FREE Estimating Software Offer!

As A Buyer Of A Building News 1995 Costbook You Are Eligible To Receive — AT NO COST! —

CONSTRUCTION ESTIMATOR

The Complete, Stand Alone, Easy-To-Use, Construction Estimating System

- This program provides the basic tools to develop accurate, comprehensive and organized cost estimates.
- The reliable cost data can be used to supplement your own records and provides a basis for accurate estimating.

- IBM Compatible — 5¼" diskette.
- Immediate on-screen access to data through "Look-up" windows. "Pull-down" menus.
- Easy to learn and master — very user-friendly.
- Different markups can be applied to each estimate.

Act NOW And Receive — FREE — An Electronic Database With Over 1,000 Lines Of Building News Costbook Data

---

Order Form

NAME _______________________________

COMPANY _______________________________

ADDRESS _______________________________

CITY, STATE, ZIP _______________________________

TELEPHONE _______________________________

Simply fill out this form, read and sign the statement below and send this complete page to:

BUILDING NEWS BOOKSTORE
3055 Overland Avenue
Los Angeles, CA 90034

I purchased my 1994 Costbook at: _______________________________
STORE   CITY   STATE

If you have received the Bonus Software through our mail offering, there is no need to order another. You may copy and share the software and data with co-workers and associates.

Bonus Software Qualifications

The CONSTRUCTION™ ESTIMATOR is a software product developed, produced and distributed by CPR International, Inc., 3195 Adelina St., Suite A, Berkeley, CA 94703, (510) 654-8338. The electronic, cost data provided with the CONSTRUCTION ESTIMATOR is derived from the Building News Construction Costbooks.

CPR International has no affiliation with BNI Publications Inc. (Building News). All questions regarding the software should be directed to CPR International, Inc.

BNI Publications Inc., its authors and editors, do not warrant or guarantee the correctness of the data or information contained in the costbooks. BNI Publications Inc., and its authors and editors, do hereby disclaim any responsibility or liability in connection with the use of data in electronic or printed form published by BNI Publications Inc., or other information in the database or costbooks.

I have read and agree to the above information.

____
NAME - PLEASE SIGN

____
NAME - PLEASE PRINT

BNi® *Building News*

TH
6010
.M454
1995

BNi. Building News

BNi. Building News
MECHANICAL/ELECTRICAL
1995 COSTBOOK
FIFTH EDITION

Co-Sponsored By

Consulting-Specifying Engineer.

BNi. Building News
Los Angeles • Anaheim
Boston • Washington, D.C.

BNi. Building News

EDITOR-IN-CHIEF
William D. Mahoney, P.E.

CONTRIBUTING EDITORS
Kenneth M. Randall
Edward B. Wetherill, CSI
Roger E. Fasteson

TECHNICAL SERVICES
Anthony Jackson
Ramon Lopez

DESIGN
Robert O. Wright

COVER DESIGN
Hannus Design Associates

BNI Publications, Inc.

LOS ANGELES
3055 OVERLAND AVENUE
LOS ANGELES, CA 90034
(310) 202-7775

ANAHEIM
1612 S. CLEMENTINE STREET
ANAHEIM, CA 92802

BOSTON
77 WEXFORD STREET
NEEDHAM HEIGHTS, MA 02194
(617) 455-1466

WASHINGTON, D.C.
502 MAPLE AVENUE
WEST VIENNA, VA 22180

ISBN 1-55701-112-5

PREFACE

For the past 49 years, Building News has been dedicated to providing construction professionals with timely and reliable information. Based on this experience, our staff has researched and compiled thousands of up-to-the-minute costs for the **Building News 1995 Costbooks.** This book is an essential reference for contractors, engineers, architects, facilities managers — any construction professional who must provide an estimate on any type of building project.

Whether working up a preliminary estimate or submitting a formal bid, the costs listed here can quickly and easily be tailored to your needs. All costs are based on national averages, while a table of modifiers is provided for regional adjustments. Overhead and profit are included in all costs.

Complete man-hour tables follow the unit costs to provide data on typical durations of specific tasks. This information can be used to schedule projects as well as to determine specific labor costs based on local labor rates.

All data is categorized according to the MASTERFORMAT of the Construction Specifications Institute (CSI). This industry standard provides an all-inclusive checklist to ensure that no element of a project is overlooked. In addition, to make specific items even easier to locate, there is a complete alphabetical index.

This costbook contains an appendix with reference charts and tables taken from an array of sources. Text explains the costs in certain categories and provides helpful pointers that should be taken into account with every estimate.

A section on square foot costs provides an overview of project costs for different building types — commercial, residential, etc. — with summaries of the actual projects. Square foot costs are invaluable for making budget estimates and checking prices when time is a factor. For a complete source of square foot costs refer to the **Building News Square Foot 1995 Costbook.**

The "Features in this Book" section presents a clear overview of the many features of this book. Included is an explanation of the data, sample page layout and discussion of how to best use the information in the book.

Of course, all buildings and construction projects are unique. The costs provided in this book are based on averages from well-managed projects with good labor productivity under normal working conditions (eight hours a day). Other circumstances affecting costs such as overtime, unusual working conditions, savings from buying bulk quantities for large projects, and unusual or hidden costs must be factored in as they arise.

The data provided in this book is for estimating purposes only. Check all applicable federal, state and local codes and regulations for specific requirements.

TABLE OF CONTENTS

CSI MASTERFORMAT

All data in the Costbook pages and Man-Hour tables is organized according to the CSI MASTERFORMAT — the industry standard numbering/classification system. The data is divided into the 16 divisions as shown below. The five digit numbers within each division correspond to the MASTERFORMAT Broadscope designations. Each section is further broken down into MASTERFORMAT Mediumscope designations. These numbers, in most cases, are the same as those used in architectural and engineering specifications.

The construction estimating information in this book is divided into two main sections:

Costbook Pages and Man-Hour Tables. Each is organized to the 16 divisions of the CSI MASTERFORMAT as shown below. In addition, there are estensive Construction Reference Tables, Geographic Cost Modifiers, Square Foot Tables and a detailed Index. Sample pages with graphic explanations are included before the Costbook pages, Man-Hour Tables and Square Foot Tables. These explanations along with the discussions below will provide a good understanding of what is included in this book and how it can best be used for construction estimating.

FEATURES IN THIS BOOK

The construction estimating information in this book is divided into two main sections: Costbook Pages and Man-Hour Tables. Each section is organized according to the 16 divisions of the CSI MASTERFORMAT as shown on the previous pages. In addition, there are extensive Supporting Construction Reference tables, Geographic Costs Modifiers, Square Foot tables and a detailed Index.

Sample pages with graphic explanations are included before the Costbook pages and Man-Hour tables. These explanations along with the discussions below, will provide a good understanding of what is included in this book and how it can best be used in construction estimating.

Material Costs

The material costs used in this book represent national averages for prices that a contractor would expect to pay plus an allowance for freight (if applicable), handling and storage. These costs reflect neither the lowest or highest prices, but rather a typical average cost over time. Periodic fluctuations in availability and in certain commodities (eg. copper, lumber) can significantly affect local material pricing. In the final estimating and bidding stages of a project when the highest degree of accuracy is required, it is best to check local, current prices.

Labor Costs

Labor costs include the basic wage, plus commonly applicable taxes, insurance and markups for overhead and profit. The labor rates used here to develop the costs are typical average prevailing wage rates. Rates for different trades are used where appropriate for each type of work.

Taxes and insurance which are most often applied to labor rates include employer-paid Social Security/Medicare taxes (FICA), Worker's Compensation insurance, state and federal unemployment taxes, and business insurance. Fixed government rates as well as average allowances are included in the labor costs. However, most of these items vary significantly from state to state and within states. For more specific data, local agencies and sources should be consulted.

Equipment Costs

Costs for various types and pieces of equipment are included in Division 1 - General Requirements and can be included in an estimate when required either as a total "Equipment" category or with specific appropriate trades. Costs for equipment are included when appropriate in the installation of costs in the Costbook pages.

Overhead And Profit

Included in the labor costs are allowances for overhead and profit for the contractor/employer whose workers are performing the specific tasks. No cost allowances or fees are included for management of subcontractors by the general contractor or construction manager. These costs, where appropriate, must be added to the costs as listed in the book.

The allowance for overhead is included to account for office overhead, the contractors' typical costs of doing business. These costs normally include in-house office staff salaries and benefits, office rent and operating expenses, professional fees, vehicle costs and other operating costs which are not directly applicable to specific jobs. It should be noted for this book that office overhead as included should be distinguished from project overhead, the General Requirements (CSI Division 1) which are specific to particular projects. Project overhead should be included on an item by item basis for each job.

Depending on the trade, an allowance of 10-15 percent is incorporated into the labor/installation costs to account for typical profit of the installing contractor. See Division 1, General Requirements, for a more detailed review of typical profit allowances.

Adjustments to Costs

The costs as presented in this book attempt to represent national averages. Costs, however, vary among regions, states and even between adjacent localities.

In order to more closely approximate the probable costs for specific locations throughout the U.S., a table of Geographic Cost Modifiers is provided. These adjustment factors are used to modify costs obtained from this book to help account for regional variations of construction costs. Whenever local current costs are known, whether material or equipment prices or labor rates, they should be used if more accuracy is required.

Man-Hour Tables

The man-hour data used to develop the labor costs are listed in the second main section of this book, the Man-Hour Tables. These productivities represent typical installation labor for thousands of construction items. The data takes into account all activities involved in normal construction under commonly experienced working conditions such as site movement, material handling, start-up, etc. As with the Costbook pages, these items are listed according to the CSI MASTERFORMAT.

Square Foot Tables

Included as an additional reference are Square Foot Tables which list hundreds of actual projects for dozens of building types each with associated building size and total square foot building cost. This data provides an overview of construction costs by building type. These costs are for actual projects. The variations within similar building types may be due, among other factors, to size, location, quality and specified components, materials and processes. Depending upon all such factors, specific building costs can vary significantly and may not necessarily fall within the range of costs as presented.

The data has been updated to reflect current construction costs and is a partial summary of the projects and costs presented in the **Building News Square Foot 1995 Costbook**. The source of this data is *Design Cost & Data* magazine which has over 25 years of history in tracking, collecting and publishing construction cost data.

Editors' Note: The **Building News 1995 Costbooks** are intended to provide accurate, reliable, average costs and typical productivities for thousands of common construction components. The data is developed and compiled from various industry sources, including government, manufacturers, suppliers and working professionals. The intent of the information is to provide assistance and guidelines to construction professionals in estimating. The user should be aware that local conditions, material and labor availability and cost variations, economic considerations, weather, local codes and regulations, etc., all affect the actual cost of construction. These and other such factors must be considered and incorporated into any and all construction estimates.

Sample Costbook Page

In order to best use the information in this book, please review this sample page and read the "Features In This Book" section.

CSI MASTERFORMAT Division

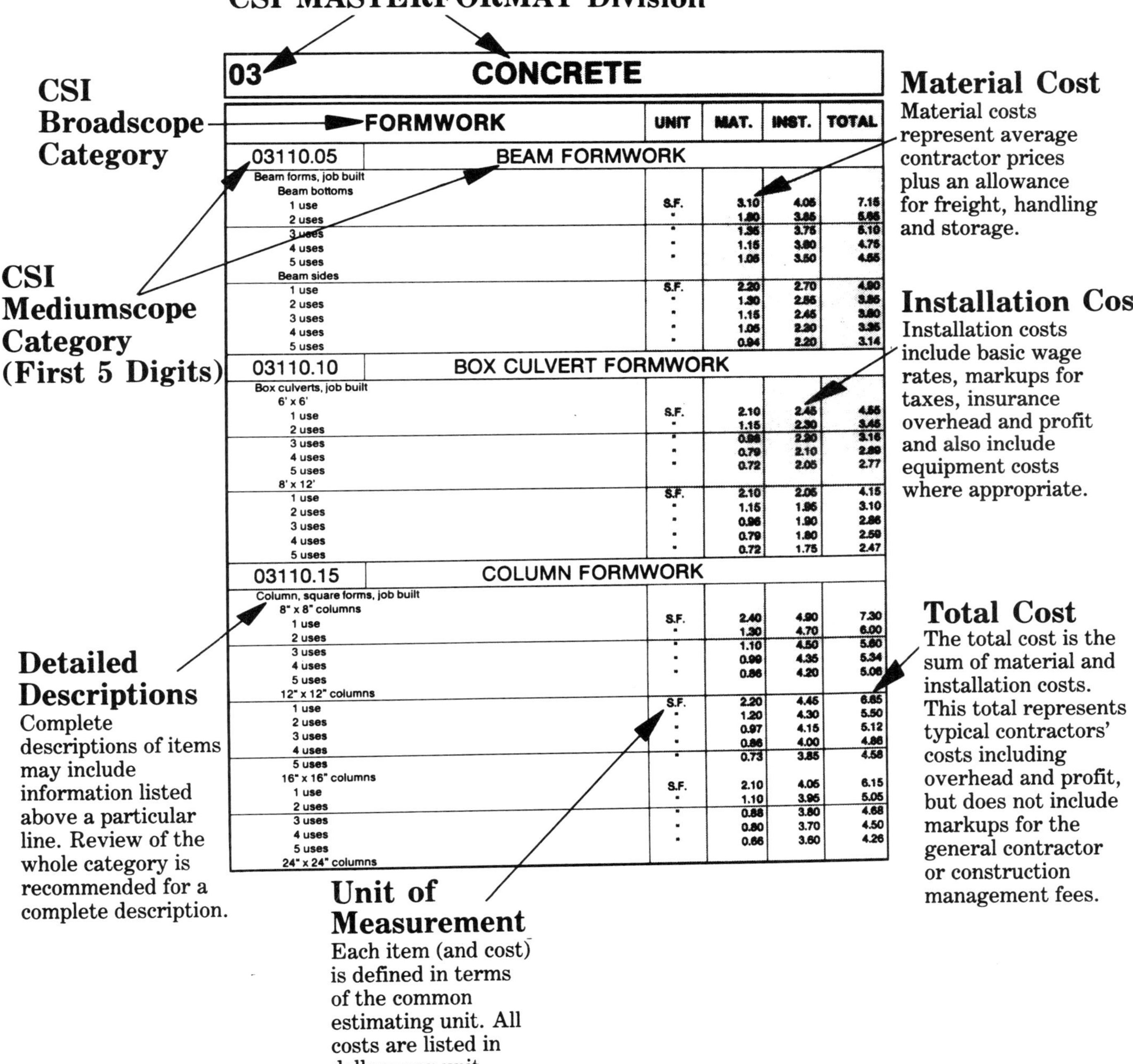

03	CONCRETE			
FORMWORK	**UNIT**	**MAT.**	**INST.**	**TOTAL**
03110.05 BEAM FORMWORK				
Beam forms, job built				
Beam bottoms				
1 use	S.F.	3.10	4.05	7.15
2 uses	"	1.80	3.85	5.65
3 uses	"	1.35	3.75	5.10
4 uses	"	1.15	3.60	4.75
5 uses	"	1.05	3.50	4.55
Beam sides				
1 use	S.F.	2.20	2.70	4.90
2 uses	"	1.30	2.55	3.85
3 uses	"	1.15	2.45	3.60
4 uses	"	1.05	2.30	3.35
5 uses	"	0.94	2.20	3.14
03110.10 BOX CULVERT FORMWORK				
Box culverts, job built				
6' x 6'				
1 use	S.F.	2.10	2.45	4.55
2 uses	"	1.15	2.30	3.45
3 uses	"	0.86	2.30	3.16
4 uses	"	0.79	2.10	2.89
5 uses	"	0.72	2.05	2.77
8' x 12'				
1 use	S.F.	2.10	2.05	4.15
2 uses	"	1.15	1.95	3.10
3 uses	"	0.96	1.90	2.86
4 uses	"	0.79	1.80	2.59
5 uses	"	0.72	1.75	2.47
03110.15 COLUMN FORMWORK				
Column, square forms, job built				
8" x 8" columns				
1 use	S.F.	2.40	4.90	7.30
2 uses	"	1.30	4.70	6.00
3 uses	"	1.10	4.50	5.60
4 uses	"	0.99	4.35	5.34
5 uses	"	0.86	4.20	5.06
12" x 12" columns				
1 use	S.F.	2.20	4.45	6.65
2 uses	"	1.20	4.30	5.50
3 uses	"	0.97	4.15	5.12
4 uses	"	0.86	4.00	4.86
5 uses	"	0.73	3.85	4.58
16" x 16" columns				
1 use	S.F.	2.10	4.05	6.15
2 uses	"	1.10	3.95	5.05
3 uses	"	0.88	3.80	4.68
4 uses	"	0.80	3.70	4.50
5 uses	"	0.66	3.60	4.26
24" x 24" columns				

CSI Broadscope Category

CSI Mediumscope Category (First 5 Digits)

Detailed Descriptions
Complete descriptions of items may include information listed above a particular line. Review of the whole category is recommended for a complete description.

Unit of Measurement
Each item (and cost) is defined in terms of the common estimating unit. All costs are listed in dollars per unit.

Material Cost
Material costs represent average contractor prices plus an allowance for freight, handling and storage.

Installation Cost
Installation costs include basic wage rates, markups for taxes, insurance overhead and profit and also include equipment costs where appropriate.

Total Cost
The total cost is the sum of material and installation costs. This total represents typical contractors' costs including overhead and profit, but does not include markups for the general contractor or construction management fees.

REQUIREMENTS	UNIT	MAT.	INST.	TOTAL
01020.10 ALLOWANCES				
Overhead				
$20,000 project				
Minimum	PCT.			15.00
Average	"			20.00
Maximum	"			40.00
$100,000 project				
Minimum	PCT.			12.00
Average	"			15.00
Maximum	"			25.00
$500,000 project				
Minimum	PCT.			10.00
Average	"			12.00
Maximum	"			20.00
$1,000,000 project				
Minimum	PCT.			6.00
Average	"			10.00
Maximum	"			12.00
Profit				
$20,000 project				
Minimum	PCT.			10.00
Average	"			15.00
Maximum	"			25.00
$100,000 project				
Minimum	PCT.			10.00
Average	"			12.00
Maximum	"			20.00
$500,000 project				
Minimum	PCT.			5.00
Average	"			10.00
Maximum	"			15.00
$1,000,000 project				
Minimum	PCT.			3.00
Average	"			8.00
Maximum	"			15.00
Professional fees				
Architectural				
$100,000 project				
Minimum	PCT.			5.00
Average	"			10.00
Maximum	"			20.00
$500,000 project				
Minimum	PCT.			5.00
Average	"			8.00
Maximum	"			12.00
$1,000,000 project				
Minimum	PCT.			3.50
Average	"			7.00
Maximum	"			10.00
Structural engineering				
Minimum	PCT.			2.00
Average	"			3.00
Maximum	"			5.00

REQUIREMENTS	UNIT	MAT.	INST.	TOTAL
01020.10 ALLOWANCES				
Mechanical engineering				
Minimum	PCT.			4.00
Average	"			5.00
Maximum	"			15.00
Electrical engineering				
Minimum	PCT.			3.00
Average	"			5.00
Maximum	"			12.00
Taxes				
Sales tax				
Minimum	PCT.			4.00
Average	"			5.00
Maximum	"			8.25
Unemployment				
Minimum	PCT.			2.00
Average	"			5.00
Maximum	"			7.00
Social security (FICA)	"			7.85
01050.10 FIELD STAFF				
Superintendent				
Minimum	YEAR			44,800
Average	"			72,800
Maximum	"			112,000
Field engineer				
Minimum	YEAR			39,200
Average	"			58,240
Maximum	"			89,600
Foreman				
Minimum	YEAR			28,000
Average	"			44,800
Maximum	"			58,800
Bookkeeper/timekeeper				
Minimum	YEAR			13,440
Average	"			21,280
Maximum	"			29,680
Watchman				
Minimum	YEAR			11,200
Average	"			14,000
Maximum	"			18,480
01310.10 SCHEDULING				
Scheduling for				
$100,000 project				
Minimum	PCT.			1.00
Average	"			2.00
Maximum	"			4.00
$500,000 project				
Minimum	PCT.			0.50
Average	"			1.00
Maximum	"			2.00
$1,000,000 project				

REQUIREMENTS	UNIT	MAT.	INST.	TOTAL
01310.10 SCHEDULING				
Minimum	PCT.			0.30
Average	"			0.80
Maximum	"			1.50
Scheduling software, not including computer				
Minimum	EA.			509.09
Average	"			2,545
Maximum	"			40,727
01330.10 SURVEYING				
Surveying				
Small crew	DAY	0.00	450.00	450.00
Average crew	"	0.00	690.00	690.00
Large crew	"	0.00	910.00	910.00
01380.10 JOB REQUIREMENTS				
Job photographs, small jobs				
Minimum	EA.			56.00
Average	"			112.00
Large projects				
Minimum	EA.			392.00
Average	"			560.00
Maximum	"			1,120
01410.10 TESTING				
Testing concrete, per test				
Minimum	EA.			12.32
Average	"			22.40
Maximum	"			33.60
Welding, per test				
Minimum	EA.			11.20
Average	"			22.40
Maximum	"			78.40
01500.10 TEMPORARY FACILITIES				
Barricades, temporary				
Highway				
Concrete	L.F.	7.30	2.25	9.55
Wood	"	2.20	0.91	3.11
Steel	"	2.35	0.76	3.11
Pedestrian barricades				
Plywood	S.F.	1.45	0.76	2.21
Chain link fence	"	1.85	0.76	2.61
Trailers, general office type, per month				
Minimum	EA.			168.00
Average	"			246.40
Maximum	"			560.00
Crew change trailers, per month				
Minimum	EA.			67.20
Average	"			78.40

REQUIREMENTS	UNIT	MAT.	INST.	TOTAL
01500.10 TEMPORARY FACILITIES				
Maximum	EA.			117.60
01505.10 MOBILIZATION				
Equipment mobilization				
Bulldozer				
Minimum	EA.			112.00
Average	"			224.00
Maximum	"			369.60
Backhoe/front-end loader				
Minimum	EA.			56.00
Average	"			112.00
Maximum	"			246.40
Crane, crawler type				
Minimum	EA.			1,120
Average	"			2,800
Maximum	"			6,160
Truck crane				
Minimum	EA.			280.00
Average	"			420.00
Maximum	"			728.00
Pile driving rig				
Minimum	EA.			5,600
Average	"			11,200
Maximum	"			19,600
01525.10 CONSTRUCTION AIDS				
Scaffolding/staging, rent per month				
Measured by lineal feet of base				
10' high	L.F.			7.84
20' high	"			13.44
30' high	"			19.60
40' high	"			22.40
50' high	"			26.88
Measured by square foot of surface				
Minimum	S.F.			0.34
Average	"			0.56
Maximum	"			1.01
Safety nets, heavy duty, per job				
Minimum	S.F.			0.22
Average	"			0.28
Maximum	"			0.56
Tarpaulins, fabric, per job				
Minimum	S.F.			0.17
Average	"			0.28
Maximum	"			0.67

REQUIREMENTS	UNIT	MAT.	INST.	TOTAL
01570.10 — SIGNS				
Construction signs, temporary				
Signs, 2' x 4'				
Minimum	EA.			21.28
Average	"			50.40
Maximum	"			179.20
Signs, 4' x 8'				
Minimum	EA.			44.80
Average	"			117.60
Maximum	"			504.00
Signs, 8' x 8'				
Minimum	EA.			56.00
Average	"			179.20
Maximum	"			1,848
01600.10 — EQUIPMENT				
Air compressor				
60 cfm				
By day	EA.			58.24
By week	"			174.72
By month	"			526.40
300 cfm				
By day	EA.			123.20
By week	"			369.60
By month	"			1,109
600 cfm				
By day	EA.			336.00
By week	"			1,008
By month	"			3,024
Air tools, per compressor, per day				
Minimum	EA.			22.40
Average	"			28.00
Maximum	"			39.20
Generators, 5 kw				
By day	EA.			50.40
By week	"			151.20
By month	"			453.60
Heaters, salamander type, per week				
Minimum	EA.			67.20
Average	"			95.20
Maximum	"			201.60
Pumps, submersible				
50 gpm				
By day	EA.			44.80
By week	"			134.40
By month	"			403.20
100 gpm				
By day	EA.			56.00
By week	"			168.00
By month	"			504.00
500 gpm				
By day	EA.			89.60

REQUIREMENTS	UNIT	MAT.	INST.	TOTAL
01600.10 — EQUIPMENT				
By week	EA.			268.80
By month	"			806.40
Diaphragm pump, by week				
Minimum	EA.			67.20
Average	"			89.60
Maximum	"			134.40
Pickup truck				
By day	EA.			78.40
By week	"			235.20
By month	"			705.60
Dump truck				
6 cy truck				
By day	EA.			224.00
By week	"			672.00
By month	"			2,016
Backhoe, track mounted				
1/2 cy capacity				
By day	EA.			448.00
By week	"			1,344
By month	"			4,032
1 cy capacity				
By day	EA.			728.00
By week	"			2,184
By month	"			6,552
Backhoe/loader, rubber tired				
1/2 cy capacity				
By day	EA.			280.00
By week	"			840.00
By month	"			2,520
3/4 cy capacity				
By day	EA.			336.00
By week	"			1,008
By month	"			3,024
Bulldozer				
75 hp				
By day	EA.			392.00
By week	"			1,176
By month	"			3,528
Cranes, crawler type				
15 ton capacity				
By day	EA.			476.00
By week	"			1,428
By month	"			4,284
25 ton capacity				
By day	EA.			588.00
By week	"			1,764
By month	"			5,292
50 ton capacity				
By day	EA.			1,008
By week	"			3,024
By month	"			9,072
100 ton capacity				

REQUIREMENTS	UNIT	MAT.	INST.	TOTAL
01600.10 EQUIPMENT				
By day	EA.			1,680
By week	"			5,040
By month	"			15,120
Truck mounted, hydraulic				
15 ton capacity				
By day	EA.			476.00
By week	"			1,428
By month	"			4,284
Loader, rubber tired				
1 cy capacity				
By day	EA.			336.00
By week	"			1,008
By month	"			3,024
01740.10 BONDS				
Performance bonds				
Minimum	PCT.			0.70
Average	"			2.00
Maximum	"			3.00

DEMOLITION	UNIT	MAT.	INST.	TOTAL
02060.10 — **BUILDING DEMOLITION**				
Cut-outs				
Concrete, elevated slabs, mesh reinforcing				
Under 5 cf	C.F.	0.00	22.70	22.70
Over 5 cf	"	0.00	18.90	18.90
Bar reinforcing				
Under 5 cf	C.F.	0.00	37.80	37.80
Over 5 cf	"	0.00	28.40	28.40
Rubbish handling				
Load in dumpster or truck				
Minimum	C.F.	0.00	0.50	0.50
Maximum	"	0.00	0.76	0.76
For use of elevators, add				
Minimum	C.F.	0.00	0.11	0.11
Maximum	"	0.00	0.23	0.23
Rubbish hauling				
Hand loaded on trucks, 2 mile trip	C.Y.	0.00	18.20	18.20
Machine loaded on trucks, 2 mile trip	"	0.00	12.35	12.35
02075.80 — **CORE DRILLING**				
Concrete				
6" thick				
3" dia.	EA.	0.00	20.65	20.65
4" dia.	"	0.00	24.05	24.05
6" dia.	"	0.00	28.90	28.90
8" dia.	"	0.00	48.10	48.10
8" thick				
3" dia.	EA.	0.00	28.90	28.90
4" dia.	"	0.00	36.10	36.10
6" dia.	"	0.00	41.30	41.30
8" dia.	"	0.00	58.00	58.00
10" thick				
3" dia.	EA.	0.00	36.10	36.10
4" dia.	"	0.00	41.30	41.30
6" dia.	"	0.00	48.10	48.10
8" dia.	"	0.00	72.00	72.00
12" thick				
3" dia.	EA.	0.00	48.10	48.10
4" dia.	"	0.00	58.00	58.00
6" dia.	"	0.00	72.00	72.00
8" dia.	"	0.00	96.50	96.50

HAZARDOUS WASTE	UNIT	MAT.	INST.	TOTAL
02080.10 ASBESTOS REMOVAL				
Enclosure using wood studs & poly, install & remove	S.F.	0.67	0.57	1.24
Trailer (change room)	DAY			61.60
Disposal suits (4 suits per man day)	"			24.64
Type C respirator mask, includes hose & filters, per man	"			12.32
Respirator mask & filter, light contamination	"			5.04
Air monitoring test, 12 tests per day				
Off job testing	DAY			672.00
On the job testing	"			896.00
Asbestos vacuum with attachments	EA.			392.00
Hydraspray piston pump	"			504.00
Negative air pressure system	"			459.20
Grade D breathing air equipment	"			1,120
Glove bag, 44" x 60" x 6 mil plastic	"			3.36
40 CY asbestos dumpster				
Weekly rental	EA.			448.00
Pick up / delivery	"			212.80
Asbestos dump fee	"			112.00
02080.12 DUCT INSULATION REMOVAL				
Remove duct insulation, duct size				
6" x 12"	L.F.	0.00	1.25	1.25
x 18"	"	0.00	1.75	1.75
x 24"	"	0.00	2.50	2.50
8" x 12"	"	0.00	1.90	1.90
x 18"	"	0.00	2.05	2.05
x 24"	"	0.00	2.85	2.85
12" x 12"	"	0.00	1.90	1.90
x 18"	"	0.00	2.50	2.50
x 24"	"	0.00	3.25	3.25
02080.15 PIPE INSULATION REMOVAL				
Removal, asbestos insulation				
2" thick, pipe				
1" to 3" dia.	L.F.	0.00	1.90	1.90
4" to 6" dia.	"	0.00	2.15	2.15
3" thick				
7" to 8" dia.	L.F.	0.00	2.25	2.25
9" to 10" dia.	"	0.00	2.40	2.40
11" to 12" dia.	"	0.00	2.50	2.50
13" to 14" dia.	"	0.00	2.65	2.65
15" to 18" dia.	"	0.00	2.85	2.85

SITE DEMOLITION	UNIT	MAT.	INST.	TOTAL
02105.43 — GAS PIPING				
Remove welded steel pipe, not including excavation				
4" dia.	L.F.	0.00	7.70	7.70
5" dia.	"	0.00	12.35	12.35
6" dia.	"	0.00	15.45	15.45
8" dia.	"	0.00	24.70	24.70
10" dia.	"	0.00	30.90	30.90
02105.45 — SANITARY PIPING				
Remove sewer pipe, not including excavation				
4" dia.	L.F.	0.00	4.95	4.95
6" dia.	"	0.00	5.60	5.60
8" dia.	"	0.00	6.15	6.15
10" dia.	"	0.00	6.50	6.50
12" dia.	"	0.00	6.85	6.85
15" dia.	"	0.00	7.25	7.25
18" dia.	"	0.00	8.25	8.25
24" dia.	"	0.00	10.30	10.30
30" dia.	"	0.00	12.35	12.35
36" dia.	"	0.00	15.45	15.45
02105.48 — WATER PIPING				
Remove water pipe, not including excavation				
4" dia.	L.F.	0.00	5.60	5.60
6" dia.	"	0.00	5.90	5.90
8" dia.	"	0.00	6.50	6.50
10" dia.	"	0.00	6.85	6.85
12" dia.	"	0.00	7.25	7.25
14" dia.	"	0.00	7.70	7.70
16" dia.	"	0.00	8.25	8.25
18" dia.	"	0.00	8.80	8.80
20" dia.	"	0.00	9.50	9.50
Remove valves				
6"	EA.	0.00	61.50	61.50
10"	"	0.00	68.50	68.50
14"	"	0.00	77.00	77.00
18"	"	0.00	100.00	100.00
02105.60 — UNDERGROUND TANKS				
Remove underground storage tank, and backfill				
50 to 250 gals	EA.	0.00	410.00	410.00
600 gals	"	0.00	410.00	410.00
1000 gals	"	0.00	620.00	620.00
4000 gals	"	0.00	990.00	990.00
5000 gals	"	0.00	990.00	990.00
10,000 gals	"	0.00	1,650	1,650
12,000 gals	"	0.00	2,060	2,060
15,000 gals	"	0.00	2,470	2,470
20,000 gals	"	0.00	3,090	3,090

SITE DEMOLITION		UNIT	MAT.	INST.	TOTAL
02105.66	SEPTIC TANKS				
Remove septic tank					
1000 gals		EA.	0.00	100.00	100.00
2000 gals		"	0.00	120.00	120.00
5000 gals		"	0.00	150.00	150.00
15,000 gals		"	0.00	1,230	1,230
25,000 gals		"	0.00	1,650	1,650
40,000 gals		"	0.00	2,470	2,470
02162.10	TRENCH SHEETING				
Closed timber, including pull and salvage, excavation					
8' deep		S.F.	1.90	4.05	5.95
20' deep		"	1.90	5.80	7.70

EARTHWORK		UNIT	MAT.	INST.	TOTAL
02210.10	HAULING MATERIAL				
Haul material by 10 cy dump truck, round trip distance					
1 mile		C.Y.	0.00	2.55	2.55
2 mile		"	0.00	3.05	3.05
5 mile		"	0.00	4.15	4.15
10 mile		"	0.00	4.55	4.55
20 mile		"	0.00	5.05	5.05
30 mile		"	0.00	6.05	6.05
Site grading, cut & fill, sandy clay, 200' haul, 75 hp dozer		"	0.00	1.80	1.80
Spread topsoil by equipment on site		"	0.00	2.00	2.00
Site grading (cut and fill to 6") less than 1 acre					
75 hp dozer		C.Y.	0.00	3.05	3.05
1.5 cy backhoe/loader		"	0.00	4.55	4.55
02210.30	BULK EXCAVATION				
Hydraulic excavator					
1 cy capacity					
Light material		C.Y.	0.00	2.05	2.05
Medium material		"	0.00	2.45	2.45
Wet material		"	0.00	3.10	3.10
Blasted rock		"	0.00	3.55	3.55
Wheel mounted front-end loader					
7/8 cy capacity					
Light material		C.Y.	0.00	1.60	1.60
Medium material		"	0.00	1.85	1.85
Wet material		"	0.00	2.15	2.15
Blasted rock		"	0.00	2.60	2.60

EARTHWORK	UNIT	MAT.	INST.	TOTAL
02210.30 — BULK EXCAVATION				
Track mounted front-end loader				
1-1/2 cy capacity				
Light material	C.Y.	0.00	1.10	1.10
Medium material	"	0.00	1.15	1.15
Wet material	"	0.00	1.30	1.30
Blasted rock	"	0.00	1.45	1.45
02220.40 — BUILDING EXCAVATION				
Structural excavation, unclassified earth				
3/8 cy backhoe	C.Y.	0.00	8.60	8.60
3/4 cy backhoe	"	0.00	6.45	6.45
1 cy backhoe	"	0.00	5.40	5.40
Foundation backfill and compaction by machine	"	0.00	12.90	12.90
02220.50 — UTILITY EXCAVATION				
Trencher, sandy clay, 8" wide trench				
18" deep	L.F.	0.00	1.00	1.00
24" deep	"	0.00	1.15	1.15
36" deep	"	0.00	1.30	1.30
Trench backfill, 95% compaction				
Tamp by hand	C.Y.	0.00	14.20	14.20
Vibratory compaction	"	0.00	11.35	11.35
Trench backfilling, with borrow sand, place & compact	"	4.50	11.35	15.85
02220.60 — TRENCHING				
Trenching and continuous footing excavation				
By gradall				
1 cy capacity				
Light soil	C.Y.	0.00	1.85	1.85
Medium soil	"	0.00	2.00	2.00
Heavy/wet soil	"	0.00	2.15	2.15
Loose rock	"	0.00	2.35	2.35
Blasted rock	"	0.00	2.50	2.50
By hydraulic excavator				
1/2 cy capacity				
Light soil	C.Y.	0.00	2.15	2.15
Medium soil	"	0.00	2.35	2.35
Heavy/wet soil	"	0.00	2.60	2.60
Loose rock	"	0.00	2.85	2.85
Blasted rock	"	0.00	3.25	3.25
Hand excavation				
Bulk, wheeled 100'				
Normal soil	C.Y.	0.00	25.20	25.20
Sand or gravel	"	0.00	22.70	22.70
Medium clay	"	0.00	32.40	32.40
Heavy clay	"	0.00	45.40	45.40
Loose rock	"	0.00	56.50	56.50
Trenches, up to 2' deep				
Normal soil	C.Y.	0.00	28.40	28.40
Sand or gravel	"	0.00	25.20	25.20
Medium clay	"	0.00	37.80	37.80

EARTHWORK	UNIT	MAT.	INST.	TOTAL
02220.60 TRENCHING				
Heavy clay	C.Y.	0.00	56.50	56.50
Loose rock	"	0.00	75.50	75.50
Trenches, to 6' deep				
Normal soil	C.Y.	0.00	32.40	32.40
Sand or gravel	"	0.00	28.40	28.40
Medium clay	"	0.00	45.40	45.40
Heavy clay	"	0.00	75.50	75.50
Loose rock	"	0.00	110.00	110.00
Backfill trenches				
With compaction				
By hand	C.Y.	0.00	18.90	18.90
By 60 hp tracked dozer	"	0.00	1.15	1.15
By 200 hp tracked dozer	"	0.00	0.72	0.72
By small front-end loader	"	0.00	1.30	1.30
Backfill trenches, sand bedding, no compaction				
By hand	C.Y.	5.60	18.90	24.50
By small front-end loader	"	5.60	1.85	7.45
02220.90 HAND EXCAVATION				
Excavation				
To 2' deep				
Normal soil	C.Y.	0.00	25.20	25.20
Sand and gravel	"	0.00	22.70	22.70
Medium clay	"	0.00	28.40	28.40
Heavy clay	"	0.00	32.40	32.40
Loose rock	"	0.00	37.80	37.80
To 6' deep				
Normal soil	C.Y.	0.00	32.40	32.40
Sand and gravel	"	0.00	28.40	28.40
Medium clay	"	0.00	37.80	37.80
Heavy clay	"	0.00	45.40	45.40
Loose rock	"	0.00	56.50	56.50
Compaction of backfill around structures or in trench				
By hand with air tamper	C.Y.	0.00	16.20	16.20
By hand with vibrating plate tamper	"	0.00	15.10	15.10
1 ton roller	"	0.00	22.75	22.75
Miscellaneous hand labor				
Trim slopes, sides of excavation	S.F.	0.00	0.04	0.04
Trim bottom of excavation	"	0.00	0.04	0.05
Excavation around obstructions and services	C.Y.	0.00	75.50	75.50
02240.05 SOIL STABILIZATION				
Straw bale secured with rebar	L.F.	1.10	0.76	1.86
Filter barrier, 18" high filter fabric	"	1.10	2.25	3.35
Sediment fence, 36" fabric with 6" mesh	"	2.50	2.85	5.35

UTILITIES	UNIT	MAT.	INST.	TOTAL
02605.30 MANHOLES				
Precast sections, 48" dia.				
Base section	EA.	110.00	100.00	210.00
1'0" riser	"	31.40	82.50	113.90
1'4" riser	"	38.10	88.00	126.10
2'8" riser	"	56.00	95.00	151.00
4'0" riser	"	110.00	100.00	210.00
2'8" cone top	"	68.50	120.00	188.50
Precast manholes, 48" dia.				
4' deep	EA.	220.00	250.00	470.00
6' deep	"	340.00	310.00	650.00
7' deep	"	390.00	350.00	740.00
8' deep	"	450.00	410.00	860.00
10' deep	"	500.00	490.00	990.00
Cast-in-place, 48" dia., with frame and cover				
5' deep	EA.	370.00	620.00	990.00
6' deep	"	490.00	710.00	1,200
8' deep	"	710.00	820.00	1,530
10' deep	"	830.00	990.00	1,820
Brick manholes, 48" dia. with cover, 8" thick				
4' deep	EA.	490.00	280.00	770.00
6' deep	"	610.00	310.00	920.00
8' deep	"	780.00	350.00	1,130
10' deep	"	970.00	400.00	1,370
12' deep	"	1,220	470.00	1,690
14' deep	"	1,480	560.00	2,040
Inverts for manholes				
Single channel	EA.	73.00	110.00	183.00
Triple channel	"	84.00	140.00	224.00
Frames and covers, 24" diameter				
300 lb	EA.	240.00	22.70	262.70
400 lb	"	250.00	25.20	275.20
500 lb	"	290.00	32.40	322.40
Watertight, 350 lb	"	300.00	75.50	375.50
For heavy equipment, 1200 lb	"	660.00	110.00	770.00
Steps for manholes				
7" x 9"	EA.	8.50	4.55	13.05
8" x 9"	"	10.90	5.05	15.95
Curb inlet, 4' throat, cast in place				
12"-30" pipe	EA.	240.00	620.00	860.00
36"-48" pipe	"	260.00	710.00	970.00
Raise exist frame and cover, repaving	"	0.00	250.00	250.00
02610.10 CAST IRON FLANGED PIPE				
Cast iron flanged sections				
4" pipe, with one bolt set				
3' section	EA.	110.00	11.20	121.20
4' section	"	120.00	12.35	132.35
5' section	"	130.00	13.70	143.70
6' section	"	140.00	15.45	155.45
8' section	"	160.00	17.65	177.65
10' section	"	180.00	24.70	204.70

UTILITIES		UNIT	MAT.	INST.	TOTAL
02610.10	**CAST IRON FLANGED PIPE**				
12' section		EA.	200.00	41.10	241.10
15' section		"	230.00	61.50	291.50
18' section		"	260.00	82.50	342.50
6" pipe, with one bolt set					
3' section		EA.	140.00	12.35	152.35
4' section		"	150.00	14.50	164.50
5' section		"	160.00	16.45	176.45
6' section		"	180.00	19.00	199.00
8' section		"	200.00	27.40	227.40
10' section		"	220.00	30.90	250.90
12' section		"	250.00	41.10	291.10
15' section		"	280.00	61.50	341.50
18' section		"	340.00	88.00	428.00
8" pipe, with one bolt set					
3' section		EA.	190.00	15.45	205.45
4' section		"	210.00	17.65	227.65
5' section		"	220.00	20.55	240.55
6' section		"	240.00	24.70	264.70
8' section		"	270.00	35.30	305.30
10' section		"	300.00	41.10	341.10
12' section		"	330.00	61.50	391.50
15' section		"	390.00	82.50	472.50
18' section		"	450.00	100.00	550.00
10" pipe, with one bolt set					
3' section		EA.	240.00	15.85	255.85
4' section		"	280.00	18.15	298.15
5' section		"	300.00	21.30	321.30
6' section		"	320.00	25.70	345.70
8' section		"	360.00	37.40	397.40
10' section		"	390.00	44.10	434.10
12' section		"	450.00	68.50	518.50
15' section		"	490.00	88.00	578.00
18' section		"	570.00	120.00	690.00
12" pipe, with one bolt set					
3' section		EA.	320.00	17.15	337.15
4' section		"	350.00	19.90	369.90
5' section		"	380.00	23.75	403.75
6' section		"	400.00	28.10	428.10
8' section		"	460.00	41.10	501.10
10' section		"	510.00	47.50	557.50
12' section		"	570.00	77.00	647.00
15' section		"	640.00	100.00	740.00
18' section		"	730.00	140.00	870.00
02610.11	**CAST IRON FITTINGS**				
Mechanical joint, with 2 bolt kits					
90 deg bend					
4"		EA.	64.00	15.10	79.10
6"		"	80.50	17.45	97.95
8"		"	110.00	22.70	132.70
10"		"	170.00	32.40	202.40

UTILITIES	UNIT	MAT.	INST.	TOTAL
02610.11 CAST IRON FITTINGS				
12"	EA.	240.00	45.40	285.40
14"	"	340.00	56.50	396.50
16"	"	370.00	75.50	445.50
45 deg bend				
4"	EA.	52.50	15.10	67.60
6"	"	75.00	17.45	92.45
8"	"	110.00	22.70	132.70
10"	"	140.00	32.40	172.40
12"	"	200.00	45.40	245.40
14"	"	240.00	56.50	296.50
16"	"	310.00	75.50	385.50
Tee, with 3 bolt kits				
4" x 4"	EA.	97.50	22.70	120.20
6" x 6"	"	130.00	28.40	158.40
8" x 8"	"	180.00	37.80	217.80
10" x 10"	"	260.00	56.50	316.50
12" x 12"	"	340.00	75.50	415.50
Wye, with 3 bolt kits				
6" x 6"	EA.	170.00	28.40	198.40
8" x 8"	"	250.00	37.80	287.80
10" x 10"	"	360.00	56.50	416.50
12" x 12"	"	500.00	75.50	575.50
Reducer, with 2 bolt kits				
6" x 4"	EA.	64.00	28.40	92.40
8" x 6"	"	87.50	37.80	125.30
10" x 8"	"	150.00	56.50	206.50
12" x 10"	"	160.00	75.50	235.50
Flanged, 90 deg bend, 125 lb.				
4"	EA.	41.40	18.90	60.30
6"	"	64.00	22.70	86.70
8"	"	93.00	28.40	121.40
10"	"	160.00	37.80	197.80
12"	"	230.00	56.50	286.50
14"	"	460.00	75.50	535.50
16"	"	690.00	75.50	765.50
Tee				
4"	EA.	80.50	28.40	108.90
6"	"	120.00	32.40	152.40
8"	"	170.00	37.80	207.80
10"	"	320.00	45.40	365.40
12"	"	440.00	56.50	496.50
14"	"	1,010	75.50	1,086
16"	"	1,460	110.00	1,570
02610.13 GATE VALVES				
Gate valve, (AWWA) mechanical joint, with adjustable box				
4" valve	EA.	340.00	41.10	381.10
6" valve	"	400.00	49.40	449.40
8" valve	"	590.00	61.50	651.50
10" valve	"	900.00	72.50	972.50
12" valve	"	1,230	88.00	1,318

UTILITIES	UNIT	MAT.	INST.	TOTAL
02610.13 GATE VALVES				
14" valve	EA.	2,800	100.00	2,900
16" valve	"	3,920	110.00	4,030
18" valve	"	5,040	120.00	5,160
Flanged, with box, post indicator (AWWA)				
4" valve	EA.	290.00	49.40	339.40
6" valve	"	360.00	56.00	416.00
8" valve	"	560.00	68.50	628.50
10" valve	"	840.00	82.50	922.50
12" valve	"	1,140	100.00	1,240
14" valve	"	2,690	120.00	2,810
16" valve	"	3,700	150.00	3,850
02610.15 WATER METERS				
Water meter, displacement type				
1"	EA.	130.00	32.00	162.00
1-1/2"	"	470.00	35.60	505.60
2"	"	710.00	40.10	750.10
02610.17 CORPORATION STOPS				
Stop for flared copper service pipe				
3/4"	EA.	15.70	16.00	31.70
1"	"	22.40	17.80	40.20
1-1/4"	"	62.50	21.35	83.85
1-1/2"	"	75.00	26.70	101.70
2"	"	110.00	32.00	142.00
02610.40 DUCTILE IRON PIPE				
Ductile iron pipe, cement lined, slip-on joints				
4"	L.F.	6.70	3.45	10.15
6"	"	8.40	3.65	12.05
8"	"	11.00	3.85	14.85
10"	"	15.10	4.10	19.20
12"	"	18.65	4.95	23.60
14"	"	23.50	6.15	29.65
16"	"	27.40	6.85	34.25
18"	"	30.80	7.70	38.50
20"	"	34.70	8.80	43.50
Mechanical joint pipe				
4"	L.F.	7.40	4.75	12.15
6"	"	9.35	5.15	14.50
8"	"	12.30	5.60	17.90
10"	"	16.15	6.15	22.30
12"	"	20.50	8.25	28.75
14"	"	25.80	9.50	35.30
16"	"	28.20	11.20	39.40
18"	"	31.80	12.35	44.15
20"	"	36.60	13.70	50.30
Fittings, mechanical joint				
90 degree elbow				
4"	EA.	95.00	15.10	110.10
6"	"	130.00	17.45	147.45

UTILITIES	UNIT	MAT.	INST.	TOTAL
02610.40 **DUCTILE IRON PIPE**				
8"	EA.	180.00	22.70	202.70
10"	"	260.00	32.40	292.40
12"	"	350.00	45.40	395.40
14"	"	540.00	56.50	596.50
16"	"	670.00	75.50	745.50
18"	"	1,010	90.50	1,101
20"	"	1,120	110.00	1,230
45 degree elbow				
4"	EA.	98.50	15.10	113.60
6"	"	110.00	17.45	127.45
8"	"	160.00	22.70	182.70
10"	"	220.00	32.40	252.40
12"	"	320.00	45.40	365.40
14"	"	450.00	56.50	506.50
16"	"	540.00	75.50	615.50
18"	"	780.00	110.00	890.00
20"	"	930.00	110.00	1,040
Tee				
4"x3"	EA.	130.00	28.40	158.40
4"x4"	"	150.00	28.40	178.40
6"x3"	"	170.00	32.40	202.40
6"x4"	"	170.00	32.40	202.40
6"x6"	"	190.00	32.40	222.40
8"x4"	"	240.00	37.80	277.80
8"x6"	"	270.00	37.80	307.80
8"x8"	"	250.00	37.80	287.80
10"x4"	"	310.00	45.40	355.40
10"x6"	"	350.00	45.40	395.40
10"x8"	"	360.00	45.40	405.40
10"x10"	"	400.00	45.40	445.40
12"x4"	"	490.00	56.50	546.50
12"x6"	"	380.00	56.50	436.50
12"x8"	"	410.00	56.50	466.50
12"x10"	"	470.00	56.50	526.50
12"x12"	"	500.00	60.50	560.50
14"x4"	"	620.00	65.00	685.00
14"x6"	"	660.00	65.00	725.00
14"x8"	"	670.00	65.00	735.00
14"x10"	"	690.00	65.00	755.00
14"x12"	"	710.00	70.00	780.00
14"x14"	"	700.00	70.00	770.00
16"x4"	"	800.00	75.50	875.50
16"x6"	"	810.00	75.50	885.50
16"x8"	"	720.00	75.50	795.50
16"x10"	"	730.00	75.50	805.50
16"x12"	"	720.00	75.50	795.50
16"x14"	"	760.00	75.50	835.50
16"x16"	"	770.00	75.50	845.50
18"x6"	"	920.00	82.50	1,003
18"x8"	"	940.00	82.50	1,023
18"x10"	"	960.00	82.50	1,043
18"x12"	"	970.00	82.50	1,053

UTILITIES		UNIT	MAT.	INST.	TOTAL
02610.40	**DUCTILE IRON PIPE**				
18"x14"		EA.	1,090	82.50	1,173
18"x16"		"	1,080	82.50	1,163
18"x18"		"	1,220	82.50	1,303
20"x6"		"	1,180	90.50	1,271
20"x8"		"	1,190	90.50	1,281
20"x10"		"	1,210	90.50	1,301
20"x12"		"	1,230	90.50	1,321
20"x14"		"	1,270	90.50	1,361
20"x16"		"	1,480	90.50	1,571
20"x18"		"	1,550	90.50	1,641
20"x20"		"	1,580	90.50	1,671
Cross					
4"x3"		EA.	170.00	37.80	207.80
4"x4"		"	180.00	37.80	217.80
6"x3"		"	190.00	45.40	235.40
6"x4"		"	200.00	45.40	245.40
6"x6"		"	220.00	45.40	265.40
8"x4"		"	260.00	50.50	310.50
8"x6"		"	280.00	50.50	330.50
8"x8"		"	310.00	50.50	360.50
10"x4"		"	360.00	56.50	416.50
10"x6"		"	380.00	56.50	436.50
10"x8"		"	410.00	56.50	466.50
10"x10"		"	490.00	56.50	546.50
12"x4"		"	460.00	65.00	525.00
12"x6"		"	520.00	65.00	585.00
12"x8"		"	500.00	65.00	565.00
12"x10"		"	580.00	70.00	650.00
12"x12"		"	630.00	70.00	700.00
14"x4"		"	590.00	75.50	665.50
14"x6"		"	670.00	75.50	745.50
14"x8"		"	700.00	75.50	775.50
14"x10"		"	750.00	75.50	825.50
14"x12"		"	810.00	82.50	892.50
14"x14"		"	880.00	82.50	962.50
16"x4"		"	770.00	90.50	860.50
16"x6"		"	800.00	90.50	890.50
16"x8"		"	840.00	90.50	930.50
16"x10"		"	900.00	90.50	990.50
16"x12"		"	940.00	90.50	1,031
16"x14"		"	1,010	90.50	1,101
16"x16"		"	1,080	90.50	1,171
18"x6"		"	1,000	100.00	1,100
18"x8"		"	1,030	100.00	1,130
18"x10"		"	1,080	100.00	1,180
18"x12"		"	1,130	100.00	1,230
18"x14"		"	1,340	100.00	1,440
18"x16"		"	1,430	100.00	1,530
18"x18"		"	1,510	100.00	1,610
20"x6"		"	1,200	110.00	1,310
20"x8"		"	1,230	110.00	1,340
20"x10"		"	1,290	110.00	1,400

UTILITIES	UNIT	MAT.	INST.	TOTAL
02610.40 **DUCTILE IRON PIPE**				
20"x12"	EA.	1,340	110.00	1,450
20"x14"	"	1,410	110.00	1,520
20"x16"	"	1,640	110.00	1,750
20"x18"	"	1,750	110.00	1,860
20"x20"	"	1,850	110.00	1,960
02610.60 **PLASTIC PIPE**				
PVC, class 150 pipe				
4" dia.	L.F.	3.00	3.10	6.10
6" dia.	"	5.70	3.35	9.05
8" dia.	"	9.05	3.55	12.60
10" dia.	"	12.90	3.85	16.75
12" dia.	"	19.05	4.10	23.15
Schedule 40 pipe				
1-1/2" dia.	L.F.	1.70	1.35	3.05
2" dia.	"	1.95	1.40	3.35
2-1/2" dia.	"	2.20	1.50	3.70
3" dia.	"	3.15	1.60	4.75
4" dia.	"	4.50	1.90	6.40
6" dia.	"	7.55	2.25	9.80
90 degree elbows				
1"	EA.	0.90	3.80	4.70
1-1/2"	"	1.70	3.80	5.50
2"	"	1.90	4.10	6.00
2-1/2"	"	5.05	4.55	9.60
3"	"	6.15	5.05	11.20
4"	"	13.45	5.65	19.10
6"	"	38.10	7.55	45.65
45 degree elbows				
1"	EA.	1.10	3.80	4.90
1-1/2"	"	1.90	3.80	5.70
2"	"	2.35	4.10	6.45
2-1/2"	"	5.05	4.55	9.60
3"	"	6.15	5.05	11.20
4"	"	15.10	5.65	20.75
6"	"	39.20	7.55	46.75
Tees				
1"	EA.	1.25	4.55	5.80
1-1/2"	"	1.90	4.55	6.45
2"	"	2.70	5.05	7.75
2-1/2"	"	4.95	5.65	10.60
3"	"	6.70	6.50	13.20
4"	"	16.80	7.55	24.35
6"	"	38.10	9.05	47.15
Couplings				
1"	EA.	0.67	3.80	4.47
1-1/2"	"	0.78	3.80	4.58
2"	"	1.35	4.10	5.45
2-1/2"	"	2.70	4.55	7.25
3"	"	5.05	5.05	10.10
4"	"	8.95	5.65	14.60

UTILITIES	UNIT	MAT.	INST.	TOTAL
02610.60 — PLASTIC PIPE				
6"	EA.	14.00	7.55	21.55
Drainage pipe				
PVC schedule 80				
1" dia.	L.F.	1.00	1.35	2.35
1-1/2" dia.	"	1.25	1.35	2.60
ABS, 2" dia.	"	1.55	1.40	2.95
2-1/2" dia.	"	2.25	1.50	3.75
3" dia.	"	2.65	1.60	4.25
4" dia.	"	3.60	1.90	5.50
6" dia.	"	5.95	2.25	8.20
8" dia.	"	7.95	3.25	11.20
10" dia.	"	10.65	3.85	14.50
12" dia.	"	17.35	4.10	21.45
90 degree elbows				
1"	EA.	1.80	3.80	5.60
1-1/2"	"	2.25	3.80	6.05
2"	"	2.70	4.10	6.80
2-1/2"	"	6.45	4.55	11.00
3"	"	6.65	5.05	11.70
4"	"	11.75	5.65	17.40
6"	"	25.80	7.55	33.35
45 degree elbows				
1"	EA.	2.90	3.80	6.70
1-1/2"	"	3.70	3.80	7.50
2"	"	4.60	4.10	8.70
2-1/2"	"	8.60	4.55	13.15
3"	"	9.20	5.05	14.25
4"	"	17.35	5.65	23.00
6"	"	40.30	7.55	47.85
Tees				
1"	EA.	1.90	4.55	6.45
1-1/2"	"	6.05	4.55	10.60
2"	"	7.40	5.05	12.45
2-1/2"	"	8.60	5.65	14.25
3"	"	9.40	6.50	15.90
4"	"	17.90	7.55	25.45
6"	"	35.30	9.05	44.35
Couplings				
1"	EA.	1.55	3.80	5.35
1-1/2"	"	2.70	3.80	6.50
2"	"	4.00	4.10	8.10
2-1/2"	"	8.40	4.55	12.95
3"	"	8.65	5.05	13.70
4"	"	8.95	5.65	14.60
6"	"	15.10	7.55	22.65
Pressure pipe				
PVC, class 200 pipe				
3/4"	L.F.	0.99	1.15	2.14
1"	"	1.10	1.20	2.30
1-1/4"	"	1.25	1.25	2.50
1-1/2"	"	1.50	1.35	2.85
2"	"	1.60	1.40	3.00

UTILITIES	UNIT	MAT.	INST.	TOTAL
02610.60 PLASTIC PIPE				
2-1/2"	L.F.	2.35	1.50	3.85
3"	"	2.95	1.60	4.55
4"	"	5.15	1.90	7.05
6"	"	9.75	2.25	12.00
8"	"	14.80	3.55	18.35
90 degree elbows				
3/4"	EA.	0.49	3.80	4.29
1"	"	0.62	3.80	4.42
1-1/4"	"	0.99	3.80	4.79
1-1/2"	"	1.20	3.80	5.00
2"	"	2.35	4.10	6.45
2-1/2"	"	3.40	4.55	7.95
3"	"	7.65	5.05	12.70
4"	"	19.70	5.65	25.35
6"	"	37.50	7.55	45.05
8"	"	74.00	11.35	85.35
45 degree elbows				
3/4"	EA.	0.74	3.80	4.54
1"	"	0.92	3.80	4.72
1-1/4"	"	1.35	3.80	5.15
1-1/2"	"	1.60	3.80	5.40
2"	"	2.30	4.10	6.40
2-1/2"	"	3.75	4.55	8.30
3"	"	8.60	5.05	13.65
4"	"	16.15	5.65	21.80
6"	"	39.80	7.55	47.35
8"	"	80.50	11.35	91.85
Tees				
3/4"	EA.	0.62	4.55	5.17
1"	"	0.81	4.55	5.36
1-1/4"	"	1.15	4.55	5.70
1-1/2"	"	1.60	4.55	6.15
2"	"	2.35	5.05	7.40
2-1/2"	"	3.70	5.65	9.35
3"	"	11.35	6.50	17.85
4"	"	16.25	7.55	23.80
6"	"	56.50	9.05	65.55
8"	"	120.00	12.60	132.60
Couplings				
3/4"	EA.	0.37	3.80	4.17
1"	"	0.56	3.80	4.36
1-1/4"	"	0.74	3.80	4.54
1-1/2"	"	0.81	3.80	4.61
2"	"	1.10	4.10	5.20
2-1/2"	"	2.35	4.55	6.90
3"	"	3.70	5.05	8.75
4"	"	5.20	5.05	10.25
6"	"	16.25	5.65	21.90
8"	"	29.10	7.55	36.65

UTILITIES	UNIT	MAT.	INST.	TOTAL

02610.90 — VITRIFIED CLAY PIPE

	UNIT	MAT.	INST.	TOTAL
Vitrified clay pipe, extra strength				
6" dia.	L.F.	2.80	5.60	8.40
8" dia.	"	3.35	5.90	9.25
10" dia.	"	5.15	6.15	11.30
12" dia.	"	7.40	8.25	15.65
15" dia.	"	13.45	12.35	25.80
18" dia.	"	20.15	13.70	33.85
24" dia.	"	37.00	17.65	54.65
30" dia.	"	62.50	24.70	87.20
36" dia.	"	89.50	35.30	124.80

02630.10 — TAPPING SADDLES & SLEEVES

	UNIT	MAT.	INST.	TOTAL
Tapping saddle, tap size to 2"				
4" saddle	EA.	35.80	11.35	47.15
6" saddle	"	42.60	14.20	56.80
8" saddle	"	49.30	18.90	68.20
10" saddle	"	58.00	22.70	80.70
12" saddle	"	69.50	32.40	101.90
14" saddle	"	78.50	45.40	123.90
Tapping sleeve				
4x4	EA.	390.00	15.10	405.10
6x4	"	500.00	17.45	517.45
6x6	"	520.00	17.45	537.45
8x4	"	530.00	22.70	552.70
8x6	"	550.00	22.70	572.70
10x4	"	840.00	49.40	889.40
10x6	"	1,120	49.40	1,169
10x8	"	1,140	49.40	1,189
10x10	"	1,180	51.50	1,232
12x4	"	1,210	51.50	1,262
12x6	"	1,250	56.00	1,306
12x8	"	1,290	61.50	1,352
12x10	"	1,340	68.50	1,409
12x12	"	1,400	77.00	1,477
Tapping valve, mechanical joint				
4" valve	EA.	360.00	150.00	510.00
6" valve	"	430.00	210.00	640.00
8" valve	"	650.00	310.00	960.00
10" valve	"	1,030	410.00	1,440
12" valve	"	1,680	620.00	2,300
Tap hole in pipe				
4" hole	EA.	0.00	28.40	28.40
6" hole	"	0.00	45.40	45.40
8" hole	"	0.00	75.50	75.50
10" hole	"	0.00	90.50	90.50
12" hole	"	0.00	110.00	110.00

02640.15 — VALVE BOXES

	UNIT	MAT.	INST.	TOTAL
Valve box, adjustable, for valves up to 20"				
3' deep	EA.	95.00	7.55	102.55
4' deep	"	95.00	9.05	104.05
5' deep	"	95.00	11.35	106.35

UTILITIES	UNIT	MAT.	INST.	TOTAL
02640.19 **THRUST BLOCKS**				
Thrust block, 3000# concrete				
1/4 c.y.	EA.	62.50	48.30	110.80
1/2 c.y.	"	87.50	58.00	145.50
3/4 c.y.	"	100.00	96.50	196.50
1 c.y.	"	130.00	190.00	320.00
02645.10 **FIRE HYDRANTS**				
Standard, 3 way post, 6" mechanical joint				
2' deep	EA.	900.00	410.00	1,310
4' deep	"	960.00	490.00	1,450
6' deep	"	1,080	620.00	1,700
8' deep	"	1,210	710.00	1,920
02665.10 **CHILLED WATER SYSTEMS**				
Chilled water pipe, 2" thick insulation, w/casing				
Align and tack weld on sleepers				
1-1/2" dia.	L.F.	10.00	1.10	11.10
3" dia.	"	15.95	1.75	17.70
4" dia.	"	18.80	2.45	21.25
6" dia.	"	21.15	3.10	24.25
8" dia.	"	29.70	3.55	33.25
10" dia.	"	37.40	4.10	41.50
12" dia.	"	44.20	4.95	49.15
14" dia.	"	61.00	5.35	66.35
16" dia.	"	77.50	6.15	83.65
Align and tack weld on trench bottom				
18" dia.	L.F.	79.50	6.85	86.35
20" dia.	"	110.00	7.70	117.70
Preinsulated fittings				
Align and tack weld on sleepers				
Elbows				
1-1/2"	EA.	200.00	20.00	220.00
3"	"	260.00	32.00	292.00
4"	"	330.00	40.10	370.10
6"	"	470.00	53.50	523.50
8"	"	670.00	64.00	734.00
Tees				
1-1/2"	EA.	320.00	21.35	341.35
3"	"	440.00	35.60	475.60
4"	"	550.00	45.80	595.80
6"	"	740.00	64.00	804.00
8"	"	990.00	80.00	1,070
Reducers				
3"	EA.	320.00	26.70	346.70
4"	"	540.00	32.00	572.00
6"	"	730.00	40.10	770.10
8"	"	760.00	53.50	813.50
Anchors, not including concrete				
4"	EA.	140.00	40.10	180.10
6"	"	210.00	40.10	250.10

UTILITIES	UNIT	MAT.	INST.	TOTAL
02665.10 CHILLED WATER SYSTEMS				
Align and tack weld on trench bottom				
Elbows				
10"	EA.	870.00	77.00	947.00
12"	"	1,050	88.00	1,138
14"	"	1,230	95.00	1,325
16"	"	1,410	100.00	1,510
18"	"	1,560	110.00	1,670
20"	"	1,880	120.00	2,000
Tees				
10"	EA.	1,380	77.00	1,457
12"	"	1,780	88.00	1,868
14"	"	1,880	95.00	1,975
16"	"	1,990	100.00	2,090
18"	"	2,220	110.00	2,330
20"	"	2,580	120.00	2,700
Reducers				
10"	EA.	1,060	51.50	1,112
12"	"	1,200	56.00	1,256
14"	"	1,410	61.50	1,472
16"	"	1,650	68.50	1,719
18"	"	1,880	77.00	1,957
20"	"	1,930	88.00	2,018
Anchors, not including concrete				
10"	EA.	310.00	51.50	361.50
12"	"	340.00	56.00	396.00
14"	"	390.00	61.50	451.50
16"	"	550.00	68.50	618.50
18"	"	770.00	77.00	847.00
20"	"	1,040	88.00	1,128
02670.10 WELLS				
Domestic water, drilled and cased				
4" dia.	L.F.	8.50	120.00	128.50
6" dia.	"	8.90	130.00	138.90
8" dia.	"	11.20	170.00	181.20
02685.10 GAS DISTRIBUTION				
Gas distribution lines				
Polyethylene, 60 psi coils				
1-1/4" dia.	L.F.	0.78	2.15	2.93
1-1/2" dia.	"	1.05	2.30	3.35
2" dia.	"	1.30	2.65	3.95
3" dia.	"	2.45	3.20	5.65
30' pipe lengths				
3" dia.	L.F.	2.80	3.55	6.35
4" dia.	"	4.35	4.00	8.35
6" dia.	"	5.80	5.35	11.15
8" dia.	"	10.90	6.40	17.30
Steel, schedule 40, plain end				
1" dia.	L.F.	2.45	2.65	5.10
2" dia.	"	4.05	2.90	6.95

UTILITIES	UNIT	MAT.	INST.	TOTAL
02685.10 GAS DISTRIBUTION				
3" dia.	L.F.	7.50	3.20	10.70
4" dia.	"	10.30	8.25	18.55
5" dia.	"	20.15	8.80	28.95
6" dia.	"	22.95	10.30	33.25
8" dia.	"	33.60	11.20	44.80
Natural gas meters, direct digital reading, threaded				
250 cfh @ 5 lbs	EA.	97.50	64.00	161.50
425 cfh @ 10 lbs	"	240.00	64.00	304.00
800 cfh @ 20 lbs	"	340.00	80.00	420.00
1000 cfh @ 25 lbs	"	1,030	80.00	1,110
1,400 cfh @ 100 lbs	"	2,240	110.00	2,350
2,300 cfh @ 100 lbs	"	3,250	160.00	3,410
5,000 cfh @ 100 lbs	"	4,820	320.00	5,140
Gas pressure regulators				
Threaded				
3/4"	EA.	41.90	40.10	82.00
1"	"	43.70	53.50	97.20
1-1/4"	"	45.90	53.50	99.40
1-1/2"	"	290.00	53.50	343.50
2"	"	300.00	64.00	364.00
Flanged				
3"	EA.	860.00	80.00	940.00
4"	"	1,200	110.00	1,310
02690.10 STORAGE TANKS				
Oil storage tank, underground				
Steel				
500 gals	EA.	490.00	150.00	640.00
1,000 gals	"	880.00	210.00	1,090
4,000 gals	"	2,350	410.00	2,760
5,000 gals	"	3,190	620.00	3,810
10,000 gals	"	4,590	1,230	5,820
Fiberglass, double wall				
550 gals	EA.	1,820	210.00	2,030
1,000 gals	"	2,330	210.00	2,540
2,000 gals	"	2,690	310.00	3,000
4,000 gals	"	3,640	620.00	4,260
6,000 gals	"	4,200	820.00	5,020
8,000 gals	"	5,710	1,230	6,940
10,000 gals	"	7,280	1,540	8,820
12,000 gals	"	8,790	2,060	10,850
15,000 gals	"	10,300	2,740	13,040
20,000 gals	"	12,660	3,090	15,750
Above ground				
Steel				
275 gals	EA.	200.00	120.00	320.00
500 gals	"	660.00	210.00	870.00
1,000 gals	"	970.00	250.00	1,220
1,500 gals	"	1,530	310.00	1,840
2,000 gals	"	2,000	410.00	2,410
5,000 gals	"	5,380	620.00	6,000

UTILITIES	UNIT	MAT.	INST.	TOTAL

02690.10   STORAGE TANKS

	UNIT	MAT.	INST.	TOTAL
Fill cap	EA.	24.65	32.00	56.65
Vent cap	"	24.65	32.00	56.65
Level indicator	"	59.50	32.00	91.50

02695.80   STEAM METERS

	UNIT	MAT.	INST.	TOTAL
In-line turbine, direct reading, 300 lb, flanged				
2"	EA.	2,800	40.10	2,840
3"	"	3,020	53.50	3,074
4"	"	3,360	64.00	3,424
Threaded, 2"				
5" line	EA.	5,380	320.00	5,700
6" line	"	5,490	320.00	5,810
8" line	"	5,590	320.00	5,910
10" line	"	5,940	320.00	6,260
12" line	"	6,050	320.00	6,370
14" line	"	6,270	320.00	6,590
16" line	"	6,610	320.00	6,930

SEWERAGE AND DRAINAGE	UNIT	MAT.	INST.	TOTAL

02720.10   CATCH BASINS

	UNIT	MAT.	INST.	TOTAL
Standard concrete catch basin				
Cast in place, 3'8" x 3'8", 6" thick wall				
2' deep	EA.	200.00	310.00	510.00
3' deep	"	260.00	310.00	570.00
4' deep	"	350.00	410.00	760.00
5' deep	"	400.00	410.00	810.00
6' deep	"	460.00	490.00	950.00
4'x4', 8" thick wall, cast in place				
2' deep	EA.	210.00	310.00	520.00
3' deep	"	290.00	310.00	600.00
4' deep	"	390.00	410.00	800.00
5' deep	"	450.00	410.00	860.00
6' deep	"	490.00	490.00	980.00
Frames and covers, cast iron				
Round				
24" dia.	EA.	170.00	56.50	226.50
26" dia.	"	190.00	56.50	246.50
28" dia.	"	230.00	56.50	286.50
Rectangular				
23"x23"	EA.	150.00	56.50	206.50
27"x20"	"	180.00	56.50	236.50
24"x24"	"	180.00	56.50	236.50

SEWERAGE AND DRAINAGE	UNIT	MAT.	INST.	TOTAL
02720.10 **CATCH BASINS**				
26"x26"	EA.	200.00	56.50	256.50
Curb inlet frames and covers				
27"x27"	EA.	370.00	56.50	426.50
24"x36"	"	220.00	56.50	276.50
24"x25"	"	210.00	56.50	266.50
24"x22"	"	180.00	56.50	236.50
20"x22"	"	240.00	56.50	296.50
Airfield catch basin frame and grating, galvanized				
2'x4'	EA.	440.00	56.50	496.50
2'x2'	"	300.00	56.50	356.50
02720.40 **STORM DRAINAGE**				
Concrete pipe				
Plain, bell and spigot joint, class II				
6" pipe	L.F.	3.00	5.60	8.60
8" pipe	"	3.45	6.15	9.60
10" pipe	"	3.60	6.50	10.10
12" pipe	"	4.35	6.85	11.20
15" pipe	"	5.45	7.25	12.70
18" pipe	"	7.40	7.70	15.10
21" pipe	"	9.25	8.25	17.50
24" pipe	"	11.55	8.80	20.35
Reinforced, class III, tongue and groove joint				
12" pipe	L.F.	6.70	6.85	13.55
15" pipe	"	8.05	7.25	15.30
18" pipe	"	10.20	7.70	17.90
21" pipe	"	13.45	8.25	21.70
24" pipe	"	16.35	8.80	25.15
27" pipe	"	19.05	9.50	28.55
30" pipe	"	21.40	10.30	31.70
36" pipe	"	32.50	11.20	43.70
42" pipe	"	41.40	12.35	53.75
48" pipe	"	58.00	13.70	71.70
54" pipe	"	76.00	15.45	91.45
60" pipe	"	89.50	17.65	107.15
66" pipe	"	100.00	20.55	120.55
72" pipe	"	130.00	24.70	154.70
Flared end-section, concrete				
12" pipe	L.F.	40.00	6.85	46.85
15" pipe	"	47.00	7.25	54.25
18" pipe	"	52.50	7.70	60.20
24" pipe	"	65.00	8.80	73.80
30" pipe	"	79.50	10.30	89.80
36" pipe	"	110.00	11.20	121.20
42" pipe	"	120.00	12.35	132.35
48" pipe	"	130.00	13.70	143.70
54" pipe	"	140.00	15.45	155.45
Corrugated metal pipe, coated, paved invert				
16 ga.				
8" pipe	L.F.	5.70	4.10	9.80
10" pipe	"	7.75	4.25	12.00
12" pipe	"	8.90	4.40	13.30

SEWERAGE AND DRAINAGE	UNIT	MAT.	INST.	TOTAL
02720.40 STORM DRAINAGE				
15" pipe	L.F.	10.95	4.75	15.70
18" pipe	"	13.25	5.15	18.40
21" pipe	"	15.70	5.60	21.30
24" pipe	"	18.50	6.15	24.65
30" pipe	"	26.40	6.85	33.25
36" pipe	"	33.60	7.70	41.30
12 ga., 48" pipe	"	54.50	8.80	63.30
10 ga.				
60" pipe	L.F.	78.50	10.30	88.80
72" pipe	"	100.00	12.35	112.35
Galvanized or aluminum, plain				
16 ga.				
8" pipe	L.F.	5.05	4.10	9.15
10" pipe	"	6.70	4.25	10.95
12" pipe	"	7.85	4.40	12.25
15" pipe	"	10.30	4.75	15.05
18" pipe	"	12.30	5.15	17.45
24" pipe	"	17.35	6.15	23.50
30" pipe	"	24.65	6.85	31.50
36" pipe	"	31.40	7.70	39.10
12 ga., 48" pipe	"	51.50	8.80	60.30
10 ga., 60" pipe	"	73.00	10.30	83.30
Galvanized or aluminum, coated oval arch				
16 ga.				
17" x 13"	L.F.	15.25	5.60	20.85
21" x 15"	"	18.45	6.15	24.60
14 ga.				
28" x 20"	L.F.	28.80	6.85	35.65
35" x 24"	"	36.90	8.80	45.70
12 ga.				
42" x 29"	L.F.	50.50	10.30	60.80
57" x 38"	"	67.00	12.35	79.35
64" x 43"	"	78.50	13.00	91.50
Oval arch culverts, plain				
16 ga.				
17" x 13"	L.F.	8.95	5.60	14.55
21" x 15"	"	10.65	6.15	16.80
14 ga.				
28" x 20"	L.F.	24.00	6.85	30.85
35" x 24"	"	30.00	8.80	38.80
12 ga.				
57" x 38"	L.F.	44.80	10.30	55.10
64" x 43"	"	50.50	12.35	62.85
71" x 47"	"	78.50	13.00	91.50
Nestable corrugated metal pipe				
16 ga.				
10" pipe	L.F.	7.60	4.25	11.85
12" pipe	"	10.15	4.40	14.55
15" pipe	"	12.70	4.75	17.45
18" pipe	"	15.00	5.15	20.15
24" pipe	"	20.75	6.15	26.90
30" pipe	"	24.25	6.85	31.10

SEWERAGE AND DRAINAGE	UNIT	MAT.	INST.	TOTAL
02720.40 STORM DRAINAGE				
14 ga., 36" pipe	L.F.	28.80	7.70	36.50
Headwalls, cast in place, 30 deg wingwall				
12" pipe	EA.	240.00	72.50	312.50
15" pipe	"	290.00	72.50	362.50
18" pipe	"	360.00	82.50	442.50
24" pipe	"	540.00	82.50	622.50
30" pipe	"	650.00	96.50	746.50
36" pipe	"	710.00	140.00	850.00
42" pipe	"	880.00	140.00	1,020
48" pipe	"	940.00	190.00	1,130
54" pipe	"	1,060	240.00	1,300
60" pipe	"	1,230	290.00	1,520
4" cleanout for storm drain				
4" pipe	EA.	320.00	28.40	348.40
6" pipe	"	390.00	28.40	418.40
8" pipe	"	550.00	28.40	578.40
Connect new drain line				
To existing manhole	EA.	61.50	75.50	137.00
To new manhole	"	50.50	45.40	95.90
02730.10 SANITARY SEWERS				
Clay				
6" pipe	L.F.	3.70	4.10	7.80
8" pipe	"	4.60	4.40	9.00
10" pipe	"	6.60	4.75	11.35
12" pipe	"	8.75	5.15	13.90
PVC				
4" pipe	L.F.	0.84	3.10	3.94
6" pipe	"	1.75	3.25	5.00
8" pipe	"	2.35	3.45	5.80
10" pipe	"	3.80	3.65	7.45
12" pipe	"	5.70	3.85	9.55
Cleanout				
4" pipe	EA.	4.25	28.40	32.65
6" pipe	"	9.05	28.40	37.45
8" pipe	"	27.70	28.40	56.10
Connect new sewer line				
To existing manhole	EA.	50.50	75.50	126.00
To new manhole	"	33.60	45.40	79.00
02740.10 DRAINAGE FIELDS				
Perforated PVC pipe, for drain field				
4" pipe	L.F.	0.86	2.75	3.61
6" pipe	"	1.60	2.95	4.55
02740.50 SEPTIC TANKS				
Septic tank, precast concrete				
1000 gals	EA.	500.00	210.00	710.00
2000 gals	"	950.00	310.00	1,260
5000 gals	"	4,140	620.00	4,760

SEWERAGE AND DRAINAGE	UNIT	MAT.	INST.	TOTAL
02740.50 SEPTIC TANKS				
25,000 gals	EA.	16,800	2,470	19,270
40,000 gals	"	24,080	4,110	28,190
Leaching pit, precast concrete, 72" diameter				
3' deep	EA.	370.00	150.00	520.00
6' deep	"	470.00	180.00	650.00
8' deep	"	580.00	210.00	790.00
02760.10 PIPELINE RESTORATION				
Relining existing water main				
6" dia.	L.F.	3.90	17.45	21.35
8" dia.	"	4.50	18.40	22.90
10" dia.	"	5.05	19.40	24.45
12" dia.	"	5.50	20.55	26.05
14" dia.	"	5.95	21.85	27.80
16" dia.	"	6.25	23.30	29.55
18" dia.	"	6.60	24.95	31.55
20" dia.	"	7.50	26.90	34.40
24" dia.	"	8.00	29.10	37.10
36" dia.	"	8.50	34.90	43.40
48" dia.	"	9.70	38.80	48.50
72" dia.	"	13.20	43.70	56.90
Replacing in line gate valves				
6" valve	EA.	450.00	230.00	680.00
8" valve	"	700.00	290.00	990.00
10" valve	"	1,050	350.00	1,400
12" valve	"	1,830	440.00	2,270
16" valve	"	4,140	500.00	4,640
18" valve	"	6,270	580.00	6,850
20" valve	"	8,620	700.00	9,320
24" valve	"	12,320	870.00	13,190
36" valve	"	33,600	1,160	34,760

POWER & COMMUNICATIONS	UNIT	MAT.	INST.	TOTAL
02780.20 HIGH VOLTAGE CABLE				
High voltage XLP copper cable, shielded, 5000v				
#6 awg	L.F.	0.97	0.48	1.45
#4 awg	"	1.10	0.59	1.69
#2 awg	"	1.45	0.71	2.16
#1 awg	"	1.55	0.78	2.33
#1/0 awg	"	1.85	0.89	2.74
#2/0 awg	"	3.10	1.10	4.20
#3/0 awg	"	3.15	1.25	4.40

POWER & COMMUNICATIONS	UNIT	MAT.	INST.	TOTAL
02780.20 HIGH VOLTAGE CABLE				
#4/0 awg	L.F.	3.30	1.35	4.65
#250 awg	"	3.85	1.60	5.45
#300 awg	"	4.40	1.80	6.20
#350 awg	"	4.95	2.00	6.95
#500 awg	"	7.05	2.70	9.75
#750 awg	"	10.40	2.95	13.35
Ungrounded, 15,000v				
#1 awg	L.F.	2.35	1.15	3.50
#1/0 awg	"	2.80	1.25	4.05
#2/0 awg	"	3.15	1.35	4.50
#3/0 awg	"	3.70	1.50	5.20
#4/0 awg	"	4.05	1.70	5.75
#250 awg	"	4.55	1.80	6.35
#300 awg	"	5.15	2.00	7.15
#350 awg	"	5.75	2.30	8.05
#500 awg	"	7.55	2.95	10.50
#750 awg	"	11.20	3.60	14.80
#1000 awg	"	16.50	4.55	21.05
Aluminum cable, shielded, 5000v				
#6 awg	L.F.	1.20	0.41	1.61
#4 awg	"	1.35	0.48	1.83
#2 awg	"	1.45	0.56	2.01
#1 awg	"	1.60	0.63	2.23
#1/0 awg	"	1.75	0.71	2.46
#2/0 awg	"	1.95	0.74	2.69
#3/0 awg	"	2.25	0.78	3.03
#4/0 awg	"	2.50	0.89	3.39
#250 awg	"	2.65	0.96	3.61
#300 awg	"	2.90	1.15	4.05
#350 awg	"	3.10	1.25	4.35
#500 awg	"	3.75	1.35	5.10
#750 awg	"	4.85	1.65	6.50
#1000 awg	"	5.50	1.85	7.35
Ungrounded, 15,000v				
#1 awg	L.F.	2.20	0.78	2.98
#1/0 awg	"	2.35	0.93	3.28
#2/0 awg	"	2.50	1.00	3.50
#3/0 awg	"	2.75	1.05	3.80
#4/0 awg	"	3.00	1.10	4.10
#250 awg	"	3.20	1.15	4.35
#300 awg	"	3.45	1.20	4.65
#350 awg	"	3.70	1.35	5.05
#500 awg	"	4.35	1.60	5.95
#750 awg	"	5.70	1.90	7.60
#1000 awg	"	7.55	2.35	9.90
Indoor terminations, 5000v				
#6 - #4	EA.	36.30	5.80	42.10
#2 - #2/0	"	42.30	5.80	48.10
#3/0 - #250	"	54.50	5.80	60.30
#300 - #750	"	60.50	100.00	160.50
#1000	"	72.50	140.00	212.50
In-line splice, 5000v				

POWER & COMMUNICATIONS	UNIT	MAT.	INST.	TOTAL
02780.20 — HIGH VOLTAGE CABLE				
#6 - #4/0	EA.	78.50	140.00	218.50
#250 - #500	"	84.50	370.00	454.50
#750 - #1000	"	120.00	480.00	600.00
T-splice, 5000v				
#2 - #4/0	EA.	78.50	440.00	518.50
#250 - #500	"	84.50	740.00	824.50
#750 - #1000	"	120.00	930.00	1,050
Indoor terminations, 15,000v				
#2 - #2/0	EA.	48.40	130.00	178.40
#3/0 - #500	"	60.50	200.00	260.50
#750 - #1000	"	72.50	230.00	302.50
In-line splice, 15,000v				
#2 - #4/0	EA.	84.50	330.00	414.50
#250 - #500	"	120.00	440.00	560.00
#750 - #1000	"	150.00	670.00	820.00
T-splice, 15,000v				
#4	EA.	84.50	670.00	754.50
#250 - #500	"	110.00	1,110	1,220
#750 - #1000	"	150.00	1,670	1,820
Compression lugs, 15,000v				
#4	EA.	5.70	14.85	20.55
#2	"	6.60	19.80	26.40
#1	"	7.20	19.80	27.00
#1/0	"	11.00	24.75	35.75
#2/0	"	11.55	24.75	36.30
#3/0	"	12.95	31.60	44.55
#4/0	"	14.30	31.60	45.90
#250	"	16.70	35.30	52.00
#300	"	19.35	35.30	54.65
#350	"	20.45	43.00	63.45
#500	"	30.20	46.40	76.60
#750	"	48.10	56.00	104.10
#1000	"	70.50	70.50	141.00
Compression splices, 15,000v				
#4	EA.	6.45	24.75	31.20
#2	"	7.05	27.00	34.05
#1	"	8.05	33.30	41.35
#1/0	"	8.50	37.10	45.60
#2/0	"	9.20	43.00	52.20
#3/0	"	10.00	46.40	56.40
#4/0	"	10.90	52.00	62.90
#250	"	11.95	56.00	67.95
#350	"	13.55	64.50	78.05
#500	"	20.05	74.00	94.05
#750	"	32.90	93.00	125.90
02780.40 — SUPPORTS & CONNECTORS				
Cable supports for conduit				
1-1/2"	EA.	40.40	12.90	53.30
2"	"	56.00	12.90	68.90
2-1/2"	"	62.50	14.85	77.35

POWER & COMMUNICATIONS	**UNIT**	**MAT.**	**INST.**	**TOTAL**
02780.40 SUPPORTS & CONNECTORS				
3"	EA.	81.00	14.85	95.85
3-1/2"	"	110.00	18.55	128.55
4"	"	130.00	18.55	148.55
5"	"	230.00	24.75	254.75
6"	"	490.00	27.00	517.00
Split bolt connectors				
#10	EA.	1.25	7.40	8.65
#8	"	1.45	7.40	8.85
#6	"	1.60	7.40	9.00
#4	"	1.90	14.85	16.75
#3	"	2.70	14.85	17.55
#2	"	3.05	14.85	17.90
#1/0	"	3.95	24.75	28.70
#2/0	"	6.30	24.75	31.05
#3/0	"	9.60	24.75	34.35
#4/0	"	10.85	24.75	35.60
#250	"	11.20	37.10	48.30
#350	"	19.80	37.10	56.90
#500	"	26.00	37.10	63.10
#750	"	44.00	56.00	100.00
#1000	"	60.50	56.00	116.50
Single barrel lugs				
#6	EA.	0.44	9.30	9.74
#1/0	"	0.85	18.55	19.40
#250	"	2.05	24.75	26.80
#350	"	2.70	24.75	27.45
#500	"	5.20	24.75	29.95
#600	"	5.55	33.30	38.85
#800	"	6.30	33.30	39.60
#1000	"	7.55	33.30	40.85
Double barrel lugs				
#1/0	EA.	1.70	33.30	35.00
#250	"	4.95	47.90	52.85
#350	"	7.05	47.90	54.95
#600	"	10.70	70.50	81.20
#800	"	12.20	70.50	82.70
#1000	"	12.60	70.50	83.10
Three barrel lugs				
#2/0	EA.	13.65	47.90	61.55
#250	"	26.10	70.50	96.60
#350	"	42.70	70.50	113.20
#600	"	46.80	99.00	145.80
#800	"	75.50	99.00	174.50
#1000	"	110.00	99.00	209.00
Four barrel lugs				
#250	EA.	29.10	100.00	129.10
#350	"	48.40	100.00	148.40
#600	"	53.00	130.00	183.00
#800	"	83.00	130.00	213.00
Compression conductor adapters				
#6	EA.	3.75	11.00	14.75
#4	"	3.95	12.90	16.85

POWER & COMMUNICATIONS	UNIT	MAT.	INST.	TOTAL
02780.40 SUPPORTS & CONNECTORS				
#2	EA.	4.15	16.50	20.65
#1	"	4.75	16.50	21.25
#1/0	"	4.95	19.80	24.75
#250	"	9.45	29.70	39.15
#350	"	11.20	31.60	42.80
#500	"	14.55	40.70	55.25
#750	"	19.95	42.40	62.35
Terminal blocks, 2 screw				
3 circuit	EA.	10.80	7.40	18.20
6 circuit	"	14.90	7.40	22.30
8 circuit	"	17.40	7.40	24.80
10 circuit	"	20.15	11.00	31.15
12 circuit	"	22.75	11.00	33.75
18 circuit	"	30.80	11.00	41.80
24 circuit	"	38.60	12.90	51.50
36 circuit	"	54.00	12.90	66.90
Compression splice				
#8 awg	EA.	1.85	14.15	16.00
#6 awg	"	2.20	10.25	12.45
#4 awg	"	2.30	10.25	12.55
#2 awg	"	3.55	19.80	23.35
#1 awg	"	5.05	19.80	24.85
#1/0 awg	"	6.15	19.80	25.95
#2/0 awg	"	6.55	31.60	38.15
#3/0 awg	"	7.65	31.60	39.25
#4/0 awg	"	8.10	31.60	39.70
#250 awg	"	8.60	50.50	59.10
#300 awg	"	9.35	50.50	59.85
#350 awg	"	9.55	52.00	61.55
#400 awg	"	12.95	52.00	64.95
#500 awg	"	15.15	56.00	71.15
#600 awg	"	22.95	56.00	78.95
#750 awg	"	24.35	64.50	88.85
#1000 awg	"	32.00	64.50	96.50

SITE IMPROVEMENTS	UNIT	MAT.	INST.	TOTAL
02810.40 LAWN IRRIGATION				
Residential system, complete				
Minimum	ACRE			10,304
Maximum	"			19,040
Commercial system, complete				
Minimum	ACRE			15,120
Maximum	"			23,520

<table>
<tr><td colspan="5"><h1>02   SITEWORK</h1></td></tr>
</table>

SITE IMPROVEMENTS	UNIT	MAT.	INST.	TOTAL
02810.40 LAWN IRRIGATION				
Components				
Pop-up head	EA.	8.95	14.20	23.15
Impact head	"	8.95	14.20	23.15
Rotary impact head	"	24.65	28.40	53.05
Shrub head	"	2.35	14.20	16.55
Hose bibb	"	4.35	28.40	32.75

FORMWORK	UNIT	MAT.	INST.	TOTAL

03110.20 — ELEVATED SLAB FORMWORK

	UNIT	MAT.	INST.	TOTAL
Elevated slab formwork				
Slab, with drop panels				
1 use	S.F.	2.50	2.30	4.80
5 uses	"	0.90	2.00	2.90
Floor slab, hung from steel beams				
1 use	S.F.	2.00	2.25	4.25
5 uses	"	0.75	1.95	2.70
Floor slab, with pans or domes				
1 use	S.F.	3.35	2.65	6.00
5 uses	"	1.70	2.25	3.95
Equipment curbs, 12" high				
1 use	L.F.	1.90	2.90	4.80
5 uses	"	0.81	2.40	3.21

03110.25 — EQUIPMENT PAD FORMWORK

	UNIT	MAT.	INST.	TOTAL
Equipment pad, job built				
1 use	S.F.	2.35	3.60	5.95
2 uses	"	1.40	3.40	4.80
3 uses	"	1.15	3.20	4.35
4 uses	"	0.86	3.05	3.91
5 uses	"	0.73	2.90	3.63

03110.35 — FOOTING FORMWORK

	UNIT	MAT.	INST.	TOTAL
Wall footings, job built, continuous				
1 use	S.F.	1.10	2.90	4.00
5 uses	"	0.50	2.40	2.90
Column footings, spread				
1 use	S.F.	1.20	3.60	4.80
5 uses	"	0.47	2.90	3.37

03110.55 — SLAB/MAT FORMWORK

	UNIT	MAT.	INST.	TOTAL
Edge forms				
6" high				
1 use	L.F.	1.75	2.65	4.40
5 uses	"	0.50	2.25	2.75
12" high				
1 use	L.F.	1.60	2.90	4.50
5 uses	"	0.46	2.40	2.86
Formwork for openings				
1 use	S.F.	2.35	5.80	8.15
5 uses	"	0.74	4.15	4.89

03110.65 — WALL FORMWORK

	UNIT	MAT.	INST.	TOTAL
Wall forms, exterior, job built				
Up to 8' high wall				
1 use	S.F.	2.05	2.90	4.95
5 uses	"	0.75	2.40	3.15
Over 8' high wall				
1 use	S.F.	2.25	3.60	5.85
5 uses	"	0.92	2.90	3.82

FORMWORK

03110.65	WALL FORMWORK	UNIT	MAT.	INST.	TOTAL
Column pier and pilaster					
1 use		S.F.	2.25	5.80	8.05
5 uses		"	1.00	4.15	5.15
Interior wall forms					
Up to 8' high					
1 use		S.F.	2.05	2.65	4.70
5 uses		"	0.74	2.25	2.99
Over 8' high					
1 use		S.F.	2.25	3.20	5.45
5 uses		"	0.93	2.65	3.58

REINFORCEMENT

03210.20	ELEVATED SLAB REINFORCING	UNIT	MAT.	INST.	TOTAL
Elevated slab					
#3 - #4		TON	650.00	410.00	1,060
#5 - #6		"	590.00	360.00	950.00
Galvanized					
#3 - #4		TON	1,150	410.00	1,560
#5 - #6		"	1,090	360.00	1,450

03210.25	EQUIP. PAD REINFORCING	UNIT	MAT.	INST.	TOTAL
Equipment pad					
#3 - #4		TON	650.00	660.00	1,310
#5 - #6		"	590.00	600.00	1,190
#7 - #8		"	560.00	550.00	1,110
#9 - #10		"	560.00	500.00	1,060
#11 - #12		"	560.00	470.00	1,030

03210.35	FOOTING REINFORCING	UNIT	MAT.	INST.	TOTAL
Footings					
Grade 50					
#3 - #4		TON	650.00	550.00	1,200
#5 - #6		"	590.00	470.00	1,060
Grade 60					
#3 - #4		TON	670.00	550.00	1,220
#5 - #6		"	610.00	470.00	1,080

03210.55	SLAB/MAT REINFORCING	UNIT	MAT.	INST.	TOTAL
Bars, slabs					
#3 - #4		TON	650.00	550.00	1,200
#5 - #6		"	590.00	470.00	1,060
#7 - #8		"	560.00	410.00	970.00

REINFORCEMENT		UNIT	MAT.	INST.	TOTAL
03210.55	**SLAB/MAT REINFORCING**				
Galvanized					
#3 - #4		TON	1,150	550.00	1,700
#5 - #6		"	1,090	470.00	1,560
03210.65	**WALL REINFORCING**				
Walls					
#3 - #4		TON	650.00	470.00	1,120
#5 - #6		"	590.00	410.00	1,000
Galvanized					
#3 - #4		TON	1,150	470.00	1,620
#5 - #6		"	1,090	410.00	1,500
Masonry wall (horizontal)					
#3 - #4		TON	650.00	1,310	1,960
#5 - #6		"	590.00	1,090	1,680
Galvanized					
#3 - #4		TON	1,150	1,310	2,460
#5 - #6		"	1,090	1,090	2,180
Masonry wall (vertical)					
#3 - #4		TON	650.00	1,640	2,290
#5 - #6		"	590.00	1,310	1,900
Galvanized					
#3 - #4		TON	1,150	1,640	2,790
#5 - #6		"	1,090	1,310	2,400

PLACING CONCRETE		UNIT	MAT.	INST.	TOTAL
03380.20	**ELEVATED SLAB CONCRETE**				
Elevated slab					
2500# or 3000# concrete					
By crane		C.Y.	57.50	21.50	79.00
By pump		"	57.50	16.55	74.05
By hand buggy		"	57.50	22.70	80.20
5000# concrete					
By crane		C.Y.	66.00	21.50	87.50
By pump		"	66.00	16.55	82.55
By hand buggy		"	66.00	22.70	88.70
03380.25	**EQUIPMENT PAD CONCRETE**				
Equipment pad					
2500# or 3000# concrete					
By chute		C.Y.	57.50	7.55	65.05
By pump		"	57.50	30.70	88.20

PLACING CONCRETE	UNIT	MAT.	INST.	TOTAL
03380.25 EQUIPMENT PAD CONCRETE				
By crane	C.Y.	57.50	35.80	93.30
3500# or 4000# concrete				
By chute	C.Y.	61.50	7.55	69.05
By pump	"	61.50	30.70	92.20
By crane	"	61.50	35.80	97.30
5000# concrete				
By chute	C.Y.	66.00	7.55	73.55
By pump	"	66.00	30.70	96.70
By crane	"	66.00	35.80	101.80
03380.35 FOOTING CONCRETE				
Continuous footing				
2500# or 3000# concrete				
By chute	C.Y.	57.50	7.55	65.05
By pump	"	57.50	26.90	84.40
By crane	"	57.50	30.70	88.20
5000# concrete				
By chute	C.Y.	66.00	7.55	73.55
By pump	"	66.00	26.90	92.90
By crane	"	66.00	30.70	96.70
Spread footing				
2500# or 3000# concrete				
Under 5 cy				
By chute	C.Y.	57.50	7.55	65.05
By pump	"	57.50	28.70	86.20
By crane	"	57.50	33.10	90.60
5000# concrete				
Under 5 c.y.				
By chute	C.Y.	66.00	7.55	73.55
By pump	"	66.00	28.70	94.70
By crane	"	66.00	33.10	99.10
03380.55 SLAB/MAT CONCRETE				
Slab on grade				
2500# or 3000# concrete				
By chute	C.Y.	57.50	5.65	63.15
By crane	"	57.50	17.90	75.40
By pump	"	57.50	15.35	72.85
By hand buggy	"	57.50	15.10	72.60
5000# concrete				
By chute	C.Y.	66.00	5.65	71.65
By crane	"	66.00	17.90	83.90
By pump	"	66.00	15.35	81.35
By hand buggy	"	66.00	15.10	81.10
03380.65 WALL CONCRETE				
Walls				
2500# or 3000# concrete				
To 4'				
By chute	C.Y.	57.50	6.50	64.00

PLACING CONCRETE	UNIT	MAT.	INST.	TOTAL
03380.65 — WALL CONCRETE				
By crane	C.Y.	57.50	35.80	93.30
By pump	"	57.50	33.10	90.60
To 8'				
By crane	C.Y.	57.50	39.10	96.60
By pump	"	57.50	35.80	93.30
Filled block (CMU)				
3000# concrete, by pump				
4" wide	S.F.	0.21	1.55	1.76
6" wide	"	0.48	1.80	2.28
8" wide	"	0.75	2.15	2.90
10" wide	"	1.00	2.55	3.55
12" wide	"	1.30	3.05	4.35
Pilasters, 3000# concrete	C.F.	2.90	43.00	45.90
Wall cavity, 2" thick, 3000# concrete	S.F.	0.54	1.45	1.99

MORTAR AND GROUT	UNIT	MAT.	INST.	TOTAL
04100.10 MASONRY GROUT				
Grout, non shrink, non-metallic, trowelable	C.F.	6.85	0.82	7.67
Grout-filled individual CMU cells				
4" wide	L.F.	0.13	0.62	0.75
8" wide	"	0.40	0.62	1.02
04150.50 MASONRY FLASHING				
Through-wall flashing				
5 oz. coated copper	S.F.	2.60	2.35	4.95
0.030" elastomeric	"	0.55	1.85	2.40

UNIT MASONRY	UNIT	MAT.	INST.	TOTAL
04210.10 BRICK MASONRY				
Standard size brick, running bond				
Face brick, red (6.4/sf)				
Veneer	S.F.	1.95	4.70	6.65
Cavity wall	"	1.95	4.00	5.95
9" solid wall	"	4.20	8.05	12.25
Chimney, standard brick, including flue				
16" x 16"	L.F.	15.10	28.10	43.20
16" x 20"	"	17.90	28.10	46.00
16" x 24"	"	21.30	28.10	49.40
20" x 20"	"	22.40	35.20	57.60
20" x 24"	"	29.10	35.20	64.30
20" x 32"	"	37.00	40.20	77.20
04220.10 CONCRETE MASONRY UNITS				
Hollow, load bearing				
4"	S.F.	1.15	2.10	3.25
8"	"	1.90	2.35	4.25
Solid, load bearing				
4"	S.F.	1.55	2.10	3.65
8"	"	2.15	2.35	4.50
Back-up block, 8" x 16"				
4"	S.F.	1.10	1.65	2.75
8"	"	1.60	1.85	3.45
Steel angles and plates				
Minimum	Lb.	0.28	0.40	0.68
Maximum	"	0.31	0.70	1.01
Various size angle lintels				
1/4" stock				
3" x 3"	L.F.	1.50	1.75	3.25
3" x 3-1/2"	"	1.60	1.75	3.35

UNIT MASONRY	UNIT	MAT.	INST.	TOTAL
04220.10 — CONCRETE MASONRY UNITS				
3/8" stock				
3" x 4"	L.F.	2.70	1.75	4.45
4" x 4"	"	2.95	1.75	4.70
1/2" stock				
6" x 4"	L.F.	4.95	1.75	6.70
04295.10 — PARGING/MASONRY PLASTER				
Parging				
1/2" thick	S.F.	0.36	1.85	2.21
3/4" thick	"	0.47	2.35	2.82

MASONRY RESTORATION	UNIT	MAT.	INST.	TOTAL
04520.10 — RESTORATION AND CLEANING				
Masonry cleaning				
Washing brick				
Smooth surface	S.F.	0.27	0.47	0.74
Rough surface	"	0.37	0.63	1.00
Pointing masonry				
Brick	S.F.	0.84	1.10	1.94
Concrete block	"	0.37	0.80	1.17
Brick removal and replacement				
Minimum	EA.	0.37	3.50	3.87
Average	"	0.49	4.70	5.19
Maximum	"	0.95	14.05	15.00
04550.10 — REFRACTORIES				
Flue liners				
Rectangular				
8" x 12"	L.F.	4.80	4.70	9.50
12" x 12"	"	6.85	5.10	11.95
12" x 18"	"	10.40	5.60	16.00
16" x 16"	"	11.75	6.25	18.00
18" x 18"	"	15.00	6.70	21.70
20" x 20"	"	23.50	7.05	30.55
24" x 24"	"	32.50	8.05	40.55
Round				
18" dia.	L.F.	24.65	6.70	31.35
24" dia.	"	40.30	8.05	48.35

METAL FASTENING	UNIT	MAT.	INST.	TOTAL
05050.10 STRUCTURAL WELDING				
Welding				
Single pass				
1/8"	L.F.	0.35	1.60	1.95
3/16"	"	0.69	2.10	2.79
1/4"	"	0.93	2.65	3.58
Miscellaneous steel shapes				
Plain	Lb.	0.69	0.06	0.75
Galvanized	"	1.05	0.11	1.16
Plates				
Plain	Lb.	0.64	0.08	0.72
Galvanized	"	0.95	0.13	1.08
05050.90 METAL ANCHORS				
Anchor bolts				
3/8" x				
8" long	EA.			3.02
10" long	"			3.25
12" long	"			3.47
1/2" x				
8" long	EA.			4.14
10" long	"			4.26
12" long	"			4.59
18" long	"			4.93
Add 25% for galvanized anchor bolts				
05050.95 METAL LINTELS				
Lintels, steel				
Plain	Lb.	0.69	0.79	1.48
Galvanized	"	1.00	0.79	1.79

COLD FORMED FRAMING	UNIT	MAT.	INST.	TOTAL
05410.10 METAL FRAMING				
Furring channel, galvanized				
Beams and columns, 3/4"				
12" o.c.	S.F.	0.34	3.15	3.49
16" o.c.	"	0.28	2.90	3.18
Walls, 3/4"				
12" o.c.	S.F.	0.34	1.60	1.94
16" o.c.	"	0.28	1.30	1.58
Stud, load bearing				

COLD FORMED FRAMING	UNIT	MAT.	INST.	TOTAL
05410.10 — METAL FRAMING				
16" o.c.				
16 ga.				
2-1/2"	S.F.	1.05	1.40	2.45
3-5/8"	"	1.30	1.40	2.70
24" o.c.				
16 ga.				
2-1/2"	S.F.	0.78	1.20	1.98
3-5/8"	"	1.00	1.20	2.20

FASTENERS AND ADHESIVES

	UNIT	MAT.	INST.	TOTAL
06050.10 ACCESSORIES				
Anchors				
Bolts, threaded two ends, with nuts and washers				
1/2" dia.				
4" long	EA.	0.90	1.80	2.70
7-1/2" long	"	1.10	1.80	2.90
3/4" dia.				
7-1/2" long	EA.	1.75	1.80	3.55
15" long	"	2.25	1.80	4.05
Bolts, carriage				
1/4 x 4	EA.	0.91	2.90	3.81
5/16 x 6	"	0.96	3.05	4.01
3/8 x 6	"	1.10	3.05	4.15
1/2 x 6	"	1.25	3.05	4.30
Joist and beam hangers				
18 ga.				
2 x 4	EA.	0.45	2.90	3.35
2 x 6	"	0.56	2.90	3.46
2 x 8	"	0.67	2.90	3.57
2 x 10	"	0.78	3.20	3.98
2 x 12	"	0.90	3.60	4.50
Strap ties, 14 ga., 1-3/8" wide				
12" long	EA.	0.94	2.40	3.34
18" long	"	1.05	2.65	3.70
24" long	"	1.35	2.90	4.25
36" long	"	2.00	3.20	5.20

ROUGH CARPENTRY

	UNIT	MAT.	INST.	TOTAL
06110.10 BLOCKING				
Steel construction				
Walls				
2x4	L.F.	0.41	1.95	2.36
2x6	"	0.63	2.25	2.88
2x8	"	0.87	2.40	3.27
2x10	"	1.30	2.65	3.95
2x12	"	1.70	2.90	4.60
Ceilings				
2x4	L.F.	0.41	2.25	2.66
2x6	"	0.63	2.65	3.28
2x8	"	0.87	2.90	3.77
2x10	"	1.30	3.20	4.50
2x12	"	1.70	3.60	5.30
Wood construction				

ROUGH CARPENTRY	UNIT	MAT.	INST.	TOTAL
06110.10 **BLOCKING**				
Walls				
2x4	L.F.	0.41	1.60	2.01
2x6	"	0.63	1.80	2.43
2x8	"	0.87	1.95	2.82
2x10	"	1.30	2.05	3.35
2x12	"	1.70	2.25	3.95
Ceilings				
2x4	L.F.	0.41	1.80	2.21
2x6	"	0.63	2.05	2.68
2x8	"	0.87	2.25	3.12
2x10	"	1.30	2.40	3.70
2x12	"	1.70	2.65	4.35
06110.20 **CEILING FRAMING**				
Ceiling joists				
12" o.c.				
2x4	S.F.	0.49	0.69	1.18
2x8	"	1.05	0.76	1.81
16" o.c.				
2x4	S.F.	0.41	0.56	0.97
2x8	"	0.87	0.60	1.47
Sister joists for ceilings				
2x4	L.F.	0.41	2.05	2.46
2x6	"	0.63	2.40	3.03
2x8	"	0.87	2.90	3.77
2x10	"	1.30	3.60	4.90
2x12	"	1.70	4.85	6.55
06110.30 **FLOOR FRAMING**				
Floor joists				
12" o.c.				
2x6	S.F.	0.74	0.58	1.32
2x8	"	1.05	0.59	1.64
2x10	"	1.55	0.60	2.15
2x12	"	2.05	0.63	2.68
16" o.c.				
2x6	S.F.	0.63	0.48	1.11
2x8	"	0.87	0.49	1.36
2x10	"	1.30	0.50	1.80
2x12	"	1.70	0.52	2.22
Sister joists for floors				
2x4	L.F.	0.41	1.80	2.21
2x6	"	0.62	2.05	2.67
2x8	"	0.87	2.40	3.27
2x10	"	1.30	2.90	4.20
2x12	"	1.70	3.60	5.30
3x6	"	1.60	2.90	4.50
3x8	"	2.10	3.20	5.30
3x10	"	2.60	3.60	6.20
3x12	"	3.15	4.15	7.30
4x6	"	2.05	2.90	4.95

ROUGH CARPENTRY	UNIT	MAT.	INST.	TOTAL
06110.30 FLOOR FRAMING				
4x8	L.F.	2.80	3.20	6.00
4x10	"	3.50	3.60	7.10
4x12	"	4.15	4.15	8.30
06110.40 FURRING				
Furring, wood strips				
Walls				
On masonry or concrete walls				
1x2 furring				
12" o.c.	S.F.	0.18	0.91	1.09
16" o.c.	"	0.15	0.83	0.98
1x3 furring				
12" o.c.	S.F.	0.25	0.91	1.16
16" o.c.	"	0.20	0.83	1.03
On wood walls				
1x2 furring				
12" o.c.	S.F.	0.18	0.64	0.82
16" o.c.	"	0.15	0.58	0.73
1x3 furring				
12" o.c.	S.F.	0.25	0.64	0.89
16" o.c.	"	0.20	0.58	0.78
Ceilings				
On masonry or concrete ceilings				
1x2 furring				
12" o.c.	S.F.	0.18	1.60	1.78
16" o.c.	"	0.15	1.45	1.60
1x3 furring				
12" o.c.	S.F.	0.25	1.60	1.85
16" o.c.	"	0.20	1.45	1.65
On wood ceilings				
1x2 furring				
12" o.c.	S.F.	0.18	1.05	1.23
16" o.c.	"	0.15	0.96	1.12
1x3				
12" o.c.	S.F.	0.25	1.05	1.30
16" o.c.	"	0.20	0.96	1.17
06110.50 ROOF FRAMING				
Roof framing				
Rafters, gable end				
0-2 pitch (flat to 2-in-12)				
12" o.c.				
2x4	S.F.	0.49	0.60	1.09
2x6	"	0.75	0.63	1.38
2x8	"	1.05	0.66	1.71
16" o.c.				
2x6	S.F.	0.63	0.52	1.15
2x8	"	0.87	0.54	1.41
Sister rafters				
2x4	L.F.	0.41	2.05	2.46

ROUGH CARPENTRY	UNIT	MAT.	INST.	TOTAL
06110.50 ROOF FRAMING				
2x6	L.F.	0.63	2.40	3.03
2x8	"	0.87	2.90	3.77
2x10	"	1.30	3.60	4.90
2x12	"	1.70	4.85	6.55
06110.60 SLEEPERS				
Sleepers, over concrete				
12" o.c.				
1x2	S.F.	0.18	0.66	0.84
1x3	"	0.25	0.69	0.94
2x4	"	0.49	0.80	1.29
2x6	"	0.74	0.85	1.59
16" o.c.				
1x2	S.F.	0.15	0.58	0.73
1x3	"	0.20	0.58	0.78
2x4	"	0.41	0.69	1.10
2x6	"	0.63	0.72	1.35
06110.65 SOFFITS				
Soffit framing				
2x3	L.F.	0.30	2.05	2.35
2x4	"	0.41	2.25	2.66
2x6	"	0.63	2.40	3.03
2x8	"	0.87	2.65	3.52
06110.70 WALL FRAMING				
Framing wall, studs				
12" o.c.				
2x3	S.F.	0.36	0.54	0.90
2x4	"	0.49	0.54	1.03
16" o.c.				
2x3	S.F.	0.30	0.45	0.75
2x4	"	0.41	0.45	0.86
24" o.c.				
2x3	S.F.	0.24	0.39	0.63
2x4	"	0.32	0.39	0.71
Plates, top or bottom				
2x3	L.F.	0.30	0.85	1.15
2x4	"	0.41	0.91	1.32
Headers, door or window				
2x6				
Single				
3' long	EA.	1.90	14.50	16.40
6' long	"	3.75	18.10	21.85
Double				
3' long	EA.	3.75	16.10	19.85
6' long	"	7.55	20.70	28.25
2x8				
Single				
4' long	EA.	3.50	18.10	21.60

ROUGH CARPENTRY	UNIT	MAT.	INST.	TOTAL
06110.70 — WALL FRAMING				
8' long	EA.	7.00	22.30	29.30
Double				
4' long	EA.	7.00	20.70	27.70
8' long	"	14.00	26.30	40.30
2x10				
Single				
5' long	EA.	6.50	22.30	28.80
10' long	"	13.00	29.00	42.00
Double				
5' long	EA.	13.00	24.15	37.15
10' long	"	26.00	29.00	55.00
06115.10 — FLOOR SHEATHING				
Sub-flooring, plywood, CDX				
1/2" thick	S.F.	0.45	0.36	0.81
5/8" thick	"	0.54	0.41	0.95
3/4" thick	"	0.64	0.48	1.12
Structural plywood				
1/2" thick	S.F.	0.49	0.36	0.85
Board type subflooring				
1x6				
Minimum	S.F.	0.74	0.64	1.38
Maximum	"	1.00	0.72	1.72
1x8				
Minimum	S.F.	0.88	0.61	1.49
Maximum	"	1.00	0.68	1.68
1x10				
Minimum	S.F.	1.05	0.58	1.63
Maximum	"	1.10	0.64	1.74
Underlayment				
Hardboard, 1/4" tempered	S.F.	0.36	0.36	0.72
Plywood, CDX				
3/8" thick	S.F.	0.39	0.36	0.75
1/2" thick	"	0.45	0.39	0.84
5/8" thick	"	0.54	0.41	0.95
3/4" thick	"	0.64	0.45	1.09
06115.20 — ROOF SHEATHING				
Sheathing				
Plywood, CDX				
3/8" thick	S.F.	0.39	0.37	0.76
1/2" thick	"	0.45	0.39	0.84
5/8" thick	"	0.54	0.41	0.95
06115.30 — WALL SHEATHING				
Sheathing				
Plywood, CDX				
3/8" thick	S.F.	0.39	0.43	0.82
1/2" thick	"	0.45	0.45	0.90

ROUGH CARPENTRY		UNIT	MAT.	INST.	TOTAL
06115.30	**WALL SHEATHING**				
5/8" thick		S.F.	0.54	0.48	1.02
3/4" thick		"	0.64	0.53	1.17
06220.10	**MILLWORK**				
Countertop, laminated plastic					
25" x 7/8" thick					
Minimum		L.F.	12.30	7.25	19.55
Average		"	19.80	9.65	29.45
Maximum		"	31.40	11.60	43.00
25" x 1-1/4" thick					
Minimum		L.F.	13.45	9.65	23.10
Average		"	20.15	11.60	31.75
Maximum		"	33.60	14.50	48.10
Add for cutouts		EA.	0.00	18.10	18.10
Backsplash, 4" high, 7/8" thick		L.F.	11.75	5.80	17.55
Base cabinets, 34-1/2" high, 24" deep, hardwood, no tops					
Minimum		L.F.	39.20	11.60	50.80
Average		"	55.00	14.50	69.50
Maximum		"	89.50	19.30	108.80
Wall cabinets					
Minimum		L.F.	28.00	9.65	37.65
Average		"	42.60	11.60	54.20
Maximum		"	78.50	14.50	93.00

MOISTURE PROTECTION

MOISTURE PROTECTION	UNIT	MAT.	INST.	TOTAL
07150.10 DAMPROOFING				
Silicone dampproofing, sprayed on				
Concrete surface				
1 coat	S.F.	0.32	0.13	0.45
2 coats	"	0.56	0.17	0.73
Concrete block				
1 coat	S.F.	0.32	0.15	0.47
2 coats	"	0.56	0.21	0.77
Brick				
1 coat	S.F.	0.38	0.17	0.55
2 coats	"	0.62	0.23	0.85
07160.10 BITUMINOUS DAMPPROOFING				
Building paper, asphalt felt				
15 lb	S.F.	0.07	0.91	0.98
30 lb	"	0.12	0.94	1.07
07190.10 VAPOR BARRIERS				
Vapor barrier, polyethylene				
2 mil	S.F.	0.01	0.11	0.12
6 mil	"	0.02	0.11	0.13
8 mil	"	0.03	0.13	0.16
10 mil	"	0.04	0.13	0.17

INSULATION

INSULATION	UNIT	MAT.	INST.	TOTAL
07210.10 BATT INSULATION				
Ceiling, fiberglass, unfaced				
3-1/2" thick, R11	S.F.	0.25	0.27	0.52
6" thick, R19	"	0.41	0.30	0.71
9" thick, R30	"	0.64	0.35	0.99
Suspended ceiling, unfaced				
3-1/2" thick, R11	S.F.	0.25	0.25	0.50
6" thick, R19	"	0.41	0.28	0.69
9" thick, R30	"	0.64	0.32	0.96
Crawl space, unfaced				
3-1/2" thick, R11	S.F.	0.25	0.35	0.60
6" thick, R19	"	0.41	0.38	0.79
9" thick, R30	"	0.64	0.41	1.05
Wall, fiberglass				
Paper backed				
2" thick, R7	S.F.	0.17	0.24	0.41
3" thick, R8	"	0.18	0.25	0.43

INSULATION	UNIT	MAT.	INST.	TOTAL
07210.10	**BATT INSULATION**			
4" thick, R11	S.F.	0.28	0.27	0.55
6" thick, R19	"	0.43	0.28	0.71
Foil backed, 1 side				
2" thick, R7	S.F.	0.36	0.24	0.60
3" thick, R11	"	0.38	0.25	0.63
4" thick, R14	"	0.43	0.27	0.70
6" thick, R21	"	0.45	0.28	0.73
Foil backed, 2 sides				
2" thick, R7	S.F.	0.43	0.27	0.70
3" thick, R11	"	0.53	0.28	0.81
4" thick, R14	"	0.62	0.30	0.92
6" thick, R21	"	0.66	0.32	0.98
Unfaced				
2" thick, R7	S.F.	0.18	0.24	0.42
3" thick, R9	"	0.20	0.25	0.45
4" thick, R11	"	0.25	0.27	0.52
6" thick, R19	"	0.41	0.28	0.69
Mineral wool batts				
Paper backed				
2" thick, R6	S.F.	0.35	0.24	0.59
4" thick, R12	"	0.43	0.25	0.68
6" thick, R19	"	0.49	0.28	0.77
Fasteners, self adhering, attached to ceiling deck				
2-1/2" long	EA.	0.11	0.38	0.49
4-1/2" long	"	0.13	0.41	0.54
Capped, self-locking washers for fastening insulation	"	0.11	0.23	0.34
07210.20	**BOARD INSULATION**			
Insulation, rigid				
Fiberglass, roof				
0.75" thick, R2.78	S.F.	0.22	0.21	0.43
1.06" thick, R4.17	"	0.45	0.22	0.67
1.31" thick, R5.26	"	0.67	0.23	0.90
1.63" thick, R6.67	"	0.84	0.24	1.08
2.25" thick, R8.33	"	0.78	0.25	1.03
Perlite board, roof				
1.00" thick, R2.78	S.F.	0.45	0.19	0.64
1.50" thick, R4.17	"	0.69	0.20	0.89
2.00" thick, R5.92	"	0.85	0.21	1.06
2.50" thick, R6.67	"	1.05	0.22	1.27
3.00" thick, R8.33	"	1.30	0.23	1.53
4.00" thick, R10.00	"	1.45	0.24	1.69
5.25" thick, R14.29	"	1.60	0.25	1.85
07210.60	**LOOSE FILL INSULATION**			
Blown-in type				
Fiberglass				
5" thick, R11	S.F.	0.20	0.19	0.39
6" thick, R13	"	0.25	0.23	0.48
9" thick, R19	"	0.28	0.32	0.60
Rockwool, attic application				

INSULATION	UNIT	MAT.	INST.	TOTAL
07210.60 — LOOSE FILL INSULATION				
6" thick, R13	S.F.	0.36	0.23	0.59
8" thick, R19	"	0.44	0.28	0.72
10" thick, R22	"	0.53	0.35	0.88
12" thick, R26	"	0.68	0.38	1.06
15" thick, R30	"	0.83	0.45	1.28
Poured type				
Fiberglass				
1" thick, R4	S.F.	0.18	0.14	0.32
2" thick, R8	"	0.34	0.16	0.50
3" thick, R12	"	0.50	0.19	0.69
4" thick, R16	"	0.67	0.23	0.90
Mineral wool				
1" thick, R3	S.F.	0.20	0.14	0.34
2" thick, R6	"	0.38	0.16	0.54
3" thick, R9	"	0.58	0.19	0.77
4" thick, R12	"	0.67	0.23	0.90
Vermiculite or perlite				
2" thick, R4.8	S.F.	0.45	0.16	0.61
3" thick, R7.2	"	0.65	0.19	0.84
4" thick, R9.6	"	0.84	0.23	1.07
Masonry, poured vermiculite or perlite				
4" block	S.F.	0.34	0.11	0.45
6" block	"	0.43	0.14	0.57
8" block	"	0.66	0.16	0.82
10" block	"	0.78	0.17	0.95
12" block	"	0.87	0.19	1.06
07210.70 — SPRAYED INSULATION				
Foam, sprayed on				
Polystyrene				
1" thick, R4	S.F.	0.34	0.23	0.57
2" thick, R8	"	0.67	0.30	0.97
Urethane				
1" thick, R7.7	S.F.	0.39	0.23	0.62
2" thick, R15.4	"	0.67	0.30	0.97
07250.10 — FIREPROOFING				
Sprayed on				
1" thick				
On beams	S.F.	0.37	0.50	0.87
On columns	"	0.39	0.45	0.84
On decks				
Flat surface	S.F.	0.37	0.23	0.60
Fluted surface	"	0.40	0.28	0.68
1-1/2" thick				
On beams	S.F.	0.56	0.65	1.21
On columns	"	0.62	0.57	1.19
On decks				
Flat surface	S.F.	0.56	0.28	0.84
Fluted surface	"	0.59	0.38	0.97

SHINGLES AND TILES	UNIT	MAT.	INST.	TOTAL
07310.10 ASPHALT SHINGLES				
Standard asphalt shingles, strip shingles				
210 lb/square	SQ.	37.00	27.90	64.90
235 lb/square	"	39.20	31.00	70.20
240 lb/square	"	40.30	34.90	75.20
260 lb/square	"	62.50	39.90	102.40
300 lb/square	"	64.50	46.50	111.00
385 lb/square	"	89.50	56.00	145.50
Roll roofing, mineral surface				
90 lb	SQ.	19.25	19.95	39.20
110 lb	"	31.10	23.25	54.35
140 lb	"	32.30	27.90	60.20
07310.30 METAL SHINGLES				
Aluminum, .020" thick				
Plain	SQ.	150.00	56.00	206.00
Colors	"	180.00	56.00	236.00
Steel, galvanized				
26 ga.				
Plain	SQ.	160.00	56.00	216.00
Colors	"	200.00	56.00	256.00
Porcelain enamel, 22 ga.				
Minimum	SQ.	480.00	70.00	550.00
Average	"	550.00	70.00	620.00
Maximum	"	620.00	70.00	690.00
Replacement shingles				
Small jobs	EA.	5.90	9.30	15.20
Large jobs	S.F.	3.90	4.65	8.55
07310.70 WOOD SHINGLES				
Wood shingles, on roofs				
White cedar, #1 shingles				
4" exposure	SQ.	160.00	93.00	253.00
5" exposure	"	140.00	70.00	210.00
On walls				
White cedar, #1 shingles				
4" exposure	SQ.	160.00	140.00	300.00
5" exposure	"	140.00	110.00	250.00
07310.80 WOOD SHAKES				
Shakes, hand split, 24" red cedar, on roofs				
5" exposure	SQ.	130.00	140.00	270.00
9" exposure	"	120.00	93.00	213.00

MEMBRANE ROOFING	UNIT	MAT.	INST.	TOTAL
07510.10 BUILT-UP ASPHALT ROOFING				
Built-up roofing, asphalt felt, including gravel				
2 ply	SQ.	25.90	70.00	95.90
3 ply	"	36.50	93.00	129.50
4 ply	"	48.20	110.00	158.20
07530.10 SINGLE-PLY ROOFING				
Elastic sheet roofing				
Neoprene, 1/16" thick	S.F.	1.40	0.35	1.75
EPDM rubber				
45 mil	S.F.	0.77	0.35	1.12
60 mil	"	1.05	0.35	1.40
PVC				
45 mil	S.F.	1.05	0.35	1.40
60 mil	"	1.55	0.35	1.90
Flashing				
Pipe flashing, 90 mil thick				
1" pipe	EA.	14.55	7.00	21.55
2" pipe	"	15.70	7.00	22.70
3" pipe	"	15.90	7.35	23.25
4" pipe	"	17.35	7.35	24.70
5" pipe	"	18.50	7.75	26.25
6" pipe	"	20.25	7.75	28.00
8" pipe	"	22.40	8.20	30.60
10" pipe	"	26.60	9.30	35.90
12" pipe	"	31.90	9.30	41.20
Neoprene flashing, 60 mil thick strip				
6" wide	L.F.	0.88	2.35	3.23
12" wide	"	1.75	3.50	5.25
18" wide	"	2.55	4.65	7.20
24" wide	"	3.30	7.00	10.30
Adhesives				
Mastic sealer, applied at joints only				
1/4" bead	L.F.	0.06	0.14	0.20
Ballast, 3/4" through 1-1/2" dia. river gravel, 100lb/sf	S.F.	0.21	27.90	28.11
Walkway for membrane roofs, 1/2" thick	"	1.25	0.93	2.18

FLASHING AND SHEET METAL	UNIT	MAT.	INST.	TOTAL
07610.10 METAL ROOFING				
Sheet metal roofing, copper, 16 oz, batten seam	SQ.	450.00	190.00	640.00
Standing seam	"	440.00	170.00	610.00
Aluminum roofing, natural finish				
Corrugated, on steel frame				

FLASHING AND SHEET METAL	UNIT	MAT.	INST.	TOTAL
07610.10 — METAL ROOFING				
.0175" thick	SQ.	67.00	80.00	147.00
.032" thick	"	160.00	80.00	240.00
Corrugated galvanized steel roofing, on steel frame				
28 ga.	SQ.	78.50	80.00	158.50
22 ga.	"	110.00	80.00	190.00
07620.10 — FLASHING AND TRIM				
Counter flashing				
Aluminum, .032"	S.F.	0.93	2.80	3.73
Stainless steel, .015"	"	2.70	2.80	5.50
16 oz.	"	2.70	2.80	5.50
Base flashing				
Aluminum, .040"	S.F.	1.45	2.35	3.80
Stainless steel, .018"	"	3.25	2.35	5.60
16 oz.	"	2.70	2.35	5.05
24 oz.	"	4.25	2.35	6.60
Scupper outlets				
10" x 10" x 4"	EA.	16.80	7.00	23.80
22" x 4" x 4"	"	20.70	7.00	27.70
8" x 8" x 5"	"	16.80	7.00	23.80
Drainage boots, roof, cast iron				
2 x 3	L.F.	29.60	3.50	33.10
3 x 4	"	37.00	3.50	40.50
4 x 5	"	53.00	3.70	56.70
4 x 6	"	50.50	3.70	54.20
5 x 7	"	60.50	4.00	64.50
Pitch pocket, copper, 16 oz.				
4 x 4	EA.	51.50	7.00	58.50
6 x 6	"	57.00	7.00	64.00
8 x 8	"	60.50	7.00	67.50
8 x 10	"	65.00	7.00	72.00
8 x 12	"	78.50	7.00	85.50
Reglets, copper 10 oz.	L.F.	2.00	1.85	3.85
Stainless steel, .020"	"	1.40	1.85	3.25
07620.20 — GUTTERS AND DOWNSPOUTS				
Copper gutter and downspout				
Downspouts, 16 oz. copper				
Round				
3" dia.	L.F.	4.55	1.85	6.40
4" dia.	"	5.80	1.85	7.65
Rectangular, corrugated				
2" x 3"	L.F.	4.35	1.75	6.10
3" x 4"	"	5.70	1.75	7.45
Rectangular, flat surface				
2" x 3"	L.F.	4.75	1.85	6.60
3" x 4"	"	6.80	1.85	8.65
Lead-coated copper downspouts				
Round				
3" dia.	L.F.	5.60	1.75	7.35

FLASHING AND SHEET METAL

FLASHING AND SHEET METAL	UNIT	MAT.	INST.	TOTAL
07620.20 GUTTERS AND DOWNSPOUTS				
4" dia.	L.F.	6.70	2.00	8.70
Rectangular, corrugated				
2" x 3"	L.F.	4.15	1.85	6.00
3" x 4"	"	4.95	1.85	6.80
Rectangular, plain				
2" x 3"	L.F.	3.60	1.85	5.45
3" x 4"	"	4.25	1.85	6.10
Gutters, 16 oz. copper				
Half round				
4" wide	L.F.	4.20	2.80	7.00
5" wide	"	5.05	3.10	8.15
Type K				
4" wide	L.F.	4.60	2.80	7.40
5" wide	"	5.05	3.10	8.15
Lead-coated copper gutters				
Half round				
4" wide	L.F.	3.90	2.80	6.70
6" wide	"	6.15	3.10	9.25
Type K				
4" wide	L.F.	5.40	2.80	8.20
5" wide	"	5.80	3.10	8.90
Aluminum gutter and downspout				
Downspouts				
2" x 3"	L.F.	0.64	1.85	2.49
3" x 4"	"	0.85	2.00	2.85
4" x 5"	"	0.95	2.15	3.10
Round				
3" dia.	L.F.	1.05	1.85	2.90
4" dia.	"	1.35	2.00	3.35
Gutters, stock units				
4" wide	L.F.	1.05	2.95	4.00
5" wide	"	1.15	3.10	4.25
Galvanized steel gutter and downspout				
Downspouts, round corrugated				
3" dia.	L.F.	0.90	1.85	2.75
4" dia.	"	1.00	1.85	2.85
5" dia.	"	1.25	2.00	3.25
6" dia.	"	1.55	2.00	3.55
Rectangular				
2" x 3"	L.F.	0.90	1.85	2.75
3" x 4"	"	1.40	1.75	3.15
4" x 4"	"	1.80	1.75	3.55
Gutters, stock units				
5" wide				
Plain	L.F.	1.00	3.10	4.10
Painted	"	1.10	3.10	4.20
6" wide				
Plain	L.F.	1.35	3.30	4.65
Painted	"	1.55	3.30	4.85

FLASHING AND SHEET METAL	UNIT	MAT.	INST.	TOTAL
07700.10 MANUFACTURED SPECIALTIES				
Ceiling access doors				
Swing up model, metal frame				
Steel door				
2'6" x 2'6"	EA.	290.00	27.90	317.90
2'6" x 3'0"	"	310.00	27.90	337.90
Aluminum door				
2'6" x 2'6"	EA.	310.00	27.90	337.90
2'6" x 3'0"	"	330.00	27.90	357.90
Swing down model, metal frame				
Steel door				
2'6" x 2'6"	EA.	290.00	27.90	317.90
2'6" x 3'0"	"	310.00	27.90	337.90
Aluminum door				
2'6" x 2'6"	EA.	300.00	27.90	327.90
2'6" x 3'0"	"	320.00	27.90	347.90
Gravity ventilators, with curb, base, damper and screen				
Stationary siphon				
6" dia.	EA.	20.70	18.60	39.30
12" dia.	"	32.20	18.60	50.80
24" dia.	"	140.00	27.90	167.90
36" dia.	"	260.00	27.90	287.90
Wind driven spinner				
6" dia.	EA.	32.50	18.60	51.10
12" dia.	"	44.20	18.60	62.80
24" dia.	"	160.00	27.90	187.90
36" dia.	"	350.00	27.90	377.90
Stationary mushroom				
16" dia.	EA.	210.00	27.90	237.90
30" dia.	"	480.00	34.90	514.90
36" dia.	"	620.00	46.50	666.50
42" dia.	"	910.00	56.00	966.00

JOINT SEALERS	UNIT	MAT.	INST.	TOTAL
07920.10 CAULKING				
Caulk exterior, two component				
1/4 x 1/2	L.F.	0.25	1.45	1.70
3/8 x 1/2	"	0.37	1.60	1.97
1/2 x 1/2	"	0.48	1.80	2.28
Caulk interior, single component				
1/4 x 1/2	L.F.	0.16	1.40	1.56
3/8 x 1/2	"	0.24	1.50	1.74
1/2 x 1/2	"	0.30	1.70	2.00

SPECIALTIES	UNIT	MAT.	INST.	TOTAL
10165.10 **TOILET PARTITIONS**				
Toilet partition, plastic laminate				
Ceiling mounted	EA.	620.00	96.50	716.50
Floor mounted	"	560.00	72.50	632.50
Metal				
Ceiling mounted	EA.	450.00	96.50	546.50
Floor mounted	"	430.00	72.50	502.50
Wheel chair partition, plastic laminate				
Ceiling mounted	EA.	780.00	96.50	876.50
Floor mounted	"	730.00	72.50	802.50
Painted metal				
Ceiling mounted	EA.	620.00	96.50	716.50
Floor mounted	"	550.00	72.50	622.50
Urinal screen, plastic laminate				
Wall hung	EA.	290.00	36.20	326.20
Floor mounted	"	250.00	36.20	286.20
Porcelain enameled steel, floor mounted	"	320.00	36.20	356.20
Painted metal, floor mounted	"	220.00	36.20	256.20
Stainless steel, floor mounted	"	410.00	36.20	446.20
Metal toilet partitions				
Toilet partitions, front door and side divider, floor mounted				
Porcelain enameled steel	EA.	690.00	72.50	762.50
Painted steel	"	410.00	72.50	482.50
Stainless steel	"	780.00	72.50	852.50
10185.10 **SHOWER STALLS**				
Shower receptors				
Precast, terrazzo				
32" x 32"	EA.	210.00	26.70	236.70
32" x 48"	"	290.00	32.00	322.00
Concrete				
32" x 32"	EA.	110.00	26.70	136.70
48" x 48"	"	160.00	35.60	195.60
Shower door, trim and hardware				
Economy, 24" wide, chrome frame, tempered glass	EA.	160.00	32.00	192.00
Porcelain enameled steel, flush	"	290.00	32.00	322.00
Baked enameled steel, flush	"	170.00	32.00	202.00
Aluminum frame, tempered glass, 48" wide, sliding	"	370.00	40.10	410.10
Folding	"	350.00	40.10	390.10
Aluminum frame and tempered glass, molded plastic				
Complete with receptor and door				
32" x 32"	EA.	470.00	80.00	550.00
36" x 36"	"	530.00	80.00	610.00
40" x 40"	"	590.00	91.50	681.50
Shower compartment, precast concrete receptor				
Single entry type				
Porcelain enameled steel	EA.	1,460	320.00	1,780
Baked enameled steel	"	840.00	320.00	1,160
Stainless steel	"	1,680	320.00	2,000
Double entry type				
Porcelain enameled steel	EA.	4,140	400.00	4,540
Baked enameled steel	"	2,690	400.00	3,090

SPECIALTIES	UNIT	MAT.	INST.	TOTAL
10185.10 — SHOWER STALLS				
Stainless steel	EA.	4,200	400.00	4,600
10270.40 — ACCESS & PEDESTAL FLOOR				
Panels, no covering, 2'x2'				
Plain	S.F.	5.50	0.36	5.86
Perforated	"	8.05	14.50	22.55
Pedestals				
For 6" to 12" clearance	EA.	4.50	2.90	7.40
Stringers				
2'	L.F.	0.76	1.40	2.16
6'	"	0.76	0.96	1.73
Accessories				
Ramp assembly	S.F.	28.00	1.15	29.15
Elevated floor assembly	"	44.30	1.05	45.35
Handrail	L.F.	33.30	14.50	47.80
Fascia plate	"	15.70	7.25	22.95
For carpet tiles, add	S.F.			3.75
For vinyl flooring, add	"			4.76
RF shielding components, floor liner				
Hot rolled steel sheet				
14 ga.	S.F.	1.85	0.72	2.57
11 ga.	"	1.95	2.25	4.20
10290.10 — PEST CONTROL				
Termite control				
Under slab spraying				
Minimum	S.F.	0.11	0.06	0.17
Average	"	0.13	0.11	0.24
Maximum	"	0.19	0.23	0.42
10500.10 — LOCKERS				
Locker bench, floor mounted, laminated maple				
4'	EA.	66.50	24.15	90.65
6'	"	94.50	24.15	118.65
Wardrobe locker, single tier type				
12" x 15" x 72"	EA.	110.00	29.00	139.00
18" x 15" x 72"	"	140.00	30.50	170.50
12" x 18" x 72"	"	120.00	32.20	152.20
18" x 18" x 72"	"	150.00	34.10	184.10
10520.10 — FIRE PROTECTION				
Portable fire extinguishers				
Water pump tank type				
2.5 gal.				
Red enameled galvanized	EA.	73.50	15.10	88.60
Red enameled copper	"	85.00	15.10	100.10
Polished copper	"	130.00	15.10	145.10

SPECIALTIES	UNIT	MAT.	INST.	TOTAL
10520.10 FIRE PROTECTION				
Carbon dioxide type, red enamel steel				
Squeeze grip with hose and horn				
2.5 lb	EA.	39.70	15.10	54.80
5 lb	"	81.50	17.45	98.95
10 lb	"	140.00	22.70	162.70
15 lb	"	160.00	28.40	188.40
20 lb	"	180.00	28.40	208.40
Wheeled type				
125 lb	EA.	1,070	45.40	1,115
250 lb	"	1,100	45.40	1,145
500 lb	"	2,020	45.40	2,065
Dry chemical, pressurized type				
Red enameled steel				
2.5 lb	EA.	21.30	15.10	36.40
5 lb	"	32.50	17.45	49.95
10 lb	"	45.90	22.70	68.60
20 lb	"	80.50	28.40	108.90
30 lb	"	220.00	28.40	248.40
Chrome plated steel, 2.5 lb	"	47.00	15.10	62.10
Other type extinguishers				
2.5 gal, stainless steel, pressurized water tanks	EA.	45.90	15.10	61.00
Soda and acid type	"	90.00	15.10	105.10
Cartridge operated, water type	"	150.00	15.10	165.10
Loaded stream, water type	"	72.00	15.10	87.10
Foam type	"	60.00	15.10	75.10
40 gal, wheeled foam type	"	3,140	45.40	3,185
Fire extinguisher cabinets				
Enameled steel				
8" x 12" x 27"	EA.	120.00	45.40	165.40
8" x 16" x 38"	"	150.00	45.40	195.40
Aluminum				
8" x 12" x 27"	EA.	150.00	45.40	195.40
8" x 16" x 38"	"	180.00	45.40	225.40
8" x 12" x 27"	"	260.00	45.40	305.40
Stainless steel				
8" x 16" x 38"	EA.	330.00	45.40	375.40
10670.10 SHELVING				
Metal storage shelving, baked enamel				
7 shelf unit, 72" or 84" high				
10" shelf	L.F.	29.10	29.00	58.10
18" shelf	"	34.70	34.10	68.80
4 shelf unit, 40" high				
10" shelf	L.F.	24.65	24.15	48.80
18" shelf	"	32.90	30.50	63.40
3 shelf unit, 32" high				
10" shelf	L.F.	20.80	14.50	35.30
18" shelf	"	24.35	17.05	41.40
Single shelf unit, attached to masonry				
10" shelf	L.F.	7.45	4.85	12.30
18" shelf	"	9.50	5.90	15.40

SPECIALTIES	UNIT	MAT.	INST.	TOTAL
10750.10 TELEPHONE ENCLOSURES				
Telephone enclosure, wall mounted, shelf, 28" x 30" x 15"	EA.	670.00	72.50	742.50
Directory shelf, stainless steel, 3 binders	"	560.00	48.30	608.30
10800.10 BATH ACCESSORIES				
Ash receiver, wall mounted, aluminum	EA.	110.00	14.50	124.50
Grab bar, 1-1/2" dia., stainless steel, wall mounted				
24" long	EA.	32.00	14.50	46.50
36" long	"	45.70	15.25	60.95
42" long	"	52.50	16.10	68.60
48" long	"	63.00	17.05	80.05
52" long	"	68.50	18.10	86.60
1" dia., stainless steel				
12" long	EA.	25.10	12.60	37.70
18" long	"	27.40	13.15	40.55
24" long	"	29.70	14.50	44.20
30" long	"	33.10	15.25	48.35
36" long	"	37.70	16.10	53.80
48" long	"	38.80	17.05	55.85
Hand dryer, surface mounted, 110 volt	"	360.00	36.20	396.20
Medicine cabinet, 16 x 22, baked enamel, steel, lighted	"	57.00	11.60	68.60
With mirror, lighted	"	100.00	19.30	119.30
Mirror, 1/4" plate glass, up to 10 sf	S.F.	5.50	2.90	8.40
Mirror, stainless steel frame				
18"x24"	EA.	91.50	9.65	101.15
18"x32"	"	95.00	11.60	106.60
18"x36"	"	100.00	14.50	114.50
24"x30"	"	110.00	14.50	124.50
24"x36"	"	130.00	16.10	146.10
24"x48"	"	230.00	24.15	254.15
24"x60"	"	240.00	29.00	269.00
30"x30"	"	130.00	29.00	159.00
30"x72"	"	370.00	36.20	406.20
48"x72"	"	380.00	48.30	428.30
With shelf, 18"x24"	"	130.00	11.60	141.60
Sanitary napkin dispenser, stainless steel, wall mounted	"	310.00	19.30	329.30
Shower rod, 1" diameter				
Chrome finish over brass	EA.	51.50	14.50	66.00
Stainless steel	"	49.70	14.50	64.20
Soap dish, stainless steel, wall mounted	"	63.00	19.30	82.30
Toilet tissue dispenser, stainless, wall mounted				
Single roll	EA.	30.80	7.25	38.05
Double roll	"	48.00	8.25	56.25
Towel dispenser, stainless steel				
Flush mounted	EA.	89.50	16.10	105.60
Surface mounted	"	84.00	14.50	98.50
Combination towel dispenser and waste receptacle	"	280.00	19.30	299.30
Towel bar, stainless steel				
18" long	EA.	26.30	11.60	37.90
24" long	"	30.80	13.15	43.95
30" long	"	33.10	14.50	47.60
36" long	"	36.60	16.10	52.70
Toothbrush and tumbler holder	"	22.30	9.65	31.95

SPECIALTIES		UNIT	MAT.	INST.	TOTAL
10800.10	BATH ACCESSORIES				
Waste receptacle, stainless steel, wall mounted		EA.	190.00	24.15	214.15
BLANK LINE					

ARCHITECTURAL EQUIPMENT

	UNIT	MAT.	INST.	TOTAL
11010.10 MAINTENANCE EQUIPMENT				
Vacuum cleaning system				
3 valves				
1.5 hp	EA.	830.00	320.00	1,150
2.5 hp	"	1,140	410.00	1,550
5 valves	"	1,540	580.00	2,120
7 valves	"	1,770	720.00	2,490
11060.10 THEATER EQUIPMENT				
Roll out stage, steel frame, wood floor				
Manual	S.F.	22.80	1.80	24.60
Electric	"	24.50	2.90	27.40
Portable stages				
8" high	S.F.	7.30	1.45	8.75
18" high	"	8.40	1.60	10.00
36" high	"	8.95	1.70	10.65
48" high	"	9.50	1.80	11.30
Band risers				
Minimum	S.F.	19.60	1.45	21.05
Maximum	"	32.80	1.45	34.25
Chairs for risers				
Minimum	EA.	56.00	1.05	57.05
Maximum	"	100.00	1.05	101.05
11080.10 POLICE EQUIPMENT				
Firing range equipment, rifle				
3 position	EA.	8,740	970.00	9,710
4 position	"	11,200	1,450	12,650
5 position	"	14,000	1,610	15,610
6 position	"	16,800	1,700	18,500
11090.10 CHECKROOM EQUIPMENT				
Motorized checkroom equipment				
No shelf system, 6'4" height				
7'6" length	EA.	1,680	290.00	1,970
14'6" length	"	1,900	290.00	2,190
28' length	"	2,580	290.00	2,870
One shelf, 6'8" height				
7'6" length	EA.	2,020	290.00	2,310
14'6" length	"	2,460	290.00	2,750
28' length	"	3,700	290.00	3,990
Two shelves, 7'5" height				
7'6" length	EA.	2,460	290.00	2,750
14'6" length	"	3,020	290.00	3,310
28' length	"	4,260	290.00	4,550
Three shelves, 8' height				
7'6" length	EA.	2,580	580.00	3,160
14'6" length	"	3,140	580.00	3,720
28' length	"	4,480	580.00	5,060
Four shelves, 8'7" height				
7'6" length	EA.	2,580	580.00	3,160
14'6" length	"	3,140	580.00	3,720

ARCHITECTURAL EQUIPMENT	UNIT	MAT.	INST.	TOTAL
11090.10 — CHECKROOM EQUIPMENT				
28' length	EA.	4,480	580.00	5,060
200 lb				
11110.10 — LAUNDRY EQUIPMENT				
High capacity, heavy duty				
Washer extractors				
135 lb				
Standard	EA.	38,610	240.00	38,850
Pass through	"	44,580	240.00	44,820
200 lb				
Standard	EA.	43,350	240.00	43,590
Pass through	"	49,290	240.00	49,530
110 lb dryer	"	6,740	240.00	6,980
Hand operated presser	"	5,260	320.00	5,580
Mushroom press	"	3,370	320.00	3,690
Spreader feeders				
2 station	EA.	50,490	320.00	50,810
4 station	"	55,860	580.00	56,440
Delivery carts				
12 bushel	EA.	340.00	3.60	343.60
Low capacity				
Pressers				
Air operated	EA.	4,170	120.00	4,290
Hand operated	"	3,400	120.00	3,520
Extractor, low capacity	"	4,860	120.00	4,980
Ironer, 48"	"	2,260	58.00	2,318
Coin washers				
10 lb capacity	EA.	1,020	58.00	1,078
20 lb capacity	"	2,570	58.00	2,628
Coin dryer	"	430.00	36.20	466.20
Coin dry cleaner, 20 lb	"	20,160	120.00	20,280
11161.10 — LOADING DOCK EQUIPMENT				
Dock leveler, 10 ton capacity				
6' x 8'	EA.	3,430	290.00	3,720
7' x 8'	"	3,770	290.00	4,060
11170.10 — WASTE HANDLING				
Incinerator, electric				
100 lb/hr				
Minimum	EA.	14,840	300.00	15,140
Maximum	"	21,280	300.00	21,580
400 lb/hr				
Minimum	EA.	26,710	590.00	27,300
Maximum	"	42,560	590.00	43,150
1000 lb/hr				
Minimum	EA.	45,710	900.00	46,610
Maximum	"	61,600	900.00	62,500
Incinerator, medical-waste				
25 lb/hr, 2-7 x 4-0	EA.	13,650	590.00	14,240

ARCHITECTURAL EQUIPMENT	UNIT	MAT.	INST.	TOTAL
11170.10 WASTE HANDLING				
50 lb/hr, 2-11 x 4-11	EA.	14,960	590.00	15,550
75 lb/hr, 3-8 x 5-0	"	21,610	1,190	22,800
100 lb/hr, 3-8 x 6-0	"	23,740	1,190	24,930
Industrial compactor				
1 cy	EA.	7,660	330.00	7,990
3 cy	"	15,080	420.00	15,500
5 cy	"	17,450	590.00	18,040
11400.10 FOOD SERVICE EQUIPMENT				
Unit kitchens				
30" compact kitchen				
Refrigerator, with range, sink	EA.	2,460	150.00	2,610
Sink only	"	2,340	99.00	2,439
Range only	"	2,060	74.00	2,134
Cabinet for upper wall section	"	290.00	42.40	332.40
Stainless shield, for rear wall	"	97.00	11.85	108.85
Side wall	"	95.00	11.85	106.85
42" compact kitchen				
Refrigerator with range, sink	EA.	3,060	160.00	3,220
Sink only	"	2,630	150.00	2,780
Cabinet for upper wall section	"	410.00	49.50	459.50
Stainless shield, for rear wall	"	130.00	12.35	142.35
Side wall	"	51.50	12.35	63.85
54" compact kitchen				
Refrigerator, oven, range, sink	EA.	3,770	210.00	3,980
Cabinet for upper wall section	"	540.00	59.50	599.50
Stainless shield, for				
Rear wall	EA.	190.00	13.50	203.50
Side wall	"	51.50	13.50	65.00
60" compact kitchen				
Refrigerator, oven, range, sink	EA.	4,340	210.00	4,550
Cabinet for upper wall section	"	620.00	59.50	679.50
Stainless shield, for				
Rear wall	EA.	190.00	13.50	203.50
Side wall	"	51.50	13.50	65.00
72" compact kitchen				
Refrigerator, oven, range, sink	EA.	4,170	250.00	4,420
Cabinet for upper wall section	"	1,200	59.50	1,260
Stainless shield for				
Rear wall	EA.	220.00	14.85	234.85
Side wall	"	51.50	14.85	66.35
Bake oven				
Single deck				
Minimum	EA.	1,340	37.10	1,377
Maximum	"	3,140	74.00	3,214
Double deck				
Minimum	EA.	1,570	49.50	1,620
Maximum	"	5,600	74.00	5,674
Triple deck				
Minimum	EA.	2,240	49.50	2,290
Maximum	"	6,720	99.00	6,819

ARCHITECTURAL EQUIPMENT	UNIT	MAT.	INST.	TOTAL
11400.10 — FOOD SERVICE EQUIPMENT				
Convection type oven, electric, 40" x 45" x 57"				
Minimum	EA.	2,240	37.10	2,277
Maximum	"	3,920	74.00	3,994
Broiler, without oven, 69" x 26" x 39"				
Minimum	EA.	2,020	37.10	2,057
Maximum	"	3,920	49.50	3,970
Coffee urns, 10 gallons				
Minimum	EA.	4,480	99.00	4,579
Maximum	"	7,280	150.00	7,430
Fryer, with submerger				
Single				
Minimum	EA.	670.00	59.50	729.50
Maximum	"	2,020	99.00	2,119
Double				
Minimum	EA.	1,120	74.00	1,194
Maximum	"	3,810	99.00	3,909
Griddle, counter				
3' long				
Minimum	EA.	1,120	49.50	1,170
Maximum	"	3,360	59.50	3,420
5' long				
Minimum	EA.	1,680	74.00	1,754
Maximum	"	4,370	99.00	4,469
Kettles, steam, jacketed				
20 gallons				
Minimum	EA.	1,790	74.00	1,864
Maximum	"	4,260	150.00	4,410
40 gallons				
Minimum	EA.	2,130	74.00	2,204
Maximum	"	4,700	150.00	4,850
60 gallons				
Minimum	EA.	2,800	74.00	2,874
Maximum	"	6,270	150.00	6,420
Range				
Heavy duty, single oven, open top				
Minimum	EA.	1,570	37.10	1,607
Maximum	"	4,030	99.00	4,129
Fry top				
Minimum	EA.	2,020	37.10	2,057
Maximum	"	4,370	99.00	4,469
Steamers, electric				
27 kw				
Minimum	EA.	6,050	74.00	6,124
Maximum	"	11,710	99.00	11,809
18 kw				
Minimum	EA.	4,460	74.00	4,534
Maximum	"	7,710	99.00	7,809
Dishwasher, rack type				
Single tank, 190 racks/hr	EA.	9,520	150.00	9,670
Double tank				
234 racks/hr	EA.	13,650	160.00	13,810
265 racks/hr	"	15,190	200.00	15,390

ARCHITECTURAL EQUIPMENT	UNIT	MAT.	INST.	TOTAL
11400.10 — FOOD SERVICE EQUIPMENT				
Dishwasher, automatic 100 meals/hr	EA.	45,010	99.00	45,109
Disposals				
100 gal/hr	EA.	1,000	99.00	1,099
120 gal/hr	"	1,450	100.00	1,550
250 gal/hr	"	3,770	110.00	3,880
Exhaust hood for dishwasher, gutter 4 sides, s-steel				
4'x4'x2'	EA.	1,340	110.00	1,450
4'x7'x2'	"	2,240	120.00	2,360
Food preparation machines				
Vertical cutter mixers				
25 quart	EA.	5,880	99.00	5,979
40 quart	"	7,080	99.00	7,179
80 quart	"	8,110	150.00	8,260
130 quart	"	9,020	250.00	9,270
Choppers				
5 lb	EA.	2,030	74.00	2,104
16 lb	"	2,800	99.00	2,899
40 lb	"	3,880	150.00	4,030
Mixers, floor models				
20 quart	EA.	2,240	37.10	2,277
60 quart	"	6,720	37.10	6,757
80 quart	"	8,960	42.40	9,002
140 quart	"	10,080	59.50	10,140
Ice cube maker				
50 lb per day				
Minimum	EA.	900.00	300.00	1,200
Maximum	"	1,790	300.00	2,090
500 lb per day				
Minimum	EA.	3,540	490.00	4,030
Maximum	"	6,510	490.00	7,000
Ice flakers				
300 lb per day	EA.	2,400	300.00	2,700
600 lb per day	"	2,860	490.00	3,350
1000 lb per day	"	3,370	660.00	4,030
2000 lb per day	"	7,650	740.00	8,390
Refrigerated cases				
Dairy products				
Multi deck type	L.F.	510.00	19.80	529.80
For rear sliding doors, add	"			56.00
Delicatessen case, service deli				
Single deck	L.F.	470.00	150.00	620.00
Multi deck	"	520.00	190.00	710.00
Meat case				
Single deck	L.F.	340.00	170.00	510.00
Multi deck	"	370.00	190.00	560.00
Produce case				
Single deck	L.F.	350.00	170.00	520.00
Multi deck	"	400.00	190.00	590.00
Bottle coolers				
6' long				
Minimum	EA.	1,290	590.00	1,880
Maximum	"	2,710	590.00	3,300

ARCHITECTURAL EQUIPMENT	UNIT	MAT.	INST.	TOTAL
11400.10 FOOD SERVICE EQUIPMENT				
10' long				
Minimum	EA.	1,830	990.00	2,820
Maximum	"	3,600	990.00	4,590
Frozen food cases				
Chest type	L.F.	370.00	170.00	540.00
Reach-in, glass door	"	760.00	190.00	950.00
Island case, single	"	370.00	170.00	540.00
Multi deck	"	820.00	190.00	1,010
Ice storage bins				
500 lb capacity	EA.	1,020	420.00	1,440
1000 lb capacity	"	2,030	850.00	2,880
11450.10 RESIDENTIAL EQUIPMENT				
Compactor, 4 to 1 compaction	EA.	1,010	74.00	1,084
Dishwasher, built-in				
2 cycles	EA.	840.00	150.00	990.00
4 or more cycles	"	1,230	150.00	1,380
Disposal				
Garbage disposer	EA.	340.00	99.00	439.00
Heaters, electric, built-in				
Ceiling type	EA.	170.00	99.00	269.00
Wall type				
Minimum	EA.	110.00	74.00	184.00
Maximum	"	220.00	99.00	319.00
Hood for range, 2-speed, vented				
30" wide	EA.	120.00	99.00	219.00
42" wide	"	470.00	99.00	569.00
Ice maker, automatic				
30 lb per day	EA.	580.00	42.40	622.40
50 lb per day	"	1,230	150.00	1,380
Ranges electric				
Built-in, 30", 1 oven	EA.	810.00	99.00	909.00
2 oven	"	1,230	99.00	1,329
Counter top, 4 burner, standard	"	450.00	74.00	524.00
With grill	"	1,340	74.00	1,414
Free standing, 21", 1 oven	"	520.00	99.00	619.00
30", 1 oven	"	670.00	59.50	729.50
2 oven	"	2,020	59.50	2,080
Water softener				
30 grains per gallon	EA.	670.00	99.00	769.00
70 grains per gallon	"	1,230	150.00	1,380
11470.10 DARKROOM EQUIPMENT				
Dryers				
36" x 25" x 68"	EA.	6,720	160.00	6,880
48" x 25" x 68"	"	6,940	160.00	7,100
Processors, film				
Black and white	EA.	10,000	160.00	10,160
Color negatives	"	11,140	160.00	11,300
Prints	"	11,650	160.00	11,810
Transparencies	"	12,450	160.00	12,610

ARCHITECTURAL EQUIPMENT	UNIT	MAT.	INST.	TOTAL
11470.10 **DARKROOM EQUIPMENT**				
Sinks with cabinet and/or stand				
5" sink with stand				
24" x 48"	EA.	740.00	80.00	820.00
32" x 64"	"	910.00	110.00	1,020
38" x 52"	"	910.00	110.00	1,020
42" x 132"	"	1,150	160.00	1,310
48" x 52"	"	1,110	160.00	1,270
5" sink with cabinet				
24" x 48"	EA.	1,470	80.00	1,550
32" x 64"	"	1,580	110.00	1,690
38" x 52"	"	1,560	110.00	1,670
42" x 132"	"	3,240	160.00	3,400
48" x 52"	"	2,000	160.00	2,160
10" sink with stand				
24" x 48"	EA.	900.00	80.00	980.00
32" x 64"	"	1,050	110.00	1,160
38" x 52"	"	1,040	110.00	1,150
10" sink with cabinet				
24" x 48"	EA.	1,470	80.00	1,550
38" x 52"	"	1,750	110.00	1,860
11500.10 **INDUSTRIAL EQUIPMENT**				
Vehicular paint spray booth, solid back, 14'4" x 9'6"				
24' deep	EA.	8,930	290.00	9,220
26'6" deep	"	9,100	290.00	9,390
28'6" deep	"	9,310	290.00	9,600
Drive through, 14'9" x 9'6"				
24' deep	EA.	9,370	290.00	9,660
26'6" deep	"	9,600	290.00	9,890
28'6" deep	"	9,710	290.00	10,000
Water wash, paint spray booth				
5' x 11'2" x 10'8"	EA.	4,680	290.00	4,970
6' x 11'2" x 10'8"	"	4,800	290.00	5,090
8' x 11'2" x 10'8"	"	6,050	290.00	6,340
10' x 11'2" x 11'2"	"	6,740	290.00	7,030
12' x 12'2" x 11'2"	"	8,230	290.00	8,520
14' x 12'2" x 11'2"	"	9,710	290.00	10,000
16' x 12'2" x 11'2"	"	10,970	290.00	11,260
20' x 12'2" x 11'2"	"	12,540	290.00	12,830
Dry type spray booth, with paint arrestors				
5'4" x 7'2" x 6'8"	EA.	1,960	290.00	2,250
6'4" x 7'2" x 6'8"	"	2,380	290.00	2,670
8'4" x 7'2" x 9'2"	"	2,790	290.00	3,080
10'4" x 7'2" x 9'2"	"	3,190	290.00	3,480
12'4" x 7'6" x 9'2"	"	3,770	290.00	4,060
14'4" x 7'6" x 9'8"	"	4,300	290.00	4,590
16'4" x 7'7" x 9'8"	"	4,910	290.00	5,200
20'4" x 7'7" x 10'8"	"	5,510	290.00	5,800
Air compressor, electric				
1 hp				
115 volt	EA.	2,460	190.00	2,650
5 hp				

ARCHITECTURAL EQUIPMENT	UNIT	MAT.	INST.	TOTAL
11500.10 INDUSTRIAL EQUIPMENT				
115 volt	EA.	5,080	290.00	5,370
230 volt	"	4,280	290.00	4,570
Hydraulic lifts				
8,000 lb capacity	EA.	380.00	720.00	1,100
11,000 lb capacity	"	5,370	1,160	6,530
24,000 lb capacity	"	7,540	1,930	9,470
Power tools				
Band saws				
10"	EA.	500.00	24.15	524.15
14"	"	900.00	29.00	929.00
Motorized shaper	"	480.00	22.30	502.30
Motorized lathe	"	560.00	24.15	584.15
Bench saws				
9" saw	EA.	1,030	19.30	1,049
10" saw	"	1,550	20.70	1,571
12" saw	"	1,990	24.15	2,014
Electric grinders				
1/3 hp	EA.	190.00	11.60	201.60
1/2 hp	"	270.00	12.60	282.60
3/4 hp	"	320.00	12.60	332.60
11600.10 LABORATORY EQUIPMENT				
Cabinets, base				
Minimum	L.F.	89.50	24.15	113.65
Maximum	"	160.00	24.15	184.15
Full storage, 7' high				
Minimum	L.F.	130.00	24.15	154.15
Maximum	"	200.00	24.15	224.15
Wall				
Minimum	L.F.	59.50	29.00	88.50
Maximum	"	100.00	29.00	129.00
Counter tops				
Minimum	S.F.	12.30	3.60	15.90
Average	"	29.10	4.15	33.25
Maximum	"	48.20	4.85	53.05
Tables				
Open underneath	S.F.	67.00	14.50	81.50
Doors underneath	"	120.00	18.10	138.10
Medical laboratory equipment				
Analyzer				
Chloride	EA.	390.00	14.85	404.85
Blood	"	16,800	24.75	16,825
Bath, water, utility, countertop unit	"	280.00	29.70	309.70
Hot plate, lab, countertop	"	89.50	27.00	116.50
Stirrer	"	160.00	27.00	187.00
Incubator, anaerobic, 23x23x36"	"	4,480	150.00	4,630
Dry heat bath	"	340.00	49.50	389.50
Incinerator, for sterilizing	"	130.00	2.95	132.95
Meter, serum protein	"	600.00	3.70	603.70
Ph analog, general purpose	"	450.00	4.25	454.25
Refrigerator, blood bank, undercounter type 153 litres	"	3,250	49.50	3,300
5.4 cf, undercounter type	"	2,800	49.50	2,850

ARCHITECTURAL EQUIPMENT	UNIT	MAT.	INST.	TOTAL
11600.10 LABORATORY EQUIPMENT				
Refrigerator/freezer, 4.4 cf, undercounter type	EA.	310.00	49.50	359.50
Sealer, impulse, free standing, 20x12x4"	"	170.00	9.90	179.90
Timer, electric, 1-60 minutes, bench or wall mounted	"	45.90	16.50	62.40
Glassware washer - dryer, undercounter	"	5,210	370.00	5,580
Balance, torsion suspension, tabletop, 4.5 lb capacity	"	360.00	16.50	376.50
Binocular microscope, with in base illuminator	"	1,400	11.40	1,411
Centrifuge, table model, 19x16x13"	"	850.00	11.85	861.85
Clinical model, with four place head	"	440.00	6.60	446.60
11700.10 MEDICAL EQUIPMENT				
Hospital equipment, lights				
Examination, portable	EA.	710.00	24.75	734.75
Meters				
Air flow meter	EA.	41.40	16.50	57.90
Oxygen flow meters	"	28.00	12.35	40.35
Physical therapy				
Chair, hydrotherapy	EA.	200.00	4.85	204.85
Diathermy, shortwave, portable, on casters	"	1,790	11.60	1,802
Exercise bicycle, floor standing, 35" x 15"	"	520.00	9.65	529.65
Hydrocollator, 4 pack, portable, 129 x 90 x 160"	"	200.00	4.15	204.15
Lamp, infrared, mobile with variable heat control	"	270.00	22.30	292.30
Ultra violet, base mounted	"	250.00	22.30	272.30
Stimulator, galvanic-faradic, hand held	"	58.00	1.95	59.95
Ultrasound, muscle stimulator, portable, 13x13x8"	"	1,340	2.40	1,342
Whirlpool, 85 gallon	"	1,790	140.00	1,930
65 gallon capacity	"	1,520	140.00	1,660
Radiology				
Radiographic table, motor driven tilting table	EA.	19,940	2,900	22,840
Fluoroscope image/tv system	"	56,000	5,790	61,790
Processor for washing and drying radiographs				
Water filter unit, 30" x 48-1/2" x 37-1/2"	EA.	17.90	490.00	507.90
Steam sterilizers				
For heat and moisture stable materials	EA.	2,020	29.70	2,050
For fast drying after sterilization	"	1,340	37.10	1,377
Compact unit	"	1,010	37.10	1,047
Semi-automatic	"	4,030	150.00	4,180
Floor loading				
Single door	EA.	76,160	250.00	76,410
Double door	"	96,320	300.00	96,620
Utensil washer, sanitizer	"	4,260	230.00	4,490
Automatic washer/sterilizer	"	15,680	590.00	16,270
16 x 16 x 26", including generator & accessories	"	15,120	990.00	16,110
Steam generator, elec., 10 kw to 180 kw	"	14,500	590.00	15,090
Surgical scrub				
Minimum	EA.	1,570	99.00	1,669
Maximum	"	3,300	99.00	3,399
Gas sterilizers				
Automatic, free standing, 21x19x29"	EA.	2,860	300.00	3,160
Surgical lights, ceiling mounted				
Minimum	EA.	2,800	490.00	3,290
Maximum	"	4,590	590.00	5,180
Water stills				

ARCHITECTURAL EQUIPMENT	UNIT	MAT.	INST.	TOTAL
11700.10 MEDICAL EQUIPMENT				
4 liters / hr	EA.	1,440	99.00	1,539
8 liters / hr	"	2,130	99.00	2,229
19 liters / hr	"	5,880	250.00	6,130
X-ray equipment				
Mobile unit				
Minimum	EA.	5,040	150.00	5,190
Maximum	"	24,640	300.00	24,940
Minimum	"	2,800	300.00	3,100
Maximum	"	5,600	300.00	5,900
Incubators				
15 cf	EA.	2,580	150.00	2,730
29 cf	"	3,140	250.00	3,390
Infant transport, portable	"	6,720	160.00	6,880
Headwall				
Aluminum, with back frame and console	EA.	2,130	150.00	2,280
Hospital ground detection system				
Power ground module	EA.	500.00	85.00	585.00
Ground slave module	"	340.00	64.50	404.50
Master ground module	"	220.00	56.00	276.00
Remote indicator	"	240.00	59.50	299.50
X-ray indicator	"	740.00	64.50	804.50
Micro ammeter	"	910.00	74.00	984.00
Supervisory module	"	690.00	64.50	754.50
Ground cords	"	56.00	11.00	67.00
Hospital isolation monitors, 5 ma				
120v	EA.	1,400	130.00	1,530
208v	"	1,340	130.00	1,470
240v	"	1,340	130.00	1,470
Digital clock-timers separate display	"	560.00	59.50	619.50
One display	"	530.00	59.50	589.50
Remote control	"	310.00	46.40	356.40
Battery pack	"	73.00	46.40	119.40
Surgical chronometer clock and 3 timers	"	1,120	93.00	1,213
Auxilary control	"	410.00	43.00	453.00
11700.20 DENTAL EQUIPMENT				
Dental care equipment				
Drill console with accessories	EA.	8,400	490.00	8,890
Amalgamator	"	210.00	14.85	224.85
Lathe	"	220.00	9.90	229.90
Finish polisher	"	160.00	19.80	179.80
Model trimmer	"	260.00	13.50	273.50
Motor, wall mounted	"	290.00	13.50	303.50
Cleaner, ultrasonic	"	1,090	29.70	1,120
Curing unit, bench mounted	"	1,450	49.50	1,500
Oral evacuation system, dual pump	"	8,170	37.10	8,207
Sterilizer, table top, self contained	"	570.00	16.50	586.50
Dental lights				
Light, floor or ceiling mounted	EA.	1,340	150.00	1,490
X-ray unit				
Portable	EA.	2,020	74.00	2,094
Wall mounted with remote control	"	4,260	250.00	4,510

ARCHITECTURAL EQUIPMENT		UNIT	MAT.	INST.	TOTAL
11700.20	DENTAL EQUIPMENT				
Illuminator, single panel		EA.	6,120	420.00	6,540
X-ray film processor		"	1,010	250.00	1,260
Shield, portable x-ray, lead lined		"	210.00	19.80	229.80

INTERIOR		UNIT	MAT.	INST.	TOTAL
12302.10	**CASEWORK**				
Kitchen base cabinet, prefinished, 24" deep, 35" high					
12"wide		EA.	120.00	29.00	149.00
18" wide		"	140.00	29.00	169.00
24" wide		"	150.00	32.20	182.20
27" wide		"	180.00	32.20	212.20
36" wide		"	210.00	36.20	246.20
48" wide		"	240.00	36.20	276.20
Corner cabinet, 36" wide		"	160.00	36.20	196.20
Wall cabinet, 12" deep, 12" high					
30" wide		EA.	89.50	29.00	118.50
36" wide		"	95.00	29.00	124.00
15" high					
30" wide		EA.	95.00	32.20	127.20
36" wide		"	100.00	32.20	132.20
24" high					
30" wide		EA.	120.00	32.20	152.20
36" wide		"	140.00	32.20	172.20
30" high					
12" wide		EA.	86.00	36.20	122.20
18" wide		"	98.00	36.20	134.20
24" wide		"	100.00	36.20	136.20
27" wide		"	130.00	36.20	166.20
30" wide		"	140.00	41.40	181.40
36" wide		"	150.00	41.40	191.40
Corner cabinet, 30" high					
24" wide		EA.	140.00	48.30	188.30
30" wide		"	170.00	48.30	218.30
36" wide		"	190.00	48.30	238.30
Wardrobe		"	440.00	72.50	512.50
Vanity with top, laminated plastic					
24" wide		EA.	270.00	72.50	342.50
30" wide		"	330.00	72.50	402.50
36" wide		"	350.00	96.50	446.50
48" wide		"	380.00	120.00	500.00
12390.10	**COUNTER TOPS**				
Stainless steel, counter top, with backsplash		S.F.	67.00	7.25	74.25
Acid-proof, kemrock surface		"	17.90	4.85	22.75

CONSTRUCTION	UNIT	MAT.	INST.	TOTAL
13152.10 SWIMMING POOL EQUIPMENT				
Diving boards				
14' long				
Aluminum	EA.	890.00	130.00	1,020
Fiberglass	"	550.00	130.00	680.00
Ladders, heavy duty				
2 steps				
Minimum	EA.	350.00	45.40	395.40
Maximum	"	570.00	45.40	615.40
4 steps				
Minimum	EA.	390.00	56.50	446.50
Maximum	"	750.00	56.50	806.50
Lifeguard chair				
Minimum	EA.	830.00	230.00	1,060
Maximum	"	1,300	230.00	1,530
Lights, underwater				
12 volt, with transformer	EA.	260.00	56.50	316.50
110 volt				
Minimum	EA.	240.00	56.50	296.50
Maximum	"	770.00	56.50	826.50
Ground fault interrupter for 110 volt, each light	"	97.00	18.90	115.90
Pool covers				
Reinforced polyethylene	S.F.	0.44	1.75	2.19
Vinyl water tube				
Minimum	S.F.	0.49	1.75	2.24
Maximum	"	0.75	1.75	2.50
Slides with water tube				
Minimum	EA.	260.00	190.00	450.00
Maximum	"	750.00	190.00	940.00

ELEVATORS	UNIT	MAT.	INST.	TOTAL
14210.10 ELEVATORS				
Passenger elevators, electric, geared				
Based on a shaft of 6 stops and 6 openings				
50 fpm, 2000 lb	EA.	59,360	1,230	60,590
100 fpm, 2000 lb	"	61,600	1,370	62,970
150 fpm				
2000 lb	EA.	67,200	1,540	68,740
3000 lb	"	85,680	1,760	87,440
4000 lb	"	90,250	2,060	92,310
Based on a shaft of 8 stops and 8 openings				
300 fpm				
3000 lb	EA.	102,360	2,470	104,830
3500 lb	"	104,530	2,470	107,000
4000 lb	"	110,240	2,740	112,980
5000 lb	"	124,290	2,940	127,230
Freight elevators, electric				
Based on a shaft of 6 stops and 6 openings				
50 fpm				
3500 lb	EA.	119,950	1,370	121,320
4000 lb	"	120,410	1,370	121,780
5000 lb	"	121,550	1,540	123,090
100 fpm				
3500 lb	EA.	126,240	1,540	127,780
4000 lb	"	126,810	1,540	128,350
5000 lb	"	123,720	1,760	125,480
For variable voltage control, add 20%				
14300.10 ESCALATORS				
Escalators				
32" wide, floor to floor				
12' high	EA.	78,040	2,060	80,100
15' high	"	85,030	2,470	87,500
18' high	"	92,020	3,090	95,110
22' high	"	99,010	4,110	103,120
25' high	"	110,660	4,940	115,600

LIFTS	UNIT	MAT.	INST.	TOTAL
14410.10 PERSONNEL LIFTS				
Electrically operated, 1 or 2 person lift				
With attached foot platforms				
3 stops	EA.			13,524
5 stops	"			16,699

LIFTS	UNIT	MAT.	INST.	TOTAL
14410.10 PERSONNEL LIFTS				
7 stops	EA.			19,874
For each additional stop, add $1250				
Residential stair climber, per story	EA.	4,000	250.00	4,250

MATERIAL HANDLING	UNIT	MAT.	INST.	TOTAL
14580.10 PNEUMATIC SYSTEMS				
Pneumatic message tube system				
Average, 20 station job				
3" round system	E.A.	207,650	2,700	210,350
4" round system	"	259,560	2,970	262,530
6" round system	"	392,220	3,300	395,520
4" x 7" oval system	"	409,530	5,940	415,470
Trash and linen tube system				
10 stations	EA.	132,660	6,170	138,830
15 stations	"	167,270	8,230	175,500
20 stations	"	224,390	9,500	233,890
30 stations	"	294,170	11,220	305,390

HOISTS AND CRANES	UNIT	MAT.	INST.	TOTAL
14600.10 INDUSTRIAL HOISTS				
Industrial hoists, electric, light to medium duty				
500 lb	EA.	4,550	150.00	4,700
1000 lb	"	4,770	160.00	4,930
2000 lb	"	4,890	160.00	5,050
3000 lb	"	4,980	170.00	5,150
4000 lb	"	5,630	190.00	5,820
5000 lb	"	7,960	200.00	8,160
6000 lb	"	8,230	200.00	8,430
7500 lb	"	8,800	210.00	9,010
10,000 lb	"	14,850	220.00	15,070
15,000 lb	"	22,050	230.00	22,280
20,000 lb	"	26,050	250.00	26,300
25,000 lb	"	28,560	270.00	28,830

HOISTS AND CRANES	**UNIT**	**MAT.**	**INST.**	**TOTAL**
14600.10 INDUSTRIAL HOISTS				
30,000 lb	EA.	28,000	300.00	28,300
Heavy duty				
500 lb	EA.	6,710	150.00	6,860
1000 lb	"	6,910	160.00	7,070
2000 lb	"	7,690	160.00	7,850
3000 lb	"	8,110	170.00	8,280
4000 lb	"	8,570	190.00	8,760
5000 lb	"	8,910	200.00	9,110
6000 lb	"	9,140	200.00	9,340
7500 lb	"	12,570	210.00	12,780
10,000 lb	"	13,140	220.00	13,360
15,000 lb	"	14,510	230.00	14,740
20,000 lb	"	17,940	250.00	18,190
25,000 lb	"	26,050	270.00	26,320
30,000 lb	"	27,190	300.00	27,490

BASIC MATERIALS	UNIT	MAT.	INST.	TOTAL
15100.10 SPECIALTIES				
Wall penetration				
Concrete wall, 6" thick				
2" dia.	EA.	0.00	7.55	7.55
4" dia.	"	0.00	11.35	11.35
8" dia.	"	0.00	16.20	16.20
12" thick				
2" dia.	EA.	0.00	10.30	10.30
4" dia.	"	0.00	16.20	16.20
8" dia.	"	0.00	25.20	25.20
Non-destructive testing, piping systems				
X-ray of welds				
3" dia. pipe	EA.	43.50	32.00	75.50
4" dia. pipe	"	37.60	32.00	69.60
6" dia. pipe	"	31.80	32.00	63.80
8" dia. pipe	"	34.10	40.10	74.20
10" dia. pipe	"	36.50	40.10	76.60
Liquid penetration of welds				
2" dia. pipe	EA.	21.65	20.00	41.65
3" dia. pipe	"	22.25	20.00	42.25
4" dia. pipe	"	24.70	20.00	44.70
6" dia. pipe	"	24.70	20.00	44.70
8" dia. pipe	"	30.60	20.00	50.60
10" dia. pipe	"	30.60	20.00	50.60
15120.10 BACKFLOW PREVENTERS				
Backflow preventer, flanged, cast iron, with valves				
3" pipe	EA.	1,970	160.00	2,130
4" pipe	"	2,520	180.00	2,700
6" pipe	"	3,750	270.00	4,020
8" pipe	"	6,940	320.00	7,260
Threaded				
3/4" pipe	EA.	310.00	20.00	330.00
2" pipe	"	680.00	32.00	712.00
Reduced pressure assembly, bronze, threaded				
3/4"	EA.	420.00	20.00	440.00
1"	"	530.00	22.90	552.90
1-1/4"	"	770.00	26.70	796.70
1-1/2"	"	840.00	32.00	872.00
15140.10 PIPE HANGERS, HEAVY				
Hangers				
1/2" pipe, clevis pipe hanger				
Black steel	EA.	1.75	10.70	12.45
Galvanized	"	2.00	10.70	12.70
U bolt	"	0.96	3.20	4.16
3/4" pipe, clevis pipe hanger				
Black steel	EA.	1.75	10.70	12.45
Galvanized	"	2.10	10.70	12.80
U bolt	"	0.96	3.20	4.16
1" pipe, clevis pipe hanger				

BASIC MATERIALS	UNIT	MAT.	INST.	TOTAL
15140.10 PIPE HANGERS, HEAVY				
Black steel	EA.	1.90	10.70	12.60
Galvanized	"	2.20	10.70	12.90
U bolt	"	1.00	3.20	4.20
1-1/4" pipe, clevis pipe hanger				
Black steel	EA.	2.10	10.70	12.80
Galvanized	"	2.50	10.70	13.20
U bolt	"	1.15	3.20	4.35
1-1/2" pipe, clevis pipe hanger				
Black steel	EA.	2.30	10.70	13.00
Galvanized	"	2.80	10.70	13.50
U bolt	"	1.20	3.20	4.40
2" pipe, clevis pipe hanger				
Black steel	EA.	2.80	10.70	13.50
Galvanized	"	3.15	10.70	13.85
Adjustable pipe roll stand	"	36.90	40.10	77.00
U bolt	"	1.25	3.20	4.45
Adjustable roller hanger	"	6.15	10.70	16.85
2-1/2" pipe, clevis pipe hanger				
Black steel	EA.	4.15	10.70	14.85
Galvanized	"	5.35	10.70	16.05
Adjustable pipe roll stand	"	37.20	10.70	47.90
Adjustable roller hanger	"	6.70	10.70	17.40
3" pipe, clevis pipe hanger				
Black steel	EA.	4.75	10.70	15.45
Galvanized	"	8.40	10.70	19.10
Adjustable pipe roll stand	"	37.20	10.70	47.90
U bolt	"	2.10	3.20	5.30
Adjustable roller hanger	"	7.75	10.70	18.45
3-1/2" pipe, clevis pipe hanger				
Black steel	EA.	4.15	10.70	14.85
Galvanized	"	6.25	10.70	16.95
Adjustable pipe roll stand	"	37.20	10.70	47.90
U bolt	"	2.20	3.20	5.40
Adjustable roller hanger	"	8.85	10.70	19.55
4" pipe, clevis pipe hanger				
Black steel	EA.	6.55	10.70	17.25
Galvanized	"	8.30	10.70	19.00
Adjustable pipe roll stand	"	43.30	10.70	54.00
U bolt	"	2.20	3.20	5.40
Adjustable roller hanger	"	11.00	10.70	21.70
5" pipe, clevis pipe hanger				
Black steel	EA.	7.30	12.80	20.10
Galvanized	"	9.50	12.80	22.30
Adjustable pipe roll stand	"	76.50	12.80	89.30
U bolt	"	2.35	3.20	5.55
Adjustable roller hanger	"	15.35	12.80	28.15
6" pipe, clevis pipe hanger				
Black steel	EA.	10.25	12.80	23.05
Galvanized	"	15.25	12.80	28.05
Adjustable pipe roll stand	"	42.80	12.80	55.60
U bolt	"	4.20	4.00	8.20
Adjustable roller hanger	"	17.80	12.80	30.60

BASIC MATERIALS	UNIT	MAT.	INST.	TOTAL
15140.10 PIPE HANGERS, HEAVY				
8" pipe, clevis pipe hanger				
Black steel	EA.	16.75	12.80	29.55
Galvanized	"	16.90	12.80	29.70
Adjustable pipe roll stand	"	67.00	12.80	79.80
U bolt	"	5.10	4.00	9.10
Adjustable roller hanger	"	21.60	12.80	34.40
10" clevis pipe hanger				
Black steel	EA.	17.70	12.80	30.50
Galvanized	"	24.25	12.80	37.05
Adjustable pipe roll stand	"	67.00	12.80	79.80
Adjustable roller hanger	"	26.60	12.80	39.40
12" pipe, clevis pipe hanger				
Black steel	EA.	16.35	12.80	29.15
Galvanized	"	23.20	12.80	36.00
Adjustable pipe roll stand	"	44.40	12.80	57.20
Adjustable roller hanger	"	29.00	12.80	41.80
14" pipe, clevis pipe hanger				
Black steel	EA.	19.50	14.55	34.05
Galvanized	"	28.20	14.55	42.75
Threaded rod, galvanized				
3/8"	L.F.			0.29
1/2"	"			0.49
5/8"	"			0.75
3/4"	"			1.30
7/8"	"			1.75
1"	"			2.37
Hex nuts, galvanized				
3/8"	EA.			0.03
1/2"	"			0.07
5/8"	"			0.16
3/4"	"			0.21
7/8"	"			0.35
1"	"			0.55
C-clamp, steel, with lock nut				
3/8"	EA.	1.20	4.00	5.20
1/2"	"	1.50	4.00	5.50
5/8"	"	2.30	4.00	6.30
3/4"	"	2.95	4.00	6.95
7/8"	"	6.50	4.00	10.50
Angle support, medium, welded steel				
12"x18"	EA.	66.00	32.00	98.00
18"x24"	"	79.50	32.00	111.50
24"x30"	"	100.00	32.00	132.00
Heavy, welded steel				
12"x18"	EA.	93.00	32.00	125.00
18"x24"	"	130.00	32.00	162.00
24"x30"	"	150.00	32.00	182.00
15140.11 PIPE HANGERS, LIGHT				
A band, black iron				
1/2"	EA.	0.43	2.30	2.73

BASIC MATERIALS	UNIT	MAT.	INST.	TOTAL
15140.11 PIPE HANGERS, LIGHT				
1"	EA.	0.54	2.35	2.89
1-1/4"	"	0.59	2.45	3.04
1-1/2"	"	0.68	2.65	3.33
2"	"	0.76	2.90	3.66
2-1/2"	"	1.00	3.20	4.20
3"	"	1.05	3.55	4.60
4"	"	1.55	4.00	5.55
5"	"	2.05	4.25	6.30
6"	"	2.85	4.60	7.45
8"	"	4.85	5.35	10.20
Copper				
1/2"	EA.	0.49	2.30	2.79
3/4"	"	0.49	2.35	2.84
1"	"	0.52	2.35	2.87
1-1/4"	"	0.54	2.45	2.99
1-1/2"	"	0.63	2.65	3.28
2"	"	0.74	2.90	3.64
2-1/2"	"	1.05	3.20	4.25
3"	"	1.25	3.55	4.80
4"	"	1.70	4.00	5.70
Black riser friction hangers				
3/4"	EA.	2.35	2.65	5.00
1"	"	2.35	2.80	5.15
1-1/4"	"	3.00	2.90	5.90
1-1/2"	"	3.20	3.05	6.25
2"	"	3.30	3.20	6.50
2-1/2"	"	3.55	3.55	7.10
3"	"	3.65	4.00	7.65
4"	"	4.65	4.60	9.25
5"	"	6.80	4.95	11.75
6"	"	7.80	5.35	13.15
8"	"	13.30	5.85	19.15
10"	"	17.10	6.40	23.50
Short pattern black riser clamps				
1-1/2"	EA.	2.00	2.90	4.90
2"	"	2.10	3.05	5.15
3"	"	2.25	3.20	5.45
4"	"	2.50	3.55	6.05
Copper riser friction hanger				
1/2"	EA.	2.20	2.45	4.65
3/4"	"	2.25	2.55	4.80
1"	"	2.50	2.65	5.15
1-1/4"	"	2.55	2.80	5.35
1-1/2"	"	2.70	2.90	5.60
2"	"	3.20	3.05	6.25
2-1/2"	"	4.75	3.20	7.95
3"	"	5.30	3.20	8.50
4"	"	5.90	3.55	9.45
Auto grip hangers, galvanized				
1/2"	EA.	0.56	2.30	2.86
3/4"	"	0.56	2.45	3.01
1"	"	0.56	2.55	3.11

BASIC MATERIALS	UNIT	MAT.	INST.	TOTAL
15140.11 — **PIPE HANGERS, LIGHT**				
1-1/4"	EA.	0.60	2.65	3.25
1-1/2"	"	0.63	2.80	3.43
2"	"	0.66	2.90	3.56
2-1/2"	"	1.85	3.05	4.90
3"	"	2.85	3.20	6.05
4"	"	1.85	3.55	5.40
Copper				
1/2"	EA.	0.74	2.30	3.04
3/4"	"	0.74	2.45	3.19
1"	"	0.80	2.55	3.35
1-1/4"	"	0.84	2.65	3.49
1-1/2"	"	0.86	2.80	3.66
2"	"	0.93	2.90	3.83
2-1/2"	"	2.05	3.05	5.10
3"	"	2.15	3.20	5.35
4"	"	3.10	3.55	6.65
Split rings (F&M), galvanized				
3/8"	EA.	1.25	2.30	3.55
1/2"	"	1.45	2.45	3.90
3/4"	"	1.60	2.55	4.15
1"	"	2.00	2.65	4.65
1-1/4"	"	2.15	2.80	4.95
1-1/2"	"	2.50	2.90	5.40
2"	"	2.90	3.05	5.95
2-1/2"	"	8.15	3.20	11.35
3"	"	8.45	3.35	11.80
4"	"	10.25	3.55	13.80
Copper				
1/4"	EA.	1.15	2.30	3.45
3/8"	"	1.15	2.30	3.45
1/2"	"	1.20	2.45	3.65
1"	"	1.45	2.65	4.10
1-1/4"	"	1.70	2.80	4.50
1-1/2"	"	1.85	2.90	4.75
2"	"	2.15	3.05	5.20
2-1/2"	"	4.00	3.20	7.20
3"	"	5.15	3.35	8.50
4"	"	13.25	3.55	16.80
F&M plates, galvanized				
3/8"	EA.	0.94	2.30	3.24
1/2"	"	1.30	2.35	3.65
Copper				
3/8"	EA.	0.90	2.30	3.20
1/2"	"	1.00	2.35	3.35
2 hole clips, galvanized				
3/4"	EA.	0.15	2.15	2.30
1"	"	0.17	2.20	2.37
1-1/4"	"	0.21	2.30	2.51
1-1/2"	"	0.26	2.35	2.61
2"	"	0.34	2.45	2.79
2-1/2"	"	0.63	2.55	3.18
3"	"	0.91	2.65	3.56

BASIC MATERIALS		UNIT	MAT.	INST.	TOTAL
15140.11	**PIPE HANGERS, LIGHT**				
4"		EA.	1.95	2.90	4.85
Perforated strap					
3/4"					
Galvanized, 20 ga.		L.F.	0.21	1.60	1.81
Copper, 22 ga.		"	0.26	1.60	1.86
Threaded rod-couplings					
1/4"		EA.	0.71	2.00	2.71
3/4"		"	0.74	2.15	2.89
1/2"		"	0.83	2.30	3.13
5/8"		"	1.30	2.45	3.75
Reducing rod coupling, 1/2" x 3/8"		"	1.15	2.30	3.45
C-clamps					
3/4"		EA.	1.10	3.20	4.30
Top beam clamp					
3/8"		EA.	1.70	2.65	4.35
1/2"		"	2.10	2.90	5.00
Side beam connector					
3/8"		EA.	0.76	2.65	3.41
1/2"		"	1.65	2.90	4.55
Hex nuts, heavy					
1"		EA.			1.65
Heavy washers					
3/8"		EA.			0.07
1/2"		"			0.15
5/8"		"			0.28
3/4"		"			0.40
Lag rod, 3/8" x					
4"		EA.			0.56
4-1/2"		"			0.56
6"		"			0.66
8"		"			0.83
10"		"			1.14
12"		"			1.39
18"		"			1.92
6"		"			1.77
8"		"			2.03
10"		"			2.31
12"		"			2.39
Drive screws					
1-1/2" x 12"		EA.			0.15
2" x 12"		"			0.23
Wood screws					
3/4"					
#8		EA.			0.03
1"					
#8		EA.			0.04
#10		"			0.04
#12		"			0.08
1-1/4"					
#8		EA.			0.04
#10		"			0.06
#12		"			0.08

BASIC MATERIALS	UNIT	MAT.	INST.	TOTAL
15140.11 PIPE HANGERS, LIGHT				
1-1/2"				
#8	EA.			0.06
#10	"			0.07
2-1/2"				
#10	EA.			0.10
3"				
#12	EA.			0.17
4"				
#12	EA.			0.34
#14	"			0.41
J-Hooks				
1/2"	EA.	0.25	1.45	1.70
3/4"	"	0.26	1.45	1.71
1"	"	0.27	1.55	1.82
1-1/4"	"	0.28	1.55	1.83
1-1/2"	"	0.29	1.60	1.89
2"	"	0.30	1.60	1.90
3"	"	0.36	1.70	2.06
4"	"	0.36	1.70	2.06
PVC coated hangers, galvanized, 28 ga.				
1-1/2" x 12"	EA.	0.72	2.15	2.87
2" x 12"	"	0.80	2.30	3.10
3" x 12"	"	0.87	2.45	3.32
4" x 12"	"	0.95	2.65	3.60
Copper, 30 ga.				
1-1/2" x 12"	EA.	0.77	2.15	2.92
2" x 12"	"	0.92	2.30	3.22
3" x 12"	"	1.05	2.45	3.50
4" x 12"	"	1.10	2.65	3.75
2" x 24"	"	1.85	2.45	4.30
3" x 24"	"	2.10	2.65	4.75
4" x 24"	"	2.25	2.90	5.15
Milford hangers				
1/2" x 6"	EA.	0.64	2.30	2.94
1/2" x 12"	"	0.87	2.45	3.32
3/4" x 6"	"	0.71	2.35	3.06
3/4" x 12"	"	0.90	2.55	3.45
1" x 6"	"	0.75	2.45	3.20
1" x 12"	"	0.94	2.55	3.49
1-1/4" x 6"	"	0.85	2.55	3.40
1-1/4" x 12"	"	1.05	2.65	3.70
1-1/2" x 6"	"	0.94	2.65	3.59
1-1/2" x 12"	"	1.10	2.80	3.90
2" x 6"	"	1.05	2.80	3.85
2" x 12"	"	1.15	2.90	4.05
Wire hook hangers				
Black wire, 1/2" x				
4"	EA.	0.21	1.60	1.81
6"	"	0.25	1.70	1.95
8"	"	0.27	1.80	2.07
10"	"	0.35	1.80	2.15
12"	"	0.44	1.90	2.34

BASIC MATERIALS	UNIT	MAT.	INST.	TOTAL
15140.11 PIPE HANGERS, LIGHT				
3/4" x				
4"	EA.	0.26	1.70	1.96
6"	"	0.28	1.80	2.08
8"	"	0.29	1.90	2.19
10"	"	0.41	2.00	2.41
12"	"	0.43	2.15	2.58
1" x				
4"	EA.	0.26	1.80	2.06
6"	"	0.27	1.90	2.17
8"	"	0.29	2.00	2.29
10"	"	0.43	2.15	2.58
12"	"	0.46	2.30	2.76
1-1/4" x				
4"	EA.	0.28	1.90	2.18
6"	"	0.29	2.00	2.29
8"	"	0.34	2.15	2.49
10"	"	0.38	2.30	2.68
12"	"	0.46	2.45	2.91
1-1/2" x				
6"	EA.	0.32	2.15	2.47
8"	"	0.34	2.30	2.64
10"	"	0.44	2.45	2.89
12"	"	0.48	2.65	3.13
2" x				
6"	EA.	0.31	2.30	2.61
8"	"	0.41	2.45	2.86
10"	"	0.43	2.65	3.08
12"	"	0.49	2.90	3.39
Copper wire hooks				
1/2" x				
4"	EA.	0.27	1.60	1.87
6"	"	0.32	1.70	2.02
8"	"	0.38	1.80	2.18
10"	"	0.45	1.90	2.35
12"	"	0.53	2.00	2.53
3/4" x				
4"	EA.	0.29	1.70	1.99
6"	"	0.34	1.80	2.14
8"	"	0.39	1.90	2.29
10"	"	0.46	2.00	2.46
12"	"	0.55	2.15	2.70
1" x				
4"	EA.	0.31	1.80	2.11
6"	"	0.35	1.90	2.25
8"	"	0.40	2.00	2.40
10"	"	0.49	2.15	2.64
12"	"	0.57	2.30	2.87
1-1/4" x				
6"	EA.	0.40	1.90	2.30
8"	"	0.46	2.00	2.46
10"	"	0.54	2.15	2.69
12"	"	0.56	2.30	2.86

BASIC MATERIALS	UNIT	MAT.	INST.	TOTAL
15140.11 — PIPE HANGERS, LIGHT				
1-1/2" x				
6"	EA.	0.45	2.15	2.60
8"	"	0.49	2.30	2.79
10"	"	0.54	2.45	2.99
12"	"	0.59	2.65	3.24
2" x				
6"	EA.	0.48	2.30	2.78
8"	"	0.53	2.45	2.98
10"	"	0.57	2.65	3.22
12"	"	0.60	2.90	3.50
15175.60 — EXPANSION TANKS				
Expansion tank, 125 psi, galvanized steel				
20 gallon	EA.	200.00	40.10	240.10
30 gallon	"	220.00	53.50	273.50
65 gallon	"	350.00	80.00	430.00
80 gallon	"	350.00	91.50	441.50
15240.10 — VIBRATION CONTROL				
Vibration isolator, in-line, stainless connector, screwed				
1/2"	EA.	56.00	17.80	73.80
3/4"	"	65.00	18.85	83.85
1"	"	69.50	20.00	89.50
1-1/4"	"	78.50	21.35	99.85
1-1/2"	"	87.50	22.90	110.40
2"	"	110.00	24.65	134.65
2-1/2"	"	260.00	26.70	286.70
3"	"	290.00	29.10	319.10
4"	"	390.00	32.00	422.00
6"	"	580.00	35.60	615.60
Flanged				
8"	EA.	900.00	40.10	940.10
10"	"	1,510	45.80	1,556
12"	"	1,900	53.50	1,954

INSULATION	UNIT	MAT.	INST.	TOTAL
15260.10 — FIBERGLASS PIPE INSULATION				
Fiberglass insulation on 1/2" pipe				
1" thick	L.F.	1.00	1.05	2.05
1-1/2" thick	"	2.45	1.35	3.80
3/4" pipe				

INSULATION	UNIT	MAT.	INST.	TOTAL
15260.10 — FIBERGLASS PIPE INSULATION				
1" thick	L.F.	2.20	1.05	3.25
1-1/2" thick	"	2.60	1.35	3.95
1" pipe				
1" thick	L.F.	2.25	1.05	3.30
1-1/2" thick	"	2.80	1.35	4.15
2" thick	"	3.90	1.60	5.50
1-1/4" dia. pipe				
1" thick	L.F.	2.80	1.35	4.15
1-1/2" thick	"	3.35	1.45	4.80
2" thick	"	4.50	1.60	6.10
1-1/2" pipe				
1" thick	L.F.	2.90	1.35	4.25
1-1/2" thick	"	3.45	1.45	4.90
2" thick	"	5.15	1.55	6.70
2-1/2" thick	"	5.70	1.60	7.30
3-1/2" thick	"	9.05	1.70	10.75
2" pipe				
1" thick	L.F.	3.15	1.35	4.50
1-1/2" thick	"	3.70	1.45	5.15
2" thick	"	5.50	1.60	7.10
2-1/2" thick	"	6.05	1.80	7.85
3-1/2" thick	"	9.40	2.00	11.40
2-1/2" pipe				
1" thick	L.F.	3.45	1.35	4.80
1-1/2" thick	"	4.05	1.45	5.50
2" thick	"	5.95	1.60	7.55
2-1/2" thick	"	6.40	1.80	8.20
3" thick	"	7.95	2.00	9.95
3-1/2" thick	"	9.40	2.30	11.70
3" pipe				
1" thick	L.F.	3.60	1.55	5.15
1-1/2" thick	"	4.35	1.60	5.95
2" thick	"	6.05	1.80	7.85
2-1/2" thick	"	7.15	2.00	9.15
3" thick	"	8.50	2.30	10.80
3-1/2" thick	"	9.40	2.65	12.05
4" pipe				
1" thick	L.F.	4.50	1.55	6.05
1-1/2" thick	"	5.60	1.60	7.20
2" thick	"	7.85	1.80	9.65
2-1/2" thick	"	8.95	2.00	10.95
3" thick	"	10.65	2.30	12.95
3-1/2" thick	"	12.30	2.65	14.95
5" pipe				
1" thick	L.F.	4.50	1.55	6.05
2" thick	"	8.40	1.60	10.00
3" thick	"	11.20	1.90	13.10
4" thick	"	14.55	2.45	17.00
6" pipe				
1" thick	L.F.	5.60	1.70	7.30
2" thick	"	8.40	1.80	10.20
4" thick	"	19.05	2.15	21.20

INSULATION	UNIT	MAT.	INST.	TOTAL
15260.10 FIBERGLASS PIPE INSULATION				
6" thick	L.F.	31.40	2.90	34.30
8" pipe				
2" thick	L.F.	11.20	1.70	12.90
3" thick	"	15.70	1.80	17.50
4" thick	"	21.30	2.15	23.45
6" thick	"	37.00	2.90	39.90
10" pipe				
2" thick	L.F.	12.30	1.70	14.00
3" thick	"	17.90	1.80	19.70
4" thick	"	24.65	2.15	26.80
6" thick	"	41.40	2.90	44.30
12" pipe				
2" thick	L.F.	13.45	1.70	15.15
3" thick	"	20.15	1.80	21.95
4" thick	"	26.90	2.15	29.05
6" thick	"	47.00	2.90	49.90
15260.20 CALCIUM SILICATE				
Calcium silicate insulation, 6" pipe				
2" thick	L.F.	6.90	2.30	9.20
2-1/2" thick	"	8.75	2.45	11.20
3" thick	"	12.65	2.65	15.30
4" thick	"	17.25	2.90	20.15
6" thick	"	27.80	3.20	31.00
8" pipe				
2" thick	L.F.	8.05	2.45	10.50
2-1/2" thick	"	10.60	2.65	13.25
3" thick	"	15.30	2.90	18.20
4" thick	"	20.40	3.20	23.60
6" thick	"	33.40	3.55	36.95
10" pipe				
2" thick	L.F.	10.30	2.45	12.75
2-1/2" thick	"	13.15	2.65	15.80
3" thick	"	18.70	2.90	21.60
4" thick	"	25.50	3.20	28.70
6" thick	"	38.00	3.55	41.55
12" pipe				
2" thick	L.F.	11.50	2.45	13.95
2-1/2" thick	"	14.30	2.65	16.95
3" thick	"	21.05	2.90	23.95
4" thick	"	26.70	3.20	29.90
6" thick	"	46.80	3.55	50.35
15260.60 EXTERIOR PIPE INSULATION				
Fiberglass insulation, aluminum jacket				
1/2" pipe				
1" thick	L.F.	2.80	2.45	5.25
1-1/2" thick	"	3.30	2.65	5.95
3/4" pipe				
1" thick	L.F.	2.90	2.45	5.35
1-1/2" thick	"	3.40	2.65	6.05
1" pipe				

INSULATION	UNIT	MAT.	INST.	TOTAL
15260.60 EXTERIOR PIPE INSULATION				
1" thick	L.F.	3.00	2.45	5.45
1-1/2" thick	"	3.70	2.65	6.35
3-1/2" thick	"	10.10	3.55	13.65
1-1/4" pipe				
1" thick	L.F.	3.45	2.90	6.35
1-1/2" thick	"	3.90	3.05	6.95
3-1/2" thick	"	10.30	3.55	13.85
1-1/2" pipe				
1" thick	L.F.	3.60	2.90	6.50
1-1/2" thick	"	4.25	3.05	7.30
2" thick	"	5.95	3.20	9.15
2-1/2" thick	"	6.60	3.35	9.95
3-1/2" thick	"	10.30	3.55	13.85
2" pipe				
1" thick	L.F.	3.80	2.90	6.70
1-1/2" thick	"	4.60	3.05	7.65
2" thick	"	6.15	3.20	9.35
2-1/2" thick	"	6.85	3.35	10.20
3-1/2" thick	"	10.65	3.55	14.20
2-1/2" pipe				
1" thick	L.F.	4.05	2.90	6.95
1-1/2" thick	"	4.80	3.05	7.85
2" thick	"	6.85	3.20	10.05
2-1/2" thick	"	7.40	3.35	10.75
3" thick	"	8.95	3.55	12.50
3-1/2" thick	"	11.20	3.75	14.95
3" pipe				
1" thick	L.F.	4.25	3.20	7.45
1-1/2" thick	"	5.05	3.35	8.40
2" thick	"	7.10	3.55	10.65
2-1/2" thick	"	8.20	3.75	11.95
3" thick	"	9.70	4.00	13.70
3-1/2" thick	"	11.40	4.25	15.65
4" pipe				
1" thick	L.F.	5.05	3.20	8.25
1-1/2" thick	"	6.05	3.35	9.40
2" thick	"	7.60	3.55	11.15
2-1/2" thick	"	9.30	3.75	13.05
3" thick	"	11.40	4.00	15.40
3-1/2" thick	"	13.45	4.25	17.70
5" pipe				
1" thick	L.F.	5.25	3.20	8.45
2" thick	"	8.95	3.35	12.30
2-1/2" thick	"	10.20	3.55	13.75
3" thick	"	13.45	3.75	17.20
3-1/2" thick	"	14.10	4.00	18.10
6" pipe				
1" thick	L.F.	5.55	3.55	9.10
2" thick	"	9.20	3.75	12.95
2-1/2" thick	"	11.65	4.00	15.65
3" thick	"	13.65	4.25	17.90
3-1/2" thick	"	15.70	4.60	20.30

INSULATION	UNIT	MAT.	INST.	TOTAL
15260.60 **EXTERIOR PIPE INSULATION**				
4" thick	L.F.	21.30	4.95	26.25
6" thick	"	33.30	5.35	38.65
8" pipe				
2" thick	L.F.	11.20	3.55	14.75
3" thick	"	16.80	3.75	20.55
3-1/2" thick	"	22.40	4.00	26.40
4" thick	"	23.50	4.25	27.75
6" thick	"	39.20	4.60	43.80
10" pipe				
2" thick	L.F.	14.00	3.55	17.55
3" thick	"	20.15	3.75	23.90
3-1/2" thick	"	23.50	4.00	27.50
4" thick	"	28.00	4.25	32.25
6" thick	"	43.70	4.60	48.30
12" pipe				
2" thick	L.F.	15.70	3.55	19.25
3" thick	"	22.40	3.75	26.15
3-1/2" thick	"	28.00	4.00	32.00
4" thick	"	30.20	4.25	34.45
6" thick	"	48.20	4.60	52.80
Calcium silicate with aluminum jacket, 6" pipe				
2" thick	L.F.	8.40	4.00	12.40
2-1/2" thick	"	10.45	4.25	14.70
3" thick	"	14.15	4.60	18.75
4" thick	"	19.65	5.35	25.00
8" pipe				
2" thick	L.F.	9.75	4.00	13.75
2-1/2" thick	"	11.70	4.25	15.95
3" thick	"	16.45	4.60	21.05
4" thick	"	22.95	4.95	27.90
6" thick	"	34.40	5.35	39.75
10" pipe				
2" thick	L.F.	11.75	4.60	16.35
2-1/2" thick	"	13.80	4.95	18.75
3" thick	"	20.55	5.35	25.90
4" thick	"	28.60	5.85	34.45
6" thick	"	39.90	6.40	46.30
12" pipe				
2" thick	L.F.	13.20	4.60	17.80
2-1/2" thick	"	15.30	4.95	20.25
3" thick	"	21.75	5.35	27.10
4" thick	"	29.70	5.85	35.55
6" thick	"	51.00	6.40	57.40
15260.90 **PIPE INSULATION FITTINGS**				
Insulation protection saddle				
1" thick covering				
1/2" pipe	EA.	3.45	12.80	16.25
3/4" pipe	"	3.45	12.80	16.25
1" pipe	"	3.45	12.80	16.25
1-1/4" pipe	"	3.45	12.80	16.25

INSULATION	UNIT	MAT.	INST.	TOTAL
15260.90 **PIPE INSULATION FITTINGS**				
1-1/2" pipe	EA.	3.70	12.80	16.50
2" pipe	"	4.80	12.80	17.60
2-1/2" pipe	"	4.80	12.80	17.60
3" pipe	"	5.40	14.55	19.95
4" pipe	"	5.80	16.00	21.80
6" pipe	"	6.85	20.00	26.85
1-1/2" thick covering				
3/4" pipe	EA.	5.60	12.80	18.40
1" pipe	"	5.60	12.80	18.40
1-1/4" pipe	"	5.60	12.80	18.40
1-1/2" pipe	"	5.60	12.80	18.40
2" pipe	"	6.25	12.80	19.05
2-1/2" pipe	"	7.05	12.80	19.85
3" pipe	"	7.05	12.80	19.85
4" pipe	"	7.50	16.00	23.50
6" pipe	"	10.40	20.00	30.40
8" pipe	"	13.10	26.70	39.80
10" pipe	"	13.10	26.70	39.80
12" pipe	"	16.45	26.70	43.15
2" thick covering				
3/4" pipe	EA.	5.80	12.80	18.60
1" pipe	"	5.80	12.80	18.60
1-1/4" pipe	"	5.80	12.80	18.60
1-1/2" pipe	"	6.95	12.80	19.75
2" pipe	"	6.95	12.80	19.75
2-1/2" pipe	"	6.95	12.80	19.75
3" pipe	"	8.20	12.80	21.00
4" pipe	"	8.20	16.00	24.20
6" pipe	"	11.55	20.00	31.55
8" pipe	"	13.80	26.70	40.50
10" pipe	"	14.45	26.70	41.15
12" pipe	"	19.00	26.70	45.70
3" thick covering				
2" pipe	EA.	8.90	12.80	21.70
2-1/2" pipe	"	8.90	12.80	21.70
3" pipe	"	9.50	12.80	22.30
4" pipe	"	9.50	16.00	25.50
6" pipe	"	16.35	20.00	36.35
8" pipe	"	16.35	26.70	43.05
10" pipe	"	18.35	26.70	45.05
12" pipe	"	22.40	26.70	49.10
15280.10 **EQUIPMENT INSULATION**				
Equipment insulation, 2" thick, cellular glass	S.F.	1.55	2.00	3.55
Urethane, rigid, field applied jacket, plastered finish	"	1.75	4.00	5.75
Fiberglass, rigid, with vapor barrier	"	1.55	1.80	3.35
15290.10 **DUCTWORK INSULATION**				
Fiberglass duct insulation, plain blanket				
1-1/2" thick	S.F.	0.24	0.40	0.64

INSULATION	UNIT	MAT.	INST.	TOTAL
15290.10 DUCTWORK INSULATION				
2" thick	S.F.	0.28	0.53	0.81
With vapor barrier				
1-1/2" thick	S.F.	0.30	0.40	0.70
2" thick	"	0.36	0.53	0.89
Rigid with vapor barrier				
2" thick	S.F.	1.10	1.05	2.15
3" thick	"	1.85	1.30	3.15
4" thick	"	2.35	1.60	3.95
6" thick	"	3.25	2.15	5.40
Weatherproof, polystyrene, 3" thick, w/vapor barrier	"	2.40	3.20	5.60
Urethane board with vapor barrier	"	3.40	4.00	7.40

FIRE PROTECTION	UNIT	MAT.	INST.	TOTAL
15330.10 WET SPRINKLER SYSTEM				
Sprinkler head, 212 deg, brass, exposed piping	EA.	4.70	12.80	17.50
Chrome, concealed piping	"	4.65	17.80	22.45
Water motor alarm	"	63.50	53.50	117.00
Fire department inlet connection	"	98.50	64.00	162.50
Wall plate for fire dept connection	"	47.70	26.70	74.40
Swing check valve flanged iron body, 4"	"	110.00	110.00	220.00
Check valve, 6"	"	480.00	160.00	640.00
Wet pipe valve, flange to groove, 4"	"	410.00	35.60	445.60
Flange to flange				
6"	EA.	530.00	53.50	583.50
8"	"	690.00	110.00	800.00
Alarm valve, flange to flange, (wet valve)				
4"	EA.	340.00	35.60	375.60
8"	"	580.00	270.00	850.00
Inspector's test connection	"	29.10	26.70	55.80
Wall hydrant, polished brass, 2-1/2" x 2-1/2", single	"	280.00	22.90	302.90
2-way	"	760.00	22.90	782.90
3-way	"	1,110	22.90	1,133
Wet valve trim, includes retard chamber & gauges, 4"-6"	"	380.00	26.70	406.70
Retard pressure switch for wet systems	"	710.00	64.00	774.00
Air maintenance device	"	200.00	26.70	226.70
Wall hydrant non-freeze, 8" thick wall, vacuum breaker	"	18.50	16.00	34.50
12" thick wall	"	20.45	16.00	36.45
15330.50 DRY SPRINKLER SYSTEM				
Dry pipe valve, flange to flange				
4"	EA.	800.00	64.00	864.00

FIRE PROTECTION	UNIT	MAT.	INST.	TOTAL
15330.50 DRY SPRINKLER SYSTEM				
6"	EA.	1,040	80.00	1,120
Trim, 4" and 6", includes gauges	"	380.00	26.70	406.70
Field testing and flushing	"	0.00	270.00	270.00
Disinfection	"	0.00	270.00	270.00
Pressure switch double circuit, open/close contacts	"	180.00	80.00	260.00
Low air				
Supervisory unit	EA.	630.00	53.50	683.50
Pressure switch	"	200.00	26.70	226.70
15330.70 CO2 SYSTEM				
CO2 system, high pressure, 75# cylinder with				
Valve assemblies	EA.	1,200	64.00	1,264
Storage rack	"	680.00	45.80	725.80
Manifold	"	500.00	230.00	730.00
Flexible loops	"	39.20	4.00	43.20
Beam scale for cylinders	"	340.00	53.50	393.50
Mechanically control head	"	280.00	21.35	301.35
Electrically control head	"	280.00	21.35	301.35
Stop valves	"	720.00	32.00	752.00
Check valves	"	320.00	40.10	360.10
Activation station	"	380.00	32.00	412.00
Nozzles	"	60.50	26.70	87.20
Hose reel with 75' of 3/4" hose	"	2,180	160.00	2,340
Main/reserve transfer switch	"	2,520	53.50	2,574
Pressure switch	"	220.00	32.00	252.00
Heat responsive device	"	380.00	53.50	433.50
Battery and charger	"	2,130	160.00	2,290
Low pressure				
Battery and charger	EA.	2,130	160.00	2,290
Pressure switch	"	210.00	35.60	245.60
Nozzles	"	61.50	29.10	90.60
Master selector valve	"	180.00	53.50	233.50
Selector valve	"	2,460	53.50	2,514
Low pressure hose reel with 75' of 3/4" hose	"	3,220	160.00	3,380
Tank fill lines	"	730.00	40.10	770.10
Activation stations	"	360.00	26.70	386.70
Electro manual pilot panels	"	730.00	40.10	770.10
15330.90 HALON SYSTEM				
Halon fire protection system, per computer room				
Minimum	EA.			2,576
Average	"			3,304
Maximum	"			6,160

PLUMBING	UNIT	MAT.	INST.	TOTAL
15410.05 C.I. PIPE, ABOVE GROUND				
No hub pipe				
1-1/2" pipe	L.F.	4.05	2.30	6.35
2" pipe	"	4.40	2.65	7.05
3" pipe	"	6.25	3.20	9.45
4" pipe	"	8.40	5.35	13.75
6" pipe	"	12.30	6.40	18.70
8" pipe	"	18.50	10.70	29.20
10" pipe	"	29.10	12.80	41.90
No hub fittings, 1-1/2" pipe				
1/4 bend	EA.	5.05	10.70	15.75
1/8 bend	"	3.35	10.70	14.05
Sanitary tee	"	6.15	16.00	22.15
Sanitary cross	"	7.60	16.00	23.60
Plug	"			1.12
Coupling	"			6.16
Wye	"	6.15	16.00	22.15
Tapped tee	"	6.25	10.70	16.95
P-trap	"	9.50	10.70	20.20
Tapped cross	"	9.75	10.70	20.45
2" pipe				
1/4 bend	EA.	5.15	12.80	17.95
1/8 bend	"	3.45	12.80	16.25
Sanitary tee	"	6.40	21.35	27.75
Sanitary cross	"	8.05	21.35	29.40
Plug	"			1.34
Coupling	"			6.50
Wye	"	6.40	26.70	33.10
Double wye	"	8.95	26.70	35.65
2x1-1/2" wye & 1/8 bend	"	8.75	20.00	28.75
Double wye & 1/8 bend	"	11.75	26.70	38.45
Test tee less 2" plug	"	5.95	12.80	18.75
Tapped tee				
2"x2"	EA.	6.95	12.80	19.75
2"x1-1/2"	"	6.85	12.80	19.65
P-trap				
2"x2"	EA.	9.75	12.80	22.55
Tapped cross				
2"x1-1/2"	EA.	10.10	12.80	22.90
3" pipe				
1/4 bend	EA.	6.50	16.00	22.50
1/8 bend	"	5.80	16.00	21.80
Sanitary tee	"	7.60	20.00	27.60
3"x2" sanitary tee	"	7.40	20.00	27.40
3"x1-1/2" sanitary tee	"	7.50	20.00	27.50
Sanitary cross	"	11.40	26.70	38.10
3x2" sanitary cross	"	12.60	26.70	39.30
Plug	"			2.02
Coupling	"			7.84
Wye	"	7.85	26.70	34.55
3x2" wye	"	8.05	26.70	34.75
Double wye	"	11.75	26.70	38.45
3x2" double wye	"	11.40	26.70	38.10

PLUMBING	UNIT	MAT.	INST.	TOTAL
15410.05	**C.I. PIPE, ABOVE GROUND**			
3x2" wye & 1/8 bend	EA.	10.10	22.90	33.00
3x1-1/2" wye & 1/8 bend	"	10.10	22.90	33.00
Double wye & 1/8 bend	"	18.50	26.70	45.20
3x2" double wye & 1/8 bend	"	14.00	26.70	40.70
3x2" reducer	"	4.25	14.55	18.80
Test tee, less 3" plug	"	8.95	16.00	24.95
Plug	"			2.80
3x3" tapped tee	"	14.00	16.00	30.00
3x2" tapped tee	"	12.55	16.00	28.55
3x1-1/2" tapped tee	"	12.30	16.00	28.30
P-trap	"	11.75	16.00	27.75
3x2" tapped cross	"	9.50	16.00	25.50
3x1-1/2" tapped cross	"	9.40	16.00	25.40
Closet flange, 3-1/2" deep	"	8.40	8.00	16.40
4" pipe				
1/4 bend	EA.	8.50	16.00	24.50
1/8 bend	"	7.60	16.00	23.60
Sanitary tee	"	9.50	26.70	36.20
4x3" sanitary tee	"	9.30	26.70	36.00
4x2" sanitary tee	"	8.75	26.70	35.45
Sanitary cross	"	23.50	32.00	55.50
4x3" sanitary cross	"	17.35	32.00	49.35
4x2" sanitary cross	"	15.25	32.00	47.25
Plug	"			2.91
Coupling	"			8.40
Wye	"	10.65	26.70	37.35
4x3" wye	"	8.95	26.70	35.65
4x2" wye	"	6.85	26.70	33.55
Double wye	"	24.65	32.00	56.65
4x3" double wye	"	14.55	32.00	46.55
4x2" double wye	"	12.90	32.00	44.90
Wye & 1/8 bend	"	13.45	26.70	40.15
4x3" wye & 1/8 bend	"	9.95	26.70	36.65
4x2" wye & 1/8 bend	"	8.05	26.70	34.75
Double wye & 1/8 bend	"	31.40	32.00	63.40
4x3" double wye & 1/8 bend	"	21.30	32.00	53.30
4x2" double wye & 1/8 bend	"	14.55	32.00	46.55
4x3" reducer	"	5.15	16.00	21.15
4x2" reducer	"	5.05	16.00	21.05
Test tee, less 4" plug	"	11.20	16.00	27.20
Plug	"			4.48
4x2" tapped tee	"	7.85	16.00	23.85
4x1-1/2" tapped tee	"	6.70	16.00	22.70
P-trap	"	20.05	16.00	36.05
4x2" tapped cross	"	12.30	16.00	28.30
4x1-1/2" tapped cross	"	9.95	16.00	25.95
Closet flange				
3" deep	EA.	7.60	16.00	23.60
8" deep	"	14.10	16.00	30.10
6" pipe				
1/4 bend	EA.	16.25	26.70	42.95
1/8 bend	"	14.55	26.70	41.25

PLUMBING	UNIT	MAT.	INST.	TOTAL
15410.05 C.I. PIPE, ABOVE GROUND				
Sanitary tee	EA.	26.90	32.00	58.90
6x4" sanitary tee	"	18.50	32.00	50.50
Coupling	"			17.92
Wye	"	25.80	32.00	57.80
6x4" wye	"	17.35	32.00	49.35
6x3" wye	"	15.45	32.00	47.45
6x2" wye	"	14.10	32.00	46.10
Double wye	"	37.00	40.10	77.10
6x4" double wye	"	31.40	40.10	71.50
Wye & 1/8 bend	"	33.00	32.00	65.00
6x4" wye & 1/8 bend	"	22.05	32.00	54.05
6x3" wye & 1/8 bend	"	21.50	32.00	53.50
6x2" wye & 1/8 bend	"	16.90	32.00	48.90
6x4" reducer	"	10.20	17.80	28.00
6x4" reducer	"	10.10	16.00	26.10
6x3" reducer	"	10.10	17.80	27.90
Test tee				
Less 6" plug	EA.	24.65	20.00	44.65
Plug	"			6.50
P-trap	"	40.30	20.00	60.30
8" pipe				
1/4 bend	EA.	38.10	26.70	64.80
1/8 bend	"	24.65	26.70	51.35
Sanitary tee	"	62.50	40.10	102.60
8x6" sanitary tee	"	44.80	40.10	84.90
Plug	"			11.20
Coupling	"			16.80
Wye	"	51.50	32.00	83.50
8x6" wye	"	40.30	26.70	67.00
8x4" wye	"	31.40	26.70	58.10
Double wye	"	74.00	26.70	100.70
Wye & 1/8 bend	"	76.00	26.70	102.70
8x6" wye & 1/8 bend	"	49.30	26.70	76.00
8x4" wye & 1/8 bend	"	40.30	26.70	67.00
8x6" reducer	"	14.55	26.70	41.25
8x4" reducer	"	14.00	26.70	40.70
8x3" reducer	"	12.20	26.70	38.90
8x2" reducer	"	12.10	17.80	29.90
Test tee				
Less 8" plug	EA.	49.30	26.70	76.00
Plug	"			11.42
10" pipe				
1/4 bend	EA.	76.00	26.70	102.70
1/8 bend	"	51.50	26.70	78.20
Plug	"			19.04
Coupling	"			25.76
Wye	"	110.00	53.50	163.50
10x8" wye	"	96.50	53.50	150.00
10x6" wye	"	85.00	53.50	138.50
10x4" wye	"	85.00	53.50	138.50
10x8" reducer	"	31.40	26.70	58.10
10x6" reducer	"	31.40	26.70	58.10

PLUMBING	UNIT	MAT.	INST.	TOTAL
15410.05 C.I. PIPE, ABOVE GROUND				
10x4" reducer	EA.	24.10	26.70	50.80
15410.06 C.I. PIPE, BELOW GROUND				
No hub pipe				
1-1/2" pipe	L.F.	4.05	1.60	5.65
2" pipe	"	4.40	1.80	6.20
3" pipe	"	6.25	2.00	8.25
4" pipe	"	8.40	2.65	11.05
6" pipe	"	12.30	2.90	15.20
8" pipe	"	18.50	3.55	22.05
10" pipe	"	29.10	4.00	33.10
Fittings, 1-1/2"				
1/4 bend	EA.	5.05	9.15	14.20
1/8 bend	"	3.35	9.15	12.50
Plug	"			1.12
Wye	"	6.15	12.80	18.95
Wye & 1/8 bend	"	6.45	9.15	15.60
P-trap	"	9.50	9.15	18.65
2"				
1/4 bend	EA.	5.15	10.70	15.85
1/8 bend	"	3.45	10.70	14.15
Plug	"			1.34
Double wye	"	8.95	20.00	28.95
Wye & 1/8 bend	"	8.75	16.00	24.75
Double wye & 1/8 bend	"	11.75	20.00	31.75
P-trap	"	9.75	10.70	20.45
3"				
1/4 bend	EA.	6.50	12.80	19.30
1/8 bend	"	5.80	12.80	18.60
Plug	"			2.02
Wye	"	7.85	20.00	27.85
3x2" wye	"	8.05	20.00	28.05
Wye & 1/8 bend	"	10.10	20.00	30.10
Double wye & 1/8 bend	"	18.50	20.00	38.50
3x2" double wye & 1/8 bend	"	14.00	20.00	34.00
3x2" reducer	"	4.25	12.80	17.05
P-trap	"	11.75	12.80	24.55
4"				
1/4 bend	EA.	8.50	12.80	21.30
1/8 bend	"	7.60	12.80	20.40
Plug	"			2.91
Wye	"	10.65	20.00	30.65
4x3" wye	"	8.95	20.00	28.95
4x2" wye	"	6.85	20.00	26.85
Double wye	"	24.65	26.70	51.35
4x3" double wye	"	14.55	26.70	41.25
4x2" double wye	"	12.90	26.70	39.60
Wye & 1/8 bend	"	13.45	20.00	33.45
4x3" wye & 1/8 bend	"	9.95	20.00	29.95
4x2" wye & 1/8 bend	"	8.05	20.00	28.05

PLUMBING	UNIT	MAT.	INST.	TOTAL
15410.06	**C.I. PIPE, BELOW GROUND**			
Double wye & 1/8 bend	EA.	31.40	26.70	58.10
4x3" double wye & 1/8 bend	"	21.30	26.70	48.00
4x2" double wye & 1/8 bend	"	14.55	26.70	41.25
4x3" reducer	"	5.15	12.80	17.95
4x2" reducer	"	5.05	12.80	17.85
6"				
1/4 bend	EA.	16.25	20.00	36.25
1/8 bend	"	14.55	20.00	34.55
Wye & 1/8 bend	"	33.00	26.70	59.70
6x4" wye & 1/8 bend	"	22.05	26.70	48.75
6x3" wye & 1/8 bend	"	21.50	26.70	48.20
6x2" wye & 1/8 bend	"	16.90	26.70	43.60
6x3" reducer	"	10.10	14.55	24.65
P-trap	"	40.30	16.00	56.30
8"				
1/4 bend	EA.	38.10	20.00	58.10
1/8 bend	"	24.65	20.00	44.65
Plug	"			11.20
Wye	"	51.50	26.70	78.20
8x6" wye	"	40.30	20.00	60.30
8x4" wye	"	31.40	20.00	51.40
8x6" wye & 1/8 bend	"	49.30	20.00	69.30
8x4" reducer	"	14.00	16.00	30.00
8x3" reducer	"	12.20	16.00	28.20
8x2" reducer	"	12.10	16.00	28.10
10"				
1/4 bend	EA.	76.00	20.00	96.00
1/8 bend	"	51.50	20.00	71.50
Plug	"			19.04
Wye	"	110.00	40.10	150.10
10x8" wye	"	96.50	40.10	136.60
10x6" wye	"	85.00	40.10	125.10
10x4" wye	"	85.00	40.10	125.10
10x8" reducer	"	31.40	20.00	51.40
10x6" reducer	"	31.40	20.00	51.40
15410.08	**EXTRA HEAVY SOIL PIPE**			
Extra heavy soil pipe, single hub				
2" x 5'	EA.	22.40	6.40	28.80
3" x 5'	"	28.00	6.80	34.80
4" x 5'	"	42.60	7.45	50.05
6" x 5'	"	61.50	8.00	69.50
Double hub				
2" x 5'	EA.	31.40	8.00	39.40
4" x 5'	"	43.70	8.45	52.15
5" x 5'	"	58.00	8.90	66.90
6" x 5'	"	73.00	9.15	82.15
Single hub				
3" x 10'	EA.	43.70	5.35	49.05
4" x 10'	"	56.00	5.50	61.50
6" x 10'	"	84.00	5.85	89.85

PLUMBING	UNIT	MAT.	INST.	TOTAL
15410.08 EXTRA HEAVY SOIL PIPE				
8" x 10'	EA.	130.00	6.15	136.15
"Mini", single hub				
2" x 42"	EA.	18.50	6.40	24.90
3" x 42"	"	22.40	6.80	29.20
4" x 42"	"	24.10	7.45	31.55
6" x 42"	"	31.90	8.00	39.90
8" x 42"	"	44.80	8.45	53.25
Fittings, 1/4" bend				
2"	EA.	6.25	10.70	16.95
3"	"	10.35	12.80	23.15
4"	"	12.15	12.80	24.95
1/8 bend				
2"	EA.	5.15	10.70	15.85
3"	"	8.75	12.80	21.55
4"	"	12.90	12.80	25.70
5"	"	16.80	16.00	32.80
6"	"	20.70	20.00	40.70
8"	"	47.00	20.00	67.00
Long sweep				
2"	EA.	11.20	10.70	21.90
3"	"	16.80	12.80	29.60
4"	"	23.50	12.80	36.30
Straight T				
4" x 2"	EA.	20.70	20.00	40.70
4" x 3"	"	22.40	20.00	42.40
4"	"	24.10	22.90	47.00
Sanitary T				
3" x 2"	EA.	17.90	20.00	37.90
3"	"	20.15	20.00	40.15
4" x 2"	"	20.70	22.90	43.60
4" x 3"	"	22.10	22.90	45.00
4"	"	24.10	24.65	48.75
Wye				
4" x 2"	EA.	20.70	22.90	43.60
4" x 3"	"	24.10	22.90	47.00
4"	"	28.00	24.65	52.65
Combination Y and 1/8 bend, 4"	"	28.00	24.65	52.65
Double wye, 4"	"	31.40	24.65	56.05
Tapped sanitary T, 4" x 2"	"	15.70	26.70	42.40
P trap				
2"	EA.	11.75	12.80	24.55
4"	"	27.40	21.35	48.75
Dandy				
4", with 3" brass plug	EA.	21.60	17.80	39.40
6", 4" brass plug	"	39.20	20.00	59.20
15410.09 SERVICE WEIGHT PIPE				
Service weight pipe, single hub				
2" x 5'	EA.	20.15	6.40	26.55
3" x 5'	"	23.50	6.80	30.30
4" x 5'	"	28.60	7.10	35.70

PLUMBING		UNIT	MAT.	INST.	TOTAL
15410.09	**SERVICE WEIGHT PIPE**				
5" x 5'		EA.	34.40	7.65	42.05
6" x 5'		"	46.50	8.00	54.50
8" x 5'		"	66.00	8.45	74.45
10" x 5'		"	100.00	8.90	108.90
12" x 5'		"	130.00	10.00	140.00
Double hub					
2" x 5'		EA.	22.10	8.00	30.10
3" x 5'		"	24.65	8.65	33.30
4" x 5'		"	30.80	9.15	39.95
5" x 5'		"	42.60	10.00	52.60
6" x 5'		"	48.70	10.70	59.40
10" x 5'		"	100.00	11.45	111.45
12' x 5'		"	130.00	12.30	142.30
Single hub					
2" x 10'		EA.	25.80	8.00	33.80
3" x 10'		"	31.90	8.65	40.55
4" x 10'		"	39.80	9.15	48.95
5" x 10'		"	52.50	10.00	62.50
6" x 10'		"	65.00	10.70	75.70
8" x 10'		"	100.00	11.45	111.45
10" x 10'		"	170.00	12.30	182.30
12" x 10'		"	250.00	13.35	263.35
Shorty					
2" x 42"		EA.	13.45	6.40	19.85
3" x 42"		"	14.55	6.80	21.35
4" x 42"		"	17.35	7.10	24.45
5" x 42"		"	25.20	7.65	32.85
6" x 42"		"	28.00	8.00	36.00
8" x 42"		"	39.20	8.45	47.65
10" x 42"		"	54.50	8.90	63.40
Soil plug					
2"		EA.	2.70	10.70	13.40
3"		"	3.80	11.45	15.25
4"		"	4.35	12.30	16.65
5"		"	6.70	12.80	19.50
6"		"	7.05	13.35	20.40
1 / 5 bend					
3"		EA.	8.95	12.80	21.75
4"		"	12.30	14.55	26.85
1 / 6 bend					
2"		EA.	6.70	10.70	17.40
3"		"	9.20	12.80	22.00
4"		"	12.65	14.55	27.20
1 / 8 bend					
2"		EA.	3.60	10.70	14.30
3"		"	5.60	12.80	18.40
4"		"	7.85	14.55	22.40
5"		"	10.65	15.25	25.90
6"		"	12.30	16.00	28.30
8"		"	39.20	20.00	59.20
10"		"	54.00	22.90	76.90
12"		"	87.50	26.70	114.20

PLUMBING	UNIT	MAT.	INST.	TOTAL
15410.09 SERVICE WEIGHT PIPE				
1/16 bend				
2"	EA.	3.80	10.70	14.50
3"	"	5.70	12.80	18.50
4"	"	7.85	14.55	22.40
5"	"	10.65	15.25	25.90
6"	"	11.75	16.00	27.75
1/4 bend				
2"	EA.	4.50	10.70	15.20
3"	"	7.05	12.80	19.85
4"	"	9.95	14.55	24.50
5"	"	12.90	15.25	28.15
6"	"	16.25	16.00	32.25
8"	"	48.70	20.00	68.70
10"	"	68.50	22.90	91.40
12"	"	92.00	26.70	118.70
1/4 bend, long				
2" x 12"	EA.	8.40	11.45	19.85
4" x 18"	"	18.15	16.00	34.15
4" x 12"	"	16.80	16.00	32.80
Sweep				
2"	EA.	6.70	10.70	17.40
3"	"	9.50	12.80	22.30
4"	"	14.45	14.55	29.00
5"	"	21.15	15.25	36.40
6"	"	25.80	16.00	41.80
8"	"	62.50	20.00	82.50
Reducing long sweep				
3" x 2"	EA.	11.20	12.80	24.00
4" x 3"	"	13.45	16.00	29.45
Straight T				
2"	EA.	8.40	20.00	28.40
3"	"	12.90	21.35	34.25
4" x 2"	"	12.90	22.90	35.80
4" x 3"	"	15.10	24.65	39.75
4"	"	16.35	26.70	43.05
Sanitary T				
2"	EA.	6.85	20.00	26.85
3" x 2"	"	9.85	21.35	31.20
3"	"	11.15	22.90	34.05
4" x 2"	"	11.50	24.65	36.15
4" x 3"	"	13.35	26.70	40.05
4"	"	14.10	26.70	40.80
5"	"	24.40	29.10	53.50
6"	"	28.00	29.10	57.10
Wye				
2"	EA.	6.70	16.00	22.70
3" x 2"	"	9.50	17.80	27.30
3"	"	11.85	17.80	29.65
4" x 2"	"	11.60	18.85	30.45
4" x 3"	"	13.65	18.85	32.50
4"	"	15.10	18.85	33.95
5" x 2"	"	16.15	20.00	36.15

PLUMBING		UNIT	MAT.	INST.	TOTAL
15410.09	**SERVICE WEIGHT PIPE**				
5" x 3"		EA.	17.25	20.00	37.25
5" x 4"		"	19.60	20.00	39.60
5"		"	23.50	20.00	43.50
6" x 2"		"	21.30	20.00	41.30
6" x 3"		"	21.50	20.00	41.50
6" x 4"		"	22.95	21.35	44.30
6" x 5"		"	29.70	21.35	51.05
6"		"	32.50	22.90	55.40
8" x 4"		"	40.20	22.90	63.10
8" x 6"		"	49.30	24.65	73.95
8"		"	82.00	26.70	108.70
10" x 4"		"	71.50	26.70	98.20
10" x 6"		"	77.50	29.10	106.60
10" x 8"		"	110.00	32.00	142.00
Service wye					
10"		EA.	120.00	40.10	160.10
12" x 4"		"	110.00	40.10	150.10
12" x 6"		"	110.00	40.10	150.10
12" x 8"		"	130.00	45.80	175.80
12"		"	200.00	53.50	253.50
Combination wye and 1/8 bend					
2"		EA.	8.95	21.35	30.30
3" x 2"		"	10.20	22.90	33.10
3"		"	13.45	22.90	36.35
4" x 2"		"	12.90	24.65	37.55
4" x 3"		"	14.55	24.65	39.20
4"		"	16.80	26.70	43.50
5" x 4"		"	28.60	26.70	55.30
6" x 4"		"	28.90	32.00	60.90
8"		"	94.00	40.10	134.10
Straight cross, 4"		"	18.95	26.70	45.65
Sanitary cross					
2"		EA.	9.85	22.90	32.75
3"		"	15.70	24.65	40.35
3" x 2"		"	13.45	24.65	38.10
4"		"	20.50	26.70	47.20
4" x 3"		"	18.25	26.70	44.95
4" x 2"		"	17.15	29.10	46.25
Double wye					
2"		EA.	10.10	22.90	33.00
3" x 2"		"	14.00	24.65	38.65
3"		"	15.55	24.65	40.20
4" x 2"		"	16.70	26.70	43.40
4" x 3"		"	19.05	26.70	45.75
4"		"	22.40	26.70	49.10
5"		"	33.00	32.00	65.00
6" x 4"		"	37.50	40.10	77.60
Combination double wye and 1/8 bend					
2"		EA.	15.70	22.90	38.60
3" x 2"		"	17.35	24.65	42.00
3"		"	20.60	24.65	45.25
4" x 3"		"	23.50	26.70	50.20

PLUMBING	UNIT	MAT.	INST.	TOTAL
15410.09 SERVICE WEIGHT PIPE				
4"	EA.	28.00	26.70	54.70
Tapped sanitary T				
2" x 1-1/2"	EA.	8.95	22.90	31.85
2" x 2"	"	9.05	22.90	31.95
3" x 1-1/2"	"	9.85	24.65	34.50
3" x 2"	"	9.85	24.65	34.50
4" x 1-1/2"	"	11.20	26.70	37.90
4" x 2"	"	11.40	26.70	38.10
Tapped straight T				
3" x 1-1/2"	EA.	8.95	24.65	33.60
3" x 2"	"	8.95	24.65	33.60
4" x 2"	"	11.20	26.70	37.90
4" x 1-1/2"	"	15.10	26.70	41.80
Tapped Y				
4" x 2"	EA.	15.25	26.70	41.95
Tapped sanitary cross				
2" x 1-1/2"	EA.	7.75	22.90	30.65
2" x 2"	"	7.75	24.65	32.40
3" x 1-1/2"	"	10.10	24.65	34.75
3" x 2"	"	10.10	24.65	34.75
3" x 3"	"	15.25	24.65	39.90
4" x 1-1/2"	"	13.45	26.70	40.15
4" x 2"	"	13.45	26.70	40.15
Cleanout, dandy, with brass plug				
2", 1-1/2" plug	EA.	11.10	22.90	34.00
3", 2" plug	"	21.85	24.65	46.50
4", 3" plug	"	25.80	26.70	52.50
5", 4" plug	"	31.40	29.10	60.50
6", 4" plug	"	39.20	32.00	71.20
8", 6" plug	"	80.50	40.10	120.60
Reducer, 3" x 2"	"	5.80	22.90	28.70
4" x				
2"	EA.	6.05	24.65	30.70
3"	"	7.05	24.65	31.70
5" x				
2"	EA.	8.60	26.70	35.30
3"	"	8.75	26.70	35.45
4"	"	8.85	26.70	35.55
6" x				
2"	EA.	8.95	32.00	40.95
3"	"	10.10	32.00	42.10
4"	"	11.10	32.00	43.10
5"	"	11.55	32.00	43.55
8" x				
4"	EA.	19.60	40.10	59.70
6"	"	19.95	40.10	60.05
10" x				
4"	EA.	28.00	45.80	73.80
6"	"	29.10	45.80	74.90
8"	"	32.50	45.80	78.30
12" x				
4"	EA.	44.20	53.50	97.70

PLUMBING	UNIT	MAT.	INST.	TOTAL
15410.09 **SERVICE WEIGHT PIPE**				
6"	EA.	47.60	53.50	101.10
8"	"	48.20	53.50	101.70
10"	"	50.50	53.50	104.00
P trap				
2"	EA.	7.85	22.90	30.75
3"	"	10.65	24.65	35.30
4"	"	15.70	26.70	42.40
5"	"	31.40	32.00	63.40
6"	"	44.20	40.10	84.30
15410.10 **COPPER PIPE**				
Type "K" copper				
1/2"	L.F.	1.70	1.00	2.70
3/4"	"	3.45	1.05	4.50
1"	"	4.35	1.15	5.50
1-1/4"	"	5.50	1.25	6.75
1-1/2"	"	6.95	1.35	8.30
2"	"	9.50	1.45	10.95
2-1/2"	"	10.65	1.60	12.25
3"	"	13.65	1.70	15.35
4"	"	22.40	1.80	24.20
DWV, copper				
1-1/4"	L.F.	3.35	1.35	4.70
1-1/2"	"	3.80	1.45	5.25
2"	"	5.05	1.60	6.65
3"	"	9.50	1.80	11.30
4"	"	14.45	2.00	16.45
6"	"	50.50	2.30	52.80
Refrigeration tubing, copper, sealed				
1/8"	L.F.	0.50	1.30	1.80
3/16"	"	0.54	1.35	1.89
1/4"	"	0.63	1.40	2.03
5/16"	"	0.81	1.45	2.26
3/8"	"	0.90	1.55	2.45
1/2"	"	1.20	1.60	2.80
7/8"	"	2.00	1.85	3.85
1-1/8"	"	3.25	2.15	5.40
1-3/8"	"	3.90	2.45	6.35
Type "L" copper				
1/4"	L.F.	1.00	0.94	1.94
3/8"	"	1.20	0.94	2.14
1/2"	"	1.55	1.00	2.55
3/4"	"	2.00	1.05	3.05
1"	"	2.80	1.15	3.95
1-1/4"	"	3.90	1.25	5.15
1-1/2"	"	4.70	1.35	6.05
2"	"	7.00	1.45	8.45
2-1/2"	"	9.50	1.60	11.10
3"	"	12.30	1.70	14.00
3-1/2"	"	15.35	1.75	17.10
4"	"	20.15	1.80	21.95

PLUMBING	UNIT	MAT.	INST.	TOTAL
15410.10 — COPPER PIPE				
Type "M" copper				
1/2"	L.F.	1.10	1.00	2.10
3/4"	"	1.70	1.05	2.75
1"	"	2.25	1.15	3.40
1-1/4"	"	3.35	1.25	4.60
2"	"	6.50	1.45	7.95
2-1/2"	"	8.95	1.60	10.55
3"	"	10.10	1.70	11.80
4"	"	18.50	1.80	20.30
Type "K" tube, coil				
1/4" x 60'	EA.			49.28
1/2" x 60'	"			94.08
1/2" x 100'	"			134.40
3/4" x 60'	"			145.60
3/4" x 100'	"			224.00
1" x 60'	"			224.00
1" x 100'	"			336.00
1-1/4" x 60'	"			336.00
1-1/2" x 60'	"			425.60
2" x 40'	"			448.00
Type "L" tube, coil				
1/4" x 60'	EA.			44.80
3/8" x 60'	"			67.20
1/2" x 60'	"			89.60
1/2" x 100'	"			134.40
3/4" x 60'	"			134.40
3/4" x 100'	"			201.60
1" x 60'	"			224.00
1" x 100'	"			336.00
1-1/4" x 60'	"			336.00
1-1/2" x 60'	"			358.40
15410.11 — COPPER FITTINGS				
Coupling, with stop				
1/4"	EA.	0.34	10.70	11.04
3/8"	"	0.65	12.80	13.45
1/2"	"	0.78	13.95	14.73
5/8"	"	1.00	16.00	17.00
3/4"	"	1.10	17.80	18.90
1"	"	2.25	18.85	21.10
3"	"	17.90	32.00	49.90
4"	"	40.30	40.10	80.40
Reducing coupling				
1/4" x 1/8"	EA.	1.10	12.80	13.90
3/8" x 1/4"	"	1.25	13.95	15.20
1/2" x				
3/8"	EA.	1.20	16.00	17.20
1/4"	"	1.20	16.00	17.20
1/8"	"	1.20	16.00	17.20
3/4" x				
3/8"	EA.	1.55	17.80	19.35

PLUMBING		UNIT	MAT.	INST.	TOTAL
15410.11	**COPPER FITTINGS**				
1/2"		EA.	1.60	17.80	19.40
1" x					
3/8"		EA.	3.35	20.00	23.35
1" x 1/2"		"	3.45	20.00	23.45
1" x 3/4"		"	3.60	20.00	23.60
1-1/4" x					
1/2"		EA.	4.05	21.35	25.40
3/4"		"	4.15	21.35	25.50
1"		"	4.35	21.35	25.70
1-1/2" x					
1/2"		EA.	7.30	22.90	30.20
3/4"		"	7.40	22.90	30.30
1"		"	7.50	22.90	30.40
1-1/4"		"	7.60	22.90	30.50
2" x					
1/2"		EA.	10.65	26.70	37.35
3/4"		"	10.85	26.70	37.55
1"		"	11.10	26.70	37.80
1-1/4"		"	11.30	26.70	38.00
1-1/2"		"	11.40	26.70	38.10
2-1/2" x					
1"		EA.	22.40	32.00	54.40
1-1/4"		"	22.95	32.00	54.95
1-1/2"		"	22.95	32.00	54.95
2"		"	23.20	32.00	55.20
3" x					
1-1/2"		EA.	29.10	40.10	69.20
2"		"	30.20	40.10	70.30
2-1/2"		"	30.80	40.10	70.90
4" x					
2"		EA.	64.00	45.80	109.80
2-1/2"		"	65.00	45.80	110.80
3"		"	66.00	45.80	111.80
Slip coupling					
1/4"		EA.	0.28	10.70	10.98
1/2"		"	0.43	12.80	13.23
3/4"		"	0.90	16.00	16.90
1"		"	1.85	17.80	19.65
1-1/4"		"	3.25	20.00	23.25
1-1/2"		"	4.50	21.35	25.85
2"		"	6.50	26.70	33.20
2-1/2"		"	9.20	26.70	35.90
3"		"	16.80	32.00	48.80
4"		"	32.50	40.10	72.60
Coupling with drain					
1/2"		EA.	4.05	16.00	20.05
3/4"		"	5.60	17.80	23.40
1"		"	7.30	20.00	27.30
Reducer					
3/8" x 1/4"		EA.	1.25	12.80	14.05
1/2" x 3/8"		"	1.35	12.80	14.15
3/4" x					

PLUMBING		UNIT	MAT.	INST.	TOTAL
15410.11	**COPPER FITTINGS**				
1/4"		EA.	1.90	14.55	16.45
3/8"		"	2.00	14.55	16.55
1/2"		"	2.15	14.55	16.70
1" x					
1/2"		EA.	3.25	16.00	19.25
3/4"		"	3.35	16.00	19.35
1-1/4" x					
1/2"		EA.	4.60	17.80	22.40
3/4"		"	5.60	17.80	23.40
1"		"	5.70	17.80	23.50
1-1/2" x					
1/2"		EA.	6.60	20.00	26.60
3/4"		"	6.65	20.00	26.65
1"		"	6.85	20.00	26.85
1-1/4"		"	6.95	20.00	26.95
2" x					
1/2"		EA.	10.10	22.90	33.00
3/4"		"	10.20	22.90	33.10
1"		"	10.20	22.90	33.10
1-1/4"		"	10.55	22.90	33.45
1-1/2"		"	10.55	22.90	33.45
2-1/2" x					
1"		EA.	21.30	26.70	48.00
1-1/4"		"	22.40	26.70	49.10
1-1/2"		"	22.40	26.70	49.10
2"		"	23.50	26.70	50.20
3" x					
1-1/4"		EA.	28.00	32.00	60.00
1-1/2"		"	28.00	32.00	60.00
2"		"	29.10	32.00	61.10
2-1/2"		"	29.10	32.00	61.10
4" x					
2"		EA.	54.00	40.10	94.10
3"		"	56.00	40.10	96.10
Female adapters					
1/4"		EA.	4.50	12.80	17.30
3/8"		"	4.70	14.55	19.25
1/2"		"	2.25	16.00	18.25
3/4"		"	3.35	17.80	21.15
1"		"	5.60	17.80	23.40
1-1/4"		"	8.85	20.00	28.85
1-1/2"		"	13.45	20.00	33.45
2"		"	16.80	21.35	38.15
2-1/2"		"	56.00	22.90	78.90
3"		"	78.50	26.70	105.20
4"		"	120.00	32.00	152.00
Increasing female adapters					
1/8" x					
3/8"		EA.	3.35	12.80	16.15
1/2"		"	3.35	12.80	16.15
1/4" x 1/2"		"	3.15	13.95	17.10
3/8" x 1/2"		"	3.15	14.55	17.70

PLUMBING	UNIT	MAT.	INST.	TOTAL
15410.11 COPPER FITTINGS				
1/2" X				
3/4"	EA.	4.50	16.00	20.50
1"	"	8.40	16.00	24.40
3/4" X				
1"	EA.	7.85	17.80	25.65
1-1/4"	"	13.45	17.80	31.25
1" x				
1-1/4"	EA.	13.45	17.80	31.25
1-1/2"	"	15.70	17.80	33.50
1-1/4" x				
1-1/2"	EA.	15.70	20.00	35.70
2"	"	21.30	20.00	41.30
1-1/2" x 2"	"	21.85	21.35	43.20
Reducing female adapters				
3/8" x 1/4"	EA.	3.35	14.55	17.90
1/2" x				
1/4"	EA.	3.35	16.00	19.35
3/8"	"	3.45	16.00	19.45
3/4" x 1/2"	"	4.50	17.80	22.30
1" x				
1/2"	EA.	7.75	17.80	25.55
3/4"	"	7.75	17.80	25.55
1-1/4" x				
1/2"	EA.	14.55	20.00	34.55
3/4"	"	14.55	20.00	34.55
1"	"	14.55	20.00	34.55
1-1/2" x				
1"	EA.	15.70	21.35	37.05
1-1/4"	"	15.70	21.35	37.05
2" x				
1"	EA.	22.40	22.90	45.30
1-1/4"	"	23.50	22.90	46.40
1-1/2"	"	24.65	22.90	47.55
Female fitting adapters				
1/2"	EA.	3.35	16.00	19.35
3/4"	"	4.50	16.00	20.50
3/4" x 1/2"	"	7.85	16.85	24.70
1"	"	7.30	17.80	25.10
1-1/4"	"	11.20	18.85	30.05
1-1/2"	"	14.55	20.00	34.55
2"	"	15.70	21.35	37.05
Male adapters				
1/4"	EA.	4.25	14.55	18.80
3/8"	"	2.45	14.55	17.00
3"	"	54.00	26.70	80.70
4"	"	80.50	32.00	112.50
Increasing male adapters				
3/8" x 1/2"	EA.	2.25	14.55	16.80
1/2" x				
3/4"	EA.	3.35	16.00	19.35
1"	"	6.70	16.00	22.70
3/4" x				

PLUMBING	UNIT	MAT.	INST.	TOTAL
15410.11 COPPER FITTINGS				
1"	EA.	6.70	16.85	23.55
1-1/4"	"	9.50	16.85	26.35
1" x 1-1/4"	"	9.50	17.80	27.30
1-1/2" x				
3/4"	EA.	12.30	18.85	31.15
1"	"	12.55	18.85	31.40
1-1/4"	"	12.75	18.85	31.60
2" x				
1"	EA.	21.30	20.00	41.30
1-1/4"	"	21.30	20.00	41.30
1-1/2"	"	21.85	20.00	41.85
2" x 2-1/2"	"	69.50	21.35	90.85
Reducing male adapters				
1/2" x				
1/4"	EA.	2.45	16.00	18.45
3/8"	"	2.45	16.00	18.45
3/4" x 1/2"	"	4.25	16.85	21.10
1" x				
1/2"	EA.	6.70	17.80	24.50
3/4"	"	6.70	17.80	24.50
1-1/4" x				
3/4"	EA.	8.95	18.85	27.80
1"	"	9.20	18.85	28.05
1-1/2" x				
3/4"	EA.	12.90	20.00	32.90
1"	"	13.10	20.00	33.10
1-1/4'	"	13.45	20.00	33.45
2" x				
3/4"	EA.	21.85	21.35	43.20
1"	"	21.85	21.35	43.20
1-1/4"	"	22.95	21.35	44.30
1-1/2"	"	22.95	21.35	44.30
2-1/2" x				
2"	EA.	76.00	24.65	100.65
Fitting x male adapters				
1/2"	EA.	5.60	16.00	21.60
3/4"	"	6.95	16.85	23.80
1"	"	8.40	17.80	26.20
1-1/4"	"	20.15	18.85	39.00
1-1/2"	"	23.50	20.00	43.50
2"	"	40.30	21.35	61.65
90 ells				
1/8"	EA.	1.45	12.80	14.25
1/4"	"	1.45	12.80	14.25
3/8"	"	1.25	14.55	15.80
1/2"	"	0.84	16.00	16.84
3/4"	"	1.40	16.85	18.25
1"	"	3.45	17.80	21.25
1-1/4"	"	5.50	18.85	24.35
1-1/2"	"	8.05	20.00	28.05
2"	"	16.60	21.35	37.95
2-1/2"	"	23.50	24.65	48.15

PLUMBING		UNIT	MAT.	INST.	TOTAL
15410.11	**COPPER FITTINGS**				
3"		EA.	29.10	26.70	55.80
4"		"	71.50	32.00	103.50
Reducing 90 ell					
3/8" x 1/4"		EA.	2.70	14.55	17.25
1/2" x					
1/4"		EA.	4.05	16.00	20.05
3/8"		"	4.05	16.00	20.05
3/4" x 1/2"		"	3.60	16.85	20.45
1" x					
1/2"		EA.	5.60	17.80	23.40
3/4"		"	5.60	17.80	23.40
1-1/4" x					
1/2"		EA.	13.45	18.85	32.30
3/4"		"	13.45	18.85	32.30
1"		"	13.65	18.85	32.50
1-1/2" x					
1/2"		EA.	20.15	20.00	40.15
3/4"		"	20.40	20.00	40.40
1"		"	20.60	20.00	40.60
x 1-1/4"		"	20.60	20.00	40.60
2" x					
3/4"		EA.	24.65	21.35	46.00
x 1"		"	24.65	21.35	46.00
x 1-1/4"		"	24.65	21.35	46.00
x 1-1/2"		"	24.65	21.35	46.00
2-1/2" x					
1-1/2"		EA.	78.50	24.65	103.15
x 2"		"	79.50	24.65	104.15
3" x					
2"		EA.	100.00	26.70	126.70
x 2-1/2"		"	110.00	26.70	136.70
Street ells, copper					
1/4"		EA.	3.35	12.80	16.15
3/8"		"	2.35	14.55	16.90
1/2"		"	0.78	16.00	16.78
3/4"		"	1.80	16.85	18.65
1"		"	4.50	17.80	22.30
1-1/4"		"	6.60	18.85	25.45
1-1/2"		"	8.30	20.00	28.30
2"		"	16.80	21.35	38.15
2-1/2"		"	29.10	24.65	53.75
3"		"	40.30	26.70	67.00
4"		"	88.50	32.00	120.50
Female, 90 ell					
1/2"		EA.	2.35	16.00	18.35
3/4"		"	4.25	16.85	21.10
1"		"	11.20	17.80	29.00
1-1/4"		"	18.95	18.85	37.80
1-1/2"		"	21.30	20.00	41.30
2"		"	32.50	21.35	53.85
Female increasing, 90 ell					
3/8" x 1/2"		EA.	3.25	14.55	17.80

PLUMBING	UNIT	MAT.	INST.	TOTAL
15410.11 — COPPER FITTINGS				
1/2" x				
3/4"	EA.	6.70	16.00	22.70
1"	"	10.65	16.00	26.65
3/4" x 1"	"	10.65	16.85	27.50
1" x 1-1/4"	"	17.35	17.80	35.15
1-1/4" x 1-1/2"	"	17.35	18.85	36.20
Female reducing, 90 ell				
1/2" x 3/8"	EA.	4.50	16.00	20.50
3/4" x 1/2"	"	6.70	16.85	23.55
1" x				
1/2"	EA.	10.65	17.80	28.45
x 3/4"	"	10.65	17.80	28.45
1-1/4" x				
3/4"	EA.	17.35	18.85	36.20
x 1"	"	17.35	18.85	36.20
x 1/2"	"	17.35	18.85	36.20
1-1/2" x				
1"	EA.	15.70	20.00	35.70
x 1-1/4"	"	15.70	20.00	35.70
2" x 1-1/2"	"	23.50	21.35	44.85
Male, 90 ell				
1/4"	EA.	3.90	12.80	16.70
3/8"	"	4.05	14.55	18.60
1/2"	"	2.45	16.00	18.45
3/4"	"	5.40	16.85	22.25
1"	"	8.95	17.80	26.75
1-1/4"	"	14.55	18.85	33.40
1-1/2"	"	18.95	20.00	38.95
2"	"	30.20	21.35	51.55
Male, increasing 90 ell				
1/2" x				
3/4"	EA.	5.05	16.00	21.05
1"	"	9.50	16.00	25.50
3/4" x 1"	"	9.50	16.85	26.35
1" x 1-1/4"	"	17.35	17.80	35.15
1-1/4" x 1-1/2"	"	17.35	18.85	36.20
Male, reducing 90 ell				
1/2" x 3/8"	EA.	4.15	16.00	20.15
3/4" x 1/2"	"	5.05	16.85	21.90
1" x				
1/2"	EA.	9.95	17.80	27.75
3/4"	"	9.95	17.80	27.75
1-1/4" x 1"	"	17.35	18.85	36.20
Drop ear ells				
1/2"	EA.	3.80	16.00	19.80
Female drop ear ells				
1/2"	EA.	3.25	16.00	19.25
1/2" x 3/8"	"	5.50	16.00	21.50
3/4"	"	6.15	16.85	23.00
Female flanged sink ell				
1/2"	EA.	5.50	16.00	21.50
45 ells				

PLUMBING	UNIT	MAT.	INST.	TOTAL
15410.11 COPPER FITTINGS				
1/4"	EA.	3.35	12.80	16.15
3/8"	"	3.15	14.55	17.70
3"	"	21.15	26.70	47.85
4"	"	73.00	32.00	105.00
45 street ell				
1/4"	EA.	3.55	12.80	16.35
3/8"	"	2.70	14.55	17.25
1/2"	"	0.90	16.00	16.90
3/4"	"	1.55	16.85	18.40
1"	"	5.40	17.80	23.20
1-1/2"	"	7.75	20.00	27.75
2"	"	11.75	21.35	33.10
2-1/2"	"	25.80	24.65	50.45
3"	"	38.10	26.70	64.80
4"	"	130.00	32.00	162.00
Tee				
1/8"	EA.	4.25	12.80	17.05
1/4"	"	4.25	12.80	17.05
3/8"	"	2.70	14.55	17.25
3"	"	53.00	26.70	79.70
4"	"	130.00	32.00	162.00
Caps				
1/4"	EA.	0.45	12.80	13.25
3/8"	"	0.67	14.55	15.22
3"	"	21.85	26.70	48.55
4"	"	40.30	32.00	72.30
Test caps				
1/2"	EA.	0.45	16.00	16.45
3/4"	"	0.50	16.85	17.35
1"	"	1.10	17.80	18.90
1-1/4"	"	1.25	18.85	20.10
1-1/2"	"	1.35	20.00	21.35
2"	"	2.25	21.35	23.60
3"	"	4.50	26.70	31.20
Flush bushing				
1/4" x 1/8"	EA.	1.10	12.80	13.90
1/2" x				
1/4"	EA.	1.35	16.00	17.35
3/8"	"	1.35	16.00	17.35
3/4" x				
3/8"	EA.	2.35	16.85	19.20
1/2"	"	2.35	16.85	19.20
1" x				
1/2"	EA.	3.90	17.80	21.70
3/4"	"	3.90	17.80	21.70
1-1/4" x				
1/2"	EA.	4.70	18.85	23.55
3/4"	"	4.70	18.85	23.55
x 1"	"	4.70	18.85	23.55
1-1/2" x				
1/2"	EA.	5.60	20.00	25.60
1-1/4"	"	5.60	20.00	25.60

PLUMBING		UNIT	MAT.	INST.	TOTAL
15410.11	**COPPER FITTINGS**				
2" x					
1"		EA.	13.45	21.35	34.80
1-1/4"		"	13.45	21.35	34.80
1-1/2"		"	13.65	21.35	35.00
Female flush bushing					
1/2" x					
1/2" x 1/8"		EA.	3.15	16.00	19.15
1/4"		"	3.15	16.00	19.15
Union					
1/4"		EA.	7.60	12.80	20.40
3/8"		"	7.60	14.55	22.15
3"		"	240.00	26.70	266.70
Female					
1/2"		EA.	10.10	16.00	26.10
3/4"		"	13.45	16.85	30.30
Male					
1/2"		EA.	8.95	16.00	24.95
3/4"		"	11.75	16.85	28.60
1"		"	26.30	17.80	44.10
45 degree wye					
1/2"		EA.	6.70	16.00	22.70
3/4"		"	11.10	16.85	27.95
1"		"	14.55	17.80	32.35
1" x 3/4" x 3/4"		"	30.20	17.80	48.00
1-1/4"		"	30.20	18.85	49.05
1-1/4" x 1" x 1"		"	31.40	18.85	50.25
Twin ells					
1" x 3/4" x 3/4"		EA.	7.85	17.80	25.65
1" x 1" x 1"		"	7.85	17.80	25.65
1-1/4" x 1" x 1"		"	12.20	18.85	31.05
Companion flanges, 125#					
2" x 6"		EA.	76.00	21.35	97.35
2-1/2"		"	87.50	24.65	112.15
3" x 7-1/2"		"	85.00	26.70	111.70
4" x 9"		"	140.00	32.00	172.00
90 union ells, male					
1/2"		EA.	11.75	16.00	27.75
3/4"		"	15.70	16.85	32.55
1"		"	35.80	17.80	53.60
DWV fittings, coupling with stop					
1-1/4"		EA.	3.35	18.85	22.20
1-1/2"		"	3.90	20.00	23.90
1-1/2" x 1-1/4"		"	4.50	20.00	24.50
2"		"	4.35	21.35	25.70
2" x 1-1/4"		"	5.60	21.35	26.95
2" x 1-1/2"		"	5.60	21.35	26.95
3"		"	7.30	26.70	34.00
3" x 1-1/2"		"	14.55	26.70	41.25
3" x 2"		"	14.55	26.70	41.25
4"		"	15.10	32.00	47.10
Slip coupling					
1-1/2"		EA.	4.35	20.00	24.35

PLUMBING	UNIT	MAT.	INST.	TOTAL
15410.11 — COPPER FITTINGS				
2"	EA.	4.70	21.35	26.05
3"	"	6.60	26.70	33.30
90 ells				
1-1/2"	EA.	7.85	20.00	27.85
1-1/2" x 1-1/4"	"	9.95	20.00	29.95
2"	"	8.40	21.35	29.75
2" x 1-1/2"	"	12.90	21.35	34.25
3"	"	24.65	26.70	51.35
4"	"	100.00	32.00	132.00
Street, 90 elbows				
1-1/2"	EA.	6.70	20.00	26.70
2"	"	11.75	21.35	33.10
3"	"	24.65	26.70	51.35
4"	"	110.00	32.00	142.00
Female, 90 elbows				
1-1/2"	EA.	11.20	20.00	31.20
2"	"	24.65	21.35	46.00
Male, 90 elbows				
1-1/2"	EA.	11.10	20.00	31.10
2"	"	24.65	21.35	46.00
90 with side inlet				
3" x 3" x 1"	EA.	37.00	26.70	63.70
3" x 3" x 1-1/2"	"	38.10	26.70	64.80
3" x 3" x 2"	"	38.10	26.70	64.80
45 ells				
1-1/4"	EA.	4.70	18.85	23.55
1-1/2"	"	4.35	20.00	24.35
2"	"	8.75	21.35	30.10
3"	"	15.70	26.70	42.40
4"	"	45.90	32.00	77.90
Street, 45 ell				
1-1/2"	EA.	5.60	20.00	25.60
2"	"	10.10	21.35	31.45
3"	"	23.50	26.70	50.20
60 ell				
1-1/2"	EA.	7.95	20.00	27.95
2"	"	14.00	21.35	35.35
3"	"	28.00	26.70	54.70
22-1/2 ell				
1-1/2"	EA.	7.85	20.00	27.85
2"	"	10.85	21.35	32.20
3"	"	16.70	26.70	43.40
11-1/4 ell				
1-1/2"	EA.	8.05	20.00	28.05
2"	"	9.50	21.35	30.85
3"	"	19.60	26.70	46.30
Wye				
1-1/4"	EA.	15.70	18.85	34.55
1-1/2"	"	13.45	20.00	33.45
2"	"	21.30	21.35	42.65
2" x 1-1/2" x 1-1/2"	"	20.70	21.35	42.05
2" x 1-1/2" x 2"	"	20.70	21.35	42.05

PLUMBING	UNIT	MAT.	INST.	TOTAL
15410.11 COPPER FITTINGS				
2" x 1-1/2" x 2"	EA.	20.85	21.35	42.20
3"	"	54.00	26.70	80.70
3" x 3" x 1-1/2"	"	55.00	26.70	81.70
3" x 3" x 2"	"	51.50	26.70	78.20
4"	"	110.00	32.00	142.00
4" x 4" x 2"	"	110.00	32.00	142.00
4" x 4" x 3"	"	110.00	32.00	142.00
Sanitary tee				
1-1/4"	EA.	10.65	18.85	29.50
1-1/2"	"	10.10	20.00	30.10
2"	"	15.70	21.35	37.05
2" x 1-1/2" x 1-1/2"	"	15.10	21.35	36.45
2" x 1-1/2" x 2"	"	15.10	21.35	36.45
2" x 2" x 1-1/2"	"	15.10	21.35	36.45
3"	"	38.10	26.70	64.80
3" x 3" x 1-1/2"	"	35.80	26.70	62.50
3" x 3" x 2"	"	35.80	26.70	62.50
4"	"	95.00	32.00	127.00
4" x 4" x 3"	"	84.00	32.00	116.00
Female sanitary tee				
1-1/2"	EA.	26.90	20.00	46.90
Long turn tee				
1-1/2"	EA.	17.90	20.00	37.90
2"	"	28.00	21.35	49.35
3" x 1-1/2"	"	40.30	26.70	67.00
Double wye				
1-1/2"	EA.	25.80	20.00	45.80
2"	"	39.20	21.35	60.55
2" x 2" x 1-1/2" x 1-1/2"	"	32.50	21.35	53.85
3"	"	55.00	26.70	81.70
3" x 3" x 1-1/2" x 1-1/2"	"	55.00	26.70	81.70
3" x 3" x 2" x 2"	"	55.00	26.70	81.70
4" x 4" x 1-1/2" x 1-1/2"	"	77.50	32.00	109.50
Double sanitary tee				
1-1/2"	EA.	17.90	20.00	37.90
2"	"	32.40	21.35	53.75
2" x 2" x 1-1/2"	"	23.50	21.35	44.85
3"	"	49.30	26.70	76.00
3" x 3" x 1-1/2" x 1-1/2"	"	34.70	26.70	61.40
3" x 3" x 2" x 2"	"	34.70	26.70	61.40
4" x 4" x 1-1/2" x 1-1/2"	"	160.00	32.00	192.00
Long				
2" x 1-1/2"	EA.	42.60	21.35	63.95
Twin elbow				
1-1/2"	EA.	20.15	20.00	40.15
2"	"	31.40	21.35	52.75
2" x 1-1/2" x 1-1/2"	"	20.15	21.35	41.50
Spigot adapter, manoff				
1-1/2" x 2"	EA.	17.90	20.00	37.90
1-1/2" x 3"	"	24.65	20.00	44.65
2"	"	10.10	21.35	31.45
2" x 3"	"	24.65	21.35	46.00

PLUMBING	UNIT	MAT.	INST.	TOTAL
15410.11 COPPER FITTINGS				
2" x 4"	EA.	34.70	21.35	56.05
3"	"	19.05	26.70	45.75
3" x 4"	"	34.70	26.70	61.40
4"	"	34.70	32.00	66.70
No-hub adapters				
1-1/2" x 2"	EA.	10.10	20.00	30.10
2"	"	10.10	21.35	31.45
2" x 3"	"	29.10	21.35	50.45
3"	"	18.50	26.70	45.20
3" x 4"	"	29.10	26.70	55.80
4"	"	31.40	32.00	63.40
Fitting reducers				
1-1/2" x 1-1/4"	EA.	4.50	20.00	24.50
2" x 1-1/2"	"	4.35	21.35	25.70
3" x 1-1/2"	"	14.00	26.70	40.70
3" x 2"	"	14.35	26.70	41.05
Slip joint (Desanco)				
1-1/4"	EA.	7.75	18.85	26.60
1-1/2"	"	8.05	20.00	28.05
1-1/2" x 1-1/4"	"	8.05	20.00	28.05
Street x slip joint (Desanco)				
1-1/2"	EA.	10.10	20.00	30.10
1-1/2" x 1-1/4"	"	10.85	20.00	30.85
Flush bushing				
1-1/2" x 1-1/4"	EA.	5.60	20.00	25.60
2" x 1-1/2"	"	8.95	21.35	30.30
3" x 1-1/2"	"	23.50	26.70	50.20
3" x 2"	"	23.50	26.70	50.20
Male hex trap bushing				
1-1/4" x 1-1/2"	EA.	7.30	18.85	26.15
1-1/2"	"	5.60	20.00	25.60
1-1/2" x 2"	"	8.95	20.00	28.95
2"	"	6.60	21.35	27.95
Round trap bushing				
1-1/2"	EA.	6.70	20.00	26.70
2"	"	7.30	21.35	28.65
Female adapter				
1-1/4"	EA.	6.70	18.85	25.55
1-1/2"	"	6.70	20.00	26.70
1-1/2" x 2"	"	11.75	20.00	31.75
2"	"	11.75	21.35	33.10
2" x 1-1/2"	"	11.75	21.35	33.10
3"	"	26.90	26.70	53.60
Fitting x female adapter				
1-1/2"	EA.	7.75	20.00	27.75
2"	"	11.20	21.35	32.55
Male adapters				
1-1/4"	EA.	6.70	18.85	25.55
1-1/4" x 1-1/2"	"	7.85	18.85	26.70
1-1/2"	"	7.75	20.00	27.75
1-1/2" x 2"	"	9.50	20.00	29.50
2"	"	8.95	21.35	30.30

PLUMBING	UNIT	MAT.	INST.	TOTAL
15410.11 — COPPER FITTINGS				
2" x 1-1/2"	EA.	9.50	21.35	30.85
3"	"	24.65	26.70	51.35
Male x slip joint adapters				
1-1/2" x 1-1/4"	EA.	11.20	20.00	31.20
Dandy cleanout				
1-1/2"	EA.	18.50	20.00	38.50
2"	"	30.20	21.35	51.55
3"	"	70.50	26.70	97.20
End cleanout, flush pattern				
1-1/2" x 1"	EA.	13.45	20.00	33.45
2" x 1-1/2"	"	20.15	21.35	41.50
3" x 2-1/2"	"	65.00	26.70	91.70
Copper caps				
1-1/2"	EA.	7.85	20.00	27.85
2"	"	11.75	21.35	33.10
Closet flanges				
3"	EA.	20.15	26.70	46.85
4"	"	20.15	32.00	52.15
Drum traps, with cleanout				
1-1/2" x 3" x 6"	EA.	47.00	20.00	67.00
P-trap, swivel, with cleanout				
1-1/2"	EA.	29.50	20.00	49.50
P-trap, solder union				
1-1/2"	EA.	20.15	20.00	40.15
2"	"	35.80	21.35	57.15
With cleanout				
1-1/2"	EA.	21.85	20.00	41.85
2"	"	39.20	21.35	60.55
2" x 1-1/2"	"	39.20	21.35	60.55
Swivel joint, with cleanout				
1-1/2" x 1-1/4"	EA.	26.90	20.00	46.90
1-1/2"	"	35.80	20.00	55.80
2" x 1-1/2"	"	43.70	21.35	65.05
Estabrook TY, with inlets				
3", with 1-1/2" inlet	EA.	58.00	26.70	84.70
Fine thread adapters				
1/2"	EA.	1.55	16.00	17.55
1/2" x 1/2" IPS	"	1.80	16.00	17.80
1/2" x 3/4" IPS	"	2.90	16.00	18.90
1/2" x male	"	1.10	16.00	17.10
1/2" x female	"	2.15	16.00	18.15
15410.14 — BRASS I.P.S. FITTINGS				
Fittings, iron pipe size, 45 deg ell				
1/8"	EA.	4.70	12.80	17.50
1/4"	"	4.70	12.80	17.50
3/8"	"	4.70	14.55	19.25
1/2"	"	4.80	16.00	20.80
3/4"	"	7.85	16.85	24.70
1"	"	12.90	17.80	30.70
1-1/4"	"	20.15	18.85	39.00

PLUMBING	UNIT	MAT.	INST.	TOTAL
15410.14 — BRASS I.P.S. FITTINGS				
1-1/2"	EA.	24.65	20.00	44.65
2"	"	39.20	21.35	60.55
90 deg ell				
1/8"	EA.	4.50	12.80	17.30
1/4"	"	4.50	12.80	17.30
3/8"	"	4.50	14.55	19.05
1/2"	"	4.50	16.00	20.50
3/4"	"	5.60	16.85	22.45
1"	"	10.10	17.80	27.90
1-1/4"	"	14.00	18.85	32.85
1-1/2"	"	18.50	20.00	38.50
2"	"	29.10	21.35	50.45
2-1/2"	"	58.00	24.65	82.65
90 deg ell, reducing				
1/4" x 1/8"	EA.	5.60	12.80	18.40
3/8" x 1/8"	"	5.60	14.55	20.15
3/8" x 1/4"	"	5.60	14.55	20.15
1/2" x 1/4"	"	5.60	16.00	21.60
1/2" x 3/8"	"	5.60	16.00	21.60
3/4" x 1/2"	"	8.40	16.85	25.25
1" x 3/8"	"	11.10	17.80	28.90
1" x 1/2"	"	11.10	17.80	28.90
1" x 3/4"	"	11.10	17.80	28.90
1-1/4" x 3/4"	"	17.90	18.85	36.75
1-1/4" x 1"	"	17.90	18.85	36.75
1-1/2" x 1/2"	"	21.30	20.00	41.30
1-1/2" x 3/4"	"	21.30	20.00	41.30
1-1/2" x 1"	"	21.30	20.00	41.30
2" x 1"	"	31.40	21.35	52.75
2" x 1-1/4"	"	31.40	21.35	52.75
2" x 1-1/2"	"	31.40	21.35	52.75
Street ell, 45 deg				
1/2"	EA.	5.60	16.00	21.60
3/4"	"	7.75	16.85	24.60
2"	"	38.10	21.35	59.45
90 deg				
1/8"	EA.	5.50	12.80	18.30
1/4"	"	5.50	12.80	18.30
3/8"	"	5.50	14.55	20.05
1/2"	"	5.50	16.00	21.50
3/4"	"	6.70	16.85	23.55
1"	"	8.95	17.80	26.75
1-1/4"	"	17.90	18.85	36.75
1-1/2"	"	19.05	20.00	39.05
2"	"	31.40	21.35	52.75
Tee, 1/8"	"	4.60	12.80	17.40
1/4"	"	4.60	12.80	17.40
3/8"	"	4.60	14.55	19.15
1/2"	"	4.60	16.00	20.60
3/4"	"	6.25	16.85	23.10
1"	"	9.40	17.80	27.20
1-1/4"	"	15.70	18.85	34.55

PLUMBING	UNIT	MAT.	INST.	TOTAL
15410.14 BRASS I.P.S. FITTINGS				
1-1/2"	EA.	16.80	20.00	36.80
2"	"	26.90	21.35	48.25
2-1/2"	"	60.50	24.65	85.15
Tee, reducing, 3/8" x				
1/4"	EA.	5.80	14.55	20.35
1/2"	"	5.80	14.55	20.35
1/2" x				
1/4"	EA.	5.80	16.00	21.80
3/8"	"	5.80	16.00	21.80
3/4"	"	6.70	16.00	22.70
3/4" x				
1/4"	EA.	6.70	16.85	23.55
1/2"	"	6.70	16.85	23.55
1"	"	13.45	16.85	30.30
1" x				
1/2"	EA.	13.45	17.80	31.25
3/4"	"	13.45	17.80	31.25
1-1/4" x				
1/2"	EA.	17.90	18.85	36.75
1"	"	17.90	18.85	36.75
1-1/2" x				
1/2"	EA.	24.65	20.00	44.65
3/4"	"	24.65	20.00	44.65
1"	"	24.65	20.00	44.65
1-1/2"	"	26.90	20.00	46.90
2" x				
1/2"	EA.	38.10	21.35	59.45
3/4"	"	38.10	21.35	59.45
1"	"	38.10	21.35	59.45
1-1/4"	"	38.10	21.35	59.45
1-1/2"	"	38.10	21.35	59.45
2-1/2" x 2"	"	85.00	24.65	109.65
Tee, reducing				
1/2" x 3/8" x 1/2"	EA.	5.80	16.00	21.80
3/4" x 1/2" x 1/2"	"	8.95	16.85	25.80
3/4" x 1/2" x 3/4"	"	8.40	16.85	25.25
1" x 1/2" x 1/2"	"	14.55	17.80	32.35
1" x 1/2" x 3/4"	"	14.55	17.80	32.35
1" x 3/4" x 1/2"	"	14.55	17.80	32.35
1" x 3/4" x 3/4"	"	14.55	17.80	32.35
1-1/4" x 1/2" x 1-1/4"	"	17.90	18.85	36.75
1-1/4" x 1" x 1"	"	19.05	18.85	37.90
1-1/4" x 1-1/4" x 3/4"	"	17.90	18.85	36.75
1-1/2" x 1-1/4" x 1-1/4"	"	24.65	20.00	44.65
Union				
1/8"	EA.	10.10	12.80	22.90
1/4"	"	10.10	12.80	22.90
3/8"	"	10.10	14.55	24.65
1/2"	"	10.10	16.00	26.10
3/4"	"	12.30	16.85	29.15
1"	"	15.70	17.80	33.50
1-1/4"	"	21.30	18.85	40.15

PLUMBING		UNIT	MAT.	INST.	TOTAL
15410.14	**BRASS I.P.S. FITTINGS**				
1-1/2"		EA.	25.80	20.00	45.80
2"		"	38.10	21.35	59.45
Brass face bushing					
3/8" x 1/4"		EA.	4.60	14.55	19.15
1/2" x 3/8"		"	4.60	16.00	20.60
3/4" x 1/2"		"	5.80	16.85	22.65
1" x 3/4"		"	9.95	17.80	27.75
1-1/4" x 1"		"	12.30	18.85	31.15
Hex bushing, 1/4" x 1/8"		"	3.25	12.80	16.05
1/2" x					
1/4"		EA.	2.90	16.00	18.90
3/8"		"	2.90	16.00	18.90
5/8" x					
1/8"		EA.	2.90	16.00	18.90
1/4"		"	2.90	16.00	18.90
3/4" x					
1/8"		EA.	4.35	16.85	21.20
1/4"		"	4.35	16.85	21.20
3/8"		"	3.25	16.85	20.10
1/2"		"	3.25	16.85	20.10
1" x					
1/4"		EA.	5.70	17.80	23.50
3/8"		"	5.70	17.80	23.50
1/2"		"	5.15	17.80	22.95
3/4"		"	5.15	17.80	22.95
1-1/4" x					
3/8"		EA.	8.95	18.85	27.80
1/2"		"	8.95	18.85	27.80
3/4"		"	6.60	18.85	25.45
1"		"	6.60	18.85	25.45
1-1/2" x					
1/4"		EA.	11.75	20.00	31.75
1/2"		"	11.75	20.00	31.75
3/4"		"	11.75	20.00	31.75
1"		"	10.10	20.00	30.10
1-1/4"		"	10.10	20.00	30.10
2" x					
1/2"		EA.	15.70	21.35	37.05
3/4"		"	15.70	21.35	37.05
1"		"	15.70	21.35	37.05
1-1/4"		"	10.75	21.35	32.10
1-1/2"		"	10.75	21.35	32.10
2-1/2" x					
1"		EA.	31.40	24.65	56.05
2"		"	26.90	24.65	51.55
3" x					
1-1/2"		EA.	45.90	26.70	72.60
2"		"	38.10	26.70	64.80
2-1/2"		"	38.10	26.70	64.80
4" x					
2"		EA.	110.00	32.00	142.00
3"		"	80.50	32.00	112.50

PLUMBING	UNIT	MAT.	INST.	TOTAL
15410.14 BRASS I.P.S. FITTINGS				
Caps				
1/8"	EA.	3.15	12.80	15.95
1/4"	"	3.15	12.80	15.95
3/8"	"	3.15	14.55	17.70
1/2"	"	3.15	16.00	19.15
3/4"	"	3.60	16.85	20.45
1"	"	5.60	17.80	23.40
1-1/4"	"	8.40	18.85	27.25
1-1/2"	"	10.65	20.00	30.65
2"	"	17.90	21.35	39.25
Couplings				
1/8"	EA.	3.45	12.80	16.25
1/4"	"	3.45	12.80	16.25
3/8"	"	3.45	14.55	18.00
1/2"	"	3.45	16.00	19.45
3/4"	"	4.50	16.85	21.35
1"	"	6.70	17.80	24.50
1-1/4"	"	10.65	18.85	29.50
1-1/2"	"	13.45	20.00	33.45
2"	"	21.30	21.35	42.65
2-1/2"	"	32.50	24.65	57.15
Couplings, reducing, 1/4" x 1/8"	"	4.05	12.80	16.85
3/8" x				
1/8"	EA.	4.05	14.55	18.60
1/4"	"	4.05	14.55	18.60
1/2" x				
1/8"	EA.	4.50	16.00	20.50
1/4"	"	4.35	16.00	20.35
3/8"	"	4.35	16.00	20.35
3/4" x				
1/4"	EA.	6.70	16.85	23.55
3/8"	"	5.70	16.85	22.55
1/2"	"	5.70	16.85	22.55
1/2"	"	7.75	16.85	24.60
3/4"	"	7.75	16.85	24.60
1-1/4" x				
1/2"	EA.	14.55	18.85	33.40
3/4"	"	13.45	18.85	32.30
1"	"	13.45	18.85	32.30
1-1/2" x				
3/4"	EA.	20.15	20.00	40.15
1"	"	17.90	20.00	37.90
1-1/4"	"	17.90	20.00	37.90
2" x				
3/4"	EA.	26.90	21.35	48.25
1"	"	26.90	21.35	48.25
1-1/4"	"	25.80	21.35	47.15
1-1/2"	"	25.80	21.35	47.15
2-1/2" x 1-1/2"	"	40.30	24.65	64.95
Square head plug, solid				
1/8"	EA.	3.35	12.80	16.15
1/4"	"	3.35	12.80	16.15

PLUMBING	UNIT	MAT.	INST.	TOTAL
15410.14 **BRASS I.P.S. FITTINGS**				
3/8"	EA.	3.35	14.55	17.90
1/2"	"	3.35	16.00	19.35
3/4"	"	3.90	16.85	20.75
Cored				
1/2"	EA.	3.25	16.00	19.25
3/4"	"	3.35	16.85	20.20
1"	"	5.60	17.80	23.40
1-1/4"	"	6.70	18.85	25.55
1-1/2"	"	6.70	20.00	26.70
2"	"	11.20	21.35	32.55
3"	"	29.10	26.70	55.80
4"	"	51.50	32.00	83.50
Countersunk				
1/2"	EA.	3.35	16.00	19.35
3/4"	"	3.60	16.85	20.45
1-1/2"	"	9.65	20.00	29.65
2"	"	12.90	21.35	34.25
Locknut				
3/4"	EA.	3.35	16.85	20.20
1"	"	3.80	17.80	21.60
1-1/4"	"	7.50	18.85	26.35
2"	"	11.75	21.35	33.10
Close standard red nipple, 1/8"	"	1.10	12.80	13.90
1/8" x				
1-1/2"	EA.	1.25	12.80	14.05
2"	"	1.35	12.80	14.15
2-1/2"	"	1.55	12.80	14.35
3"	"	1.80	12.80	14.60
3-1/2"	"	2.15	12.80	14.95
4"	"	2.25	12.80	15.05
4-1/2"	"	2.35	12.80	15.15
5"	"	2.45	12.80	15.25
5-1/2"	"	2.90	12.80	15.70
6"	"	3.15	12.80	15.95
1/4" x close	"	1.35	12.80	14.15
1/4" x				
1-1/2"	EA.	2.00	12.80	14.80
2"	"	2.15	12.80	14.95
2-1/2"	"	2.25	12.80	15.05
3"	"	2.35	12.80	15.15
3-1/2"	"	2.60	12.80	15.40
4"	"	2.70	12.80	15.50
4-1/2"	"	2.80	12.80	15.60
5"	"	2.90	12.80	15.70
5-1/2"	"	3.25	12.80	16.05
6"	"	3.35	12.80	16.15
3/8" x close	"	1.35	14.55	15.90
3/8" x				
1-1/2"	EA.	1.45	14.55	16.00
2"	"	1.55	14.55	16.10
2-1/2"	"	1.90	14.55	16.45
3"	"	2.35	14.55	16.90

PLUMBING	UNIT	MAT.	INST.	TOTAL
15410.14 BRASS I.P.S. FITTINGS				
3-1/2"	EA.	2.60	14.55	17.15
4"	"	3.25	14.55	17.80
4-1/2"	"	3.35	14.55	17.90
5"	"	3.60	14.55	18.15
5-1/2"	"	3.70	14.55	18.25
6"	"	4.35	14.55	18.90
1/2" x close	"	1.70	16.00	17.70
1/2" x				
1-1/2"	EA.	1.90	16.00	17.90
2"	"	2.35	16.00	18.35
2-1/2"	"	2.60	16.00	18.60
3"	"	2.80	16.00	18.80
3-1/2"	"	3.15	16.00	19.15
4"	"	3.35	16.00	19.35
4-1/2"	"	3.60	16.00	19.60
5"	"	3.80	16.00	19.80
5-1/2"	"	4.05	16.00	20.05
6"	"	4.25	16.00	20.25
7-1/2"	"	4.50	16.00	20.50
8"	"	4.50	16.00	20.50
3/4" x close	"	2.45	16.85	19.30
3/4" x				
1-1/2"	EA.	2.60	16.85	19.45
2"	"	2.90	16.85	19.75
2-1/2"	"	3.15	16.85	20.00
3"	"	3.45	16.85	20.30
3-1/2"	"	3.80	16.85	20.65
4"	"	4.15	16.85	21.00
4-1/2"	"	4.35	16.85	21.20
5"	"	4.60	16.85	21.45
5-1/2"	"	5.15	16.85	22.00
6"	"	5.40	16.85	22.25
1" x close	"	3.60	17.80	21.40
1" x				
2"	EA.	5.05	17.80	22.85
2-1/2"	"	5.15	17.80	22.95
3"	"	5.40	17.80	23.20
3-1/2"	"	5.60	17.80	23.40
4"	"	6.15	17.80	23.95
4-1/2"	"	6.25	17.80	24.05
5"	"	7.30	17.80	25.10
5-1/2"	"	7.50	17.80	25.30
6"	"	8.05	17.80	25.85
1-1/4" x close	"	5.15	18.85	24.00
1-1/4" x				
2"	EA.	5.25	18.85	24.10
2-1/2"	"	5.80	18.85	24.65
3"	"	7.50	18.85	26.35
3-1/2"	"	8.20	18.85	27.05
4"	"	8.20	18.85	27.05
4-1/2"	"	8.50	18.85	27.35
5"	"	9.95	18.85	28.80

PLUMBING		UNIT	MAT.	INST.	TOTAL
15410.14	**BRASS I.P.S. FITTINGS**				
5-1/2"		EA.	11.10	18.85	29.95
6"		"	11.20	18.85	30.05
1-1/2" x close		"	5.60	20.00	25.60
1-1/2" x					
2"		EA.	5.80	20.00	25.80
2-1/2"		"	6.05	20.00	26.05
3"		"	6.60	20.00	26.60
3-1/2"		"	7.15	20.00	27.15
4-1/2"		"	9.95	20.00	29.95
5"		"	10.85	20.00	30.85
5-1/2"		"	11.85	20.00	31.85
6"		"	12.10	20.00	32.10
2" x close		"	7.85	21.35	29.20
2"					
2-1/2"		EA.	8.95	21.35	30.30
3"		"	9.40	21.35	30.75
3-1/2"		"	10.65	21.35	32.00
4"		"	10.85	21.35	32.20
4-1/2"		"	12.20	21.35	33.55
5"		"	12.90	21.35	34.25
5-1/2"		"	15.70	21.35	37.05
6"		"	16.80	21.35	38.15
2-1/2" x close		"	17.90	24.65	42.55
2-1/2" x					
3"		EA.	18.50	24.65	43.15
3-1/2"		"	19.60	24.65	44.25
4"		"	23.50	24.65	48.15
2-1/2" x 5"		"	26.90	24.65	51.55
3"					
Close		EA.	25.80	26.70	52.50
15410.15	**BRASS FITTINGS**				
Compression fittings, union					
3/8"		EA.	2.15	5.35	7.50
1/2"		"	3.25	5.35	8.60
5/8"		"	4.05	5.35	9.40
Union elbow					
3/8"		EA.	3.60	5.35	8.95
1/2"		"	5.60	5.35	10.95
5/8"		"	7.15	5.35	12.50
Union tee					
3/8"		EA.	4.70	5.35	10.05
1/2"		"	7.75	5.35	13.10
5/8"		"	10.10	5.35	15.45
Male connector					
3/8"		EA.	1.55	5.35	6.90
1/2"		"	1.10	5.35	6.45
5/8"		"	2.45	5.35	7.80
Female connector					
3/8"		EA.	2.90	5.35	8.25
1/2"		"	3.80	5.35	9.15
5/8"		"	3.80	5.35	9.15

PLUMBING		UNIT	MAT.	INST.	TOTAL
15410.15	**BRASS FITTINGS**				
Brass flare fittings, union					
3/8"		EA.	1.70	5.15	6.85
1/2"		"	2.50	5.15	7.65
5/8"		"	3.55	5.15	8.70
90 deg elbow union					
3/8"		EA.	3.75	5.15	8.90
1/2"		"	5.30	5.15	10.45
5/8"		"	7.65	5.15	12.80
Three way tee					
3/8"		EA.	5.05	8.65	13.70
1/2"		"	6.40	8.65	15.05
5/8"		"	10.10	8.65	18.75
Cross					
3/8"		EA.	6.45	11.45	17.90
1/2"		"	9.40	11.45	20.85
5/8"		"	18.80	11.45	30.25
Male connector, half union					
3/8"		EA.	2.15	5.15	7.30
1/2"		"	2.85	5.15	8.00
5/8"		"	5.80	5.15	10.95
Female connector, half union					
3/8"		EA.	2.15	5.15	7.30
1/2"		"	4.35	5.15	9.50
5/8"		"	7.40	5.15	12.55
Long forged nut					
3/8"		EA.	1.55	5.15	6.70
1/2"		"	2.70	5.15	7.85
5/8"		"	4.05	5.15	9.20
Short forged nut					
3/8"		EA.	1.05	5.15	6.20
1/2"		"	1.45	5.15	6.60
5/8"		"	1.90	5.15	7.05
Compression elbow					
1/4"		EA.	0.74	5.15	5.89
5/16"		"	1.35	5.15	6.50
3/4"		"	6.70	5.15	11.85
Nut					
1/8"		EA.			0.10
1/4"		"			0.10
5/16"		"			0.16
3/8"		"			0.18
1/2"		"			0.29
5/8"		"			0.49
3/4"		"			0.89
7/8"		"			0.85
Sleeve					
1/8"		EA.	0.06	6.40	6.46
1/4"		"	0.04	6.40	6.44
5/16"		"	0.07	6.40	6.47
3/8"		"	0.06	6.40	6.46
1/2"		"	0.11	6.40	6.51
5/8"		"	0.16	6.40	6.56

PLUMBING	UNIT	MAT.	INST.	TOTAL
15410.15 BRASS FITTINGS				
3/4"	EA.	0.31	6.40	6.71
7/8"	"	0.49	6.40	6.89
Tee				
1/4"	EA.	1.25	9.15	10.40
5/16"	"	1.80	9.15	10.95
Male tee				
5/16" x 1/8"	EA.	1.35	9.15	10.50
Female union				
1/8" x 1/8"	EA.	0.47	8.00	8.47
1/4" x 3/8"	"	0.85	8.00	8.85
3/8" x 1/4"	"	0.71	8.00	8.71
3/8" x 1/2"	"	0.87	8.00	8.87
5/8" x 1/2"	"	1.40	9.15	10.55
Male union, 1/4"				
1/4" x 1/4"	EA.	0.50	8.00	8.50
3/8"	"	0.77	8.00	8.77
1/2"	"	1.35	8.00	9.35
5/16" x				
1/8"	EA.	0.47	8.00	8.47
1/4"	"	0.52	8.00	8.52
3/8"	"	0.83	8.00	8.83
3/8" x				
1/8"	EA.	0.54	8.00	8.54
1/4"	"	0.63	8.00	8.63
1/2"	"	0.85	8.00	8.85
3/4"	"	1.05	8.00	9.05
1/2" x				
1/4"	EA.	0.81	9.15	9.96
3/8"	"	0.83	9.15	9.98
5/8" x				
3/8"	EA.	1.20	9.15	10.35
1/2"	"	1.05	9.15	10.20
3/4"	"	1.70	9.15	10.85
1/2"	"	2.70	9.15	11.85
7/8" x				
1/2"	EA.	2.60	9.15	11.75
3/4"	"	4.05	9.15	13.20
Female elbow, 1/4" x 1/4"	"	1.20	9.15	10.35
5/16" x				
1/8"	EA.	1.15	9.15	10.30
1/4"	"	1.45	9.15	10.60
3/8" x				
3/8"	EA.	1.35	9.15	10.50
1/2"	"	1.50	9.15	10.65
Male elbow, 1/8" x 1/8"	"	0.74	9.15	9.89
3/16" x 1/4"	"	1.05	9.15	10.20
1/4" x				
1/8"	EA.	0.63	9.15	9.78
1/4"	"	0.71	9.15	9.86
3/8"	"	1.10	9.15	10.25
5/16" x				
1/8"	EA.	0.85	9.15	10.00

PLUMBING		UNIT	MAT.	INST.	TOTAL
15410.15	**BRASS FITTINGS**				
1/4"		EA.	0.93	9.15	10.08
3/8"		"	1.20	9.15	10.35
3/8" x					
1/8"		EA.	0.87	9.15	10.02
1/4"		"	0.90	9.15	10.05
3/8"		"	1.20	9.15	10.35
1/2"		"	1.50	9.15	10.65
1/2" x					
1/4"		EA.	1.55	10.70	12.25
3/8"		"	1.55	10.70	12.25
1/2"		"	1.90	10.70	12.60
5/8" x					
3/8"		EA.	2.05	10.70	12.75
1/2"		"	2.05	10.70	12.75
3/4"		"	3.15	10.70	13.85
3/4" x					
1/2"		EA.	3.50	10.70	14.20
3/4"		"	3.60	10.70	14.30
7/8" x					
3/4"		EA.	4.85	10.70	15.55
Union					
1/8"		EA.	0.59	9.15	9.74
3/16"		"	0.59	9.15	9.74
1/4"		"	0.60	9.15	9.75
5/16"		"	0.76	9.15	9.91
3/8"		"	1.90	9.15	11.05
3/4"		"	4.15	9.15	13.30
7/8"		"	4.25	9.15	13.40
Reducing union					
3/8" x 1/4"		EA.	1.05	10.70	11.75
5/8" x					
3/8"		EA.	1.35	10.70	12.05
1/2"		"	2.00	10.70	12.70
15410.17	**CHROME PLATED FITTINGS**				
Fittings					
90 ell					
3/8"		EA.	10.65	8.00	18.65
1/2"		"	14.00	8.00	22.00
45 ell					
3/8"		EA.	14.00	8.00	22.00
1/2"		"	18.50	8.00	26.50
Tee					
3/8"		EA.	13.45	10.70	24.15
1/2"		"	15.70	10.70	26.40
Coupling					
3/8"		EA.	8.40	8.00	16.40
1/2"		"	8.40	8.00	16.40
Union					
3/8"		EA.	13.45	8.00	21.45
1/2"		"	14.00	8.00	22.00

PLUMBING	UNIT	MAT.	INST.	TOTAL
15410.17 CHROME PLATED FITTINGS				
Tee				
1/2" x 3/8" x 3/8"	EA.	15.70	10.70	26.40
1/2" x 3/8" x 1/2"	"	15.90	10.70	26.60
15410.18 GLASS PIPE				
Glass pipe				
1-1/2" dia.	L.F.	10.65	6.40	17.05
2" dia.	"	12.30	7.10	19.40
3" dia.	"	21.85	8.00	29.85
4" dia.	"	31.40	9.15	40.55
6" dia.	"	55.00	10.70	65.70
15410.30 PVC/CPVC PIPE				
PVC schedule 40				
1/2" pipe	L.F.	0.78	1.35	2.13
3/4" pipe	"	0.90	1.45	2.35
1" pipe	"	1.00	1.60	2.60
1-1/4" pipe	"	1.10	1.80	2.90
1-1/2" pipe	"	1.25	2.00	3.25
2" pipe	"	1.55	2.30	3.85
2-1/2" pipe	"	2.45	2.65	5.10
3" pipe	"	2.80	3.20	6.00
4" pipe	"	3.90	4.00	7.90
6" pipe	"	7.85	8.00	15.85
8" pipe	"	12.90	10.70	23.60
Fittings, 1/2"	EA.	0.29	4.00	4.29
90 deg ell	"	0.45	4.00	4.45
45 deg ell	"	0.38	4.60	4.98
Tee	"	0.29	5.35	5.64
Reducing insert	"	0.72	4.00	4.72
Threaded	"	0.28	5.35	5.63
Male adapter	"	0.30	4.00	4.30
Female adapter	"	0.22	4.00	4.22
Coupling	"	4.50	6.40	10.90
Union	"	0.27	5.35	5.62
Cap	"	5.15	6.40	11.55
Flange				
3/4"	EA.	0.30	5.35	5.65
90 deg elbow	"	0.69	5.35	6.04
45 deg elbow	"	0.39	6.40	6.79
Tee	"	0.28	4.60	4.88
Reducing insert	"	0.43	5.35	5.78
Threaded				
1"	EA.	0.50	6.40	6.90
90 deg elbow	"	0.76	6.40	7.16
45 deg elbow	"	0.69	7.10	7.79
Tee	"	0.50	6.40	6.90
Reducing insert	"	0.68	7.10	7.78
Threaded	"	0.45	8.00	8.45
Male adapter	"	0.40	8.00	8.40
Female adapter				

PLUMBING	UNIT	MAT.	INST.	TOTAL
15410.30 PVC/CPVC PIPE				
Coupling	EA.	0.38	8.00	8.38
Union	"	3.65	10.70	14.35
Cap	"	0.41	6.40	6.81
Flange	"	5.15	10.70	15.85
1-1/4"				
90 deg elbow	EA.	0.88	9.15	10.03
45 deg elbow	"	1.10	9.15	10.25
Tee	"	1.05	10.70	11.75
Reducing insert	"	0.63	10.70	11.33
Threaded	"	1.05	10.70	11.75
Male adapter	"	0.55	10.70	11.25
Female adapter	"	0.63	10.70	11.33
Coupling	"	0.54	10.70	11.24
Union	"	8.35	12.80	21.15
Cap	"	0.57	10.70	11.27
Flange	"	5.30	12.80	18.10
1-1/2"				
90 deg elbow	EA.	0.99	9.15	10.14
45 deg elbow	"	1.45	9.15	10.60
Tee	"	1.35	10.70	12.05
Reducing insert	"	0.68	10.70	11.38
Threaded	"	1.20	10.70	11.90
Male adapter	"	0.77	10.70	11.47
Female adapter	"	0.77	10.70	11.47
Coupling	"	0.58	10.70	11.28
Union	"	11.20	16.00	27.20
Cap	"	0.65	10.70	11.35
Flange	"	8.85	16.00	24.85
2"				
90 deg elbow	EA.	1.50	10.70	12.20
45 deg elbow	"	1.90	10.70	12.60
Tee	"	2.00	12.80	14.80
Reducing insert	"	1.25	12.80	14.05
Threaded	"	1.70	12.80	14.50
Male adapter	"	1.00	12.80	13.80
Female adapter	"	1.05	12.80	13.85
Coupling	"	0.92	12.80	13.72
Union	"	15.70	20.00	35.70
Cap	"	0.81	12.80	13.61
Flange	"	8.95	20.00	28.95
2-1/2"				
90 deg elbow	EA.	4.55	20.00	24.55
45 deg elbow	"	6.70	20.00	26.70
Tee	"	6.05	21.35	27.40
Reducing insert	"	1.80	21.35	23.15
Threaded	"	2.65	21.35	24.00
Male adapter	"	2.80	21.35	24.15
Female adapter	"	2.50	21.35	23.85
Coupling	"	2.80	21.35	24.15
Union	"	20.15	26.70	46.85
Cap	"	2.45	20.00	22.45
Flange	"	11.75	26.70	38.45

PLUMBING	UNIT	MAT.	INST.	TOTAL
15410.30 PVC/CPVC PIPE				
3"				
90 deg elbow	EA.	5.40	26.70	32.10
45 deg elbow	"	7.05	26.70	33.75
Tee	"	8.65	29.10	37.75
Reducing insert	"	2.65	26.70	29.35
Threaded	"	3.40	26.70	30.10
Male adapter	"	4.10	26.70	30.80
Female adapter	"	3.35	26.70	30.05
Coupling	"	3.10	26.70	29.80
Union	"	23.60	32.00	55.60
Cap	"	2.70	26.70	29.40
Flange	"	12.35	32.00	44.35
4"				
90 deg elbow	EA.	9.70	32.00	41.70
45 deg elbow	"	12.65	32.00	44.65
Tee	"	14.45	35.60	50.05
Reducing insert	"	5.90	32.00	37.90
Threaded	"	7.55	32.00	39.55
Male adapter	"	5.25	32.00	37.25
Female adapter	"	4.95	32.00	36.95
Coupling	"	4.45	32.00	36.45
Union	"	28.80	40.10	68.90
Cap	"	6.05	32.00	38.05
Flange	"	15.60	40.10	55.70
PVC schedule 80 pipe				
1-1/2" pipe	L.F.	1.00	2.00	3.00
2" pipe	"	1.25	2.30	3.55
3" pipe	"	1.95	3.20	5.15
4" pipe	"	2.90	4.00	6.90
Fittings, 1-1/2"				
90 deg elbow	EA.	2.90	10.70	13.60
45 deg elbow	"	5.15	10.70	15.85
Tee	"	8.05	16.00	24.05
Reducing insert	"	2.80	10.70	13.50
Threaded	"	3.35	10.70	14.05
Male adapter	"	5.35	10.70	16.05
Female adapter	"	5.80	10.70	16.50
Coupling	"	4.50	10.70	15.20
Union	"	11.20	16.00	27.20
Cap	"	3.15	10.70	13.85
Flange	"	7.30	16.00	23.30
2"				
90 deg elbow	EA.	3.45	12.80	16.25
45 deg elbow	"	6.65	12.80	19.45
Tee	"	10.00	20.00	30.00
Reducing insert	"	4.00	12.80	16.80
Threaded	"	4.10	12.80	16.90
Male adapter	"	7.40	12.80	20.20
Female adapter	"	10.15	12.80	22.95
2-1/2"				
90 deg elbow	EA.	6.60	20.00	26.60
45 deg elbow	"	13.85	20.00	33.85

PLUMBING		UNIT	MAT.	INST.	TOTAL
15410.30	**PVC/CPVC PIPE**				
Tee		EA.	10.90	26.70	37.60
Reducing insert		"	6.95	20.00	26.95
Threaded		"	8.70	20.00	28.70
Male adapter		"	8.80	20.00	28.80
Female adapter		"	16.00	20.00	36.00
Coupling		"	8.60	20.00	28.60
Union		"	23.90	26.70	50.60
Cap		"	10.10	20.00	30.10
Flange		"	12.90	26.70	39.60
3"					
90 deg elbow		EA.	8.95	26.70	35.65
45 deg elbow		"	17.00	26.70	43.70
Tee		"	13.65	32.00	45.65
Reducing insert		"	11.05	26.70	37.75
Threaded		"	16.05	26.70	42.75
Male adapter		"	9.75	26.70	36.45
Female adapter		"	18.05	26.70	44.75
Coupling		"	9.90	26.70	36.60
Union		"	30.60	32.00	62.60
Cap		"	12.90	26.70	39.60
Flange		"	14.55	32.00	46.55
4"					
90 deg elbow		EA.	11.30	32.00	43.30
45 deg elbow		"	30.60	32.00	62.60
Tee		"	15.05	40.10	55.15
Reducing insert		"	15.40	32.00	47.40
Threaded		"	24.80	32.00	56.80
Male adapter		"	17.35	32.00	49.35
Coupling		"	12.40	32.00	44.40
Union		"	28.80	40.10	68.90
Cap		"	15.70	32.00	47.70
Flange		"	17.90	40.10	58.00
CPVC schedule 40					
1/2" pipe		L.F.	0.67	1.35	2.02
3/4" pipe		"	0.90	1.45	2.35
1" pipe		"	1.20	1.60	2.80
1-1/4" pipe		"	1.60	1.80	3.40
1-1/2" pipe		"	1.90	2.00	3.90
2" pipe		"	2.35	2.30	4.65
Fittings, CPVC, schedule 80					
1/2", 90 deg ell		EA.	2.25	3.20	5.45
Tee		"	3.35	5.35	8.70
3/4", 90 deg ell		"	3.45	3.20	6.65
Tee		"	3.90	5.35	9.25
1", 90 deg ell		"	4.25	3.55	7.80
Tee		"	4.50	5.85	10.35
1-1/4", 90 deg ell		"	4.70	3.55	8.25
Tee		"	5.40	5.85	11.25
1-1/2", 90 deg ell		"	5.15	6.40	11.55
Tee		"	7.40	8.00	15.40
2", 90 deg ell		"	7.85	6.40	14.25
Tee		"	9.50	8.00	17.50

PLUMBING	UNIT	MAT.	INST.	TOTAL
15410.30 PVC/CPVC PIPE				
Polypropylene, acid resistant, DWV pipe				
Schedule 40				
1-1/2" pipe	L.F.	3.40	2.30	5.70
2" pipe	"	4.55	2.65	7.20
3" pipe	"	8.40	3.20	11.60
4" pipe	"	11.90	4.00	15.90
6" pipe	"	21.30	8.00	29.30
Fittings				
1-1/2"				
1/4 bend	EA.	12.25	8.00	20.25
1/8 bend	"	12.05	8.00	20.05
Sanitary tee	"	15.25	16.00	31.25
Cleanout with plug	"	31.20	16.00	47.20
Wye	"	28.50	16.00	44.50
Combination wye & 1/8 bend	"	27.60	16.00	43.60
P-trap	"	26.70	20.00	46.70
Hub adapter	"	14.60	8.00	22.60
Coupling	"	9.60	8.00	17.60
2"				
1/4 bend	EA.	15.25	9.15	24.40
1/8 bend	"	14.65	9.15	23.80
Sanitary tee	"	18.30	18.85	37.15
Cleanout with plug	"	36.50	18.85	55.35
Wye	"	38.60	18.85	57.45
Combination wye & 1/8 bend	"	36.80	18.85	55.65
P-trap	"	36.60	26.70	63.30
Hub adapter	"	15.60	10.70	26.30
Mechanical joint adapter	"	15.40	10.70	26.10
Coupling	"	11.95	10.70	22.65
3"				
1/4 bend	EA.	25.90	10.70	36.60
1/8 bend	"	26.90	10.70	37.60
Sanitary tee	"	36.60	21.35	57.95
Cleanout with plug	"	60.00	21.35	81.35
Wye	"	38.90	21.35	60.25
Combination wye & 1/8 bend	"	66.00	21.35	87.35
P-trap	"	84.50	32.00	116.50
Hub adapter	"	19.55	10.70	30.25
Mechanical joint adapter	"	19.10	10.70	29.80
Coupling	"	15.40	10.70	26.10
4"				
1/4 bend	EA.	41.10	16.00	57.10
1/8 bend	"	30.50	16.00	46.50
Sanitary tee	"	54.50	32.00	86.50
Cleanout with plug	"	81.00	32.00	113.00
Wye	"	56.50	32.00	88.50
Combination wye & 1/8 bend	"	87.50	32.00	119.50
P-trap	"	110.00	53.50	163.50
Hub adapter	"	29.10	16.00	45.10
Mechanical joint adapter	"	28.10	16.00	44.10
Coupling	"	21.75	16.00	37.75
6"				

PLUMBING	UNIT	MAT.	INST.	TOTAL
15410.30 **PVC/CPVC PIPE**				
1/4 bend	EA.	100.00	26.70	126.70
1/8 bend	"	84.50	26.70	111.20
Sanitary tee	"	150.00	53.50	203.50
Cleanout with plug	"	110.00	53.50	163.50
Wye	"	150.00	53.50	203.50
Hub adapter	"	38.50	26.70	65.20
Coupling	"	34.50	26.70	61.20
Polyethylene pipe and fittings				
SDR-21				
3" pipe	L.F.	1.10	4.00	5.10
4" pipe	"	2.00	5.35	7.35
6" pipe	"	3.90	8.00	11.90
8" pipe	"	5.70	9.15	14.85
10" pipe	"	8.75	10.70	19.45
12" pipe	"	12.20	12.80	25.00
14" pipe	"	15.70	16.00	31.70
16" pipe	"	20.15	20.00	40.15
18" pipe	"	24.65	24.65	49.30
20" pipe	"	30.20	32.00	62.20
22" pipe	"	35.80	35.60	71.40
24" pipe	"	42.60	40.10	82.70
Fittings, 3"				
90 deg elbow	EA.	59.50	16.00	75.50
45 deg elbow	"	37.80	16.00	53.80
Tee	"	34.50	26.70	61.20
45 deg wye	"	89.00	26.70	115.70
Reducer	"	19.80	20.00	39.80
Flange assembly	"	33.70	16.00	49.70
4"				
90 deg elbow	EA.	88.00	20.00	108.00
45 deg elbow	"	53.50	20.00	73.50
Tee	"	76.00	32.00	108.00
45 deg wye	"	140.00	32.00	172.00
Reducer	"	47.80	26.70	74.50
Flange assembly	"	47.80	20.00	67.80
8"				
90 deg elbow	EA.	240.00	40.10	280.10
45 deg elbow	"	130.00	40.10	170.10
Tee	"	220.00	64.00	284.00
45 deg wye	"	350.00	64.00	414.00
Reducer	"	120.00	53.50	173.50
Flange assembly	"	130.00	40.10	170.10
10"				
90 deg elbow	EA.	330.00	53.50	383.50
45 deg elbow	"	180.00	53.50	233.50
Tee	"	300.00	80.00	380.00
45 deg wye	"	470.00	80.00	550.00
Reducer	"	160.00	64.00	224.00
Flange assembly	"	150.00	53.50	203.50
12"				
90 deg elbow	EA.	540.00	64.00	604.00
45 deg elbow	"	330.00	64.00	394.00

PLUMBING	UNIT	MAT.	INST.	TOTAL
15410.30 PVC/CPVC PIPE				
Tee	EA.	420.00	110.00	530.00
45 deg wye	"	650.00	110.00	760.00
Reducer	"	250.00	80.00	330.00
Flange assembly	"	200.00	64.00	264.00
14"				
90 deg elbow	EA.	730.00	80.00	810.00
45 deg elbow	"	410.00	80.00	490.00
Tee	"	530.00	130.00	660.00
45 deg wye	"	980.00	130.00	1,110
Reducer	"	210.00	110.00	320.00
Flange assembly	"	240.00	80.00	320.00
16"				
90 deg elbow	EA.	910.00	80.00	990.00
45 deg elbow	"	540.00	80.00	620.00
Tee	"	660.00	130.00	790.00
45 deg wye	"	1,090	130.00	1,220
Reducer	"	430.00	110.00	540.00
Flange assembly	"	310.00	80.00	390.00
18"				
90 deg elbow	EA.	1,390	110.00	1,500
45 deg elbow	"	900.00	110.00	1,010
Tee	"	1,070	160.00	1,230
45 deg wye	"	1,850	160.00	2,010
Reducer	"	400.00	110.00	510.00
Flange assembly	"	540.00	110.00	650.00
20"				
90 deg elbow	EA.	1,120	110.00	1,230
45 deg elbow	"	660.00	110.00	770.00
15410.33 ABS DWV PIPE				
Schedule 40 ABS				
1-1/2" pipe	L.F.	1.00	1.60	2.60
2" pipe	"	1.25	1.80	3.05
3" pipe	"	2.35	2.30	4.65
4" pipe	"	3.45	3.20	6.65
6" pipe	"	5.50	4.00	9.50
Fittings				
1/8 bend				
1-1/2"	EA.	1.00	6.40	7.40
2"	"	1.10	8.00	9.10
3"	"	3.35	10.70	14.05
4"	"	5.60	12.80	18.40
6"	"	35.80	16.00	51.80
Tee, sanitary				
1-1/2"	EA.	1.55	10.70	12.25
2"	"	2.35	12.80	15.15
3"	"	5.50	16.00	21.50
4"	"	10.10	20.00	30.10
6"	"	58.00	26.70	84.70
Tee, sanitary reducing				
2 x 1-1/2 x 1-1/2	EA.	3.00	12.80	15.80
2 x 1-1/2 x 2	"	3.70	13.35	17.05

PLUMBING	UNIT	MAT.	INST.	TOTAL
15410.33 ABS DWV PIPE				
2 x 2 x 1-1/2	EA.	2.25	14.55	16.80
3 x 3 x 1-1/2	"	4.50	16.00	20.50
3 x 3 x 2	"	4.50	17.80	22.30
4 x 4 x 1-1/2	"	12.30	20.00	32.30
4 x 4 x 2	"	10.10	22.90	33.00
4 x 4 x 3	"	13.45	24.65	38.10
6 x 6 x 4	"	57.00	26.70	83.70
Wye				
1-1/2"	EA.	2.40	9.15	11.55
2"	"	2.80	12.80	15.60
3"	"	5.60	16.00	21.60
4"	"	10.10	20.00	30.10
6"	"	58.00	26.70	84.70
Reducer				
2 x 1-1/2	EA.	1.55	8.00	9.55
3 x 1-1/2	"	3.60	10.70	14.30
3 x 2	"	3.25	10.70	13.95
4 x 2	"	6.95	12.80	19.75
4 x 3	"	6.85	12.80	19.65
6 x 4	"	10.30	16.00	26.30
P-trap				
1-1/2"	EA.	3.60	10.70	14.30
2"	"	4.70	11.85	16.55
3"	"	21.30	13.95	35.25
4"	"	37.30	16.00	53.30
6"	"	61.00	20.00	81.00
Double sanitary, tee				
1-1/2"	EA.	4.50	12.80	17.30
2"	"	6.05	16.00	22.05
3"	"	13.45	20.00	33.45
4"	"	22.40	26.70	49.10
Long sweep, 1/4 bend				
1-1/2"	EA.	2.25	6.40	8.65
2"	"	2.35	8.00	10.35
3"	"	6.25	10.70	16.95
4"	"	12.90	16.00	28.90
Wye, standard				
1-1/2"	EA.	2.70	10.70	13.40
2"	"	2.45	12.80	15.25
3"	"	6.05	16.00	22.05
4"	"	10.65	20.00	30.65
Wye, reducing				
2 x 1-1/2 x 1-1/2	EA.	5.05	10.70	15.75
2 x 2 x 1-1/2	"	4.05	12.80	16.85
4 x 4 x 2	"	10.65	20.00	30.65
4 x 4 x 3	"	9.65	21.35	31.00
Double wye				
1-1/2"	EA.	4.75	12.80	17.55
2"	"	6.15	16.00	22.15
3"	"	20.15	20.00	40.15
4"	"	35.80	26.70	62.50
2 x 2 x 1-1/2 x 1-1/2	"	6.70	16.00	22.70

PLUMBING	UNIT	MAT.	INST.	TOTAL
15410.33 ABS DWV PIPE				
3 x 3 x 2 x 2	EA.	13.45	20.00	33.45
4 x 4 x 3 x 3	"	31.40	26.70	58.10
Combination wye and 1/8 bend				
1-1/2"	EA.	5.15	10.70	15.85
2"	"	5.80	12.80	18.60
3"	"	8.05	16.00	24.05
4"	"	15.70	20.00	35.70
2 x 2 x 1-1/2	"	5.15	12.80	17.95
3 x 3 x 1-1/2	"	10.10	16.00	26.10
3 x 3 x 2	"	6.25	16.00	22.25
4 x 4 x 2	"	12.45	20.00	32.45
4 x 4 x 3	"	15.20	20.00	35.20
15410.35 PLASTIC PIPE				
Fiberglass reinforced pipe				
2" pipe	L.F.	8.95	2.45	11.40
3" pipe	"	10.20	2.65	12.85
4" pipe	"	13.45	2.90	16.35
6" pipe	"	19.60	3.20	22.80
8" pipe	"	29.10	5.35	34.45
10" pipe	"	44.80	6.40	51.20
12" pipe	"	54.00	8.00	62.00
Fittings				
90 deg elbow, flanged				
2"	EA.	71.50	32.00	103.50
3"	"	86.00	35.60	121.60
4"	"	110.00	40.10	150.10
6"	"	200.00	53.50	253.50
8"	"	360.00	64.00	424.00
10"	"	480.00	80.00	560.00
12"	"	650.00	110.00	760.00
45 deg elbow, flanged				
2"	EA.	71.50	26.70	98.20
3"	"	86.00	32.00	118.00
4"	"	110.00	40.10	150.10
6"	"	200.00	53.50	253.50
8"	"	310.00	64.00	374.00
10"	"	410.00	80.00	490.00
12"	"	520.00	110.00	630.00
Tee, flanged				
2"	EA.	92.50	40.10	132.60
3"	"	130.00	45.80	175.80
4"	"	140.00	53.50	193.50
6"	"	240.00	64.00	304.00
8"	"	430.00	80.00	510.00
10"	"	700.00	110.00	810.00
12"	"	970.00	160.00	1,130
Wye, flanged				
2"	EA.	180.00	40.10	220.10
3"	"	250.00	45.80	295.80
4"	"	320.00	53.50	373.50
6"	"	410.00	64.00	474.00

PLUMBING	UNIT	MAT.	INST.	TOTAL
15410.35 — PLASTIC PIPE				
8"	EA.	550.00	80.00	630.00
10"	"	950.00	110.00	1,060
12"	"	1,210	160.00	1,370
Concentric reducer, flanged				
2"	EA.	69.00	26.70	95.70
4"	"	100.00	32.00	132.00
6"	"	140.00	45.80	185.80
8"	"	220.00	64.00	284.00
10"	"	350.00	80.00	430.00
12"	"	500.00	110.00	610.00
Adapter, bell x male or female				
2"	EA.	8.90	26.70	35.60
3"	"	17.75	29.10	46.85
4"	"	18.80	32.00	50.80
6"	"	40.10	45.80	85.90
8"	"	58.50	64.00	122.50
10"	"	84.00	80.00	164.00
12"	"	150.00	110.00	260.00
Nipples				
2" x 6"	EA.	3.75	3.20	6.95
2" x 12"	"	5.65	4.00	9.65
3" x 8"	"	5.65	4.95	10.60
3" x 12"	"	5.65	5.35	11.00
4" x 8"	"	5.65	5.35	11.00
4" x 12"	"	6.35	6.40	12.75
6" x 12"	"	14.55	8.00	22.55
8" x 18"	"	41.50	8.00	49.50
8" x 24"	"	48.50	9.15	57.65
10" x 18"	"	48.50	10.70	59.20
10" x 24"	"	66.00	12.80	78.80
12" x 18"	"	66.00	14.55	80.55
12" x 24"	"	77.50	16.00	93.50
Sleeve coupling				
2"	EA.	8.20	26.70	34.90
3"	"	9.00	32.00	41.00
4"	"	12.45	45.80	58.25
6"	"	30.00	64.00	94.00
8"	"	49.60	80.00	129.60
10"	"	74.00	110.00	184.00
Flanges				
2"	EA.	12.25	26.70	38.95
3"	"	16.50	32.00	48.50
4"	"	21.90	45.80	67.70
6"	"	37.50	64.00	101.50
8"	"	66.00	80.00	146.00
10"	"	93.50	110.00	203.50
12"	"	120.00	110.00	230.00
15410.70 — STAINLESS STEEL PIPE				
Stainless steel, schedule 40, threaded				
1/2" pipe	L.F.	4.05	4.60	8.65

PLUMBING		UNIT	MAT.	INST.	TOTAL
15410.70	**STAINLESS STEEL PIPE**				
3/4" pipe		L.F.	5.25	4.70	9.95
1" pipe		"	6.60	4.95	11.55
1-1/2" pipe		"	9.50	5.35	14.85
2" pipe		"	15.70	5.85	21.55
2-1/2" pipe		"	20.15	6.40	26.55
3" pipe		"	29.10	7.10	36.20
4" pipe		"	37.00	8.00	45.00
Fittings, 1/2"					
90 deg ell		EA.	5.05	40.10	45.15
45 deg ell		"	5.95	40.10	46.05
Tee		"	6.70	53.50	60.20
Cap		"	2.25	20.00	22.25
Reducer		"	7.30	26.70	34.00
Union		"	11.75	40.10	51.85
Flange		"	20.15	40.10	60.25
3/4"					
90 deg ell		EA.	6.70	40.10	46.80
45 deg ell		"	7.75	40.10	47.85
Tee		"	9.05	53.50	62.55
Cap		"	4.05	20.00	24.05
Reducer		"	8.95	26.70	35.65
Union		"	14.55	40.10	54.65
Flange		"	24.65	40.10	64.75
1"					
90 deg ell		EA.	8.40	40.10	48.50
45 deg ell		"	8.40	40.10	48.50
Tee		"	9.95	53.50	63.45
Cap		"	5.50	20.00	25.50
Reducer		"	8.75	40.10	48.85
Union		"	17.35	40.10	57.45
Flange		"	24.65	40.10	64.75
1-1/4"					
90 deg ell		EA.	11.75	40.10	51.85
45 deg ell		"	11.75	40.10	51.85
Tee		"	16.80	53.50	70.30
Cap		"	10.65	20.00	30.65
Union		"	31.90	40.10	72.00
Flange		"	28.00	40.10	68.10
1-1/2"					
90 deg ell		EA.	14.55	53.50	68.05
45 deg ell		"	15.10	53.50	68.60
Tee		"	21.15	64.00	85.15
Cap		"	13.45	26.70	40.15
Reducer		"	16.25	32.00	48.25
Union		"	38.10	45.80	83.90
Flange		"	34.70	45.80	80.50
2"					
90 deg ell		EA.	21.85	64.00	85.85
45 deg ell		"	21.85	64.00	85.85
Tee		"	25.80	110.00	135.80
Cap		"	16.80	32.00	48.80
Reducer		"	19.60	40.10	59.70

PLUMBING		UNIT	MAT.	INST.	TOTAL
15410.70	**STAINLESS STEEL PIPE**				
Union		EA.	42.60	64.00	106.60
Flange		"	32.50	64.00	96.50
Type 304, sch 10 pipe					
1" pipe		L.F.	5.95	4.00	9.95
1-1/4" pipe		"	7.60	5.35	12.95
1-1/2" pipe		"	8.30	5.85	14.15
2" pipe		"	8.85	7.10	15.95
2-1/2" pipe		"	11.75	8.00	19.75
3" pipe		"	14.00	9.15	23.15
4" pipe		"	18.50	10.70	29.20
6" pipe		"	29.10	12.80	41.90
Fittings, 1"					
90 deg elbow		EA.	5.80	64.00	69.80
45 deg elbow		"	8.20	64.00	72.20
Tee		"	17.90	110.00	127.90
Reducer		"	6.15	64.00	70.15
Flange		"	22.40	40.10	62.50
Cap		"	17.90	40.10	58.00
1-1/4"					
90 deg elbow		EA.	9.50	64.00	73.50
45 deg elbow		"	11.40	64.00	75.40
Tee		"	34.70	110.00	144.70
Reducer		"	8.95	64.00	72.95
Flange		"	28.00	40.10	68.10
Cap		"	28.00	40.10	68.10
1-1/2"					
90 deg elbow		EA.	10.65	53.50	64.15
45 deg elbow		"	11.75	53.50	65.25
Tee		"	24.65	110.00	134.65
Reducer		"	13.45	80.00	93.45
Flange		"	24.65	53.50	78.15
Cap		"	22.40	53.50	75.90
2"					
90 deg elbow		EA.	10.10	64.00	74.10
45 deg elbow		"	9.20	64.00	73.20
Tee		"	24.65	160.00	184.65
Reducer		"	13.45	110.00	123.45
Flange		"	31.40	80.00	111.40
Cap		"	22.40	80.00	102.40
2-1/2"					
90 deg elbow		EA.	16.80	64.00	80.80
45 deg elbow		"	16.80	64.00	80.80
Tee		"	54.00	160.00	214.00
Reducer		"	17.90	110.00	127.90
Flange		"	54.00	80.00	134.00
Cap		"	22.40	80.00	102.40
3"					
90 deg elbow		EA.	17.00	80.00	97.00
45 deg elbow		"	14.10	80.00	94.10
Tee		"	42.60	210.00	252.60
Reducer		"	18.50	160.00	178.50
Flange		"	50.50	110.00	160.50

PLUMBING	UNIT	MAT.	INST.	TOTAL
15410.70 STAINLESS STEEL PIPE				
Cap	EA.	23.50	110.00	133.50
4"				
90 deg elbow	EA.	35.80	80.00	115.80
45 deg elbow	"	26.90	80.00	106.90
Reducer	"	33.60	160.00	193.60
Flange	"	70.50	110.00	180.50
Cap	"	24.65	110.00	134.65
6"				
90 deg elbow	EA.	77.50	110.00	187.50
45 deg elbow	"	50.50	110.00	160.50
Tee	"	160.00	460.00	620.00
Reducer	"	47.00	160.00	207.00
Flange	"	110.00	160.00	270.00
Cap	"	51.50	160.00	211.50
Type 304 tubing				
.035 wall				
1/4"	L.F.	0.74	1.80	2.54
3/8"	"	0.88	2.00	2.88
1/2"	"	1.05	2.30	3.35
5/8"	"	1.50	2.65	4.15
3/4"	"	1.70	3.20	4.90
7/8"	"	1.80	3.55	5.35
1"	"	1.90	4.00	5.90
.049 wall				
1/4"	L.F.	0.96	1.90	2.86
3/8"	"	1.05	2.15	3.20
1/2"	"	1.10	2.45	3.55
5/8"	"	1.50	2.90	4.40
3/4"	"	1.75	3.55	5.30
7/8"	"	2.15	4.00	6.15
1"	"	2.30	4.60	6.90
.065 wall				
1/4"	L.F.	1.40	2.15	3.55
3/8"	"	1.70	2.65	4.35
1/2"	"	1.85	2.90	4.75
5/8"	"	2.00	3.55	5.55
3/4"	"	2.45	4.60	7.05
7/8"	"	2.75	5.35	8.10
1"	"	3.00	6.40	9.40
Type 316 tubing				
.035 wall				
1/4"	L.F.	0.72	1.80	2.52
3/8"	"	0.88	2.00	2.88
1/2"	"	0.92	2.30	3.22
5/8"	"	1.80	2.65	4.45
3/4"	"	2.00	3.20	5.20
7/8"	"	2.70	3.55	6.25
1"	"	3.10	4.00	7.10
.049 wall				
1/4"	L.F.	1.25	2.15	3.40
3/8"	"	1.35	2.65	4.00
1/2"	"	1.70	2.90	4.60

PLUMBING		UNIT	MAT.	INST.	TOTAL
15410.70	**STAINLESS STEEL PIPE**				
5/8"		L.F.	2.15	3.55	5.70
3/4"		"	2.25	4.60	6.85
7/8"		"	3.15	5.35	8.50
1"		"	2.15	6.40	8.55
.065 wall					
1/4"		L.F.	1.80	2.15	3.95
3/8"		"	2.35	2.65	5.00
1/2"		"	2.35	2.90	5.25
5/8"		"	2.45	3.55	6.00
3/4"		"	2.80	4.60	7.40
7/8"		"	5.60	5.35	10.95
1"		"	3.35	6.40	9.75
Fittings, 1/4"					
90 deg elbow		EA.	12.30	6.40	18.70
Union tee		"	20.70	10.70	31.40
Union		"	10.10	10.70	20.80
Male connector		"	6.70	8.00	14.70
3/8"					
90 deg elbow		EA.	17.90	8.00	25.90
Union tee		"	26.90	12.30	39.20
Union		"	14.00	12.30	26.30
Male connector		"	10.10	8.00	18.10
1/2"					
90 deg elbow		EA.	26.90	8.45	35.35
Union tee		"	3.35	13.35	16.70
Union		"	21.30	13.35	34.65
Male connector		"	14.55	8.00	22.55
5/8"					
90 deg elbow		EA.	65.00	10.70	75.70
Union tee		"	86.00	16.00	102.00
Union		"	42.60	16.00	58.60
Male connector		"	21.30	10.70	32.00
3/4"					
90 deg elbow		EA.	96.50	10.70	107.20
Union tee		"	83.00	16.00	99.00
Union		"	69.50	16.00	85.50
Male connector		"	30.20	10.70	40.90
7/8"					
90 deg elbow		EA.	89.50	11.45	100.95
Union tee		"	130.00	17.80	147.80
Union		"	69.50	17.80	87.30
Male connector		"	55.00	11.45	66.45
1"					
90 deg elbow		EA.	160.00	14.55	174.55
Union tee		"	200.00	20.00	220.00
Union		"	110.00	20.00	130.00
Male connector		"	73.00	16.00	89.00
Type 316 valves					
Gate valves					
1/4"		EA.	250.00	10.70	260.70
3/8"		"	250.00	12.80	262.80
1/2"		"	160.00	13.95	173.95

PLUMBING	UNIT	MAT.	INST.	TOTAL
15410.70 STAINLESS STEEL PIPE				
3/4"	EA.	200.00	16.00	216.00
1"	"	220.00	21.35	241.35
Globe valves				
1/4"	EA.	95.00	10.70	105.70
3/8"	"	150.00	12.80	162.80
1/2"	"	190.00	13.95	203.95
3/4"	"	220.00	16.00	236.00
1"	"	220.00	21.35	241.35
Check valves				
1/4"	EA.	73.00	10.70	83.70
3/8"	"	140.00	12.80	152.80
1/2"	"	160.00	13.95	173.95
3/4"	"	210.00	16.00	226.00
1"	"	220.00	21.35	241.35
Test and balance	"	0.00	26.70	26.70
Pipe identification	"	0.11	6.40	6.51
Disinfect	"	0.00	26.70	26.70
15410.80 STEEL PIPE				
Black steel, extra heavy pipe, threaded				
1/2" pipe	L.F.	1.10	1.30	2.40
3/4" pipe	"	1.35	1.30	2.65
1" pipe	"	1.80	1.60	3.40
1-1/2" pipe	"	2.90	1.80	4.70
2-1/2" pipe	"	6.15	4.00	10.15
3" pipe	"	8.40	5.35	13.75
4" pipe	"	13.35	6.40	19.75
5" pipe	"	17.90	8.00	25.90
6" pipe	"	22.40	8.00	30.40
8" pipe	"	33.00	10.70	43.70
10" pipe	"	45.90	12.80	58.70
12" pipe	"	58.00	16.00	74.00
Fittings, malleable iron, threaded, 1/2" pipe				
90 deg ell	EA.	1.35	10.70	12.05
45 deg ell	"	2.15	10.70	12.85
Tee	"	1.80	16.00	17.80
Reducing tee	"	3.90	16.00	19.90
Cap	"	1.35	6.40	7.75
Coupling	"	1.65	12.80	14.45
Union	"	7.60	10.70	18.30
Nipple, 4" long	"	1.35	10.70	12.05
3/4" pipe				
90 deg ell	EA.	1.80	10.70	12.50
45 deg ell	"	2.90	16.00	18.90
Tee	"	2.95	16.00	18.95
Reducing tee	"	4.35	10.70	15.05
Cap	"	1.80	6.40	8.20
Coupling	"	1.95	10.70	12.65
Union	"	8.60	10.70	19.30
Nipple, 4" long	"	1.95	10.70	12.65
1" pipe				

PLUMBING		UNIT	MAT.	INST.	TOTAL
15410.80	**STEEL PIPE**				
90 deg ell		EA.	2.90	12.80	15.70
45 deg ell		"	3.25	12.80	16.05
Tee		"	4.35	17.80	22.15
Reducing tee		"	5.60	17.80	23.40
Cap		"	2.45	6.40	8.85
Coupling		"	3.35	12.80	16.15
Union		"	10.30	12.80	23.10
Nipple, 4" long		"	2.35	12.80	15.15
1-1/2" pipe					
90 deg ell		EA.	5.55	16.00	21.55
45 deg ell		"	6.45	16.00	22.45
Tee		"	8.00	22.90	30.90
Reducing tee		"	12.40	22.90	35.30
Cap		"	3.55	8.00	11.55
Coupling		"	4.95	16.00	20.95
Union		"	13.45	16.00	29.45
Nipple, 4" long		"	3.45	16.00	19.45
2-1/2" pipe					
90 deg ell		EA.	22.40	40.10	62.50
45 deg ell		"	28.00	40.10	68.10
Tee		"	30.20	53.50	83.70
Reducing tee		"	39.20	53.50	92.70
Cap		"	11.75	20.00	31.75
Coupling		"	18.50	53.50	72.00
Union		"	50.50	53.50	104.00
Nipple, 4" long		"	11.75	53.50	65.25
3" pipe					
90 deg ell		EA.	32.50	53.50	86.00
45 deg ell		"	37.00	53.50	90.50
Tee		"	41.40	80.00	121.40
Reducing tee		"	66.00	80.00	146.00
Cap		"	17.90	26.70	44.60
Coupling		"	25.80	53.50	79.30
Union		"	83.00	53.50	136.50
Nipple, 4" long		"	14.55	53.50	68.05
4" pipe					
90 deg ell		EA.	65.00	64.00	129.00
45 deg ell		"	66.00	64.00	130.00
Tee		"	95.00	110.00	205.00
Reducing tee		"	100.00	110.00	210.00
Cap		"	26.90	110.00	136.90
Coupling		"	50.50	32.00	82.50
Union		"	87.50	110.00	197.50
Nipple, 4" long		"	17.90	110.00	127.90
6" pipe					
90 deg ell		EA.	180.00	64.00	244.00
45 deg ell		"	220.00	64.00	284.00
Tee		"	260.00	110.00	370.00
Reducing tee		"	250.00	110.00	360.00
Cap		"	110.00	32.00	142.00
8" pipe					
90 deg ell		EA.	190.00	130.00	320.00

PLUMBING	UNIT	MAT.	INST.	TOTAL
15410.80 STEEL PIPE				
45 deg ell	EA.	240.00	130.00	370.00
Tee	"	280.00	200.00	480.00
Reducing tee	"	260.00	170.00	430.00
Cap	"	31.40	64.00	95.40
10" pipe				
90 deg ell	EA.	220.00	160.00	380.00
45 deg ell	"	250.00	160.00	410.00
Tee	"	320.00	200.00	520.00
Reducing tee	"	300.00	80.00	380.00
Cap	"	31.40	80.00	111.40
12" pipe				
90 deg ell	EA.	250.00	200.00	450.00
45 deg ell	"	290.00	200.00	490.00
Tee	"	370.00	270.00	640.00
Reducing tee	"	340.00	270.00	610.00
Cap	"	32.50	110.00	142.50
Butt welded, 1/2" pipe				
90 deg ell	EA.	8.85	10.70	19.55
45 deg ell	"	12.90	10.70	23.60
Tee	"	35.80	16.00	51.80
3/4" pipe				
90 deg ell	EA.	8.85	10.70	19.55
45 deg. ell	"	13.45	10.70	24.15
Tee	"	37.00	16.00	53.00
1" pipe				
90 deg ell	EA.	8.85	12.80	21.65
45 deg ell	"	13.45	12.80	26.25
Tee	"	37.00	17.80	54.80
1-1/2" pipe				
90 deg ell	EA.	12.55	16.00	28.55
45 deg. ell	"	13.35	16.00	29.35
Tee	"	33.00	22.90	55.90
Reducing tee	"	22.05	22.90	44.95
Cap	"	11.75	12.80	24.55
2-1/2" pipe				
90 deg. ell	EA.	25.80	32.00	57.80
45 deg. ell	"	25.80	32.00	57.80
Tee	"	98.50	45.80	144.30
Reducing tee	"	21.30	45.80	67.10
Cap	"	26.90	16.00	42.90
3" pipe				
90 deg ell	EA.	24.65	40.10	64.75
45 deg. ell	"	60.50	40.10	100.60
Tee	"	24.65	53.50	78.15
Reducing tee	"	25.80	53.50	79.30
Cap	"	13.45	26.70	40.15
4" pipe				
90 deg ell	EA.	49.30	53.50	102.80
45 deg. ell	"	40.30	53.50	93.80
Tee	"	88.50	80.00	168.50
Reducing tee	"	32.50	80.00	112.50
Cap	"	14.55	26.70	41.25

PLUMBING	UNIT	MAT.	INST.	TOTAL
15410.80 **STEEL PIPE**				
6" pipe				
90 deg. ell	EA.	150.00	64.00	214.00
45 deg. ell	"	110.00	64.00	174.00
Tee	"	240.00	110.00	350.00
Reducing tee	"	220.00	110.00	330.00
Cap	"	25.80	32.00	57.80
8" pipe				
90 deg. ell	EA.	240.00	110.00	350.00
45 deg. ell	"	180.00	110.00	290.00
Tee	"	550.00	160.00	710.00
Reducing tee	"	450.00	160.00	610.00
Cap	"	39.20	64.00	103.20
10" pipe				
90 deg ell	EA.	530.00	110.00	640.00
45 deg. ell	"	360.00	110.00	470.00
Tee	"	970.00	160.00	1,130
Reducing tee	"	190.00	160.00	350.00
Cap	"	61.50	80.00	141.50
12" pipe				
90 deg. ell	EA.	880.00	130.00	1,010
45 deg. ell	"	580.00	130.00	710.00
Tee	"	1,230	230.00	1,460
Reducing tee	"	440.00	230.00	670.00
Cap	"	75.00	80.00	155.00
90 deg. ell	"	2.25	10.70	12.95
45 deg. ell	"	4.50	10.70	15.20
Tee	"	2.90	16.00	18.90
Reducing tee	"	5.50	16.00	21.50
3/4" pipe				
90 deg. ell	EA.	2.35	10.70	13.05
45 deg. ell	"	2.90	10.70	13.60
Tee	"	3.65	16.00	19.65
Reducing tee	"	4.75	16.00	20.75
1" pipe				
90 deg. ell	EA.	2.85	12.80	15.65
45 deg. ell	"	3.90	12.80	16.70
Tee	"	3.55	17.80	21.35
Reducing tee	"	4.70	17.80	22.50
1-1/2" pipe				
90 deg. ell	EA.	5.55	16.00	21.55
45 deg. ell	"	7.90	16.00	23.90
Tee	"	8.10	22.90	31.00
Reducing tee	"	10.95	22.90	33.85
2-1/2" pipe				
90 deg. ell	EA.	17.75	32.00	49.75
45 deg. ell	"	21.05	32.00	53.05
Tee	"	25.30	45.80	71.10
Reducing tee	"	29.30	45.80	75.10
3" pipe				
90 deg. ell	EA.	28.70	40.10	68.80
45 deg. ell	"	32.90	40.10	73.00
Tee	"	38.10	64.00	102.10

PLUMBING	UNIT	MAT.	INST.	TOTAL
15410.80 STEEL PIPE				
Reducing tee	EA.	44.10	64.00	108.10
4" pipe				
90 deg. ell	EA.	51.50	53.50	105.00
45 deg. ell	"	64.00	53.50	117.50
Tee	"	73.50	80.00	153.50
Reducing tee	"	88.00	80.00	168.00
6" pipe				
90 deg. ell	EA.	120.00	53.50	173.50
45 deg. ell	"	140.00	53.50	193.50
Tee	"	170.00	80.00	250.00
Reducing tee	"	200.00	80.00	280.00
8" pipe				
90 deg. ell	EA.	250.00	110.00	360.00
45 deg. ell	"	280.00	110.00	390.00
Tee	"	350.00	160.00	510.00
Reducing tee	"	380.00	160.00	540.00
15410.82 GALVANIZED STEEL PIPE				
Galvanized pipe				
1/2" pipe	L.F.	1.25	3.20	4.45
3/4" pipe	"	1.55	4.00	5.55
1" pipe	"	2.25	4.60	6.85
1-1/4" pipe	"	2.70	5.35	8.05
1-1/2" pipe	"	3.25	6.40	9.65
2" pipe	"	4.70	8.00	12.70
2-1/2" pipe	"	5.80	10.70	16.50
3" pipe	"	10.10	11.45	21.55
4" pipe	"	15.70	13.35	29.05
6" pipe	"	21.85	26.70	48.55
90 degree ell, 150 lb malleable iron, galvanized				
1/2"	EA.	1.10	6.40	7.50
3/4"	"	1.35	8.00	9.35
1"	"	2.70	8.45	11.15
1-1/4"	"	4.50	9.40	13.90
1-1/2"	"	4.70	10.70	15.40
2"	"	6.70	12.80	19.50
2-1/2"	"	11.20	20.00	31.20
3"	"	15.70	24.65	40.35
4"	"	24.65	26.70	51.35
5"	"	58.00	32.00	90.00
6"	"	70.50	32.00	102.50
45 degree ell, 150 lb m.i., galv.				
1/2"	EA.	1.35	6.40	7.75
3/4"	"	2.25	8.00	10.25
1"	"	3.35	8.45	11.80
1-1/4"	"	4.50	9.40	13.90
1-1/2"	"	5.60	10.70	16.30
2"	"	6.70	12.80	19.50
2-1/2"	"	11.20	20.00	31.20
3"	"	16.80	24.65	41.45
4"	"	25.80	32.00	57.80
5"	"	67.00	32.00	99.00

PLUMBING	UNIT	MAT.	INST.	TOTAL
15410.82 GALVANIZED STEEL PIPE				
6"	EA.	78.50	40.10	118.60
Tees, straight, 150 lb m.i., galv.				
1/2"	EA.	1.55	8.00	9.55
3/4"	"	2.70	9.15	11.85
1"	"	3.80	10.70	14.50
1-1/4"	"	4.50	12.80	17.30
1-1/2"	"	5.60	16.00	21.60
2"	"	8.95	20.00	28.95
2-1/2"	"	15.70	26.70	42.40
3"	"	16.80	32.00	48.80
4"	"	33.60	40.10	73.70
5"	"	67.00	45.80	112.80
6"	"	89.50	53.50	143.00
Tees, reducing, out, 150 lb m.i., galv.				
1/2"	EA.	3.15	8.00	11.15
3/4"	"	3.35	9.15	12.50
1"	"	4.50	10.70	15.20
1-1/4"	"	5.60	12.80	18.40
1-1/2"	"	6.70	16.00	22.70
2"	"	14.00	20.00	34.00
2-1/2"	"	15.70	26.70	42.40
3"	"	20.15	32.00	52.15
4"	"	29.10	40.10	69.20
5"	"	98.50	45.80	144.30
6"	"	76.00	53.50	129.50
Couplings, straight, 150 lb m.i., galv.				
1/2"	EA.	1.80	6.40	8.20
3/4"	"	2.25	7.10	9.35
1"	"	3.35	8.00	11.35
1-1/4"	"	4.50	9.15	13.65
1-1/2"	"	5.05	10.70	15.75
2"	"	6.15	12.80	18.95
2-1/2"	"	7.85	20.00	27.85
3"	"	9.50	26.70	36.20
4"	"	21.30	29.10	50.40
5"	"	35.80	32.00	67.80
6"	"	49.30	32.00	81.30
Couplings, reducing, 150 lb m.i., galv				
1/2"	EA.	1.80	6.40	8.20
3/4"	"	2.25	7.10	9.35
1"	"	3.35	8.00	11.35
1-1/4"	"	4.50	9.15	13.65
1-1/2"	"	5.05	10.70	15.75
2"	"	6.15	12.80	18.95
2-1/2"	"	7.85	20.00	27.85
3"	"	9.50	26.70	36.20
4"	"	21.30	29.10	50.40
5"	"	35.80	32.00	67.80
6"	"	49.30	32.00	81.30
Caps, 150 lb m.i., galv.				
1/2"	EA.	1.10	3.20	4.30
3/4"	"	1.35	3.35	4.70

PLUMBING		UNIT	MAT.	INST.	TOTAL
15410.82	**GALVANIZED STEEL PIPE**				
1"		EA.	2.00	3.55	5.55
1-1/4"		"	2.45	3.75	6.20
1-1/2"		"	3.35	4.00	7.35
2"		"	3.60	4.60	8.20
2-1/2"		"	4.50	5.85	10.35
3"		"	6.70	8.00	14.70
4"		"	11.20	10.00	21.20
5"		"	23.50	12.30	35.80
6"		"	43.70	16.00	59.70
Unions, 150 lb m.i., galv.					
1/2"		EA.	3.35	8.00	11.35
3/4"		"	3.60	9.15	12.75
1"		"	4.50	10.70	15.20
1-1/4"		"	5.60	12.80	18.40
1-1/2"		"	6.70	16.00	22.70
2"		"	8.95	17.80	26.75
2-1/2"		"	15.70	21.35	37.05
3"		"	24.65	26.70	51.35
Nipples, galvanized steel, 4" long					
1/2"		EA.	1.35	4.00	5.35
3/4"		"	1.80	4.25	6.05
1"		"	2.45	4.60	7.05
1-1/4"		"	2.80	4.95	7.75
1-1/2"		"	3.35	5.35	8.70
2"		"	3.90	5.85	9.75
2-1/2"		"	4.70	6.40	11.10
3"		"	6.15	8.00	14.15
4"		"	7.75	10.70	18.45
90 degree reducing ell, 150 lb m.i., galv.					
3/4" x 1/2"		EA.	2.00	6.40	8.40
1" x 3/4"		"	2.80	7.10	9.90
1-1/4" x 1"		"	4.50	8.00	12.50
1-1/4" x 3/4"		"	4.70	9.15	13.85
1-1/4" x 1/2"		"	4.70	10.70	15.40
1-1/2" x 1-1/4"		"	4.95	10.70	15.65
1-1/2" x 1"		"	5.15	10.70	15.85
1-1/2" x 3/4"		"	5.40	10.70	16.10
2" x 1-1/2"		"	6.70	13.95	20.65
2" x 1-1/4"		"	7.05	13.95	21.00
2" x 1"		"	7.05	13.95	21.00
2" x 3/4"		"	7.05	13.95	21.00
2-1/2" x 2"		"	13.45	20.00	33.45
2-1/2" x 1-1/2"		"	14.00	20.00	34.00
3" x 2-1/2"		"	20.15	26.70	46.85
3" x 2"		"	20.15	26.70	46.85
4" x 3"		"	35.80	29.10	64.90
Square head plug (C.I.)					
1/2"		EA.	1.00	3.55	4.55
3/4"		"	2.25	4.00	6.25
1"		"	2.35	4.25	6.60
1-1/4"		"	2.45	4.60	7.05
1-1/2"		"	3.10	4.95	8.05

PLUMBING	UNIT	MAT.	INST.	TOTAL
15410.82 GALVANIZED STEEL PIPE				
2"	EA.	3.90	5.35	9.25
2-1/2"	"	5.15	7.10	12.25
3"	"	7.50	8.00	15.50
4"	"	10.90	10.70	21.60
5"	"	17.90	12.80	30.70
6"	"	24.65	16.00	40.65
Screwed flanges, galv.				
1"	EA.	9.20	16.00	25.20
1-1/4"	"	10.10	17.80	27.90
1-1/2"	"	11.40	20.00	31.40
2"	"	12.20	20.00	32.20
2-1/2"	"	12.90	21.35	34.25
3"	"	16.80	29.10	45.90
4"	"	22.40	40.10	62.50
5"	"	28.00	53.50	81.50
6"	"	35.80	53.50	89.30
15430.23 CLEANOUTS				
Cleanout, wall				
2"	EA.	51.50	21.35	72.85
3"	"	58.00	21.35	79.35
4"	"	77.00	26.70	103.70
6"	"	130.00	32.00	162.00
8"	"	160.00	40.10	200.10
Floor				
2"	EA.	56.50	26.70	83.20
3"	"	71.50	26.70	98.20
4"	"	79.00	32.00	111.00
6"	"	110.00	40.10	150.10
8"	"	200.00	45.80	245.80
15430.24 GREASE TRAPS				
Grease traps, cast iron, 3" pipe				
35 gpm, 70 lb capacity	EA.	1,850	320.00	2,170
50 gpm, 100 lb capacity	"	2,350	400.00	2,750
15430.25 HOSE BIBBS				
Hose bibb				
1/2"	EA.	4.75	10.70	15.45
3/4"	"	5.05	10.70	15.75
15430.60 VALVES				
Gate valve, 125 lb, bronze, soldered				
1/2"	EA.	10.10	8.00	18.10
3/4"	"	12.30	8.00	20.30
1"	"	16.25	10.70	26.95
1-1/2"	"	28.70	12.80	41.50
2"	"	40.10	16.00	56.10
2-1/2"	"	78.50	20.00	98.50

PLUMBING	UNIT	MAT.	INST.	TOTAL
15430.60 **VALVES**				
Threaded				
1/4", 125 lb	EA.	8.40	12.80	21.20
1/2"				
125 lb	EA.	11.75	12.80	24.55
150 lb	"	16.35	12.80	29.15
300 lb	"	28.00	12.80	40.80
3/4"				
125 lb	EA.	14.00	12.80	26.80
150 lb	"	19.05	12.80	31.85
300 lb	"	37.00	12.80	49.80
1"				
125 lb	EA.	17.80	12.80	30.60
150 lb	"	24.65	12.80	37.45
300 lb	"	50.50	16.00	66.50
1-1/2"				
125 lb	EA.	28.00	16.00	44.00
150 lb	"	42.60	16.00	58.60
300 lb	"	95.00	17.80	112.80
2"				
125 lb	EA.	39.20	22.90	62.10
150 lb	"	59.50	22.90	82.40
300 lb	"	130.00	26.70	156.70
Cast iron, flanged				
2", 150 lb	EA.	190.00	26.70	216.70
2-1/2"				
125 lb	EA.	180.00	26.70	206.70
150 lb	"	280.00	26.70	306.70
250 lb	"	360.00	26.70	386.70
3"				
125 lb	EA.	210.00	32.00	242.00
150 lb	"	290.00	32.00	322.00
250 lb	"	450.00	32.00	482.00
4"				
125 lb	EA.	260.00	45.80	305.80
150 lb	"	350.00	45.80	395.80
250 lb	"	600.00	45.80	645.80
6"				
125 lb	EA.	390.00	64.00	454.00
250 lb	"	1,010	64.00	1,074
8"				
125 lb	EA.	780.00	80.00	860.00
250 lb	"	1,900	80.00	1,980
OS&Y, flanged				
2"				
125 lb	EA.	130.00	26.70	156.70
250 lb	"	310.00	26.70	336.70
2-1/2"				
125 lb	EA.	150.00	26.70	176.70
250 lb	"	360.00	32.00	392.00
3"				
125 lb	EA.	150.00	32.00	182.00
250 lb	"	390.00	32.00	422.00

PLUMBING	UNIT	MAT.	INST.	TOTAL
15430.60 VALVES				
4"				
125 lb	EA.	210.00	53.50	263.50
250 lb	"	600.00	53.50	653.50
6"				
125 lb	EA.	330.00	64.00	394.00
250 lb	"	970.00	64.00	1,034
Check valve, bronze, soldered, 125 lb				
1/2"	EA.	13.40	8.00	21.40
3/4"	"	15.70	8.00	23.70
1"	"	20.15	10.70	30.85
1-1/4"	"	23.50	12.80	36.30
1-1/2"	"	32.50	12.80	45.30
2"	"	44.80	16.00	60.80
Threaded				
1/2"				
125 lb	EA.	10.90	10.70	21.60
150 lb	"	17.95	10.70	28.65
200 lb	"	18.15	10.70	28.85
3/4"				
125 lb	EA.	13.85	12.80	26.65
150 lb	"	22.60	12.80	35.40
200 lb	"	22.60	12.80	35.40
1"				
125 lb	EA.	18.40	16.00	34.40
150 lb	"	29.60	16.00	45.60
200 lb	"	29.60	16.00	45.60
Flow check valve, cast iron, threaded				
1"	EA.	31.40	12.80	44.20
1-1/4"	"	33.40	16.00	49.40
1-1/2"				
125 lb	EA.	30.60	16.00	46.60
150 lb	"	50.50	16.00	66.50
200 lb	"	50.50	17.80	68.30
2"				
125 lb	EA.	44.80	17.80	62.60
150 lb	"	74.00	17.80	91.80
200 lb	"	74.00	20.00	94.00
2-1/2"				
125 lb	EA.	100.00	26.70	126.70
250 lb	"	330.00	32.00	362.00
3"				
125 lb	EA.	120.00	32.00	152.00
250 lb	"	410.00	40.10	450.10
4"				
125 lb	EA.	180.00	45.80	225.80
250 lb	"	520.00	53.50	573.50
6"				
125 lb	EA.	260.00	64.00	324.00
250 lb	"	910.00	64.00	974.00
Vertical check valve, bronze, 125 lb, threaded				
1/2"	EA.	21.45	12.80	34.25
3/4"	"	29.10	14.55	43.65

PLUMBING	UNIT	MAT.	INST.	TOTAL
15430.60 — VALVES				
1"	EA.	34.90	16.00	50.90
1-1/4"	"	44.00	17.80	61.80
1-1/2"	"	50.50	20.00	70.50
2"	"	82.50	22.90	105.40
Cast iron, flanged				
2-1/2"	EA.	300.00	32.00	332.00
3"	"	330.00	40.10	370.10
4"	"	450.00	53.50	503.50
6	"	780.00	64.00	844.00
8"	"	1,460	80.00	1,540
10"	"	2,240	110.00	2,350
12"	"	2,800	130.00	2,930
Globe valve, bronze, soldered, 125 lb				
1/2"	EA.	11.20	9.15	20.35
3/4"	"	13.45	10.00	23.45
1"	"	24.65	10.70	35.35
1-1/4"	"	40.30	11.45	51.75
1-1/2"	"	54.00	13.35	67.35
2"	"	89.50	16.00	105.50
Threaded				
1/2"				
125 lb	EA.	11.20	10.70	21.90
150 lb	"	33.60	10.70	44.30
300 lb	"	42.60	10.70	53.30
3/4"				
125 lb	EA.	13.45	12.80	26.25
150 lb	"	35.80	12.80	48.60
300 lb	"	44.80	12.80	57.60
1"				
125 lb	EA.	24.65	16.00	40.65
150 lb	"	78.50	16.00	94.50
300 lb	"	89.50	16.00	105.50
1-1/4"				
125 lb	EA.	44.80	16.00	60.80
150 lb	"	110.00	16.00	126.00
300 lb	"	120.00	16.00	136.00
1-1/2"				
125 lb	EA.	54.00	17.80	71.80
150 lb	"	120.00	17.80	137.80
300 lb	"	140.00	17.80	157.80
2"				
125 lb	EA.	89.50	21.35	110.85
150 lb	"	220.00	21.35	241.35
300 lb	"	280.00	21.35	301.35
Cast iron flanged				
2-1/2"				
125 lb	EA.	500.00	32.00	532.00
250 lb	"	920.00	32.00	952.00
3"				
125 lb	EA.	590.00	40.10	630.10
250 lb	"	990.00	40.10	1,030
4"				

PLUMBING	UNIT	MAT.	INST.	TOTAL
15430.60 — VALVES				
125 lb	EA.	840.00	53.50	893.50
250 lb	"	1,390	53.50	1,444
6"				
125 lb	EA.	1,510	64.00	1,574
250 lb	"	1,860	64.00	1,924
8"				
125 lb	EA.	2,490	80.00	2,570
250 lb	"	2,910	80.00	2,990
Butterfly valve, cast iron, wafer type				
2"				
150 lb	EA.	73.00	22.90	95.90
200 lb	"	78.50	26.70	105.20
2-1/2"				
150 lb	EA.	78.50	26.70	105.20
200 lb	"	84.00	29.10	113.10
3"				
150 lb	EA.	78.50	32.00	110.50
200 lb	"	89.50	35.60	125.10
4"				
150 lb	EA.	95.00	45.80	140.80
200 lb	"	100.00	53.50	153.50
6"				
150 lb	EA.	170.00	64.00	234.00
200 lb	"	180.00	64.00	244.00
8"				
150 lb	EA.	200.00	71.00	271.00
200 lb	"	240.00	80.00	320.00
10"				
150 lb	EA.	250.00	80.00	330.00
200 lb	"	340.00	110.00	450.00
Ball valve, bronze, 250 lb, threaded				
1/2"	EA.	6.90	12.80	19.70
3/4"	"	10.65	12.80	23.45
1"	"	11.75	16.00	27.75
1-1/4"	"	20.15	17.80	37.95
1-1/2"	"	31.40	20.00	51.40
2"	"	35.80	22.90	58.70
Angle valve, bronze, 150 lb, threaded				
1/2"	EA.	39.20	11.45	50.65
3/4"	"	50.50	12.80	63.30
1"	"	73.00	12.80	85.80
1-1/4"	"	85.00	16.00	101.00
1-1/2"	"	110.00	17.80	127.80
Balancing valve, with meter connections, circuit setter				
1/2"	EA.	37.90	12.80	50.70
3/4"	"	40.30	14.55	54.85
1"	"	51.50	16.00	67.50
1-1/4"	"	71.50	17.80	89.30
1-1/2"	"	84.00	21.35	105.35
2"	"	120.00	26.70	146.70
2-1/2"	"	240.00	32.00	272.00
3"	"	350.00	40.10	390.10

PLUMBING	UNIT	MAT.	INST.	TOTAL
15430.60 VALVES				
4"	EA.	500.00	53.50	553.50
Balancing valve, straight type				
1/2"	EA.	9.90	12.80	22.70
3/4"	"	11.75	12.80	24.55
Angle type				
1/2"	EA.	13.40	12.80	26.20
3/4"	"	17.65	12.80	30.45
Square head cock, 125 lb, bronze body				
1/2"	EA.	7.90	10.70	18.60
3/4"	"	9.20	12.80	22.00
1"	"	13.15	14.55	27.70
1-1/4"	"	17.95	16.00	33.95
Radiator temp control valve, with control and sensor				
1/2" valve	EA.	52.50	20.00	72.50
1" valve	"	61.50	20.00	81.50
Pressure relief valve, 1/2", bronze				
Low pressure	EA.	11.95	12.80	24.75
High pressure	"	13.15	12.80	25.95
Press and temperature relief valve				
Bronze, 3/4"	EA.	43.90	12.80	56.70
Cast iron, 3/4"				
High pressure	EA.	25.20	12.80	38.00
Temperature relief	"	32.60	12.80	45.40
Pressure & temp relief valve	"	32.60	12.80	45.40
Pressure reducing valve, bronze, threaded, 250 lb				
1/2"	EA.	210.00	20.00	230.00
3/4"	"	210.00	20.00	230.00
1"	"	240.00	20.00	260.00
1-1/4"	"	280.00	22.90	302.90
1-1/2"	"	320.00	26.70	346.70
Pressure regulating valve, bronze, class 300				
1"	EA.	450.00	20.00	470.00
1-1/2"	"	590.00	24.65	614.65
2"	"	700.00	32.00	732.00
3"	"	770.00	45.80	815.80
4"	"	1,030	64.00	1,094
5"	"	1,430	80.00	1,510
6"	"	1,440	110.00	1,550
Solar water temperature regulating valve				
3/4"	EA.	270.00	26.70	296.70
1"	"	280.00	32.00	312.00
1-1/4"	"	300.00	35.60	335.60
1-1/2"	"	340.00	40.10	380.10
2"	"	420.00	45.80	465.80
2-1/2"	"	840.00	80.00	920.00
Tempering valve, threaded				
3/4"	EA.	46.60	10.70	57.30
1"	"	120.00	12.80	132.80
1-1/4"	"	140.00	16.00	156.00
1-1/2"	"	190.00	16.00	206.00
2"	"	230.00	20.00	250.00
2-1/2"	"	290.00	26.70	316.70

PLUMBING	UNIT	MAT.	INST.	TOTAL
15430.60 VALVES				
3"	EA.	360.00	32.00	392.00
4"	"	830.00	45.80	875.80
Thermostatic mixing valve, threaded				
1/2"	EA.	55.50	11.45	66.95
3/4"	"	58.50	12.80	71.30
1"	"	210.00	13.95	223.95
1-1/2"	"	220.00	16.00	236.00
2"	"	290.00	20.00	310.00
Sweat connection				
1/2"	EA.	65.50	11.45	76.95
3/4"	"	75.50	12.80	88.30
Mixing valve, sweat connection				
1/2"	EA.	25.70	11.45	37.15
3/4"	"	28.50	12.80	41.30
Liquid level gauge, aluminum body				
3/4"	EA.	210.00	12.80	222.80
125 psi, pvc body				
3/4"	EA.	210.00	12.80	222.80
150 psi, crs body				
3/4"	EA.	160.00	12.80	172.80
1"	"	180.00	12.80	192.80
175 psi, bronze body, 1/2"	"	330.00	11.45	341.45
15430.65 VACUUM BREAKERS				
Vacuum breaker, atmospheric, threaded connection				
3/4"	EA.	21.30	12.80	34.10
1"	"	32.50	12.80	45.30
Anti-siphon, brass				
3/4"	EA.	23.50	12.80	36.30
1"	"	32.50	12.80	45.30
1-1/4"	"	52.50	16.00	68.50
1-1/2"	"	62.50	17.80	80.30
2"	"	89.50	20.00	109.50
Air eliminators, purger, cast iron, threaded				
1"	EA.	17.90	12.80	30.70
1-1/4"	"	19.05	16.00	35.05
1-1/2"	"	37.00	16.00	53.00
2"	"	42.60	20.00	62.60
2-1/2"	"	110.00	40.10	150.10
3"	"	110.00	53.50	163.50
Airtrol fitting, 3/4"	"	24.65	12.80	37.45
Air eliminator, air vents, 1/4"	"	15.00	12.80	27.80
Air vent for hot water	"	12.45	11.45	23.90
15430.68 STRAINERS				
Strainer, Y pattern, 125 psi, cast iron body, threaded				
3/4"	EA.	10.10	11.45	21.55
1"	"	12.70	12.80	25.50
1-1/4"	"	17.35	16.00	33.35
1-1/2"	"	21.75	16.00	37.75
2"	"	33.10	20.00	53.10

PLUMBING	UNIT	MAT.	INST.	TOTAL
15430.68 STRAINERS				
250 psi, brass body, threaded				
3/4"	EA.	24.75	12.80	37.55
1"	"	35.20	12.80	48.00
1-1/4"	"	44.50	16.00	60.50
1-1/2"	"	64.00	16.00	80.00
2"	"	110.00	20.00	130.00
Cast iron body, threaded				
3/4"	EA.	16.35	12.80	29.15
1"	"	20.80	12.80	33.60
1-1/4"	"	27.80	16.00	43.80
1-1/2"	"	36.80	16.00	52.80
2"	"	46.70	20.00	66.70
15430.70 DRAINS, ROOF & FLOOR				
Floor drain, cast iron, with cast iron top				
2"	EA.	44.80	26.70	71.50
3"	"	68.00	26.70	94.70
4"	"	89.50	26.70	116.20
6"	"	160.00	32.00	192.00
Roof drain, cast iron				
2"	EA.	120.00	26.70	146.70
3"	"	130.00	26.70	156.70
4"	"	150.00	26.70	176.70
5"	"	210.00	32.00	242.00
6"	"	210.00	32.00	242.00
15430.80 TRAPS				
Bucket trap, threaded				
3/4"	EA.	78.50	20.00	98.50
1"	"	280.00	21.35	301.35
1-1/4"	"	330.00	24.65	354.65
1-1/2"	"	500.00	29.10	529.10
Inverted bucket steam trap, threaded				
3/4"	EA.	73.00	20.00	93.00
1"	"	240.00	20.00	260.00
1-1/4"	"	330.00	17.80	347.80
1-1/2"	"	350.00	26.70	376.70
With stainless interior				
1/2"	EA.	56.00	20.00	76.00
3/4"	"	89.50	20.00	109.50
1"	"	180.00	20.00	200.00
1-1/4"	"	220.00	22.90	242.90
Brass interior				
3/4"	EA.	89.50	20.00	109.50
1"	"	240.00	21.35	261.35
1-1/4"	"	290.00	22.90	312.90
Cast steel body, threaded, high temperature				
3/4"	EA.	340.00	20.00	360.00
1"	"	470.00	22.90	492.90
1-1/4	"	710.00	24.65	734.65
1-1/2"	"	1,330	26.70	1,357

PLUMBING	UNIT	MAT.	INST.	TOTAL
15430.80 TRAPS				
2"	EA.	1,330	32.00	1,362
Float trap, 15 psi				
3/4"	EA.	78.50	20.00	98.50
1"	"	95.00	21.35	116.35
1-1/4"	"	110.00	22.90	132.90
1-1/2"	"	150.00	26.70	176.70
2"	"	260.00	32.00	292.00
30 psi				
3/4"	EA.	89.50	20.00	109.50
1"	"	100.00	21.35	121.35
1-1/4"	"	180.00	26.70	206.70
1-1/2"	"	290.00	32.00	322.00
75 psi				
3/4"	EA.	110.00	20.00	130.00
1"	"	120.00	21.35	141.35
1-1/4"	"	170.00	22.90	192.90
1-1/2"	"	250.00	26.70	276.70
125 psi				
3/4"	EA.	100.00	20.00	120.00
1"	"	110.00	21.35	131.35
1-1/4	"	160.00	22.90	182.90
1-1/2	"	290.00	26.70	316.70
Float and thermostatic trap, 15 psi				
3/4"	EA.	73.00	20.00	93.00
1"	"	67.00	21.35	88.35
1-1/4"	"	100.00	22.90	122.90
1-1/2"	"	150.00	26.70	176.70
2"	"	260.00	32.00	292.00
30 psi				
3/4"	EA.	95.00	20.00	115.00
1"	"	89.50	21.35	110.85
1-1/4"	"	110.00	22.90	132.90
1-1/2"	"	180.00	26.70	206.70
75 psi				
3/4"	EA.	110.00	20.00	130.00
1"	"	120.00	21.35	141.35
1-1/4"	"	180.00	22.90	202.90
1-1/2"	"	300.00	26.70	326.70
Steam trap, cast iron body, threaded, 125 psi				
3/4"	EA.	110.00	20.00	130.00
1"	"	120.00	21.35	141.35
1-1/4"	"	180.00	22.90	202.90
1-1/2"	"	290.00	26.70	316.70
Thermostatic trap, low pressure, angle type, 25 psi				
1/2"	EA.	31.40	20.00	51.40
3/4"	"	54.00	20.00	74.00
1"	"	69.50	21.35	90.85
50 psi				
1/2"	EA.	51.50	20.00	71.50
3/4"	"	67.00	20.00	87.00
1"	"	80.50	21.35	101.85
Cast iron body, threaded, 125 psi				

PLUMBING	UNIT	MAT.	INST.	TOTAL
15430.80 TRAPS				
3/4"	EA.	77.50	20.00	97.50
1"	"	94.00	22.90	116.90
1-1/4"	"	120.00	24.65	144.65
1-1/2"	"	170.00	26.70	196.70
Low pressure, 25 psi, swivel type, 1/2"	"	33.70	20.00	53.70
Straightway type, 3/4"	"	46.60	20.00	66.60
Vertical type, 1/2"	"	31.30	20.00	51.30
Medium pressure, 50 psi, angle type, 1/2"	"	42.80	20.00	62.80
High pressure, 125 psi, angle type				
1/2"	EA.	56.00	20.00	76.00
3/4"	"	73.00	20.00	93.00
1"	"	110.00	22.90	132.90
Straightway type				
1/2"	EA.	58.00	20.00	78.00
Thermo disc trap				
3/4"	EA.	88.50	20.00	108.50
1"	"	100.00	21.35	121.35
Drip pan ell, cast iron				
2-1/2"	EA.	93.00	40.10	133.10
3"	"	100.00	45.80	145.80
4"	"	140.00	53.50	193.50
5"	"	180.00	64.00	244.00
6"	"	280.00	80.00	360.00
8"	"	310.00	110.00	420.00
Steel				
2-1/2"	EA.	480.00	40.10	520.10
3"	"	500.00	45.80	545.80
4"	"	650.00	53.50	703.50
5"	"	1,050	64.00	1,114
6"	"	1,050	80.00	1,130
8"	"	1,500	110.00	1,610

PLUMBING FIXTURES	UNIT	MAT.	INST.	TOTAL
15440.10 BATHS				
Bath tub, 5' long				
Minimum	EA.	330.00	110.00	440.00
Average	"	720.00	160.00	880.00
Maximum	"	1,230	320.00	1,550
6' long				
Minimum	EA.	380.00	110.00	490.00
Average	"	770.00	160.00	930.00
Maximum	"	1,340	320.00	1,660

PLUMBING FIXTURES	UNIT	MAT.	INST.	TOTAL
15440.10 BATHS				
Square tub, whirlpool, 4'x4'				
Minimum	EA.	1,140	160.00	1,300
Average	"	1,570	320.00	1,890
Maximum	"	3,250	400.00	3,650
5'x5'				
Minimum	EA.	1,160	160.00	1,320
Average	"	1,790	320.00	2,110
Maximum	"	3,360	400.00	3,760
6'x6'				
Minimum	EA.	1,350	160.00	1,510
Average	"	1,960	320.00	2,280
Maximum	"	4,030	400.00	4,430
For trim and rough-in				
Minimum	EA.	100.00	110.00	210.00
Average	"	150.00	160.00	310.00
Maximum	"	220.00	320.00	540.00
15440.12 DISPOSALS & ACCESSORIES				
Continuous feed				
Minimum	EA.	67.00	64.00	131.00
Average	"	120.00	80.00	200.00
Maximum	"	240.00	110.00	350.00
Batch feed, 1/2 hp				
Minimum	EA.	250.00	64.00	314.00
Average	"	350.00	80.00	430.00
Maximum	"	580.00	110.00	690.00
Hot water dispenser				
Minimum	EA.	100.00	64.00	164.00
Average	"	130.00	80.00	210.00
Maximum	"	290.00	110.00	400.00
Epoxy finish faucet	"	170.00	64.00	234.00
Lock stop assembly	"	36.90	40.10	77.00
Mounting gasket	"	4.25	26.70	30.95
Tailpipe gasket	"	0.63	26.70	27.33
Stopper assembly	"	14.60	32.00	46.60
Switch assembly, on/off	"	16.75	53.50	70.25
Tailpipe gasket washer	"	0.31	16.00	16.31
Stop gasket	"	1.50	17.80	19.30
Tailpipe flange	"	0.17	16.00	16.17
Tailpipe	"	1.95	20.00	21.95
15440.15 FAUCETS				
Kitchen				
Minimum	EA.	88.00	53.50	141.50
Average	"	140.00	64.00	204.00
Maximum	"	180.00	80.00	260.00
Bath				
Minimum	EA.	140.00	53.50	193.50
Average	"	210.00	64.00	274.00
Maximum	"	240.00	80.00	320.00
Lavatory, domestic				
Minimum	EA.	90.50	53.50	144.00

PLUMBING FIXTURES	UNIT	MAT.	INST.	TOTAL
15440.15 FAUCETS				
Average	EA.	190.00	64.00	254.00
Maximum	"	290.00	80.00	370.00
Hospital, patient rooms				
Minimum	EA.	220.00	80.00	300.00
Average	"	320.00	110.00	430.00
Maximum	"	410.00	160.00	570.00
Operating room				
Minimum	EA.	260.00	80.00	340.00
Average	"	340.00	110.00	450.00
Maximum	"	500.00	160.00	660.00
Washroom				
Minimum	EA.	130.00	53.50	183.50
Average	"	210.00	64.00	274.00
Maximum	"	310.00	80.00	390.00
Handicapped				
Minimum	EA.	160.00	64.00	224.00
Average	"	230.00	80.00	310.00
Maximum	"	350.00	110.00	460.00
Shower				
Minimum	EA.	110.00	53.50	163.50
Average	"	200.00	64.00	264.00
Maximum	"	310.00	80.00	390.00
For trim and rough-in				
Minimum	EA.	47.00	64.00	111.00
Average	"	70.50	80.00	150.50
Maximum	"	110.00	160.00	270.00
15440.18 HYDRANTS				
Wall hydrant				
8" thick	EA.	67.00	53.50	120.50
12" thick	"	78.50	64.00	142.50
18" thick	"	89.50	71.00	160.50
24" thick	"	100.00	80.00	180.00
Ground hydrant				
2' deep	EA.	170.00	40.10	210.10
4' deep	"	180.00	45.80	225.80
6' deep	"	210.00	53.50	263.50
8' deep	"	250.00	80.00	330.00
15440.20 LAVATORIES				
Lavatory, counter top, porcelain enamel on cast iron				
Minimum	EA.	120.00	64.00	184.00
Average	"	180.00	80.00	260.00
Maximum	"	270.00	110.00	380.00
Wall hung, china				
Minimum	EA.	140.00	64.00	204.00
Average	"	190.00	80.00	270.00
Maximum	"	470.00	110.00	580.00
Handicapped				
Minimum	EA.	280.00	80.00	360.00
Average	"	320.00	110.00	430.00
Maximum	"	490.00	160.00	650.00

PLUMBING FIXTURES	UNIT	MAT.	INST.	TOTAL
15440.20 LAVATORIES				
For trim and rough-in				
Minimum	EA.	110.00	80.00	190.00
Average	"	140.00	110.00	250.00
Maximum	"	240.00	160.00	400.00
15440.30 SHOWERS				
Shower, fiberglass, 36"x34"x84"				
Minimum	EA.	460.00	230.00	690.00
Average	"	640.00	320.00	960.00
Maximum	"	920.00	320.00	1,240
Steel, 1 piece, 36"x36"				
Minimum	EA.	420.00	230.00	650.00
Average	"	640.00	320.00	960.00
Maximum	"	750.00	320.00	1,070
Receptor, molded stone, 36"x36"				
Minimum	EA.	140.00	110.00	250.00
Average	"	240.00	160.00	400.00
Maximum	"	370.00	270.00	640.00
For trim and rough-in				
Minimum	EA.	110.00	150.00	260.00
Average	"	140.00	180.00	320.00
Maximum	"	190.00	320.00	510.00
15440.40 SINKS				
Service sink, 24"x29"				
Minimum	EA.	420.00	80.00	500.00
Average	"	530.00	110.00	640.00
Maximum	"	710.00	160.00	870.00
Kitchen sink, single, stainless steel, single bowl				
Minimum	EA.	120.00	64.00	184.00
Average	"	160.00	80.00	240.00
Maximum	"	230.00	110.00	340.00
Double bowl				
Minimum	EA.	160.00	80.00	240.00
Average	"	190.00	110.00	300.00
Maximum	"	250.00	160.00	410.00
Porcelain enamel, cast iron, single bowl				
Minimum	EA.	120.00	64.00	184.00
Average	"	150.00	80.00	230.00
Maximum	"	200.00	110.00	310.00
Double bowl				
Minimum	EA.	170.00	80.00	250.00
Average	"	190.00	110.00	300.00
Maximum	"	260.00	160.00	420.00
Mop sink, 24"x36"x10"				
Minimum	EA.	310.00	64.00	374.00
Average	"	400.00	80.00	480.00
Maximum	"	490.00	110.00	600.00
Washing machine box				
Minimum	EA.	110.00	80.00	190.00
Average	"	140.00	110.00	250.00
Maximum	"	190.00	160.00	350.00

PLUMBING FIXTURES	UNIT	MAT.	INST.	TOTAL
15440.40 **SINKS**				
For trim and rough-in				
Minimum	EA.	160.00	110.00	270.00
Average	"	240.00	160.00	400.00
Maximum	"	290.00	210.00	500.00
15440.50 **URINALS**				
Urinal, flush valve, floor mounted				
Minimum	EA.	440.00	80.00	520.00
Average	"	540.00	110.00	650.00
Maximum	"	710.00	160.00	870.00
Wall mounted				
Minimum	EA.	130.00	80.00	210.00
Average	"	330.00	110.00	440.00
Maximum	"	420.00	160.00	580.00
For trim and rough-in				
Minimum	EA.	82.50	80.00	162.50
Average	"	130.00	160.00	290.00
Maximum	"	150.00	210.00	360.00
15440.60 **WATER CLOSETS**				
Water closet flush tank, floor mounted				
Minimum	EA.	170.00	80.00	250.00
Average	"	430.00	110.00	540.00
Maximum	"	840.00	160.00	1,000
Handicapped				
Minimum	EA.	250.00	110.00	360.00
Average	"	440.00	160.00	600.00
Maximum	"	670.00	320.00	990.00
Bowl, with flush valve, floor mounted				
Minimum	EA.	300.00	80.00	380.00
Average	"	340.00	110.00	450.00
Maximum	"	650.00	160.00	810.00
Wall mounted				
Minimum	EA.	300.00	80.00	380.00
Average	"	360.00	110.00	470.00
Maximum	"	680.00	160.00	840.00
For trim and rough-in				
Minimum	EA.	100.00	80.00	180.00
Average	"	120.00	110.00	230.00
Maximum	"	150.00	160.00	310.00
15440.70 **WATER HEATERS**				
Water heater, electric				
6 gal	EA.	190.00	53.50	243.50
10 gal	"	200.00	53.50	253.50
15 gal	"	210.00	53.50	263.50
20 gal	"	240.00	64.00	304.00
30 gal	"	250.00	64.00	314.00
40 gal	"	260.00	64.00	324.00
52 gal	"	290.00	80.00	370.00
66 gal	"	450.00	80.00	530.00
80 gal	"	490.00	80.00	570.00

PLUMBING FIXTURES	UNIT	MAT.	INST.	TOTAL
15440.70 — WATER HEATERS				
100 gal	EA.	600.00	110.00	710.00
120 gal	"	770.00	110.00	880.00
Oil fired				
20 gal	EA.	770.00	160.00	930.00
50 gal	"	1,110	230.00	1,340
15440.90 — MISCELLANEOUS FIXTURES				
Electric water cooler				
Floor mounted	EA.	460.00	110.00	570.00
Wall mounted	"	560.00	110.00	670.00
Wash fountain				
Wall mounted	EA.	770.00	160.00	930.00
Circular, floor supported	"	1,460	320.00	1,780
Deluge shower and eye wash	"	550.00	160.00	710.00
15440.95 — FIXTURE CARRIERS				
Water fountain, wall carrier				
Minimum	EA.	89.50	32.00	121.50
Average	"	110.00	40.10	150.10
Maximum	"	130.00	53.50	183.50
Lavatory, wall carrier				
Minimum	EA.	89.50	32.00	121.50
Average	"	110.00	40.10	150.10
Maximum	"	130.00	53.50	183.50
Sink, industrial, wall carrier				
Minimum	EA.	95.00	32.00	127.00
Average	"	120.00	40.10	160.10
Maximum	"	140.00	53.50	193.50
Toilets, water closets, wall carrier				
Minimum	EA.	120.00	32.00	152.00
Average	"	130.00	40.10	170.10
Maximum	"	180.00	53.50	233.50
Floor support				
Minimum	EA.	56.00	26.70	82.70
Average	"	67.00	32.00	99.00
Maximum	"	78.50	40.10	118.60
Urinals, wall carrier				
Minimum	EA.	73.00	32.00	105.00
Average	"	95.00	40.10	135.10
Maximum	"	120.00	53.50	173.50
Floor support				
Minimum	EA.	61.50	26.70	88.20
Average	"	89.50	32.00	121.50
Maximum	"	110.00	40.10	150.10
15450.30 — PUMPS				
In-line pump, bronze, centrifugal				
5 gpm, 20' head	EA.	430.00	20.00	450.00
20 gpm, 40' head	"	800.00	20.00	820.00
50 gpm				

PLUMBING FIXTURES	**UNIT**	**MAT.**	**INST.**	**TOTAL**
15450.30 PUMPS				
50' head	EA.	1,060	40.10	1,100
100' head	"	1,590	40.10	1,630
70 gpm, 100' head	"	1,590	53.50	1,644
100 gpm, 80' head	"	1,590	53.50	1,644
250 gpm, 150' head	"	4,990	80.00	5,070
Cast iron, centrifugal				
50 gpm, 200' head	EA.	2,320	40.10	2,360
100 gpm				
100' head	EA.	2,140	53.50	2,194
200' head	"	2,490	53.50	2,544
200 gpm				
100' head	EA.	2,370	80.00	2,450
200' head	"	3,210	80.00	3,290
Centrifugal, close coupled, c.i., single stage				
50 gpm, 100' head	EA.	940.00	40.10	980.10
100 gpm, 100' head	"	1,140	53.50	1,194
Base mounted				
50 gpm, 100' head	EA.	1,530	40.10	1,570
100 gpm, 50' head	"	1,690	53.50	1,744
200 gpm, 100' head	"	2,160	80.00	2,240
300 gpm, 175' head	"	2,800	80.00	2,880
Suction diffuser, flanged, strainer				
3" inlet, 2-1/2" outlet	EA.	240.00	40.10	280.10
3" outlet	"	250.00	40.10	290.10
4" inlet				
3" outlet	EA.	290.00	53.50	343.50
4" outlet	"	340.00	53.50	393.50
6" inlet				
4" outlet	EA.	390.00	64.00	454.00
5" outlet	"	480.00	64.00	544.00
6" Outlet	"	490.00	64.00	554.00
8" inlet				
6" outlet	EA.	530.00	80.00	610.00
8" outlet	"	930.00	80.00	1,010
10" inlet				
8" outlet	EA.	1,250	110.00	1,360
Vertical turbine				
Single stage, C.I., 3550 rpm, 200 gpm, 50'head	EA.	3,030	110.00	3,140
Multi stage, 3550 rpm				
50 gpm, 100' head	EA.	2,940	80.00	3,020
100 gpm				
100' head	EA.	3,170	80.00	3,250
200 gpm				
50' head	EA.	3,380	110.00	3,490
100' head	"	3,440	110.00	3,550
Bronze				
Single stage, 3550 rpm, 100 gpm, 50' head	EA.	2,870	80.00	2,950
Multi stage, 3550 rpm, 50 gpm, 100' head	"	3,050	80.00	3,130
100 gpm				
100' head	EA.	3,050	80.00	3,130
200 gpm				
50' head	EA.	3,050	110.00	3,160

PLUMBING FIXTURES	UNIT	MAT.	INST.	TOTAL
15450.30		PUMPS		
100' head	EA.	3,320	110.00	3,430
Sump pump, bronze, 1750 rpm, 25 gpm				
20' head	EA.	2,920	400.00	3,320
150' head	"	4,230	530.00	4,760
50 gpm				
100' head	EA.	3,050	400.00	3,450
100 gpm				
50' head	EA.	3,050	400.00	3,450
Condensate pump, simplex				
1000 sf EDR, 2 gpm	EA.	740.00	270.00	1,010
2000 sf EDR, 3 gpm	"	750.00	270.00	1,020
4000 sf EDR, 6 gpm	"	750.00	290.00	1,040
6000 sf EDR, 9 gpm	"	760.00	290.00	1,050
Duplex, bronze				
8000 sf EDR, 12 gpm	EA.	1,020	290.00	1,310
10,000 sf EDR, 15 gpm	"	1,030	400.00	1,430
15,000 sf EDR, 23 gpm	"	1,290	460.00	1,750
20,000 sf EDR, 30 gpm	"	1,530	640.00	2,170
25,000 sf EDR, 38 gpm	"	1,550	640.00	2,190
30,000 sf EDR, 45 gpm	"	2,320	710.00	3,030
40,000 sf EDR, 60 gpm	"	2,320	400.00	2,720
50,000 sf EDR, 75 gpm	"	2,320	460.00	2,780
75,000 sf EDR, 112 gpm	"	3,320	530.00	3,850
100,000 sf EDR, 150 gpm	"	3,430	800.00	4,230
15450.40		STORAGE TANKS		
Hot water storage tank, cement lined				
10 gallon	EA.	2,680	110.00	2,790
70 gallon	"	2,670	160.00	2,830
200 gallon	"	2,760	230.00	2,990
900 gallon	"	3,160	400.00	3,560
1100 gallon	"	3,650	400.00	4,050
2000 gallon	"	7,290	400.00	7,690
15480.10		SPECIAL SYSTEMS		
Air compressor, air cooled, two stage				
5.0 cfm, 175 psi	EA.	2,300	640.00	2,940
10 cfm, 175 psi	"	2,800	710.00	3,510
20 cfm, 175 psi	"	3,860	760.00	4,620
50 cfm, 125 psi	"	5,630	840.00	6,470
80 cfm, 125 psi	"	8,060	920.00	8,980
Single stage, 125 psi				
1.0 cfm	EA.	2,120	460.00	2,580
1.5 cfm	"	2,150	460.00	2,610
2.0 cfm	"	2,200	460.00	2,660
Automotive compressor, hose reel, air and water, 50' hose	"	990.00	270.00	1,260
Lube equipment, 3 reel, with pumps	"	5,150	1,280	6,430
Tire changer				
Truck	EA.	8,060	460.00	8,520
Passenger car	"	1,900	250.00	2,150
Air hose reel, includes, 50' hose	"	500.00	250.00	750.00

PLUMBING FIXTURES

PLUMBING FIXTURES	UNIT	MAT.	INST.	TOTAL
15480.10 — SPECIAL SYSTEMS				
Hose reel, 5 reel, motor oil, gear oil, lube, air & water	EA.	4,700	1,280	5,980
Water hose reel, 50' hose	"	500.00	250.00	750.00
Pump, air operated, for motor or gear oil, fits 55 gal drum	"	660.00	32.00	692.00
For chassis lube	"	1,080	32.00	1,112
Fuel dispensing pump, lighted dial, one product				
One hose	EA.	2,350	270.00	2,620
Two hose	"	4,140	270.00	4,410
Two products, two hose	"	4,370	270.00	4,640

HEATING & VENTILATING

HEATING & VENTILATING	UNIT	MAT.	INST.	TOTAL
15555.10 — BOILERS				
Cast iron, gas fired, hot water				
115 mbh	EA.	1,340	1,030	2,370
175 mbh	"	1,570	1,120	2,690
235 mbh	"	2,350	1,230	3,580
940 mbh	"	6,940	2,470	9,410
1600 mbh	"	9,520	3,090	12,610
3000 mbh	"	20,160	4,110	24,270
6000 mbh	"	44,800	6,170	50,970
Steam				
115 mbh	EA.	1,940	1,030	2,970
175 mbh	"	2,470	1,120	3,590
235 mbh	"	2,700	1,230	3,930
940 mbh	"	6,830	2,470	9,300
1600 mbh	"	12,320	3,090	15,410
3000 mbh	"	23,520	4,110	27,630
6000 mbh	"	43,680	6,170	49,850
Electric, hot water				
115 mbh	EA.	3,920	620.00	4,540
175 mbh	"	4,700	620.00	5,320
235 mbh	"	6,380	620.00	7,000
940 mbh	"	11,200	1,230	12,430
1600 mbh	"	16,800	2,470	19,270
3000 mbh	"	29,120	3,090	32,210
6000 mbh	"	43,680	4,110	47,790
Steam				
115 mbh	EA.	5,040	620.00	5,660
175 mbh	"	5,940	620.00	6,560
235 mbh	"	7,170	620.00	7,790
940 mbh	"	16,800	1,230	18,030
1600 mbh	"	28,000	2,470	30,470
3000 mbh	"	32,480	3,090	35,570
6000 mbh	"	50,400	4,110	54,510

HEATING & VENTILATING	UNIT	MAT.	INST.	TOTAL
15555.10 BOILERS				
Oil fired, hot water				
115 mbh	EA.	1,900	820.00	2,720
175 mbh	"	2,350	950.00	3,300
235 mbh	"	3,080	1,120	4,200
940 mbh	"	9,070	2,060	11,130
1600 mbh	"	15,740	2,470	18,210
3000 mbh	"	26,880	3,090	29,970
6000 mbh	"	47,040	6,170	53,210
Steam				
115 mbh	EA.	2,000	820.00	2,820
175 mbh	"	2,410	950.00	3,360
235 mbh	"	3,180	1,120	4,300
940 mbh	"	10,230	2,060	12,290
1600 mbh	"	16,700	2,470	19,170
3000 mbh	"	28,930	3,090	32,020
6000 mbh	"	47,600	6,170	53,770
15610.10 FURNACES				
Electric, hot air				
40 mbh	EA.	530.00	160.00	690.00
60 mbh	"	620.00	170.00	790.00
80 mbh	"	680.00	180.00	860.00
100 mbh	"	920.00	190.00	1,110
125 mbh	"	1,010	190.00	1,200
160 mbh	"	1,470	200.00	1,670
200 mbh	"	2,700	210.00	2,910
400 mbh	"	4,820	210.00	5,030
Gas fired hot air				
40 mbh	EA.	530.00	160.00	690.00
60 mbh	"	560.00	170.00	730.00
80 mbh	"	650.00	180.00	830.00
100 mbh	"	680.00	190.00	870.00
125 mbh	"	740.00	190.00	930.00
160 mbh	"	880.00	200.00	1,080
200 mbh	"	1,570	210.00	1,780
400 mbh	"	6,720	210.00	6,930
Oil fired hot air				
40 mbh	EA.	650.00	160.00	810.00
60 mbh	"	810.00	170.00	980.00
80 mbh	"	880.00	180.00	1,060
100 mbh	"	1,030	190.00	1,220
125 mbh	"	1,150	190.00	1,340
160 mbh	"	1,230	200.00	1,430
200 mbh	"	2,820	210.00	3,030
400 mbh	"	9,760	210.00	9,970

REFRIGERATION	UNIT	MAT.	INST.	TOTAL
15670.10		**CONDENSING UNITS**		
Air cooled condenser, single circuit				
3 ton	EA.	1,340	53.50	1,394
5 ton	"	2,240	53.50	2,294
7.5 ton	"	3,580	150.00	3,730
20 ton	"	7,560	160.00	7,720
25 ton	"	8,820	160.00	8,980
30 ton	"	10,160	160.00	10,320
40 ton	"	14,000	230.00	14,230
50 ton	"	17,020	230.00	17,250
60 ton	"	19,600	200.00	19,800
With low ambient dampers				
3 ton	EA.	1,570	80.00	1,650
5 ton	"	2,460	80.00	2,540
7.5 ton	"	3,810	160.00	3,970
20 ton	"	7,840	210.00	8,050
25 ton	"	9,070	210.00	9,280
30 ton	"	10,420	210.00	10,630
40 ton	"	14,670	270.00	14,940
50 ton	"	17,700	290.00	17,990
60 ton	"	20,270	290.00	20,560
Dual circuit				
10 ton	EA.	3,360	160.00	3,520
15 ton	"	4,930	230.00	5,160
20 ton	"	8,400	230.00	8,630
25 ton	"	9,630	230.00	9,860
30 ton	"	11,200	230.00	11,430
40 ton	"	16,020	270.00	16,290
50 ton	"	17,700	270.00	17,970
60 ton	"	18,370	270.00	18,640
80 ton	"	22,960	360.00	23,320
100 ton	"	26,880	360.00	27,240
120 ton	"	31,920	360.00	32,280
With low ambient dampers				
15 ton	EA.	5,490	230.00	5,720
20 ton	"	8,960	230.00	9,190
25 ton	"	10,190	230.00	10,420
30 ton	"	10,980	230.00	11,210
40 ton	"	16,580	270.00	16,850
50 ton	"	18,260	270.00	18,530
60 ton	"	18,930	270.00	19,200
80 ton	"	24,080	360.00	24,440
100 ton	"	28,000	360.00	28,360
120 ton	"	33,380	360.00	33,740
15680.10		**CHILLERS**		
Chiller, reciprocal				
Air cooled, remote condenser, starter				
20 ton	EA.	15,680	410.00	16,090
25 ton	"	17,700	410.00	18,110
30 ton	"	18,700	410.00	19,110
40 ton	"	27,660	620.00	28,280

REFRIGERATION	UNIT	MAT.	INST.	TOTAL
15680.10 CHILLERS				
50 ton	EA.	30,130	690.00	30,820
60 ton	"	34,500	730.00	35,230
80 ton	"	42,560	1,120	43,680
100 ton	"	50,400	1,230	51,630
120 ton	"	57,120	1,370	58,490
150 ton	"	72,800	1,540	74,340
180 ton	"	83,440	1,760	85,200
200 ton	"	91,840	2,060	93,900
Water cooled, with starter				
20 ton	EA.	13,440	410.00	13,850
25 ton	"	14,900	410.00	15,310
30 ton	"	17,920	620.00	18,540
40 ton	"	24,640	620.00	25,260
50 ton	"	26,880	690.00	27,570
60 ton	"	29,120	730.00	29,850
80 ton	"	33,600	1,120	34,720
100 ton	"	39,200	1,230	40,430
120 ton	"	45,920	1,370	47,290
150 ton	"	57,120	1,540	58,660
180 ton	"	59,360	1,760	61,120
200 ton	"	62,720	2,060	64,780
Packaged, air cooled, with starter				
20 ton	EA.	15,010	310.00	15,320
25 ton	"	16,240	310.00	16,550
30 ton	"	18,930	310.00	19,240
40 ton	"	21,730	310.00	22,040
50 ton	"	24,530	410.00	24,940
60 ton	"	28,900	410.00	29,310
80 ton	"	33,710	620.00	34,330
100 ton	"	39,540	620.00	40,160
120 ton	"	44,910	620.00	45,530
Heat recovery, air cooled, with starter				
40 ton	EA.	25,200	620.00	25,820
50 ton	"	29,120	620.00	29,740
60 ton	"	31,920	820.00	32,740
75 ton	"	36,510	1,230	37,740
100 ton	"	41,660	1,230	42,890
Water cooled, with starter				
40 ton	EA.	24,860	620.00	25,480
50 ton	"	29,680	620.00	30,300
60 ton	"	32,700	820.00	33,520
75 ton	"	39,760	1,230	40,990
100 ton	"	44,910	1,370	46,280
Centrifugal, single bundle condenser, with starter				
80 ton	EA.	58,800	1,760	60,560
130 ton	"	61,940	2,060	64,000
160 ton	"	62,940	2,240	65,180
180 ton	"	67,420	2,470	69,890
230 ton	"	70,450	2,740	73,190
280 ton	"	77,620	3,090	80,710
360 ton	"	86,580	3,090	89,670
460 ton	"	106,740	4,110	110,850

REFRIGERATION		UNIT	MAT.	INST.	TOTAL
15680.10	CHILLERS				
560 ton		EA.	11,680	4,410	16,090
670 ton		"	14,240	4,940	19,180
15710.10	COOLING TOWERS				
Cooling tower, propeller type					
100 ton		EA.	6,850	410.00	7,260
200 ton		"	11,420	620.00	12,040
300 ton		"	17,140	1,030	18,170
400 ton		"	22,850	1,230	24,080
600 ton		"	34,270	1,760	36,030
800 ton		"	45,700	2,470	48,170
1000 ton		"	54,840	3,090	57,930
Centrifugal					
100 ton		EA.	6,850	410.00	7,260
200 ton		"	13,710	620.00	14,330
300 ton		"	20,560	1,030	21,590
400 ton		"	27,420	1,230	28,650
600 ton		"	41,130	1,760	42,890
800 ton		"	54,840	2,470	57,310
1000 ton		"	63,970	3,090	67,060

HEAT TRANSFER		UNIT	MAT.	INST.	TOTAL
15780.10	COMPUTER ROOM A/C				
Air cooled, alarm, high efficiency filter, elec. heat					
3 ton		EA.	8,110	250.00	8,360
5 ton		"	8,700	270.00	8,970
7.5 ton		"	14,820	320.00	15,140
10 ton		"	15,290	400.00	15,690
15 ton		"	16,930	460.00	17,390
Steam heat					
3 ton		EA.	9,410	250.00	9,660
5 ton		"	10,000	270.00	10,270
7.5 ton		"	15,880	320.00	16,200
10 ton		"	16,350	400.00	16,750
15 ton		"	18,110	460.00	18,570
Hot water heat					
3 ton		EA.	9,410	250.00	9,660
5 ton		"	10,000	270.00	10,270
7.5 ton		"	15,880	320.00	16,200
10 ton		"	16,350	400.00	16,750
15 ton		"	18,110	460.00	18,570
Air cooled condenser, low ambient damper					
3 ton		EA.	2,120	64.00	2,184

HEAT TRANSFER	UNIT	MAT.	INST.	TOTAL
15780.10 COMPUTER ROOM A/C				
5 ton	EA.	2,350	80.00	2,430
7.5 ton	"	3,410	160.00	3,570
10 ton	"	3,410	230.00	3,640
15 ton	"	3,760	190.00	3,950
Water cooled, high efficiency filter, alarm, elec. heat				
3 ton	EA.	9,290	230.00	9,520
5 ton	"	10,000	270.00	10,270
7.5 ton	"	15,880	400.00	16,280
10 ton	"	16,460	460.00	16,920
15 ton	"	19,290	530.00	19,820
Steam heat				
3 ton	EA.	10,580	230.00	10,810
5 ton	"	12,100	270.00	12,370
7.5 ton	"	17,050	400.00	17,450
10 ton	"	17,640	460.00	18,100
15 ton	"	20,460	530.00	20,990
Hot water heat				
3 ton	EA.	10,580	230.00	10,810
5 ton	"	11,290	270.00	11,560
7.5 ton	"	17,050	400.00	17,450
10 ton	"	17,640	460.00	18,100
15 ton	"	20,460	530.00	20,990
Chilled water, alarm, high eff. filter, elec. heat				
7.5 ton	EA.	7,410	290.00	7,700
10 ton	"	7,760	360.00	8,120
15 ton	"	8,820	400.00	9,220
Steam heat				
7.5 ton	EA.	8,470	290.00	8,760
10 ton	"	8,820	360.00	9,180
15 ton	"	9,880	400.00	10,280
Hot water heat				
7.5 ton	EA.	8,470	290.00	8,760
10 ton	"	8,820	360.00	9,180
15 ton	"	9,880	400.00	10,280
15780.20 ROOFTOP UNITS				
Packaged, single zone rooftop unit, with roof curb				
2 ton	EA.	2,350	320.00	2,670
3 ton	"	2,580	320.00	2,900
4 ton	"	3,470	400.00	3,870
5 ton	"	3,810	530.00	4,340
7.5 ton	"	5,320	640.00	5,960
15820.10 DEHUMIDIFIERS				
Dessicant dehumidifier, 1125 cfm	EA.			11,760
15830.10 RADIATION UNITS				
Baseboard radiation unit				
1.7 mbh/lf	L.F.	41.40	12.80	54.20
2.1 mbh/lf	"	54.50	16.00	70.50

HEAT TRANSFER	UNIT	MAT.	INST.	TOTAL
15830.10 — RADIATION UNITS				
Enclosure only				
Two tier	L.F.	21.30	5.35	26.65
Three tier	"	28.00	5.35	33.35
Copper element only, 3/4" dia.				
Two tier	L.F.	26.90	8.00	34.90
Three tier	"	39.20	10.70	49.90
Fin-tube, 16 ga, sloping cover, 1-1/4" steel				
One tier	L.F.	30.80	10.70	41.50
Two tier	"	47.60	12.80	60.40
2" steel				
Two tier	L.F.	50.50	12.80	63.30
Three tier	"	70.50	16.00	86.50
1-1/4" copper				
Two tier	L.F.	52.50	10.70	63.20
18 ga flat cover, 1-1/4" steel				
One tier	L.F.	17.90	10.70	28.60
Two tier	"	32.50	12.80	45.30
Three tier	"	51.50	16.00	67.50
2" steel				
One tier	L.F.	22.40	10.70	33.10
Two tier	"	37.00	12.80	49.80
Three tier	"	52.50	16.00	68.50
1-1/4" copper				
One tier	L.F.	20.15	10.70	30.85
Two tier	"	35.80	12.80	48.60
Three tier	"	47.00	16.00	63.00
15830.20 — FAN COIL UNITS				
Fan coil unit, 2 pipe, complete				
200 cfm ceiling hung	EA.	560.00	110.00	670.00
Floor mounted	"	530.00	80.00	610.00
300 cfm, ceiling hung	"	600.00	130.00	730.00
Floor mounted	"	570.00	110.00	680.00
400 cfm, ceiling hung	"	630.00	150.00	780.00
Floor mounted	"	600.00	110.00	710.00
500 cfm, ceiling hung	"	730.00	160.00	890.00
Floor mounted	"	700.00	120.00	820.00
600 cfm, ceiling hung	"	930.00	180.00	1,110
Floor mounted	"	860.00	150.00	1,010
800 cfm, ceiling hung	"	1,080	200.00	1,280
Floor mounted	"	860.00	150.00	1,010
1000 cfm, ceiling hung	"	1,230	230.00	1,460
Floor mounted	"	1,350	170.00	1,520
1200 cfm ceiling hung	"	1,400	270.00	1,670
Floor mounted	"	1,470	200.00	1,670
15830.70 — UNIT HEATERS				
Steam unit heater, horizontal				
12,500 btuh, 200 cfm	EA.	190.00	53.50	243.50
17,000 btuh, 300 cfm	"	220.00	53.50	273.50
40,000 btuh, 500 cfm	"	260.00	53.50	313.50

HEAT TRANSFER	UNIT	MAT.	INST.	TOTAL
15830.70 — UNIT HEATERS				
60,000 btuh, 700 cfm	EA.	310.00	53.50	363.50
70,000 btuh, 1000 cfm	"	410.00	80.00	490.00
Vertical				
12,500 btuh, 200 cfm	EA.	210.00	53.50	263.50
17,000 btuh, 300 cfm	"	340.00	53.50	393.50
40,000 btuh, 500 cfm	"	440.00	53.50	493.50
60,000 btuh, 700 cfm	"	500.00	53.50	553.50
70,000 btuh, 1000 cfm	"	550.00	53.50	603.50
Gas unit heater, horizontal				
27,400 btuh	EA.	410.00	130.00	540.00
38,000 btuh	"	470.00	130.00	600.00
56,000 btuh	"	530.00	130.00	660.00
82,200 btuh	"	590.00	130.00	720.00
103,900 btuh	"	620.00	200.00	820.00
125,700 btuh	"	650.00	200.00	850.00
133,200 btuh	"	590.00	200.00	790.00
149,000 btuh	"	760.00	200.00	960.00
172,000 btuh	"	810.00	200.00	1,010
190,000 btuh	"	880.00	200.00	1,080
225,000 btuh	"	970.00	200.00	1,170
Hot water unit heater, horizontal				
12,500 btuh, 200 cfm	EA.	210.00	53.50	263.50
17,000 btuh, 300 cfm	"	270.00	53.50	323.50
25,000 btuh, 500 cfm	"	290.00	53.50	343.50
30,000 btuh, 700 cfm	"	340.00	53.50	393.50
50,000 btuh, 1000 cfm	"	380.00	80.00	460.00
60,000 btuh, 1300 cfm	"	480.00	80.00	560.00
Vertical				
12,500 btuh, 200 cfm	EA.	250.00	53.50	303.50
17,000 btuh, 300 cfm	"	320.00	53.50	373.50
25,000 btuh, 500 cfm	"	350.00	53.50	403.50
30,000 btuh, 700 cfm	"	390.00	53.50	443.50
50,000 btuh, 1000 cfm	"	440.00	53.50	493.50
60,000 btuh, 1300 cfm	"	480.00	53.50	533.50
Cabinet unit heaters, ceiling, exposed, hot water				
200 cfm	EA.	660.00	110.00	770.00
300 cfm	"	690.00	130.00	820.00
400 cfm	"	760.00	150.00	910.00
600 cfm	"	780.00	170.00	950.00
800 cfm	"	1,090	200.00	1,290
1000 cfm	"	1,180	230.00	1,410
1200 cfm	"	1,280	270.00	1,550
2000 cfm	"	2,290	360.00	2,650

AIR HANDLING	UNIT	MAT.	INST.	TOTAL
15855.10 AIR HANDLING UNITS				
Air handling unit, medium pressure, single zone				
1500 cfm	EA.	2,240	200.00	2,440
3000 cfm	"	2,800	360.00	3,160
4000 cfm	"	3,140	400.00	3,540
5000 cfm	"	3,580	430.00	4,010
6000 cfm	"	4,030	460.00	4,490
7000 cfm	"	4,480	490.00	4,970
8500 cfm	"	5,260	530.00	5,790
10,500 cfm	"	5,940	640.00	6,580
12,500 cfm	"	7,620	710.00	8,330
15,500 cfm	"	12,320	920.00	13,240
17,500 cfm	"	14,560	1,070	15,630
20,500 cfm	"	16,800	1,280	18,080
25,000 cfm	"	19,040	1,600	20,640
31,500 cfm	"	23,520	2,140	25,660
Rooftop air handling units				
4950 cfm	EA.	7,060	360.00	7,420
7370 cfm	"	8,060	460.00	8,520
9790 cfm	"	9,520	530.00	10,050
14,300 cfm	"	13,440	460.00	13,900
21,725 cfm	"	19,040	460.00	19,500
33,000 cfm	"	26,880	530.00	27,410
15870.20 EXHAUST FANS				
Belt drive roof exhaust fans				
640 cfm, 2618 fpm	EA.	670.00	40.10	710.10
940 cfm, 2604 fpm	"	870.00	40.10	910.10
1050 cfm, 3325 fpm	"	770.00	40.10	810.10
1170 cfm, 2373 fpm	"	1,120	40.10	1,160
2440 cfm, 4501 fpm	"	880.00	40.10	920.10
2760 cfm, 4950 fpm	"	970.00	40.10	1,010
3890 cfm, 6769 fpm	"	1,110	40.10	1,150
2380 cfm, 3382 fpm	"	1,230	40.10	1,270
2880 cfm, 3859 fpm	"	1,290	40.10	1,330
3200 cfm, 4173 fpm	"	1,300	53.50	1,354
3660 cfm, 3437 fpm	"	1,320	53.50	1,374
4070 cfm, 3694 fpm	"	1,680	53.50	1,734
5030 cfm, 3251 fpm	"	1,180	53.50	1,234
5830 cfm, 6932 fpm	"	1,620	64.00	1,684
6380 cfm, 3817 fpm	"	1,620	64.00	1,684
8460 cfm, 6721 fpm	"	1,570	64.00	1,634
10,970 cfm, 5906 fpm	"	1,930	80.00	2,010
12,470 cfm, 6620 fpm	"	2,200	110.00	2,310
7000 cfm, 3449 fpm	"	1,570	80.00	1,650
13,000 cfm, 5456 fpm	"	2,320	80.00	2,400
11,250 cfm, 4854 fpm	"	2,130	80.00	2,210
18,490 cfm, 7405 fpm	"	3,100	150.00	3,250
11,300 cfm, 3232 fpm	"	2,020	140.00	2,160
18,330 cfm, 4488 fpm	"	3,360	140.00	3,500
21,720 cfm, 5131 fpm	"	3,530	140.00	3,670
31,110 cfm, 6965 fpm	"	3,890	160.00	4,050
Direct drive fans				

AIR HANDLING	UNIT	MAT.	INST.	TOTAL
15870.20 EXHAUST FANS				
60 to 390 cfm	EA.	550.00	40.10	590.10
145 to 590 cfm	"	660.00	40.10	700.10
295 to 860 cfm	"	800.00	40.10	840.10
235 to 1300 cfm	"	860.00	40.10	900.10
415 to 1630 cfm	"	970.00	40.10	1,010
590 to 2045 cfm	"	1,120	40.10	1,160
805 cfm, 3235 fpm	"	740.00	40.10	780.10
1455 cfm, 4360 fpm	"	800.00	40.10	840.10
1385 cfm, 3655 fpm	"	820.00	40.10	860.10
2260 cfm, 4930 fpm	"	860.00	40.10	900.10
1720 cfm, 3870 fpm	"	830.00	40.10	870.10
2700 cfm, 5220 fpm	"	870.00	40.10	910.10
Terminal blenders and cooling				
400 cfm	EA.	300.00	64.00	364.00
800 cfm	"	320.00	64.00	384.00
1200 cfm	"	390.00	80.00	470.00
2000 cfm	"	450.00	80.00	530.00

AIR DISTRIBUTION	UNIT	MAT.	INST.	TOTAL
15890.10 METAL DUCTWORK				
Rectangular duct				
Galvanized steel				
Minimum	Lb.	1.80	2.90	4.70
Average	"	2.00	3.55	5.55
Maximum	"	2.15	5.35	7.50
Aluminum				
Minimum	Lb.	4.70	6.40	11.10
Average	"	5.15	8.00	13.15
Maximum	"	5.60	10.70	16.30
Fittings				
Minimum	EA.	2.70	10.70	13.40
Average	"	5.40	16.00	21.40
Maximum	"	13.45	32.00	45.45
For work				
10-20' high, add per pound, $.30				
30-50', add per pound, $.50				
15890.30 FLEXIBLE DUCTWORK				
Flexible duct, 1.25" fiberglass				
5" dia.	L.F.	1.35	1.60	2.95
6" dia.	"	1.45	1.80	3.25
7" dia.	"	1.70	1.90	3.60

AIR DISTRIBUTION	UNIT	MAT.	INST.	TOTAL
15890.30 — FLEXIBLE DUCTWORK				
8" dia.	L.F.	2.00	2.00	4.00
10" dia.	"	2.25	2.30	4.55
12" dia.	"	2.80	2.45	5.25
14" dia.	"	3.35	2.65	6.00
16" dia.	"	3.90	2.90	6.80
Flexible duct connector, 3" wide fabric	"	1.35	5.35	6.70
15895.10 — ROOF CURBS				
8" high, insulated, with liner and raised can				
15" x 15"	EA.	58.00	16.00	74.00
17" x 17"	"	62.50	16.00	78.50
19" x 19"	"	66.00	16.00	82.00
21" x 21"	"	73.00	16.00	89.00
25" x 25"	"	79.50	20.00	99.50
28" x 28"	"	84.00	21.35	105.35
32" x 32"	"	95.00	22.90	117.90
36" x 36"	"	110.00	22.90	132.90
40" x 40"	"	120.00	22.90	142.90
44" x 44"	"	130.00	24.65	154.65
48" x 48"	"	300.00	24.65	324.65
52" x 52"	"	360.00	26.70	386.70
56" x 56"	"	460.00	26.70	486.70
60" x 60"	"	550.00	32.00	582.00
64" x 64"	"	660.00	32.00	692.00
68" x 68"	"	760.00	35.60	795.60
72" x 72"	"	870.00	40.10	910.10
15910.10 — DAMPERS				
Horizontal parallel aluminum backdraft damper				
12" x 12"	EA.	29.40	8.00	37.40
16" x 16"	"	41.20	9.15	50.35
20" x 20"	"	53.00	11.45	64.45
24" x 24"	"	64.50	16.00	80.50
28" x 28"	"	89.50	17.80	107.30
32" x 32"	"	120.00	20.00	140.00
36" x 36"	"	150.00	22.90	172.90
40" x 40"	"	190.00	26.70	216.70
44" x 44"	"	220.00	29.10	249.10
48" x 48"	"	260.00	32.00	292.00
"Up", parallel dampers				
12" x 12"	EA.	51.50	8.00	59.50
16" x 16"	"	73.00	9.15	82.15
20" x 20"	"	80.00	11.45	91.45
24" x 24"	"	89.50	16.00	105.50
28" x 28"	"	130.00	17.80	147.80
32" x 32"	"	150.00	20.00	170.00
36" x 36"	"	150.00	22.90	172.90
40" x 40"	"	200.00	26.70	226.70
44" x 44"	"	250.00	29.10	279.10
48" x 48"	"	300.00	32.00	332.00
"Down", parallel dampers				
12" x 12"	EA.	51.50	8.00	59.50

AIR DISTRIBUTION	UNIT	MAT.	INST.	TOTAL
15910.10 DAMPERS				
16" x 16"	EA.	73.00	9.15	82.15
20" x 20"	"	80.00	11.45	91.45
24" x 24"	"	89.50	16.00	105.50
28" x 28"	"	130.00	17.80	147.80
32" x 32"	"	150.00	20.00	170.00
36" x 36"	"	150.00	22.90	172.90
40" x 40"	"	210.00	26.70	236.70
44" x 44"	"	250.00	29.10	279.10
48" x 48"	"	300.00	32.00	332.00
Fire damper, 1.5 hr rating				
12" x 12"	EA.	17.65	16.00	33.65
16" x 16"	"	28.20	16.00	44.20
20" x 20"	"	30.60	16.00	46.60
24" x 24"	"	35.30	16.00	51.30
28" x 28"	"	42.30	22.90	65.20
32" x 32"	"	49.40	26.70	76.10
36" x 36"	"	59.00	32.00	91.00
40" x 40"	"	70.50	35.60	106.10
44" x 44"	"	84.50	40.10	124.60
48" x 48"	"	110.00	45.80	155.80
15940.10 DIFFUSERS				
Ceiling diffusers, round, baked enamel finish				
6" dia.	EA.	29.40	10.70	40.10
8" dia.	"	35.30	13.35	48.65
10" dia.	"	41.20	13.35	54.55
12" dia.	"	53.00	13.35	66.35
14" dia.	"	66.00	14.55	80.55
16" dia.	"	76.50	14.55	91.05
18" dia.	"	88.00	16.00	104.00
20" dia.	"	110.00	16.00	126.00
Rectangular				
6x6"	EA.	29.10	10.70	39.80
9x9"	"	34.70	16.00	50.70
12x12"	"	51.50	16.00	67.50
15x15"	"	65.00	16.00	81.00
18x18"	"	82.50	16.00	98.50
21x21"	"	110.00	20.00	130.00
24x24"	"	140.00	20.00	160.00
Lay in, flush mounted, perforated face, with grid				
6x6/24x24	EA.	47.00	12.80	59.80
8x8/24x24	"	47.00	12.80	59.80
9x9/24x24	"	47.00	12.80	59.80
10x10/24x24	"	49.30	12.80	62.10
12x12/24x24	"	51.50	12.80	64.30
15x15/24x24	"	51.50	12.80	64.30
18x6/24x24	"	51.50	12.80	64.30
18x18/24x24	"	54.00	12.80	66.80
Two-way slot diffuser with balancing damper, 4'	"	56.00	32.00	88.00

AIR DISTRIBUTION	UNIT	MAT.	INST.	TOTAL
15940.20 RELIEF VENTILATORS				
Intake ventilator, aluminum, with screen, no curbs				
12" x 12"	EA.	110.00	26.70	136.70
16" x 16"	"	150.00	32.00	182.00
20" x 20"	"	250.00	32.00	282.00
30" x 30"	"	390.00	45.80	435.80
36" x 36"	"	590.00	53.50	643.50
42" x 42"	"	810.00	53.50	863.50
48" x 48"	"	960.00	64.00	1,024
15940.40 REGISTERS AND GRILLES				
Lay in flush mounted, perforated face, return				
6x6 / 24x24	EA.	24.70	12.80	37.50
8x8 / 24x24	"	24.70	12.80	37.50
9x9 / 24x24	"	24.70	12.80	37.50
10x10 / 24x24	"	24.70	12.80	37.50
12x12 / 24x24	"	24.70	12.80	37.50
Rectangular, ceiling return, single deflection				
10x10	EA.	11.20	16.00	27.20
12x12	"	13.45	16.00	29.45
14x14	"	16.80	16.00	32.80
16x8	"	13.45	16.00	29.45
16x16	"	13.45	16.00	29.45
18x8	"	15.70	16.00	31.70
20x20	"	26.90	16.00	42.90
24x12	"	38.10	16.00	54.10
24x18	"	50.50	16.00	66.50
36x24	"	100.00	17.80	117.80
36x30	"	150.00	17.80	167.80
Wall, return air register				
12x12	EA.	22.35	8.00	30.35
16x16	"	32.90	8.00	40.90
18x18	"	38.80	8.00	46.80
20x20	"	45.90	8.00	53.90
24x24	"	63.50	8.00	71.50
Ceiling, return air grille				
6x6	EA.	8.25	10.70	18.95
8x8	"	9.40	12.80	22.20
10x10	"	10.60	12.80	23.40
Ceiling, exhaust grille, aluminum egg crate				
6x6	EA.	10.60	10.70	21.30
8x8	"	10.60	12.80	23.40
10x10	"	11.75	12.80	24.55
12x12	"	15.30	16.00	31.30
14x14	"	20.00	16.00	36.00
16x16	"	23.50	16.00	39.50
18x18	"	28.20	16.00	44.20
15940.80 PENTHOUSE LOUVERS				
Penthouse louvers				
12" high, extruded aluminum, 4" louver				
6' perimeter	EA.	270.00	80.00	350.00
8' perimeter	"	360.00	80.00	440.00

AIR DISTRIBUTION

	UNIT	MAT.	INST.	TOTAL
15940.80 PENTHOUSE LOUVERS				
10' perimeter	EA.	450.00	80.00	530.00
12' perimeter	"	650.00	80.00	730.00
14' perimeter	"	800.00	110.00	910.00
16' perimeter	"	920.00	130.00	1,050
18' perimeter	"	1,080	180.00	1,260
20' perimeter	"	1,340	210.00	1,550
16" high x 4' perimeter	"	220.00	80.00	300.00
6' perimeter	"	310.00	80.00	390.00
8' perimeter	"	400.00	80.00	480.00
10' perimeter	"	490.00	80.00	570.00
12' perimeter	"	730.00	80.00	810.00
14' perimeter	"	900.00	110.00	1,010
16' perimeter	"	1,010	130.00	1,140
18' perimeter	"	1,230	180.00	1,410
20' perimeter	"	1,570	210.00	1,780
22' perimeter	"	1,680	270.00	1,950
24' perimeter	"	1,790	360.00	2,150
20" high x 4' perimeter	"	290.00	80.00	370.00
6' perimeter	"	340.00	80.00	420.00
8' perimeter	"	450.00	80.00	530.00
10' perimeter	"	560.00	80.00	640.00
12' perimeter	"	820.00	80.00	900.00
14' perimeter	"	1,010	110.00	1,120
16' perimeter	"	1,230	130.00	1,360
18' perimeter	"	1,400	180.00	1,580
20' perimeter	"	1,680	210.00	1,890
22' perimeter	"	1,800	270.00	2,070
24' perimeter	"	2,000	360.00	2,360
24" high x 4' perimeter	"	310.00	80.00	390.00
6' perimeter	"	380.00	80.00	460.00
8' perimeter	"	500.00	80.00	580.00
10' perimeter	"	620.00	80.00	700.00
12' perimeter	"	910.00	80.00	990.00
16' perimeter	"	1,340	130.00	1,470
18' perimeter	"	1,570	180.00	1,750
20' perimeter	"	1,890	210.00	2,100
22' perimeter	"	2,090	270.00	2,360
24' perimeter	"	2,240	360.00	2,600

CONTROLS

	UNIT	MAT.	INST.	TOTAL
15950.10 HVAC CONTROLS				
Pressure gauge, direct reading gage cock and siphon	EA.	48.20	20.00	68.20
Control valve, 1", modulating				

CONTROLS	UNIT	MAT.	INST.	TOTAL
15950.10 HVAC CONTROLS				
2-way	EA.	490.00	26.70	516.70
3-way	"	560.00	40.10	600.10
Self contained control valve with sensing element, 3/4"	"	89.50	20.00	109.50
Instrument air system 2-2 hp compressors, rcvr refrigerant dryer	"			4,144
Thermostat primary control device	"			89.60
Humidistat primary control device	"			67.20
Timers primary control device, indoor/outdoor, 24 hour	"			134.40
Thermometer, dir. reading, 3 dial	"			67.20
Control dampers, round				
6" dia.	EA.	49.30	12.80	62.10
8" dia	"	73.00	12.80	85.80
10" dia	"	96.50	12.80	109.30
12" dia	"	120.00	12.80	132.80
12" dia	"	180.00	16.00	196.00
18" dia	"	200.00	16.00	216.00
20" dia	"	260.00	16.00	276.00
Rectangular, parallel blade standard leakage				
12" x 12"	EA.	37.00	16.00	53.00
16" x 16"	"	55.00	16.00	71.00
20" x 20"	"	65.00	16.00	81.00
28" x 28"	"	84.00	20.00	104.00
32" x 32"	"	96.50	20.00	116.50
36" x 36"	"	110.00	26.70	136.70
40" x 40"	"	130.00	32.00	162.00
44" x 44"	"	170.00	40.10	210.10
48" x 48"	"	190.00	45.80	235.80
48" x 52"	"	210.00	53.50	263.50
48" x 56"	"	230.00	53.50	283.50
48" x 60"	"	240.00	53.50	293.50
48" x 64"	"	240.00	53.50	293.50
48" x 68"	"	260.00	53.50	313.50
48" x 72"	"	290.00	53.50	343.50
Low leakage				
12" x 12"	EA.	73.00	16.00	89.00
16" x 16"	"	93.00	16.00	109.00
20" x 20"	"	120.00	16.00	136.00
24" x 24"	"	160.00	16.00	176.00
28" x 28"	"	180.00	20.00	200.00
32" x 32"	"	210.00	22.90	232.90
36" x 36"	"	240.00	26.70	266.70
40" x 40"	"	370.00	32.00	402.00
44" x 44"	"	390.00	40.10	430.10
48" x 48"	"	410.00	45.80	455.80
48" x 56"	"	480.00	53.50	533.50
48" x 60"	"	500.00	53.50	553.50
48" x 64"	"	530.00	53.50	583.50
48" x 68"	"	550.00	53.50	603.50
48" x 72"	"	650.00	53.50	703.50
Rectangular, opposed horizontal blade				
12" x 12"	EA.	48.20	16.00	64.20
16" x 16"	"	67.00	16.00	83.00
20" x 20"	"	78.50	16.00	94.50

CONTROLS		UNIT	MAT.	INST.	TOTAL
15950.10	HVAC CONTROLS				
24" x 24"		EA.	89.50	16.00	105.50
28" x 28"		"	120.00	20.00	140.00
32" x 32"		"	130.00	21.35	151.35
36" x 36"		"	150.00	26.70	176.70
40" x 40"		"	190.00	32.00	222.00
44" x 44"		"	240.00	40.10	280.10
48" x 48"		"	290.00	45.80	335.80
48" x 52"		"	300.00	45.80	345.80
48" x 56"		"	310.00	53.50	363.50
48" x 60"		"	330.00	53.50	383.50
48" x 64"		"	350.00	53.50	403.50
48" x 68"		"	390.00	53.50	443.50
48" x 72"		"	410.00	53.50	463.50

BASIC MATERIALS	UNIT	MAT.	INST.	TOTAL
16050.30 — BUS DUCT				
Bus duct, 100a, plug-in				
10', 600v	EA.	170.00	100.00	270.00
With ground	"	230.00	160.00	390.00
10', 277/480v	"	220.00	100.00	320.00
With ground	"	270.00	160.00	430.00
Cable tap box	"	93.00	93.00	186.00
End closure	"	22.40	14.85	37.25
Edgewise hanger	"	9.50	27.00	36.50
Flatwise hanger	"	9.75	27.00	36.75
Outside elbow	"	140.00	29.70	169.70
Inside elbow	"	140.00	29.70	169.70
Outside tee	"	180.00	40.80	220.80
Inside tee	"	180.00	40.80	220.80
Outlet cover	"	9.75	14.85	24.60
Wall flange	"	28.00	14.85	42.85
Circuit breakers, with enclosure				
1 pole				
15a-60a	EA.	170.00	37.10	207.10
70a-100a	"	200.00	46.40	246.40
2 pole				
15a-60a	EA.	250.00	40.80	290.80
70a-100a	"	310.00	48.30	358.30
3 pole				
15a-60a	EA.	290.00	43.00	333.00
70a-100a	"	340.00	56.00	396.00
Bus duct, copper feeder duct, 277/480v, 4 wire				
800a	L.F.	170.00	14.85	184.85
1000a	"	200.00	18.55	218.55
1200a	"	230.00	19.80	249.80
1350a	"	270.00	22.85	292.85
1600a	"	320.00	27.00	347.00
2000a	"	400.00	29.70	429.70
2500a	"	490.00	31.60	521.60
3000a	"	630.00	35.30	665.30
Weatherproof				
800a	L.F.	210.00	16.50	226.50
1000a	"	240.00	19.80	259.80
1350a	"	320.00	24.75	344.75
1600a	"	390.00	27.00	417.00
2000a	"	480.00	30.90	510.90
2500a	"	590.00	33.30	623.30
3000a	"	750.00	36.20	786.20
4000a	"	960.00	56.00	1,016
5000a	"	1,160	67.50	1,228
Plug-in feeder duct, 277/480v, 4 wire				
400a	L.F.	110.00	14.85	124.85
600a	"	130.00	16.50	146.50
800a	"	190.00	18.55	208.55
1000a	"	210.00	18.55	228.55
1200a	"	260.00	19.80	279.80
1350a	"	280.00	22.85	302.85
1600a	"	340.00	27.00	367.00

BASIC MATERIALS	UNIT	MAT.	INST.	TOTAL
16050.30 BUS DUCT				
2000a	L.F.	410.00	29.70	439.70
2500a	"	510.00	31.60	541.60
3000a	"	640.00	35.30	675.30
Ground bus				
225a	L.F.			19.45
400a	"			20.35
600a	"			20.90
800a	"			22.45
1000a	"			23.45
1200a	"			29.00
1350a	"			30.20
1600a	"			40.60
2000a	"			49.40
2500a	"			69.00
3000a	"			91.50
4000a	"			110.00
5000a	"			140.00
Copper flanged ends, 277/480v, 4 wire				
225a	EA.	190.00	93.00	283.00
400a	"	230.00	100.00	330.00
600a	"	300.00	110.00	410.00
800a	"	340.00	110.00	450.00
1000a	"	390.00	120.00	510.00
1200a	"	420.00	120.00	540.00
1350a	"	440.00	120.00	560.00
1600a	"	510.00	130.00	640.00
2000a	"	590.00	130.00	720.00
2500a	"	590.00	130.00	720.00
3000a	"	850.00	140.00	990.00
4000a	"	1,140	160.00	1,300
5000a	"	1,320	170.00	1,490
Bus duct, copper elbows, 277/480v-4w				
225a-1000a	EA.	360.00	78.00	438.00
1200a-3000a	"	510.00	93.00	603.00
4000a-5000a	"	600.00	110.00	710.00
Tees, 277/480v-4w				
225a-1000a	EA.	430.00	82.50	512.50
1200a-3000a	"	600.00	95.50	695.50
4000a-5000a	"	700.00	110.00	810.00
Crosses, 277/480v-4w				
225a-1000a	EA.	500.00	82.50	582.50
1200a-3000a	"	700.00	95.50	795.50
4000a-5000a	"	800.00	110.00	910.00
Copper end closures, 277/480v-4w				
225a-1000a	EA.	89.50	33.30	122.80
1200a-3000a	"	120.00	44.30	164.30
4000a-5000a	"	150.00	62.00	212.00
Tap boxes, 277/480v-4w				
225a	EA.	410.00	130.00	540.00
400a	"	630.00	160.00	790.00
600a	"	900.00	270.00	1,170
800a	"	920.00	300.00	1,220

BASIC MATERIALS		UNIT	MAT.	INST.	TOTAL
16050.30	**BUS DUCT**				
1000a		EA.	930.00	370.00	1,300
1200a		"	940.00	410.00	1,350
1350a		"	950.00	480.00	1,430
1600a		"	960.00	520.00	1,480
2000a		"	1,030	630.00	1,660
2500a		"	1,120	850.00	1,970
3000a		"	1,160	1,040	2,200
4000a		"	1,240	1,410	2,650
5000a		"	1,460	1,670	3,130
Circuit breaker, adapter cubicle					
225a		EA.	2,090	56.00	2,146
200a		"	2,460	59.50	2,520
600a		"	3,650	63.00	3,713
800a		"	4,170	67.50	4,238
1000a		"	4,830	70.50	4,901
1200a		"	5,790	74.00	5,864
1600a		"	7,050	78.00	7,128
2000a		"	8,250	82.50	8,333
Transformer taps, 1 phase 277/480v					
600a		EA.	360.00	270.00	630.00
800a		"	380.00	300.00	680.00
1000a		"	460.00	370.00	830.00
1200a		"	500.00	410.00	910.00
1350a		"	530.00	480.00	1,010
1600a		"	600.00	520.00	1,120
2000a		"	690.00	630.00	1,320
2500a		"	830.00	850.00	1,680
3000a		"	980.00	1,040	2,020
4000a		"	1,160	1,410	2,570
5000a		"	1,450	1,710	3,160
3 phase, 480v, 3 wire					
600a		EA.	820.00	270.00	1,090
800a		"	880.00	300.00	1,180
1000a		"	1,000	370.00	1,370
1200a		"	1,090	410.00	1,500
1350a		"	1,140	480.00	1,620
1600a		"	1,380	520.00	1,900
2000a		"	1,550	630.00	2,180
2500a		"	1,800	850.00	2,650
3000a		"	2,090	1,040	3,130
4000a		"	2,410	1,410	3,820
5000a		"	2,780	1,710	4,490
3 phase, 4 wire, 277/480v					
600a		EA.	890.00	270.00	1,160
800a		"	950.00	300.00	1,250
1000a		"	1,090	370.00	1,460
1200a		"	1,200	410.00	1,610
1350a		"	1,260	480.00	1,740
1600a		"	1,470	520.00	1,990
2000a		"	1,680	630.00	2,310
2500a		"	1,960	850.00	2,810
3000a		"	2,290	1,040	3,330

BASIC MATERIALS	UNIT	MAT.	INST.	TOTAL
16050.30 BUS DUCT				
4000a	EA.	2,660	1,510	4,170
5000a	"	3,020	1,710	4,730
Transformer connection, 4 wire, 277/480v				
600a	EA.	1,840	100.00	1,940
800a	"	1,920	110.00	2,030
1000a	"	2,000	110.00	2,110
1200a	"	2,060	110.00	2,170
1350a	"	2,090	120.00	2,210
1600a	"	2,250	120.00	2,370
2000a	"	2,340	130.00	2,470
2500a	"	2,640	130.00	2,770
3000a	"	2,900	140.00	3,040
4000a	"	3,830	160.00	3,990
5000a	"	5,130	170.00	5,300
Unfused reducers, 3 wire, 480v, 3 phase				
400a	EA.	250.00	93.00	343.00
600a	"	280.00	140.00	420.00
800a	"	340.00	170.00	510.00
1000a	"	410.00	190.00	600.00
1200a	"	690.00	200.00	890.00
1350a	"	880.00	210.00	1,090
1600a	"	950.00	230.00	1,180
2000a	"	1,280	240.00	1,520
2500a	"	1,580	250.00	1,830
3000a	"	1,920	270.00	2,190
4000a	"	2,490	320.00	2,810
5000a	"	3,000	400.00	3,400
Circuit breaker reducers, 4 wire, 277/480v				
400a	EA.	450.00	82.50	532.50
600a	"	480.00	130.00	610.00
800a	"	600.00	160.00	760.00
1000a	"	700.00	170.00	870.00
1350a	"	1,530	200.00	1,730
1600a	"	1,660	210.00	1,870
2000a	"	2,240	230.00	2,470
2500a	"	2,770	250.00	3,020
3000a	"	3,340	260.00	3,600
4000a	"	4,340	270.00	4,610
5000a	"	5,220	370.00	5,590
Expansion fittings, 4 wire, 277/480v				
225a	EA.	580.00	93.00	673.00
400a	"	630.00	140.00	770.00
600a	"	720.00	170.00	890.00
800a	"	880.00	190.00	1,070
1000a	"	990.00	200.00	1,190
1200a	"	1,180	210.00	1,390
1350a	"	1,310	220.00	1,530
1600a	"	1,580	230.00	1,810
2000a	"	1,690	250.00	1,940
2500a	"	1,830	270.00	2,100
3000a	"	2,390	320.00	2,710
4000a	"	2,770	400.00	3,170

BASIC MATERIALS	UNIT	MAT.	INST.	TOTAL
16050.30 BUS DUCT				
5000a	EA.	2,900	440.00	3,340
Wall flanges				
225a-2500a	EA.	110.00	150.00	260.00
3000a-5000a	"	170.00	230.00	400.00
Weather seals	"	220.00	37.10	257.10
Roof flanges	"	470.00	150.00	620.00
Fire barriers	"	220.00	56.00	276.00
Spring hangers	"	64.00	64.50	128.50
Sway brace collars	"	13.20	46.40	59.60
Hook-sticks				
8'	EA.			120.00
14'	"			200.00
Fusible switches, 240v, 3 phase				
30a	EA.	170.00	37.10	207.10
60a	"	210.00	46.40	256.40
100a	"	280.00	56.00	336.00
200a	"	480.00	78.00	558.00
400a	"	1,320	150.00	1,470
600a	"	1,900	230.00	2,130
208v, 4 wire				
30a	EA.	200.00	44.30	244.30
60a	"	220.00	50.50	270.50
100a	"	300.00	67.50	367.50
200a	"	530.00	100.00	630.00
400a	"	1,420	190.00	1,610
600a	"	2,080	300.00	2,380
600v				
30a	EA.	190.00	37.10	227.10
60a	"	200.00	46.40	246.40
100a	"	290.00	56.00	346.00
200a	"	500.00	78.00	578.00
400a	"	1,320	150.00	1,470
600a	"	1,920	230.00	2,150
800a	"	3,310	250.00	3,560
1000a	"	3,900	300.00	4,200
1200a	"	6,210	410.00	6,620
1600a	"	6,280	440.00	6,720
480v, 4 wire				
30a	EA.	220.00	44.30	264.30
60a	"	230.00	50.50	280.50
100a	"	330.00	67.50	397.50
200a	"	560.00	100.00	660.00
400a	"	1,420	190.00	1,610
600a	"	2,080	300.00	2,380
800a	"	3,420	310.00	3,730
1000a	"	4,040	410.00	4,450
1200a	"	6,300	440.00	6,740
1600a	"	6,590	520.00	7,110
Fusible combination starters, 600v, 3 phase				
Size 0	EA.	630.00	47.90	677.90
Size 1	"	710.00	59.50	769.50
Size 2	"	900.00	67.50	967.50

BASIC MATERIALS

16050.30   BUS DUCT

	UNIT	MAT.	INST.	TOTAL
Size 3	EA.	1,470	95.50	1,566
Circuit breaker combination starters, 600v, 3 phase				
Size 0	EA.	690.00	47.90	737.90
Size 1	"	720.00	59.50	779.50
Size 2	"	1,040	67.50	1,108
Size 3	"	1,350	95.50	1,446
Fusible combination contactors, 600v, 3 phase				
30a	EA.	650.00	47.90	697.90
60a	"	800.00	59.50	859.50
100a	"	860.00	67.50	927.50
200a	"	1,400	95.50	1,496
Circuit breaker, combination contactors, 600v, 3 phase				
30a	EA.	670.00	47.90	717.90
60a	"	690.00	59.50	749.50
100a	"	970.00	67.50	1,038
200a	"	1,280	95.50	1,376
Fusible contactor electrically held, 480v, 4 wire				
30a	EA.	650.00	47.90	697.90
60a	"	830.00	59.50	889.50
100a	"	1,190	67.50	1,258
200a	"	2,490	100.00	2,590
Mechanically held				
30a	EA.	700.00	47.90	747.90
60a	"	1,010	59.50	1,070
100a	"	1,400	69.00	1,469
200a	"	2,870	100.00	2,970
Circuit breakers, 240v, 3 phase				
15a-60a	EA.	340.00	43.00	383.00
70a-100a	"	370.00	59.50	429.50
600v, 3 phase				
15a-60a	EA.	370.00	43.00	413.00
125a-225a	"	930.00	85.00	1,015
250a-400a	"	1,920	160.00	2,080
500a-600a	"	2,730	200.00	2,930
700a-800a	"	3,300	300.00	3,600
900a-1000a	"	3,910	370.00	4,280
1200a-1600a	"	6,210	410.00	6,620
120/208v, 4 wire				
15a-60a	EA.	310.00	47.90	357.90
70a-100a	"	360.00	69.00	429.00
277/480v, 4 wire				
15a-60a	EA.	380.00	47.90	427.90
70a-100a	"	410.00	70.50	480.50
125a-225a	"	1,000	100.00	1,100
250a-400a	"	2,010	190.00	2,200
500a-600a	"	2,870	300.00	3,170
700a-800a	"	3,420	300.00	3,720
900a-1000a	"	4,040	410.00	4,450
1200a-1600a	"	6,160	520.00	6,680
600v, 3 phase, 65,000 aic.				
60a	EA.	470.00	43.00	513.00
70a-100a	"	510.00	59.50	569.50

BASIC MATERIALS	UNIT	MAT.	INST.	TOTAL
16050.30 BUS DUCT				
125a-225a	EA.	1,670	85.00	1,755
250a-400a	"	2,660	160.00	2,820
500a-600a	"	3,150	230.00	3,380
700a-800a	"	3,730	300.00	4,030
900a-1000a	"	4,230	370.00	4,600
277/480v, 4 wire, 65,000 aic.				
15a-60a	EA.	510.00	47.90	557.90
70a-100a	"	550.00	70.50	620.50
125a-225a	"	1,720	100.00	1,820
250a-400a	"	2,760	190.00	2,950
500a-600a	"	3,300	230.00	3,530
700a-800a	"	3,860	300.00	4,160
900a-1000a	"	6,300	380.00	6,680
600v, 3 phase, current limiting				
15a-60a	EA.	1,230	43.00	1,273
70a-100a	"	1,510	59.50	1,570
125a-225a	"	2,820	85.00	2,905
250a-400a	"	3,350	160.00	3,510
500a-600a	"	5,040	230.00	5,270
700a-800a	"	5,680	300.00	5,980
900a-1000a	"	6,320	370.00	6,690
277/480v, 4 wire, current limiting				
15a-60a	EA.	1,260	47.90	1,308
70a-100a	"	1,620	1,480	3,100
125a-225a	"	2,900	100.00	3,000
250a-400a	"	3,450	190.00	3,640
500a-600a	"	3,850	230.00	4,080
700a-800a	"	5,810	300.00	6,110
900a-1000a	"	6,450	410.00	6,860
Capacitors, 3 phase, 240v				
5 kvar	EA.	1,360	190.00	1,550
7.5 kvar	"	1,710	230.00	1,940
10 kvar	"	1,990	300.00	2,290
15 kvar	"	2,610	340.00	2,950
480v				
2.5 kvar	EA.	590.00	100.00	690.00
5 kvar	"	880.00	170.00	1,050
7.5 kvar	"	1,070	210.00	1,280
10 kvar	"	1,190	300.00	1,490
15 kvar	"	1,430	330.00	1,760
20 kvar	"	1,780	430.00	2,210
25 kvar	"	2,230	480.00	2,710
30 kvar	"	2,630	530.00	3,160
Transformers, 3 phase, 480v				
1.0 kva	EA.	560.00	64.50	624.50
1.5 kva	"	610.00	74.00	684.00
2 kva	"	660.00	93.00	753.00
3 kva	"	780.00	100.00	880.00
5 kva	"	1,080	150.00	1,230
7.5 kva	"	1,300	190.00	1,490
10 kva	"	1,520	200.00	1,720

BASIC MATERIALS	UNIT	MAT.	INST.	TOTAL
16110.12 — CABLE TRAY				
Cable tray, 6"	L.F.	11.60	2.20	13.80
Ventilated cover	"	4.70	1.10	5.80
Solid cover	"	3.65	1.10	4.75
Flat 90	EA.	52.00	18.55	70.55
Outside 90	"	43.90	18.55	62.45
Inside 90	"	48.00	18.55	66.55
Flat 45	"	39.70	18.55	58.25
Outside 45	"	32.90	18.55	51.45
Inside 45	"	34.30	18.55	52.85
Adjustable elbow	"	49.40	18.55	67.95
Support riser	"	66.00	18.55	84.55
Adjustable riser	"	7.30	18.55	25.85
Tee	"	67.00	64.50	131.50
Cross	"	90.50	67.50	158.00
Blind end	"	5.35	11.00	16.35
Expansion joint	"	24.65	18.55	43.20
Box connector	"	16.45	93.00	109.45
Standard dropout	"	2.85	18.55	21.40
2"	"	14.05	22.85	36.90
3"	"	24.65	29.70	54.35
4"	"	39.70	37.10	76.80
Cable tray, 9"	L.F.	12.30	2.60	14.90
Ventilated cover	"	5.25	1.50	6.75
Solid cover	"	4.25	1.50	5.75
Flat 90	EA.	57.50	19.80	77.30
Outside 90	"	47.40	19.80	67.20
Inside 90	"	49.90	19.80	69.70
Flat 45	"	41.10	19.80	60.90
Outside 45	"	34.90	19.80	54.70
Inside 45	"	34.90	19.80	54.70
Adjustable elbow	"	52.50	19.80	72.30
Support riser	"	71.00	19.80	90.80
Adjustable riser	"	7.30	19.80	27.10
Tee	"	73.50	67.50	141.00
Cross	"	95.00	67.50	162.50
Blind end	"	5.75	11.85	17.60
Expansion joint	"	34.90	19.80	54.70
Box connector	"	16.65	95.50	112.15
Standard dropout	"	3.15	19.80	22.95
2"	"	14.15	24.75	38.90
3"	"	23.65	24.75	48.40
4"	"	39.70	40.70	80.40
Cable tray, 12"	L.F.	12.60	2.95	15.55
Ventilated cover	"	4.85	1.85	6.70
Solid cover	"	4.45	1.85	6.30
Flat 90	EA.	60.00	19.80	79.80
Outside 90	"	47.50	19.80	67.30
Inside 90	"	52.00	19.80	71.80
Flat 45	"	43.60	19.80	63.40
Outside 45	"	68.50	19.80	88.30
Inside 45	"	56.00	19.80	75.80
Adjustable elbow	"	55.00	19.80	74.80

BASIC MATERIALS	UNIT	MAT.	INST.	TOTAL
16110.12 CABLE TRAY				
Support riser	EA.	71.00	19.80	90.80
Adjustable riser	"	8.20	19.80	28.00
Tee	"	77.00	74.00	151.00
Cross	"	97.50	74.00	171.50
Blind end	"	6.50	12.90	19.40
Expansion joint	"	37.40	22.85	60.25
Box connector	"	17.40	100.00	117.40
Standard dropout	"	3.40	22.85	26.25
2"	"	14.85	27.00	41.85
3"	"	24.65	33.30	57.95
4"	"	41.10	43.00	84.10
Cable tray, 18"	L.F.	13.25	3.70	16.95
Ventilated cover	"	6.85	2.20	9.05
Solid cover	"	5.75	2.20	7.95
Flat 90	EA.	70.00	27.00	97.00
Outside 90	"	75.00	27.00	102.00
Inside 90	"	57.00	27.00	84.00
Flat 45	"	49.90	27.00	76.90
Outside 45	"	70.00	27.00	97.00
Inside 45	"	58.00	27.00	85.00
Adjustable elbow	"	75.00	27.00	102.00
Support riser	"	88.00	27.00	115.00
Adjustable riser	"	7.30	27.00	34.30
Tee	"	92.50	78.00	170.50
Cross	"	140.00	78.00	218.00
Blind end	"	9.65	14.85	24.50
Expansion joint	"	48.70	27.00	75.70
Box connector	"	20.55	110.00	130.55
Standard dropout	"	5.20	24.75	29.95
2"	"	16.45	27.00	43.45
3"	"	26.00	35.30	61.30
4"	"	42.40	44.30	86.70
Cable tray, 24"	L.F.	15.10	4.55	19.65
Ventilated cover	"	9.35	2.60	11.95
Solid cover	"	6.45	2.60	9.05
Flat 90	EA.	86.00	27.00	113.00
Outside 90	"	76.00	27.00	103.00
Inside 90	"	60.50	27.00	87.50
Flat 45	"	57.00	27.00	84.00
Outside 45	"	70.50	27.00	97.50
Inside 45	"	60.50	27.00	87.50
Adjustable elbow	"	75.00	27.00	102.00
Support riser	"	90.50	27.00	117.50
Adjustable riser	"	8.15	27.00	35.15
Tee	"	110.00	82.50	192.50
Cross	"	160.00	82.50	242.50
Blind end	"	10.60	16.50	27.10
Expansion joint	"	49.90	27.00	76.90
Box connector	"	21.95	130.00	151.95
Standard dropout	"	5.50	24.75	30.25
2"	"	17.10	29.70	46.80
3"	"	27.10	36.20	63.30

BASIC MATERIALS	UNIT	MAT.	INST.	TOTAL
16110.12 CABLE TRAY				
4"	EA.	43.60	46.40	90.00
Cable tray, 36"	L.F.	17.45	5.40	22.85
Ventilated cover	"	10.75	2.95	13.70
Solid cover	"	7.45	2.95	10.40
Flat 90	EA.	100.00	31.60	131.60
Outside 90	"	87.50	31.60	119.10
Inside 90	"	76.00	31.60	107.60
Flat 45	"	67.00	31.60	98.60
Outside 45	"	77.50	31.60	109.10
Inside 45	"	67.00	31.60	98.60
Adjustable elbow	"	77.50	31.60	109.10
Support riser	"	94.00	31.60	125.60
Adjustable riser	"	8.55	31.60	40.15
Tee	"	120.00	85.00	205.00
Cross	"	180.00	110.00	290.00
Blind end	"	10.95	17.45	28.40
Expansion joint	"	51.50	27.00	78.50
Box connector	"	25.50	140.00	165.50
Standard dropout	"	6.15	27.00	33.15
2"	"	18.15	29.70	47.85
3"	"	28.00	37.10	65.10
4"	"	43.70	47.90	91.60
Reducers				
9" - 6"	EA.	39.40	18.55	57.95
12" - 9"	"	41.10	18.55	59.65
18" - 12"	"	43.60	22.85	66.45
24" - 18"	"	48.70	27.00	75.70
36" - 18"	"	59.50	29.70	89.20
36" - 24"	"	60.50	33.30	93.80
Conduit dropouts				
3/4"	EA.	5.25	12.90	18.15
1"	"	5.40	12.90	18.30
1-1/4"	"	6.30	14.85	21.15
1-1/2"	"	7.25	18.55	25.80
2"	"	8.30	19.80	28.10
2-1/2"	"	9.50	27.00	36.50
3"	"	11.00	29.70	40.70
Wall brackets				
6"	EA.	20.50	5.40	25.90
9"	"	23.65	5.40	29.05
12"	"	24.95	7.40	32.35
18"	"	29.90	9.30	39.20
24"	"	34.80	11.00	45.80
36"	"	43.70	14.85	58.55
16110.15 FIBERGLASS CABLE TRAY				
Fiberglass cable tray, 6"	L.F.	18.65	1.50	20.15
Tray cover	"	5.15	1.10	6.25
Horizontal				
90	EA.	200.00	10.25	210.25
45	"	86.00	10.25	96.25
30	"	88.50	10.25	98.75

BASIC MATERIALS	UNIT	MAT.	INST.	TOTAL
16110.15 FIBERGLASS CABLE TRAY				
Inside				
90	EA.	77.50	10.25	87.75
45	"	62.50	10.25	72.75
30	"	84.00	10.25	94.25
Horizontal tee	"	140.00	27.00	167.00
Horizontal cross	"	160.00	40.70	200.70
Splice plate	"	7.55	5.40	12.95
Floor flange	"	8.35	12.90	21.25
Panel flange	"	8.65	46.40	55.05
End plate	"	5.70	5.40	11.10
Nylon rivet	"	0.20	1.85	2.05
Barrier strip	"	2.85	1.85	4.70
Hold down clamp	"	1.25	1.85	3.10
Drop out	"	9.00	19.80	28.80
Cover stand off	"	2.40	1.85	4.25
Wall bracket	"	33.70	7.40	41.10
Sealer	"			12.30
Outside				
90	EA.	77.50	10.25	87.75
45	"	62.50	10.25	72.75
30	"	85.00	10.25	95.25
Fiberglass cable tray, 9"	L.F.	19.55	1.85	21.40
Tray cover	"	5.40	1.50	6.90
Horizontal				
90	EA.	210.00	10.25	220.25
45	"	89.50	10.25	99.75
30	"	90.50	10.25	100.75
Inside				
90	EA.	82.00	10.25	92.25
45	"	65.00	10.25	75.25
30	"	87.50	10.25	97.75
Horizontal tee	"	140.00	27.00	167.00
Horizontal cross	"	170.00	40.70	210.70
Splice plate	"	7.95	5.40	13.35
Floor flange	"	8.75	12.90	21.65
Panel flange	"	9.10	46.40	55.50
End plate	"	6.00	5.40	11.40
Nylon rivet	"	0.21	1.85	2.06
Barrier strap	"	3.00	1.85	4.85
Hold down clamp	"	1.30	1.85	3.15
Drop out	"	9.40	19.80	29.20
Cover stand off	"	2.50	1.85	4.35
Wall bracket	"	35.40	7.40	42.80
Sealer	"			12.30
Outside				
90	EA.	82.50	10.25	92.75
45	"	67.50	10.25	77.75
30	"	89.50	10.25	99.75
Fiberglass cable tray, 12"	L.F.	20.55	2.20	22.75
Tray cover	"	5.75	1.85	7.60
Horizontal				
90	EA.	220.00	12.90	232.90

BASIC MATERIALS	UNIT	MAT.	INST.	TOTAL
16110.15 FIBERGLASS CABLE TRAY				
45	EA.	94.00	12.90	106.90
30	"	96.50	12.90	109.40
Inside				
90	EA.	85.00	12.90	97.90
45	"	67.00	12.90	79.90
Inside 30	"	90.50	12.90	103.40
Horizontal tee	"	150.00	35.30	185.30
Horizontal cross	"	170.00	47.90	217.90
Splice plate	"	8.30	5.95	14.25
Floor flange	"	9.20	13.50	22.70
Panel flange	"	9.55	59.50	69.05
End plate	"	6.25	5.95	12.20
Nylon rivet	"	0.22	1.85	2.07
Barrier strip	"	3.10	1.85	4.95
Hold down clamp	"	1.40	1.85	3.25
Drop out	"	14.05	19.80	33.85
Cover stand off	"	2.60	1.85	4.45
Wall bracket	"	37.20	7.40	44.60
Sealer	"			12.30
Outside				
90	EA.	85.00	12.90	97.90
45	"	68.50	12.90	81.40
30	"	94.00	12.90	106.90
Fiberglass cable tray, 18"	L.F.	21.55	2.60	24.15
Tray cover	"	10.60	2.20	12.80
Horizontal				
90	EA.	230.00	14.85	244.85
45	"	100.00	14.85	114.85
30	"	100.00	14.85	114.85
Inside				
90	EA.	88.50	14.85	103.35
45	"	70.50	14.85	85.35
30	"	95.00	14.85	109.85
Horizontal tee	"	160.00	57.00	217.00
Horizontal cross	"	180.00	50.50	230.50
Splice plate	"	8.75	5.95	14.70
Floor flange	"	9.65	14.15	23.80
Panel flange	"	10.00	63.00	73.00
End plate	"	6.60	6.30	12.90
Nylon rivet	"	0.24	1.85	2.09
Barrier strip	"	3.30	1.85	5.15
Hold down clamp	"	1.45	1.85	3.30
Drop out	"	20.55	19.80	40.35
Cover stand off	"	2.80	1.85	4.65
Wall bracket	"	39.00	7.40	46.40
Sealer	"			12.30
Outside				
90	EA.	89.50	14.85	104.35
45	"	71.50	14.85	86.35
30	"	100.00	14.85	114.85
Fiberglass cable tray, 24"	L.F.	22.55	2.95	25.50
Tray cover	"	12.60	2.60	15.20

BASIC MATERIALS	UNIT	MAT.	INST.	TOTAL
16110.15 FIBERGLASS CABLE TRAY				
Horizontal				
90	EA.	240.00	18.55	258.55
45	"	100.00	18.55	118.55
30	"	100.00	18.55	118.55
Inside				
90	EA.	92.00	18.55	110.55
45	"	74.00	18.55	92.55
30	"	100.00	18.55	118.55
Horizontal tee	"	160.00	37.10	197.10
Horizontal cross	"	190.00	56.00	246.00
Splice plate	"	9.15	5.95	15.10
Floor flange	"	10.15	14.85	25.00
Panel flange	"	10.55	67.50	78.05
End plate	"	6.90	6.30	13.20
Nylon Rivet	"	0.25	1.85	2.10
Barrier strip	"	3.45	1.85	5.30
Hold down clamp	"	1.50	1.85	3.35
Drop out	"	26.80	19.80	46.60
Cover stand off	"	2.90	1.85	4.75
Wall bracket	"	42.70	7.40	50.10
Sealer	"			12.30
Outside				
90	EA.	92.00	18.55	110.55
45	"	75.00	18.55	93.55
30	"	100.00	18.55	118.55
Fiberglass cable tray, 30"	L.F.	25.80	3.30	29.10
Tray cover	"	17.90	2.95	20.85
Horizontal				
90	EA.	240.00	22.85	262.85
45	"	100.00	22.85	122.85
30	"	110.00	22.85	132.85
Inside				
90	EA.	94.00	22.85	116.85
45	"	76.00	22.85	98.85
30	"	100.00	22.85	122.85
Horizontal tee	"	180.00	40.70	220.70
Horizontal cross	"	210.00	59.50	269.50
Splice plate	"	9.60	5.95	15.55
Floor flange	"	10.65	14.85	25.50
Panel flange	"	11.05	70.50	81.55
End plate	"	7.25	7.40	14.65
Nylon rivet	"	0.26	1.85	2.11
Barrier strip	"	3.65	1.85	5.50
Hold down clamp	"	1.60	1.85	3.45
Dropout	"	30.80	22.85	53.65
Cover stand off	"	3.05	1.85	4.90
Wall bracket	"	48.00	7.40	55.40
Sealer	"			12.30
Outside				
90	EA.	94.00	22.85	116.85
45	"	77.50	22.85	100.35
30	"	100.00	22.85	122.85

BASIC MATERIALS	UNIT	MAT.	INST.	TOTAL
16110.15 — FIBERGLASS CABLE TRAY				
Fiberglass cable tray, 36"	L.F.	29.80	3.70	33.50
Tray cover	"	23.75	3.30	27.05
Horizontal				
90	EA.	240.00	24.75	264.75
45	"	110.00	24.75	134.75
30	"	110.00	24.75	134.75
Inside				
90	EA.	96.50	24.75	121.25
45	"	77.50	24.75	102.25
30	"	100.00	24.75	124.75
Horizontal tee	"	200.00	44.30	244.30
horizontal cross	"	240.00	63.00	303.00
Splice plate	"	10.10	5.95	16.05
Floor flange	"	11.20	16.50	27.70
Panel flange	"	11.60	78.00	89.60
End plate	"	7.65	7.40	15.05
Nylon rivet	"	0.29	1.85	2.14
Barrier strip	"	3.80	1.85	5.65
Hold down clamp	"	1.65	1.85	3.50
Drop out	"	35.30	24.75	60.05
5Cover stand off	"	3.20	1.85	5.05
Wall bracket	"	53.50	7.40	60.90
Sealer	"			12.30
Outside				
90	EA.	96.50	24.75	121.25
45	"	80.50	24.75	105.25
30	"	110.00	24.75	134.75
Reducers				
12" - 6"	EA.	120.00	11.00	131.00
12" - 9"	"	130.00	11.00	141.00
18" - 6"	"	110.00	12.90	122.90
18" - 9"	"	110.00	12.90	122.90
18" - 12"	"	120.00	14.85	134.85
24" - 6"	"	140.00	14.85	154.85
24" - 9"	"	140.00	14.85	154.85
24" - 12"	"	140.00	14.85	154.85
24" - 18"	"	140.00	16.50	156.50
30" - 9"	"	140.00	14.85	154.85
30" - 12"	"	140.00	14.85	154.85
30" - 18"	"	140.00	16.50	156.50
30" - 24"	"	150.00	18.55	168.55
36" - 9"	"	140.00	18.55	158.55
36" - 12"	"	150.00	18.55	168.55
36" - 18"	"	160.00	19.80	179.80
36" - 24"	"	160.00	22.85	182.85
36" - 30"	"	160.00	27.00	187.00
16110.20 — CONDUIT SPECIALTIES				
Rod beam clamp, 1/2"	EA.	3.40	1.85	5.25
Hanger rod				
3/8"	L.F.	0.72	1.50	2.22
1/2"	"	1.15	1.85	3.00

BASIC MATERIALS	UNIT	MAT.	INST.	TOTAL
16110.20 CONDUIT SPECIALTIES				
All thread rod				
1/4"	L.F.	0.41	1.10	1.51
3/8"	"	0.69	1.50	2.19
1/2"	"	1.15	1.85	3.00
5/8"	"	1.85	2.95	4.80
Hanger channel, 1-1/2"				
No holes	EA.	2.80	1.10	3.90
Holes	"	3.10	1.10	4.20
Channel strap				
1/2"	EA.	0.75	1.85	2.60
3/4"	"	0.88	1.85	2.73
1"	"	0.95	1.85	2.80
1-1/4"	"	1.05	2.95	4.00
1-1/2"	"	1.20	2.95	4.15
2"	"	1.40	2.95	4.35
2-1/2"	"	1.50	4.55	6.05
3"	"	1.65	4.55	6.20
3-1/2"	"	2.05	4.55	6.60
4"	"	2.30	5.40	7.70
5"	"	3.20	5.40	8.60
6"	"	3.85	5.40	9.25
Conduit penetrations, roof and wall, 8" thick				
1/2"	EA.	0.00	22.85	22.85
3/4"	"	0.00	22.85	22.85
1"	"	0.00	29.70	29.70
1-1/4"	"	0.00	29.70	29.70
1-1/2"	"	0.00	29.70	29.70
2"	"	0.00	59.50	59.50
2-1/2"	"	0.00	59.50	59.50
3"	"	0.00	59.50	59.50
3-1/2"	"	0.00	74.00	74.00
4"	"	0.00	74.00	74.00
Plastic duct bank conduit spacer, 3" separation				
2"	EA.	0.36	1.85	2.21
3"	"	0.37	1.85	2.22
4"	"	0.41	1.85	2.26
5"	"	0.49	1.85	2.34
6"	"	0.55	1.85	2.40
Intermediate, 3" separation				
2"	EA.	0.40	1.85	2.25
3"	"	0.43	1.85	2.28
4"	"	0.44	1.85	2.29
5"	"	0.50	1.85	2.35
6"	"	0.59	1.85	2.44
Base with 1-1/2" separation				
2"	EA.	0.34	1.85	2.19
3"	"	0.36	5.95	6.31
4"	"	0.41	5.95	6.36
5"	"	0.45	5.95	6.40
6"	"	0.49	5.95	6.44
Intermediate, 1-1/2" separation				
2"	EA.	0.36	5.95	6.31

BASIC MATERIALS	UNIT	MAT.	INST.	TOTAL
16110.20 CONDUIT SPECIALTIES				
3"	EA.	0.37	5.95	6.32
3-1/2"	"	0.43	5.95	6.38
4"	"	0.43	5.95	6.38
5"	"	0.48	6.40	6.88
6"	"	0.53	5.95	6.48
OD beam clamp, 1/4"	"	0.66	7.40	8.06
Threaded rod couplings				
1/4"	EA.	0.54	1.85	2.39
3/8"	"	0.62	1.85	2.47
1/2"	"	0.82	1.85	2.67
5/8"	"	2.00	1.85	3.85
3/4"	"	2.95	1.85	4.80
Hex nuts				
1/4"	EA.	0.08	1.85	1.93
3/8"	"	0.16	1.85	2.01
1/2"	"	0.25	1.85	2.10
5/8"	"	0.71	1.85	2.56
3/4"	"	0.80	1.85	2.65
Square nuts				
1/4"	EA.	0.08	1.85	1.93
3/8"	"	0.16	1.85	2.01
3/8"	"	0.25	1.85	2.10
5/8"	"	0.71	1.85	2.56
3/4"	"	0.80	1.85	2.65
Flat washers				
1/4"	EA.			0.08
3/8"	"			0.10
1/2"	"			0.16
5/8"	"			0.23
3/4"	"			0.33
Lockwashers				
1/4"	EA.			0.08
3/8"	"			0.08
1/2"	"			0.08
5/8"	"			0.11
3/4"	"			0.20
Channel closure strip	L.F.	1.30	4.95	6.25
Channel end cap	EA.	0.56	4.95	5.51
Li-channel trapeze hangers				
12" long	EA.	10.00	5.40	15.40
18" long	"	11.40	5.40	16.80
24" long	"	14.15	5.40	19.55
30" long	"	15.90	9.30	25.20
36" long	"	18.60	9.30	27.90
42" long	"	22.30	11.00	33.30
Channel spring nuts				
1/4"	EA.	0.76	2.20	2.96
3/8"	"	0.88	2.95	3.83
1/2"	"	0.96	3.70	4.66
Fireproofing, for conduit penetrations				
1/2"	EA.	1.60	18.55	20.15
3/4"	"	1.60	18.55	20.15

BASIC MATERIALS	UNIT	MAT.	INST.	TOTAL
16110.20 CONDUIT SPECIALTIES				
1"	EA.	1.65	18.55	20.20
1-1/4"	"	2.30	29.10	31.40
1-1/2"	"	2.30	27.00	29.30
2"	"	2.30	27.00	29.30
2-1/2"	"	4.65	33.30	37.95
3"	"	4.65	36.00	40.65
3-1/2"	"	5.60	46.40	52.00
4"	"	6.75	56.00	62.75
16110.21 ALUMINUM CONDUIT				
Aluminum conduit				
1/2"	L.F.	0.85	1.10	1.95
3/4"	"	1.10	1.50	2.60
1"	"	1.15	1.85	3.00
1-1/4"	"	2.10	2.20	4.30
1-1/2"	"	2.60	2.95	5.55
2"	"	3.50	3.30	6.80
2-1/2"	"	5.55	3.70	9.25
3"	"	7.25	3.95	11.20
3-1/2"	"	8.70	4.55	13.25
4"	"	10.35	5.40	15.75
5"	"	14.80	6.75	21.55
6"	"	19.50	7.40	26.90
90 deg. elbow				
1/2"	EA.	2.95	7.05	10.00
3/4"	"	4.00	9.30	13.30
1"	"	5.60	11.40	17.00
1-1/4"	"	8.90	14.15	23.05
1-1/2"	"	11.75	14.85	26.60
2"	"	17.45	16.50	33.95
2-1/2"	"	29.40	21.20	50.60
3"	"	45.50	24.75	70.25
3-1/2"	"	71.00	29.70	100.70
4"	"	84.00	33.00	117.00
5"	"	230.00	42.40	272.40
6"	"	320.00	82.50	402.50
Coupling				
1/2"	EA.	0.94	1.85	2.79
3/4"	"	1.45	2.20	3.65
1"	"	1.90	2.95	4.85
1-1/4"	"	2.35	3.30	5.65
1-1/2"	"	2.70	3.70	6.40
2"	"	3.80	3.95	7.75
2-1/2"	"	8.65	4.55	13.20
3"	"	11.25	4.55	15.80
3-1/2"	"	15.50	5.40	20.90
4"	"	18.65	5.95	24.60
5"	"	47.30	5.95	53.25
6"	"	74.00	7.05	81.05

BASIC MATERIALS	UNIT	MAT.	INST.	TOTAL
16110.22 EMT CONDUIT				
EMT conduit				
1/2"	L.F.	0.18	1.10	1.28
3/4"	"	0.25	1.50	1.75
1"	"	0.39	1.85	2.24
1-1/4"	"	0.56	2.20	2.76
1-1/2"	"	0.65	2.95	3.60
2"	"	0.83	3.30	4.13
2-1/2"	"	1.90	3.70	5.60
3"	"	2.35	4.55	6.90
3-1/2"	"	3.35	5.40	8.75
4"	"	3.85	6.75	10.60
90 deg. elbow				
1/2"	EA.	2.15	3.30	5.45
3/4"	"	2.20	3.70	5.90
1"	"	2.30	3.95	6.25
1-1/4"	"	3.20	4.55	7.75
1-1/2"	"	4.05	5.40	9.45
2"	"	6.45	7.05	13.50
2-1/2"	"	18.35	7.80	26.15
3"	"	28.60	9.00	37.60
3-1/2"	"	38.40	10.35	48.75
4"	"	45.20	10.60	55.80
Connector, steel compression				
1/2"	EA.	0.58	3.30	3.88
3/4"	"	0.83	3.30	4.13
1"	"	1.35	3.30	4.65
1-1/4"	"	2.65	3.95	6.60
1-1/2"	"	3.90	5.40	9.30
2"	"	5.60	7.05	12.65
2-1/2"	"	17.65	9.30	26.95
3"	"	24.30	10.60	34.90
3-1/2"	"	36.70	11.40	48.10
4"	"	37.60	12.35	49.95
Coupling, steel, compression				
1/2"	EA.	0.72	2.20	2.92
3/4"	"	1.00	2.20	3.20
1"	"	1.60	2.20	3.80
1-1/4"	"	2.90	3.30	6.20
1-1/2"	"	4.25	3.95	8.20
2"	"	5.75	5.40	11.15
2-1/2"	"	22.05	8.25	30.30
3"	"	27.50	9.30	36.80
3-1/2"	"	43.10	10.60	53.70
4"	"	44.10	11.40	55.50
1 hole strap, steel				
1/2"	EA.	0.10	1.50	1.60
3/4"	"	0.15	1.50	1.65
1"	"	0.19	1.50	1.69
1-1/4"	"	0.29	1.85	2.14
1-1/2"	"	0.45	1.85	2.30
2"	"	0.59	1.85	2.44
2-1/2"	"	1.20	2.20	3.40

BASIC MATERIALS	UNIT	MAT.	INST.	TOTAL
16110.22 EMT CONDUIT				
3"	EA.	1.30	2.20	3.50
3-1/2"	"	1.95	2.20	4.15
4"	"	2.45	2.20	4.65
Connector, steel set screw				
1/2"	EA.	0.36	2.60	2.96
3/4"	"	0.55	2.60	3.15
1"	"	0.91	2.60	3.51
1-1/4"	"	1.85	3.95	5.80
1-1/2"	"	2.65	5.40	8.05
2"	"	3.70	6.75	10.45
2-1/2"	"	11.05	9.00	20.05
3"	"	13.20	9.90	23.10
3-1/2"	"	17.60	11.00	28.60
4"	"	20.35	12.90	33.25
Insulated throat				
1/2"	EA.	0.50	2.60	3.10
3/4"	"	0.77	2.60	3.37
1"	"	1.30	2.60	3.90
1-1/4"	"	2.30	3.95	6.25
1-1/2"	"	3.35	5.40	8.75
2"	"	4.80	6.75	11.55
2-1/2"	"	21.35	9.00	30.35
3"	"	25.10	9.90	35.00
3-1/2"	"	33.60	11.00	44.60
4"	"	37.30	12.90	50.20
Connector, die cast set screw				
1/2"	EA.	0.19	2.20	2.39
3/4"	"	0.30	2.20	2.50
1"	"	0.49	2.20	2.69
1-1/4"	"	0.85	3.30	4.15
1-1/2"	"	1.35	3.95	5.30
2"	"	1.70	5.40	7.10
2-1/2"	"	5.55	7.40	12.95
3"	"	6.80	8.25	15.05
3-1/2"	"	9.55	9.30	18.85
4"	"	11.00	10.60	21.60
Insulated throat				
1/2"	EA.	0.31	2.20	2.51
3/4"	"	0.52	2.20	2.72
1"	"	0.77	2.20	2.97
1-1/4"	"	1.60	3.30	4.90
1-1/2"	"	2.20	3.95	6.15
2"	"	3.00	5.40	8.40
2-1/2"	"	11.50	7.40	18.90
3"	"	14.75	8.25	23.00
3-1/2"	"	19.10	10.00	29.10
4"	"	21.10	10.60	31.70
Coupling, steel set screw				
1/2"	EA.	0.48	1.50	1.98
3/4"	"	0.74	1.50	2.24
1"	"	1.10	1.50	2.60
1-1/4"	"	2.10	1.85	3.95

BASIC MATERIALS	UNIT	MAT.	INST.	TOTAL
16110.22 EMT CONDUIT				
1-1/2"	EA.	3.10	2.95	6.05
2"	"	4.10	3.95	8.05
2-1/2"	"	9.25	5.95	15.20
3"	"	10.35	7.05	17.40
3-1/2"	"	12.20	8.25	20.45
4"	"	13.95	9.30	23.25
Diecast set screw				
1/2"	EA.	0.19	1.50	1.69
3/4"	"	0.30	1.50	1.80
1"	"	0.52	1.50	2.02
1-1/4"	"	0.91	1.85	2.76
1-1/2"	"	1.35	2.95	4.30
2"	"	1.80	3.95	5.75
2-1/2"	"	4.45	5.95	10.40
3"	"	5.10	6.90	12.00
3-1/2"	"	6.15	8.25	14.40
4"	"	6.90	9.30	16.20
1 hole malleable straps				
1/2"	EA.	0.20	1.50	1.70
3/4"	"	0.26	1.50	1.76
1"	"	0.38	1.50	1.88
1-1/4"	"	0.62	1.85	2.47
1-1/2"	"	0.75	1.85	2.60
2"	"	1.45	1.85	3.30
2-1/2"	"	2.65	2.20	4.85
3"	"	3.75	2.20	5.95
3-1/2"	"	5.40	2.20	7.60
4"	"	12.05	2.20	14.25
EMT to rigid compression coupling				
1/2"	EA.	1.35	3.70	5.05
3/4"	"	2.00	3.70	5.70
1"	"	2.95	5.55	8.50
Set screw couplings				
1/2"	EA.	1.00	3.70	4.70
3/4"	"	1.45	3.70	5.15
1"	"	2.20	5.40	7.60
Set screw offset connectors				
1/2"	EA.	1.35	3.70	5.05
3/4"	"	1.90	3.70	5.60
1"	"	2.80	5.40	8.20
Compression offset connectors				
1/2"	EA.	1.55	3.70	5.25
3/4"	"	2.25	3.70	5.95
1"	"	3.45	5.40	8.85
Type "LB" set screw condulets				
1/2"	EA.	2.95	8.50	11.45
3/4"	"	3.65	11.00	14.65
1"	"	5.60	14.15	19.75
1-1/4"	"	8.35	16.50	24.85
1-1/2"	"	10.70	19.80	30.50
2"	"	17.65	22.85	40.50
2-1/2"	"	55.00	27.00	82.00

BASIC MATERIALS	UNIT	MAT.	INST.	TOTAL
16110.22 — EMT CONDUIT				
3"	EA.	72.50	37.10	109.60
3-1/2"	"	110.00	49.50	159.50
4"	"	140.00	59.50	199.50
Type "T" set screw condulets				
1/2"	EA.	4.05	11.00	15.05
3/4"	"	5.10	14.85	19.95
1"	"	6.85	16.50	23.35
1-1/4"	"	11.95	19.80	31.75
1-1/2"	"	16.20	22.85	39.05
2"	"	26.80	24.75	51.55
Type "C" set screw condulets				
1/2"	EA.	3.50	9.30	12.80
3/4"	"	4.30	11.00	15.30
1"	"	6.40	14.15	20.55
1-1/4"	"	9.65	16.50	26.15
1-1/2"	"	12.75	19.80	32.55
2"	"	24.85	14.15	39.00
Type "LL" setscrew condulets				
1/2"	EA.	3.50	9.30	12.80
3/4"	"	4.30	11.00	15.30
1"	"	6.40	14.15	20.55
1-1/4"	"	9.65	16.50	26.15
1-1/2"	"	12.65	19.80	32.45
2"	"	21.25	22.85	44.10
Type "LR" set screw condulets				
1/2"	EA.	3.00	9.30	12.30
3/4"	"	4.30	11.00	15.30
1"	"	6.40	14.15	20.55
1-1/4"	"	9.65	16.50	26.15
1-1/2"	"	12.65	19.80	32.45
2"	"	21.25	22.85	44.10
Type "LB" compression condulets				
1/2"	EA.	7.45	11.00	18.45
3/4"	"	10.95	18.55	29.50
1"	"	14.05	18.55	32.60
Type "T" compression condulets				
1/2"	EA.	9.95	14.85	24.80
3/4"	"	13.05	16.50	29.55
1"	"	20.25	22.85	43.10
Condulet covers				
1/2"	EA.	0.83	4.55	5.38
3/4"	"	1.10	4.55	5.65
1"	"	1.40	4.55	5.95
1-1/4"	"	1.75	4.55	6.30
1-1/2"	"	1.90	5.40	7.30
2"	"	3.10	5.40	8.50
2-1/2"	"	5.15	5.40	10.55
3"	"	5.35	6.75	12.10
3-1/2"	"	7.20	6.75	13.95
4"	"	7.95	6.75	14.70
Clamp type entrance caps				
1/2"	EA.	3.80	9.30	13.10

BASIC MATERIALS	UNIT	MAT.	INST.	TOTAL
16110.22 EMT CONDUIT				
3/4"	EA.	4.55	11.00	15.55
1"	"	5.55	14.85	20.40
1-1/4"	"	6.55	19.80	26.35
1-1/2"	"	11.70	22.85	34.55
2"	"	21.55	33.30	54.85
2-1/2"	"	57.00	37.10	94.10
3"	"	57.50	56.00	113.50
3-1/2"	"	72.00	64.50	136.50
4"	"	98.00	82.50	180.50
Slip fitter type entrance caps				
1/2"	EA.	2.60	9.30	11.90
3/4"	"	3.60	11.00	14.60
1"	"	4.65	14.85	19.50
1-1/4"	"	4.90	19.80	24.70
1-1/2"	"	6.85	22.85	29.70
2"	"	8.40	33.30	41.70
2-1/2"	"	31.10	37.10	68.20
3"	"	51.50	56.00	107.50
3-1/2"	"	70.00	69.50	139.50
4"	"	78.00	82.50	160.50
16110.23 FLEXIBLE CONDUIT				
Flexible conduit, steel				
3/8"	L.F.	0.20	1.10	1.30
1/2	"	0.21	1.10	1.31
3/4"	"	0.29	1.50	1.79
1"	"	0.59	1.50	2.09
1-1/4"	"	0.75	1.85	2.60
1-1/2"	"	1.00	2.20	3.20
2"	"	1.30	2.95	4.25
2-1/2"	"	1.50	3.30	4.80
3"	"	1.90	3.95	5.85
Flexible conduit, liquid tight				
3/8"	L.F.	0.85	1.10	1.95
1/2"	"	0.92	1.10	2.02
3/4"	"	1.00	1.50	2.50
1"	"	1.95	1.50	3.45
1-1/4"	"	2.65	1.85	4.50
1-1/2"	"	3.70	2.20	5.90
2"	EA.	4.65	2.95	7.60
2-1/2"	"	8.60	3.30	11.90
3"	"	11.90	3.95	15.85
4"	"	17.35	5.40	22.75
Connector, straight				
3/8"	EA.	0.53	2.95	3.48
1/2"	"	1.30	2.95	4.25
3/4"	"	1.05	3.30	4.35
1"	"	2.10	3.70	5.80
1-1/4"	"	3.25	3.95	7.20
1-1/2"	"	5.70	4.55	10.25
2"	"	7.90	5.40	13.30
2-1/2"	"	14.80	6.75	21.55

BASIC MATERIALS	UNIT	MAT.	INST.	TOTAL
16110.23 — FLEXIBLE CONDUIT				
3"	EA.	20.65	7.05	27.70
Straight insulated throat connectors				
3/8"	EA.	2.25	4.55	6.80
1/2"	"	2.30	4.55	6.85
3/4"	"	3.35	5.40	8.75
1"	"	5.15	5.40	10.55
1-1/4"	"	8.05	6.75	14.80
1-1/2"	"	11.55	7.80	19.35
2"	"	21.65	8.50	30.15
2-1/2"	"	110.00	9.90	119.90
3"	"	120.00	12.35	132.35
4"	"	140.00	15.60	155.60
90 deg connectors				
3/8"	EA.	3.15	5.50	8.65
1/2"	"	3.25	5.50	8.75
3/4"	"	4.80	6.30	11.10
1"	"	9.75	6.75	16.50
1-1/4"	"	15.15	8.50	23.65
1-1/2"	"	18.15	9.30	27.45
2"	"	26.60	9.90	36.50
2-1/2"	"	120.00	12.35	132.35
3"	"	140.00	14.15	154.15
4"	"	180.00	16.50	196.50
90 degree insulated throat connectors				
3/8"	EA.	3.80	5.40	9.20
1/2"	"	3.85	5.40	9.25
3/4"	"	5.70	6.30	12.00
1"	"	10.90	6.60	17.50
1-1/4"	"	15.80	8.50	24.30
1-1/2"	"	19.40	9.30	28.70
2"	"	29.00	9.90	38.90
2-1/2"	"	130.00	12.35	142.35
3"	"	160.00	14.15	174.15
4"	"	210.00	16.50	226.50
Flexible aluminum conduit				
3/8"	L.F.	0.18	1.10	1.28
1/2"	"	0.27	1.10	1.37
3/4"	"	0.36	1.50	1.86
1"	"	0.81	1.50	2.31
1-1/4"	"	1.05	1.85	2.90
1-1/2"	"	1.30	2.20	3.50
2"	"	1.75	2.95	4.70
2-1/2"	"	2.20	3.30	5.50
3"	"	2.55	3.95	6.50
3-1/2"	"	3.30	4.55	7.85
4"	"	3.95	5.40	9.35
Connector, straight				
3/8"	EA.	2.05	3.70	5.75
1/2"	"	2.10	3.70	5.80
3/4"	"	2.90	3.95	6.85
1"	"	4.25	4.55	8.80
1-1/4"	"	7.35	5.40	12.75

BASIC MATERIALS	UNIT	MAT.	INST.	TOTAL
16110.23 FLEXIBLE CONDUIT				
1-1/2"	EA.	10.45	6.75	17.20
2"	"	19.20	7.05	26.25
2-1/2"	"	90.50	8.25	98.75
3"	"	100.00	10.25	110.25
4"	"	120.00	12.90	132.90
Straight insulated throat connectors				
3/8"	EA.	0.56	3.30	3.86
1/2"	"	1.10	3.30	4.40
3/4"	"	1.25	3.30	4.55
1"	"	2.75	3.70	6.45
1-1/4"	"	4.50	3.95	8.45
1-1/2"	"	6.65	4.55	11.20
2"	"	10.05	5.40	15.45
2-1/2"	"	20.05	6.75	26.80
3"	"	26.00	7.05	33.05
3-1/2"	"	84.00	8.25	92.25
4"	"	110.00	10.25	120.25
90 deg connectors				
3/8"	EA.	1.05	5.40	6.45
1/2"	"	1.65	5.40	7.05
3/4"	"	2.55	5.40	7.95
1"	"	4.25	6.30	10.55
1-1/4"	"	8.60	6.75	15.35
1-1/2"	"	14.25	7.40	21.65
2"	"	18.45	7.80	26.25
2-1/2"	"	53.50	8.50	62.00
3"	"	65.50	9.90	75.40
90 deg insulated throat connectors				
3/8"	EA.	1.10	5.40	6.50
1/2"	"	1.85	5.40	7.25
3/4"	"	2.80	5.40	8.20
1"	"	4.75	6.30	11.05
1-1/4"	"	9.10	6.75	15.85
1-1/2"	"	16.30	7.40	23.70
2"	"	21.30	7.80	29.10
2-1/2"	"	57.00	8.50	65.50
3"	"	71.00	9.90	80.90
3-1/2"	"	210.00	12.35	222.35
4"	"	320.00	15.60	335.60
16110.24 GALVANIZED CONDUIT				
Galvanized rigid steel conduit				
1/2"	L.F.	0.77	1.50	2.27
3/4"	"	0.96	1.85	2.81
1"	"	1.30	2.20	3.50
1-1/4"	"	1.70	2.95	4.65
1-1/2"	"	2.05	3.30	5.35
2"	"	2.70	3.70	6.40
2-1/2"	"	4.30	5.40	9.70
3"	"	5.70	6.75	12.45
3-1/2"	"	5.75	7.05	12.80

BASIC MATERIALS	UNIT	MAT.	INST.	TOTAL
16110.24 — GALVANIZED CONDUIT				
4"	L.F.	8.45	7.80	16.25
5"	"	18.00	10.60	28.60
6"	"	25.90	14.15	40.05
90 degree ell				
1/2"	EA.	2.95	9.30	12.25
3/4"	"	3.55	11.40	14.95
1"	"	5.25	14.15	19.40
1-1/4"	"	7.60	16.50	24.10
1-1/2"	"	9.10	18.55	27.65
2"	"	13.40	19.80	33.20
2-1/2"	"	22.55	24.75	47.30
3"	"	33.70	33.00	66.70
3-1/2"	"	58.50	37.10	95.60
4"	"	68.00	49.50	117.50
5"	"	180.00	82.50	262.50
6"	"	240.00	120.00	360.00
Couplings, with set screws				
1/2"	EA.	1.45	1.85	3.30
3/4"	"	1.95	2.20	4.15
1"	"	3.10	2.95	6.05
1-1/4"	"	5.25	3.70	8.95
1-1/2"	"	6.80	4.55	11.35
2"	"	15.30	5.40	20.70
2-1/2"	"	37.90	7.05	44.95
3"	"	45.50	9.30	54.80
3-1/2"	"	64.50	10.60	75.10
4"	"	84.00	11.40	95.40
5"	"	130.00	16.50	146.50
6"	"	170.00	18.55	188.55
Split couplings				
1/2"	EA.	1.30	7.05	8.35
3/4"	"	1.65	9.30	10.95
1"	"	2.30	10.25	12.55
1-1/4"	"	4.65	11.40	16.05
1-1/2"	"	5.80	14.15	19.95
2"	"	26.90	21.20	48.10
2-1/2"	"	27.40	21.20	48.60
3"	"	40.60	27.00	67.60
3-1/2"	"	64.50	37.10	101.60
4"	"	76.00	49.50	125.50
5"	"	130.00	60.50	190.50
6"	"	180.00	76.00	256.00
Erickson couplings				
1/2"	EA.	2.15	16.50	18.65
3/4"	"	2.60	18.55	21.15
1"	"	5.25	22.85	28.10
1-1/4"	"	9.40	33.00	42.40
1-1/2"	"	12.25	37.10	49.35
2"	"	24.55	49.50	74.05
2-1/2"	"	47.90	69.00	116.90
3"	"	70.50	78.00	148.50
3-1/2"	"	130.00	93.00	223.00

BASIC MATERIALS	UNIT	MAT.	INST.	TOTAL
16110.24 GALVANIZED CONDUIT				
4"	EA.	150.00	99.00	249.00
5"	"	300.00	110.00	410.00
6"	"	400.00	120.00	520.00
Seal fittings				
1/2"	EA.	8.95	24.75	33.70
3/4"	"	10.35	29.70	40.05
1"	"	12.70	37.10	49.80
1-1/4"	"	15.40	42.40	57.80
1-1/2"	"	22.75	49.50	72.25
2"	"	29.70	59.50	89.20
2-1/2"	"	45.70	70.50	116.20
3"	"	55.00	78.00	133.00
3-1/2"	"	150.00	93.00	243.00
4"	"	230.00	110.00	340.00
5"	"	350.00	160.00	510.00
6"	"	470.00	190.00	660.00
Entrance fitting, (weather head), threaded				
1/2"	EA.	2.50	16.50	19.00
3/4"	"	3.05	18.55	21.60
1"	"	3.95	21.20	25.15
1-1/4"	"	5.10	27.00	32.10
1-1/2"	"	7.50	29.70	37.20
2"	"	13.80	33.00	46.80
2-1/2"	"	47.50	37.10	84.60
3"	"	67.00	49.50	116.50
3-1/2"	"	90.50	64.50	155.00
4"	"	110.00	93.00	203.00
5"	"	240.00	130.00	370.00
6"	"	250.00	160.00	410.00
Locknuts				
1/2"	EA.	0.09	1.85	1.94
3/4"	"	0.10	1.85	1.95
1"	"	0.19	1.85	2.04
1-1/4"	"	0.25	1.85	2.10
1-1/2"	"	0.36	2.20	2.56
2"	"	0.52	2.20	2.72
2-1/2"	"	1.35	2.95	4.30
3"	"	1.70	2.95	4.65
3-1/2"	"	2.90	2.95	5.85
4"	"	3.65	3.30	6.95
5"	"	7.75	3.30	11.05
6"	"	13.60	3.30	16.90
Plastic conduit bushings				
1/2"	EA.	0.09	4.55	4.64
3/4"	"	0.13	5.40	5.53
1"	"	0.19	7.05	7.24
1-1/4"	"	0.26	8.25	8.51
1-1/2"	"	0.35	9.30	9.65
2"	"	0.63	11.40	12.03
2-1/2"	"	1.55	18.55	20.10
3"	"	1.95	24.75	26.70
3-1/2"	"	2.05	29.70	31.75

BASIC MATERIALS	UNIT	MAT.	INST.	TOTAL
16110.24 GALVANIZED CONDUIT				
4"	EA.	2.35	33.00	35.35
5"	"	5.35	42.40	47.75
6"	"	10.25	59.50	69.75
Conduit bushings, steel				
1/2"	EA.	0.13	4.55	4.68
3/4"	"	0.18	5.40	5.58
1"	"	0.27	7.05	7.32
1-1/4"	"	0.38	8.25	8.63
1-1/2"	"	0.57	9.30	9.87
2"	"	0.88	11.40	12.28
2-1/2"	"	2.00	18.55	20.55
3"	"	2.45	24.75	27.20
3-1/2"	"	5.15	29.70	34.85
4"	"	6.25	33.00	39.25
5"	"	13.25	42.40	55.65
6"	"	23.55	59.50	83.05
Pipe cap				
1/2"	EA.	0.16	1.85	2.01
3/4"	"	0.17	1.85	2.02
1"	"	0.28	1.85	2.13
1-1/4"	"	0.48	2.95	3.43
1-1/2"	"	0.75	2.95	3.70
2"	"	0.84	2.95	3.79
2-1/2"	"	1.40	3.30	4.70
3"	"	1.80	3.30	5.10
3-1/2"	"	2.45	3.30	5.75
4"	"	3.15	3.95	7.10
5"	"	4.15	5.40	9.55
6"	"	5.25	7.40	12.65
GRS elbows, 36" radius				
2"	EA.	70.00	24.75	94.75
2-1/2"	"	95.50	30.00	125.50
3"	"	130.00	39.10	169.10
3-1/2"	"	170.00	46.40	216.40
4"	"	190.00	56.00	246.00
5"	"	310.00	93.00	403.00
6"	"	330.00	140.00	470.00
42" radius				
2"	EA.	77.50	30.00	107.50
2-1/2"	"	110.00	37.10	147.10
3"	"	140.00	46.40	186.40
3-1/2"	"	190.00	56.00	246.00
4"	"	230.00	64.50	294.50
5"	"	340.00	110.00	450.00
6"	"	360.00	150.00	510.00
48" radius				
2"	EA.	87.50	34.50	122.00
2-1/2"	"	120.00	41.80	161.80
3"	"	160.00	53.00	213.00
3-1/2"	"	220.00	64.50	284.50
4"	"	270.00	80.00	350.00
5"	"	390.00	110.00	500.00

BASIC MATERIALS	UNIT	MAT.	INST.	TOTAL
16110.24 GALVANIZED CONDUIT				
6"	EA.	410.00	160.00	570.00
Threaded couplings				
1/2"	EA.	0.97	1.85	2.82
3/4"	"	1.20	2.20	3.40
1"	"	1.70	2.95	4.65
1-1/4"	"	2.10	3.30	5.40
1-1/2"	"	2.70	3.70	6.40
2"	"	3.55	3.95	7.50
2-1/2"	"	7.90	4.55	12.45
3"	"	10.80	5.40	16.20
3-1/2"	"	14.45	5.40	19.85
4"	"	15.20	5.95	21.15
5"	"	34.40	6.75	41.15
6"	"	45.70	7.05	52.75
Threadless couplings				
1/2"	EA.	1.65	3.70	5.35
3/4"	"	2.45	4.55	7.00
1"	"	4.20	5.40	9.60
1-1/4"	"	6.35	7.05	13.40
1-1/2"	"	8.10	9.30	17.40
2"	"	18.35	11.40	29.75
2-1/2"	"	53.50	18.55	72.05
3"	"	70.50	22.85	93.35
3-1/2"	"	94.00	30.00	124.00
4"	"	120.00	37.10	157.10
5"	"	270.00	46.40	316.40
6"	"	340.00	200.00	540.00
Threadless connectors				
1/2"	EA.	1.05	3.70	4.75
3/4"	"	1.70	4.55	6.25
1"	"	2.40	5.40	7.80
1-1/4"	"	4.35	7.05	11.40
1-1/2"	"	6.05	9.30	15.35
2"	"	11.85	11.40	23.25
2-1/2"	"	37.70	18.55	56.25
3"	"	50.50	22.85	73.35
3-1/2"	"	65.00	30.00	95.00
4"	"	79.50	37.10	116.60
5"	"	200.00	46.40	246.40
6"	"	270.00	56.00	326.00
Setscrew connectors				
1/2"	EA.	0.96	2.95	3.91
3/4"	"	1.35	3.30	4.65
1"	"	2.10	3.70	5.80
1-1/4"	"	3.70	4.55	8.25
1-1/2"	"	5.40	5.40	10.80
2"	"	10.65	7.05	17.70
2-1/2"	"	31.80	9.30	41.10
3"	"	45.20	11.40	56.60
3-1/2"	"	59.00	14.15	73.15
4"	"	74.00	18.55	92.55
5"	"	130.00	22.85	152.85

BASIC MATERIALS	UNIT	MAT.	INST.	TOTAL
16110.24 GALVANIZED CONDUIT				
6"	EA.	170.00	30.00	200.00
Clamp type entrance caps				
1/2"	EA.	2.80	11.40	14.20
3/4"	"	3.30	14.15	17.45
1"	"	4.00	16.50	20.50
1-1/4"	"	4.10	18.55	22.65
1-1/2"	"	8.45	22.85	31.30
2"	"	15.80	27.00	42.80
3-1/2"	"	41.60	34.90	76.50
3"	"	56.00	41.80	97.80
3-1/2"	"	60.50	51.00	111.50
4"	"	80.50	90.00	170.50
"LB" condulets				
1/2"	EA.	3.20	11.40	14.60
3/4"	"	3.85	14.15	18.00
1"	"	5.80	16.50	22.30
1-1/4"	"	10.00	18.55	28.55
1-1/2"	"	13.05	22.85	35.90
2"	"	21.50	27.00	48.50
2-1/2"	"	45.10	37.10	82.20
3"	"	59.50	51.00	110.50
3-1/2"	"	110.00	64.50	174.50
4"	"	120.00	78.00	198.00
"T" condulets				
1/2"	EA.	4.00	14.15	18.15
3/4"	"	4.80	16.50	21.30
1"	"	7.20	18.55	25.75
1-1/4"	"	10.60	21.20	31.80
1-1/2"	"	14.10	22.85	36.95
2"	"	21.85	27.00	48.85
2-1/2"	"	47.20	41.80	89.00
3"	"	61.50	56.00	117.50
3-1/2"	"	110.00	69.00	179.00
4"	"	130.00	82.50	212.50
"X" condulets				
1/2"	EA.	5.95	16.50	22.45
3/4"	"	6.40	18.55	24.95
1"	"	10.50	21.20	31.70
1-1/4"	"	13.70	22.85	36.55
1-1/2"	"	17.70	24.75	42.45
2"	"	36.40	32.60	69.00
Blank steel condulet covers				
1/2"	EA.	0.92	3.70	4.62
3/4"	"	1.10	3.70	4.80
1"	"	1.55	3.70	5.25
1-1/4"	"	1.85	4.55	6.40
1-1/2"	"	1.95	4.55	6.50
2"	"	3.30	4.55	7.85
2-1/2"	"	5.20	5.40	10.60
3"	"	5.60	5.40	11.00
3-1/2"	"	6.15	5.40	11.55
4"	"	6.70	7.40	14.10

BASIC MATERIALS	UNIT	MAT.	INST.	TOTAL
16110.24 GALVANIZED CONDUIT				
Solid condulet gaskets				
1/2"	EA.	1.10	1.85	2.95
3/4"	"	1.10	1.85	2.95
1"	"	1.40	1.85	3.25
1-1/4"	"	1.75	2.95	4.70
1-1/2"	"	1.85	2.95	4.80
2"	"	2.05	2.95	5.00
2-1/2"	"	3.40	3.70	7.10
3"	"	3.50	3.70	7.20
3-1/2"	"	4.30	3.70	8.00
4"	"	4.65	5.40	10.05
One-hole malleable straps				
1/2"	EA.	0.17	1.50	1.67
3/4"	"	0.21	1.50	1.71
1"	"	0.30	1.50	1.80
1-1/4"	"	0.55	1.85	2.40
1-1/2"	"	0.66	1.85	2.51
2"	"	1.30	1.85	3.15
2-1/2"	"	2.70	2.20	4.90
3"	"	3.85	2.20	6.05
3-1/2"	"	5.55	2.20	7.75
4"	"	12.30	2.95	15.25
5"	"	43.40	2.95	46.35
6"	"	46.50	2.95	49.45
One-hole steel straps				
1/2"	EA.	0.12	1.50	1.62
3/4"	"	0.13	1.50	1.63
1"	"	0.24	1.50	1.74
1-1/4"	"	0.31	1.85	2.16
1-1/2"	"	0.39	1.85	2.24
2"	"	0.75	1.85	2.60
2-1/2"	"	1.10	2.20	3.30
3"	"	1.25	2.20	3.45
3-1/2"	"	1.85	2.20	4.05
4"	"	2.35	2.95	5.30
Bushed chase nipples				
1/2"	EA.	0.22	2.20	2.42
3/4"	"	0.29	2.60	2.89
1"	"	0.55	3.30	3.85
1-1/4"	"	0.88	3.70	4.58
1-1/2"	"	1.15	4.55	5.70
2"	"	1.75	5.40	7.15
2-1/2"	"	4.10	5.40	9.50
3"	"	5.90	6.75	12.65
3-1/2"	"	12.55	9.30	21.85
4"	"	20.60	11.00	31.60
Offset nipples				
1/2"	EA.	1.60	2.20	3.80
3/4"	"	1.70	2.60	4.30
1"	"	2.10	3.30	5.40
1-1/4"	"	4.55	3.95	8.50
1-1/2"	"	5.60	4.55	10.15

BASIC MATERIALS	UNIT	MAT.	INST.	TOTAL
16110.24 GALVANIZED CONDUIT				
2"	EA.	8.90	5.40	14.30
3"	"	29.00	6.75	35.75
Short elbows				
1/2"	EA.	1.10	5.40	6.50
3/4"	"	1.50	7.40	8.90
1"	"	2.60	9.30	11.90
1-1/4"	"	6.90	11.00	17.90
1-1/2"	"	10.20	12.90	23.10
2"	"	17.00	14.85	31.85
Pulling elbows, female to female				
1/2"	EA.	2.80	9.30	12.10
3/4"	"	3.25	11.00	14.25
1"	"	5.45	14.85	20.30
1-1/4"	"	8.00	19.80	27.80
1-1/2"	"	15.00	27.00	42.00
2"	"	12.30	31.60	43.90
Grounding locknuts				
1/2"	EA.	0.73	2.95	3.68
3/4"	"	0.91	2.95	3.86
1"	"	1.30	2.95	4.25
1-1/4"	"	1.40	3.30	4.70
1-1/2"	"	1.45	3.30	4.75
2"	"	2.00	3.30	5.30
2-1/2"	"	3.70	3.70	7.40
3"	"	4.65	3.70	8.35
3-1/2"	"	7.55	3.70	11.25
4"	"	10.15	5.40	15.55
Insulated grounding metal bushings				
1/2"	EA.	2.10	7.05	9.15
3/4"	"	2.70	8.25	10.95
1"	"	3.65	9.30	12.95
1-1/4"	"	4.40	11.40	15.80
1-1/2"	"	5.05	14.15	19.20
2"	"	5.95	16.50	22.45
2-1/2"	"	8.10	24.75	32.85
3"	"	9.35	30.00	39.35
3-1/2"	"	13.55	34.90	48.45
4"	"	16.90	39.10	56.00
5"	"	33.90	58.00	91.90
6"	"	51.00	64.50	115.50
Nipples				
1/2" x				
4"	EA.	1.30	5.40	6.70
6"	"	1.70	5.40	7.10
8"	"	2.90	5.40	8.30
10"	"	3.35	5.40	8.75
12"	"	3.85	5.40	9.25
3/4" x				
4"	EA.	1.45	5.40	6.85
6"	"	1.95	5.40	7.35
8"	"	3.25	5.40	8.65
10"	"	3.90	5.40	9.30

BASIC MATERIALS	UNIT	MAT.	INST.	TOTAL
16110.24 — GALVANIZED CONDUIT				
12"	EA.	4.35	5.40	9.75
1" x				
4"	EA.	2.10	5.40	7.50
6"	"	2.60	5.40	8.00
8"	"	4.25	5.40	9.65
10"	"	5.30	5.40	10.70
12"	"	6.05	5.40	11.45
1-1/4" x				
4"	EA.	2.55	9.30	11.85
6"	"	3.40	9.30	12.70
8"	"	5.50	9.30	14.80
10"	"	6.90	9.30	16.20
12"	"	8.10	9.30	17.40
1-1/2" x				
4"	EA.	3.40	9.30	12.70
6"	"	4.70	9.30	14.00
8"	"	6.90	9.30	16.20
10"	"	8.30	9.30	17.60
12"	"	9.05	9.30	18.35
2" x				
4"	EA.	4.45	9.30	13.75
6"	"	5.70	9.30	15.00
8"	"	8.60	9.30	17.90
10"	"	10.30	9.30	19.60
12"	"	11.75	9.30	21.05
2-1/2" x				
6"	EA.	12.45	11.10	23.55
8"	"	16.35	11.10	27.45
10"	"	19.00	11.10	30.10
12"	"	22.10	11.10	33.20
3" x				
6"	EA.	14.50	11.10	25.60
8"	"	19.45	11.10	30.55
10"	"	25.60	11.10	36.70
12"	"	27.90	11.10	39.00
3-1/2" x				
6"	EA.	17.75	11.10	28.85
8"	"	22.25	11.10	33.35
10"	"	26.90	11.10	38.00
12"	"	31.40	11.10	42.50
4" x				
8"	EA.	25.10	14.85	39.95
10"	"	31.00	14.85	45.85
12"	"	37.00	14.85	51.85
5" x				
8"	EA.	41.80	14.85	56.65
10"	"	47.30	14.85	62.15
12"	"	59.50	14.85	74.35
6" x				
8"	EA.	52.00	14.85	66.85
10"	"	65.00	14.85	79.85
12"	"	72.00	14.85	86.85

BASIC MATERIALS	UNIT	MAT.	INST.	TOTAL
16110.25 PLASTIC CONDUIT				
PVC conduit, schedule 40				
1/2"	L.F.	0.21	1.10	1.31
3/4"	"	0.27	1.10	1.37
1"	"	0.39	1.50	1.89
1-1/4"	"	0.53	1.50	2.03
1-1/2"	"	0.64	1.85	2.49
2"	"	0.83	1.85	2.68
2-1/2"	"	1.35	2.20	3.55
3"	"	1.60	2.20	3.80
3-1/2"	"	2.05	2.95	5.00
4"	"	2.25	2.95	5.20
5"	"	3.30	3.30	6.60
6"	"	4.30	3.70	8.00
Couplings				
1/2"	EA.	0.19	1.85	2.04
3/4"	"	0.20	1.85	2.05
1"	"	0.31	1.85	2.16
1-1/4"	"	0.40	2.20	2.60
1-1/2"	"	0.54	2.20	2.74
2"	"	0.74	2.20	2.94
2-1/2"	"	1.35	2.20	3.55
3"	"	2.15	2.95	5.10
3-1/2"	"	2.40	2.95	5.35
4"	"	3.30	3.70	7.00
5"	"	8.20	3.70	11.90
6"	"	10.55	3.70	14.25
90 degree elbows				
1/2"	EA.	0.72	3.70	4.42
3/4"	"	0.77	4.55	5.32
1"	"	0.95	4.55	5.50
1-1/4"	"	1.35	5.40	6.75
1-1/2"	"	1.90	7.05	8.95
2"	"	2.70	8.25	10.95
2-1/2"	"	4.80	9.30	14.10
3"	"	8.40	11.40	19.80
3-1/2"	"	11.70	14.15	25.85
4"	"	17.40	18.55	35.95
5"	"	25.50	22.85	48.35
6"	"	46.30	27.00	73.30
Terminal adapters				
1/2"	EA.	0.25	3.70	3.95
3/4"	"	0.41	3.70	4.11
1"	"	0.49	3.70	4.19
1-1/4"	"	0.65	5.95	6.60
1-1/2"	"	0.80	5.95	6.75
2"	"	1.10	5.95	7.05
2-1/2"	"	1.90	8.25	10.15
3"	"	2.70	8.25	10.95
3-1/2"	"	3.45	8.25	11.70
4"	"	4.50	14.15	18.65
5"	"	9.00	14.15	23.15
6"	"	10.80	14.15	24.95

BASIC MATERIALS	UNIT	MAT.	INST.	TOTAL
16110.25 — PLASTIC CONDUIT				
End bells				
1"	EA.	1.35	3.70	5.05
1-1/4"	"	1.45	5.95	7.40
1-1/2"	"	1.50	5.95	7.45
2"	"	2.25	5.95	8.20
2-1/2"	"	2.45	8.25	10.70
3"	"	2.75	8.25	11.00
3-1/2"	"	2.85	8.25	11.10
4"	"	3.25	14.15	17.40
5"	"	4.95	14.15	19.10
6"	"	5.85	14.15	20.00
LB conduit body				
1/2"	EA.	1.70	7.05	8.75
3/4"	"	2.35	7.05	9.40
1	"	2.60	7.05	9.65
1-1/4"	"	3.90	11.40	15.30
1-1/2"	"	4.95	11.40	16.35
2"	"	7.15	11.40	18.55
2-1/2"	"	25.30	16.50	41.80
3"	"	26.10	19.80	45.90
3-1/2"	"	27.20	22.85	50.05
4"	"	28.50	27.00	55.50
Direct burial, conduit				
2"	L.F.	0.41	1.85	2.26
3"	"	0.76	2.20	2.96
4"	"	1.50	2.95	4.45
5"	"	2.15	3.30	5.45
6"	"	3.00	3.70	6.70
Encased burial conduit				
2"	L.F.	0.43	1.85	2.28
3"	"	0.69	2.20	2.89
4"	"	1.05	2.95	4.00
5"	"	1.55	3.30	4.85
6"	"	2.00	3.70	5.70
"EB" and "DB" duct, 90 degree elbows				
1-1/2"	EA.	5.25	5.40	10.65
2"	"	5.90	8.50	14.40
3"	"	7.50	14.15	21.65
3-1/2"	"	12.85	16.50	29.35
4"	"	13.75	19.80	33.55
5"	"	16.65	24.75	41.40
6"	"	26.80	33.30	60.10
45 degree elbows				
1-1/2"	EA.	4.15	8.50	12.65
2"	"	4.25	8.50	12.75
3"	"	5.35	14.15	19.50
3-1/2"	"	7.60	16.50	24.10
4"	"	7.75	19.80	27.55
5"	"	11.50	24.75	36.25
6"	"	39.20	33.30	72.50
Couplings				
1-1/2"	EA.	0.67	2.20	2.87

BASIC MATERIALS	UNIT	MAT.	INST.	TOTAL
16110.25 — PLASTIC CONDUIT				
2"	EA.	0.88	2.20	3.08
3"	"	1.05	2.95	4.00
3-1/2"	"	1.15	2.95	4.10
4"	"	2.70	3.70	6.40
5"	"	3.95	3.70	7.65
6"	"	5.40	5.95	11.35
Bell ends				
1-1/2"	EA.	2.60	5.95	8.55
2"	"	3.30	5.95	9.25
3"	"	3.70	8.25	11.95
3-1/2"	"	3.85	8.25	12.10
4"	"	4.15	14.15	18.30
5"	"	5.60	14.15	19.75
6"	"	6.05	14.15	20.20
Female adapters, 1-1/2"	"	0.78	7.40	8.18
5 degree couplings				
1-1/2"	EA.	2.75	2.60	5.35
2"	"	3.10	2.60	5.70
3"	"	4.00	3.70	7.70
4"	"	2.15	5.40	7.55
5"	"	2.75	5.40	8.15
6"	"	4.20	5.40	9.60
45 degree elbows				
1/2"	EA.	0.53	4.55	5.08
3/4"	"	0.65	5.40	6.05
1"	"	0.90	5.40	6.30
1-1/4"	"	1.30	6.75	8.05
1-1/2"	"	1.80	8.50	10.30
2"	"	2.65	9.90	12.55
2-1/2"	"	4.90	11.00	15.90
3"	"	8.45	14.15	22.60
3-1/2"	"	9.50	16.50	26.00
4"	"	13.35	22.85	36.20
5"	"	21.35	27.00	48.35
6"	"	31.10	33.30	64.40
Female adapters				
1/2"	EA.	0.26	4.55	4.81
3/4"	"	0.43	4.55	4.98
1"	"	0.54	4.55	5.09
1-1/4"	"	0.67	7.40	8.07
1-1/2"	"	0.76	7.40	8.16
2"	"	1.05	7.40	8.45
2-1/2"	"	1.85	9.90	11.75
3"	"	3.05	9.90	12.95
3-1/2"	"	3.80	9.90	13.70
4"	"	4.85	16.50	21.35
5"	"	10.85	16.50	27.35
6"	"	12.35	16.50	28.85
Expansion couplings				
1/2"	EA.	8.75	4.55	13.30
3/4"	"	8.95	4.55	13.50
1"	"	9.10	5.40	14.50

BASIC MATERIALS	UNIT	MAT.	INST.	TOTAL
16110.25 PLASTIC CONDUIT				
1-1/4"	EA.	9.20	7.40	16.60
1-1/2"	"	9.35	7.40	16.75
2"	"	9.90	7.40	17.30
2-1/2"	"	14.20	11.00	25.20
3"	"	18.10	11.00	29.10
3-1/2"	"	20.85	11.00	31.85
4"	"	25.50	16.50	42.00
5"	"	37.30	16.50	53.80
6"	"	49.70	16.50	66.20
Plugs				
2"	EA.	0.54	7.40	7.94
3"	"	0.69	11.00	11.69
3-1/2"	"	0.85	11.00	11.85
4"	"	1.05	16.50	17.55
5"	"	1.35	16.50	17.85
6"	"	2.70	18.55	21.25
PVC cement				
1 pint	EA.			5.22
1 quart	"			7.59
1 gallon	"			23.15
Type "T" condulets				
1/2"	EA.	2.95	11.00	13.95
3/4"	"	3.35	11.00	14.35
1"	"	3.80	11.00	14.80
1-1/4"	"	4.75	18.55	23.30
1-1/2"	"	5.95	18.55	24.50
2"	"	8.00	18.55	26.55
EB & DB female adapters				
2"	EA.	1.05	9.30	10.35
3"	"	3.95	14.15	18.10
3-1/2"	"	4.85	22.85	27.70
4"	"	5.25	27.00	32.25
5"	"	11.55	37.10	48.65
6"	"	13.05	59.50	72.55
16110.27 PLASTIC COATED CONDUIT				
Rigid steel conduit, plastic coated				
1/2"	L.F.	2.30	1.85	4.15
3/4"	"	2.65	2.20	4.85
1"	"	3.45	2.95	6.40
1-1/4"	"	4.35	3.70	8.05
1-1/2"	"	5.30	4.55	9.85
2"	"	6.95	5.40	12.35
2-1/2"	"	10.60	7.05	17.65
3"	"	13.35	8.25	21.60
3-1/2"	"	16.20	9.30	25.50
4"	"	19.80	11.40	31.20
5"	"	34.00	14.15	48.15
90 degree elbows				
1/2"	EA.	9.05	11.40	20.45
3/4"	"	9.20	14.15	23.35
1"	"	10.70	16.50	27.20

BASIC MATERIALS	UNIT	MAT.	INST.	TOTAL
16110.27 PLASTIC COATED CONDUIT				
1-1/4"	EA.	13.40	18.55	31.95
1-1/2"	"	16.20	22.85	39.05
2"	"	22.55	29.70	52.25
2-1/2"	"	43.60	42.40	86.00
3"	"	44.10	49.50	93.60
3-1/2"	"	94.50	60.50	155.00
4"	"	100.00	74.00	174.00
5"	"	230.00	93.00	323.00
Couplings				
1/2"	EA.	2.60	2.20	4.80
3/4"	"	2.80	2.95	5.75
1"	"	3.75	3.30	7.05
1-1/4"	"	4.35	3.95	8.30
1-1/2"	"	5.65	4.55	10.20
2"	"	7.60	5.40	13.00
2-1/2"	"	19.00	6.75	25.75
3"	"	22.20	7.05	29.25
3-1/2"	"	30.50	7.40	37.90
4"	"	37.40	8.25	45.65
5"	"	110.00	9.30	119.30
1 hole conduit straps				
3/4"	EA.	3.00	1.85	4.85
1"	"	4.35	1.85	6.20
1-1/4"	"	5.50	2.20	7.70
1-1/2"	"	6.65	2.20	8.85
2"	"	7.65	2.20	9.85
3"	"	11.90	2.95	14.85
3-1/2"	"	13.30	2.95	16.25
4"	"	18.15	3.70	21.85
"L.B." condulets with covers				
1/2"	EA.	23.35	18.55	41.90
3/4"	"	25.90	18.55	44.45
1"	"	34.70	22.85	57.55
1-1/4"	"	50.50	27.00	77.50
1-1/2"	"	61.50	32.60	94.10
2"	"	90.50	37.10	127.60
2-1/2"	"	160.00	51.00	211.00
3"	"	200.00	64.50	264.50
3-1/2"	"	290.00	80.00	370.00
4"	"	330.00	93.00	423.00
"T" condulets with covers				
1/2"	EA.	27.00	21.20	48.20
3/4"	"	30.10	22.85	52.95
1"	"	40.30	24.75	65.05
1-1/4"	"	57.00	30.00	87.00
1-1/2"	"	71.00	34.90	105.90
2"	"	100.00	39.10	139.10
2-1/2"	"	170.00	56.00	226.00
3-1/2"	"	320.00	82.50	402.50
4"	"	350.00	99.00	449.00
5"	"	370.00	120.00	490.00

BASIC MATERIALS	UNIT	MAT.	INST.	TOTAL

16110.28 STEEL CONDUIT

	UNIT	MAT.	INST.	TOTAL
Intermediate metal conduit (IMC)				
1/2"	L.F.	0.73	1.10	1.83
3/4"	"	0.86	1.50	2.36
1"	"	1.20	1.85	3.05
1-1/4"	"	1.55	2.20	3.75
1-1/2"	"	1.80	2.95	4.75
2"	"	2.50	3.30	5.80
2-1/2"	"	3.90	4.45	8.35
3"	"	5.25	5.40	10.65
3-1/2"	"	6.95	6.75	13.70
4"	"	8.25	7.05	15.30
90 degree ell				
1/2"	EA.	2.35	9.30	11.65
3/4"	"	3.10	11.40	14.50
1"	"	4.50	14.15	18.65
1-1/4"	"	6.25	16.50	22.75
1-1/2"	"	7.90	18.55	26.45
2"	"	11.45	21.20	32.65
2-1/2"	"	19.80	24.75	44.55
3"	"	30.30	33.00	63.30
3-1/2"	"	53.50	42.40	95.90
4"	"	62.00	49.50	111.50
Couplings				
1/2"	EA.	0.94	1.85	2.79
3/4"	"	1.15	2.20	3.35
1"	"	1.65	2.95	4.60
1-1/4"	"	2.05	3.30	5.35
1-1/2"	"	2.60	3.70	6.30
2"	"	3.45	3.95	7.40
2-1/2"	"	7.70	4.55	12.25
3"	"	10.50	5.40	15.90
3-1/2"	"	14.05	5.40	19.45
4"	"	14.80	5.95	20.75

16110.32 FLEXIBLE WIRING SYSTEMS

	UNIT	MAT.	INST.	TOTAL
Single circuit cables				
5'	EA.	10.00	2.20	12.20
10'	"	14.25	3.70	17.95
15'	"	17.65	5.40	23.05
20'	"	24.75	7.40	32.15
25'	"	31.90	9.90	41.80
30'	"	39.00	11.00	50.00
40'	"	53.50	14.85	68.35
Two circuit cables				
5'	EA.	13.60	2.20	15.80
10'	"	15.95	3.70	19.65
15'	"	20.00	5.40	25.40
20'	"	28.20	7.40	35.60
25'	"	36.30	9.90	46.20
30'	"	44.30	11.00	55.30
40'	"	60.50	14.85	75.35

BASIC MATERIALS	UNIT	MAT.	INST.	TOTAL
16110.32 FLEXIBLE WIRING SYSTEMS				
Two wire switch and receptacle cables				
5'	EA.	5.60	2.20	7.80
10'	"	11.35	3.70	15.05
15'	"	14.55	5.40	19.95
20'	"	25.10	7.40	32.50
25'	"	33.30	9.90	43.20
30'	"	45.00	11.00	56.00
40'	"	57.50	12.90	70.40
Three wire switch				
5'	EA.	6.45	2.20	8.65
10'	"	12.95	3.70	16.65
15'	"	17.00	5.40	22.40
20'	"	25.10	7.40	32.50
25'	"	33.30	9.90	43.20
30'	"	45.00	11.00	56.00
40'	"	57.50	12.90	70.40
Distribution boxes				
2 circuit	EA.	10.50	19.80	30.30
3 circuit	"	12.10	24.75	36.85
4 circuit	"	18.25	29.70	47.95
6 circuit	"	24.30	40.70	65.00
12 circuit	"	37.50	82.50	120.00
18 circuit	"	59.00	110.00	169.00
Tap boxes				
1 single pole switch	EA.	21.20	14.85	36.05
2 single pole switches	"	23.35	19.80	43.15
1 3 way switch	"	24.70	16.50	41.20
1 4 way switch	"	30.70	19.80	50.50
1 receptacle	"	19.05	14.85	33.90
2 receptacles	"	23.85	19.80	43.65
4 receptacles	"	29.80	29.70	59.50
1 clock	"	23.85	14.85	38.70
2 clocks	"	27.60	19.80	47.40
4 clocks	"	31.10	29.70	60.80
Dust cap	"	3.10	3.70	6.80
Cable coupler	"	7.40	7.40	14.80
Reversing connector	"	10.50	11.00	21.50
16110.35 WIREMOLD				
Wiremold raceway with fittings, surface mounted				
#200	L.F.	0.49	1.10	1.59
#500	"	0.59	1.10	1.69
#700	"	0.65	1.50	2.15
#800	"	1.05	1.50	2.55
Fittings, #200, 90 degree flat elbow	EA.	0.95	1.85	2.80
Internal elbow	"	1.90	1.85	3.75
Extension adapter	"	3.35	2.20	5.55
#200, #500, #700				
Single pole switch and box	EA.	6.35	14.85	21.20
Duplex receptacle with box	"	7.35	11.40	18.75
#500, #700				

BASIC MATERIALS	UNIT	MAT.	INST.	TOTAL
16110.35 WIREMOLD				
90 deg. flat elbow	EA.	0.82	2.95	3.77
Internal elbow	"	1.05	2.95	4.00
Junction box	"	5.40	4.95	10.35
Fixture box	"	5.45	4.95	10.40
Shallow switch and receptacle box	"	4.45	2.95	7.40
#800				
90 deg. flat elbow	EA.	1.00	2.95	3.95
Internal elbow	"	1.00	2.95	3.95
Junction box	"	3.15	4.55	7.70
#1500 series				
Raceway	L.F.	1.05	1.85	2.90
Wire clip	EA.	0.25	1.50	1.75
2 hole clip	"	0.30	2.20	2.50
Flat elbow	"	3.60	7.40	11.00
Internal elbow	"	3.70	7.40	11.10
External elbow	"	4.00	7.40	11.40
Adapter fitting	"	4.05	1.85	5.90
Duplex receptacle box	"	7.55	5.40	12.95
#2000 series				
Raceway	L.F.	1.10	1.85	2.95
Wire clip	EA.	0.29	1.50	1.79
Coupling	"	0.35	1.85	2.20
Supporting clip	"	0.66	1.50	2.16
Entrance end	"	3.40	5.40	8.80
Flat elbow	"	3.20	5.40	8.60
Tee	"	6.40	7.40	13.80
Single pole switch	"	8.40	7.40	15.80
#2100 series				
Raceway	L.F.	0.71	1.85	2.56
Cover	"	0.55	0.37	0.92
Wireclip	EA.	0.30	1.50	1.80
Coupling	"	0.41	1.85	2.26
Entrance end	"	3.35	7.40	10.75
Flat elbow	"	3.65	7.40	11.05
Tee	"	4.55	7.40	11.95
Device box	"	5.20	5.40	10.60
#2220 series				
Raceway	L.F.	2.20	3.70	5.90
Cover	"	0.43	0.74	1.17
Wire cup	EA.	0.52	1.50	2.02
Coupling	"	1.20	1.85	3.05
Entrance end	"	8.00	7.40	15.40
Flat elbow	"	14.80	7.40	22.20
Wall box connection	"	5.95	7.40	13.35
External elbow	"	7.85	5.40	13.25
#2600 series				
Raceway	L.F.	2.45	3.70	6.15
Wireclip	EA.	0.37	1.50	1.87
Bushing	"	0.59	1.50	2.09
Flat elbow	"	5.15	7.40	12.55
Junction box	"	6.95	7.40	14.35
#3000 series				

BASIC MATERIALS	UNIT	MAT.	INST.	TOTAL
16110.35 WIREMOLD				
Raceway	L.F.	2.65	3.70	6.35
Wire clip	EA.	0.71	1.50	2.21
Coupling	"	1.35	1.50	2.85
Supporting clip	"	1.45	1.85	3.30
Device bracket	"	3.10	2.20	5.30
Entrance end	"	4.35	7.40	11.75
Flat elbow	"	9.75	9.30	19.05
Tee	"	14.25	12.90	27.15
Device box	"	6.00	7.40	13.40
Circuit breaker housing	"	14.30	9.30	23.60
Panel connection	"	5.40	18.55	23.95
Conduit connector	"	19.10	9.30	28.40
#4000 series				
Raceway	L.F.	3.15	3.70	6.85
Cover	"	1.90	1.85	3.75
Divider	"	0.55	1.85	2.40
Wire clip	EA.	1.20	1.85	3.05
Coupling	"	3.85	2.20	6.05
Device bracket	"	3.60	2.20	5.80
Entrance end	"	11.45	11.40	22.85
Flat elbow	"	13.10	9.30	22.40
Tee	"	23.20	11.40	34.60
Internal elbow	"	7.40	11.40	18.80
External elbow	"	16.80	11.40	28.20
Panel connector	"	8.30	27.00	35.30
#6000 series				
Raceway	L.F.	4.95	5.50	10.45
Cover	"	2.05	1.85	3.90
Wire cup	EA.	1.35	1.85	3.20
Coupling	"	4.40	3.70	8.10
Device bracket	"	4.85	3.70	8.55
Entrance end	"	13.00	16.50	29.50
Flat elbow	"	17.35	14.85	32.20
Tee	"	26.50	18.55	45.05
Internal elbow	"	16.45	14.85	31.30
External elbow	"	16.65	16.50	33.15
Panel connector	"	7.70	33.30	41.00
Telepower poles				
# 21TP2	EA.	84.00	27.00	111.00
# 21TP4	"	93.00	37.10	130.10
# 30TP4	"	160.00	37.10	197.10
Fittings				
Telephone	EA.	6.05	7.40	13.45
T-bar clamp	"	4.30	7.40	11.70
C-hanger clamp	"	7.00	7.40	14.40
Tap-off fitting	"	4.20	7.40	11.60
Take-off connector	"	28.80	27.00	55.80
16110.40 UNDERFLOOR DUCT				
Underfloor blank duct, insert duct				
7/8"	L.F.	5.35	1.85	7.20

BASIC MATERIALS	UNIT	MAT.	INST.	TOTAL
16110.40 — UNDERFLOOR DUCT				
1-3/8"	L.F.	5.85	1.85	7.70
1-7/8"	"	5.95	1.85	7.80
Box opening plugs	EA.	2.05	5.40	7.45
Duct end plugs	"	2.35	5.40	7.75
Sleeve couplings	"	5.50	12.90	18.40
Expansion couplings	"	31.30	12.90	44.20
Vertical elbow	"	26.40	12.90	39.30
Offset elbow	"	21.10	12.90	34.00
Horizontal elbow	"	51.00	12.90	63.90
Adjustable elbow	"	9.35	12.90	22.25
Cabinet connector	"	7.80	44.30	52.10
Y-take off	"	23.50	22.85	46.35
Underfloor duct leveling legs	"	2.35	5.40	7.75
Conduit adapters				
1/2"	EA.	9.35	9.30	18.65
3/4"	"	9.50	9.30	18.80
1"	"	9.60	9.30	18.90
1-1/4"	"	10.65	11.00	21.65
2"	"	14.85	12.90	27.75
Reducer bushings				
1-1/4" x 3/4"	EA.	4.70	7.40	12.10
1-1/4" x 1"	"	5.50	7.40	12.90
2" x 1-1/2"	"	6.45	9.30	15.75
Support couplers				
1 standard	EA.	10.95	9.30	20.25
2 standard	"	13.30	11.00	24.30
3 standard	"	16.45	12.90	29.35
Supports				
1 duct	EA.	11.65	5.40	17.05
2 duct	"	13.30	6.30	19.60
3 duct	"	14.00	7.05	21.05
4 duct	"	18.00	9.30	27.30
5 duct	"	19.65	11.00	30.65
Single level junction box				
1 standard	EA.	130.00	29.70	159.70
2 standard	"	190.00	56.00	246.00
3 standard	"	370.00	110.00	480.00
4 standard	"	480.00	150.00	630.00
Two level junction boxes				
1 standard	EA.	150.00	37.10	187.10
2 standard	"	290.00	74.00	364.00
Sealing compound	"			2.95
Insert adapters	"	7.05	5.40	12.45
Ellipsoids	"	7.80	12.90	20.70
Insert closing cap	"	0.81	5.40	6.21
Marker Screws	"	2.20	5.40	7.60
Access boxes	"	160.00	37.10	197.10
Closing caps	"	0.81	5.40	6.21
Afterset markers	"	2.15	5.40	7.55
Cell markers	"	4.70	5.40	10.10
Tie down straps	"	4.35	5.40	9.75
Plastic grommets	"	2.20	14.85	17.05

BASIC MATERIALS	UNIT	MAT.	INST.	TOTAL
16110.40 **UNDERFLOOR DUCT**				
Metal grommets	EA.	15.05	14.85	29.90
Receptacle				
Duplex, 20a	EA.	32.90	18.55	51.45
Single				
30a	EA.	43.70	18.55	62.25
50a	"	48.00	22.85	70.85
Double duplex	"	34.20	22.85	57.05
Single, 20a	"	30.10	16.50	46.60
Double single	"	35.60	18.55	54.15
Twist lock	"	32.90	16.50	49.40
1 conduit opening	"	31.50	16.50	48.00
2 conduit openings	"	35.60	18.55	54.15
1 bushed opening	"	31.50	16.50	48.00
2 bushed openings	"	35.60	18.55	54.15
Amphenol connector				
1"	EA.	52.00	18.55	70.55
2"	"	59.00	19.80	78.80
5"	"	70.00	22.85	92.85
Standpipes				
Aluminum	EA.	39.40	9.30	48.70
Brass	"	41.40	9.30	50.70
Abandonment plates				
Aluminum	EA.	11.10	9.30	20.40
Brass	"	18.50	9.30	27.80
Split bell caps				
Aluminum	EA.	63.00	9.30	72.30
Brass	"	79.00	9.30	88.30
Flush floor receptacles				
Aluminum	EA.	53.50	27.00	80.50
Brass	"	58.50	27.00	85.50
Flush floor telephone				
Aluminum	EA.	37.80	18.55	56.35
Brass	"	47.20	18.55	65.75
Super underfloor duct blank duct				
1/2"	L.F.	9.35	2.20	11.55
7/8"	"	9.60	2.20	11.80
1-3/8"	"	10.30	2.20	12.50
1-7/8"	"	10.95	2.20	13.15
Box opening plugs	EA.	2.25	5.40	7.65
End plugs	"	2.35	5.40	7.75
Conduit adapters	"	9.35	27.00	36.35
Sleeve coupling	"	7.05	16.50	23.55
Expansion coupling	"	46.90	16.50	63.40
Reducing coupling	"	32.30	16.50	48.80
Vertical elbow	"	33.20	16.50	49.70
Offset elbow	"	34.20	16.50	50.70
Horizontal elbow	"	66.50	16.50	83.00
Adjustable elbow	"	19.60	16.50	36.10
Cabinet connector	"	14.85	56.00	70.85
Super underfloor duct Y-take off	"	66.00	27.00	93.00
Leveling legs	"	2.75	5.40	8.15
Support couplers				

BASIC MATERIALS	UNIT	MAT.	INST.	TOTAL
16110.40 UNDERFLOOR DUCT				
1 super	EA.	13.30	9.30	22.60
2 super	"	18.00	11.00	29.00
1 super, 1 standard	"	16.45	11.00	27.45
2 super, 2 standard	"	19.50	12.90	32.40
Single level junction boxes				
1 super	EA.	180.00	56.00	236.00
2 super	"	320.00	74.00	394.00
4 super	"	490.00	130.00	620.00
Double level junction boxes				
1 super	EA.	200.00	44.30	244.30
2 super	"	290.00	56.00	346.00
16110.50 WALL DUCT				
Lay-in wall duct, 10"	L.F.	21.30	2.20	23.50
Horizontal elbow	EA.	96.00	27.00	123.00
Edgewise elbow	"	51.50	27.00	78.50
Tee	"	96.00	33.30	129.30
Cross	"	89.50	40.70	130.20
Cabinet connector	"	57.50	59.50	117.00
Reverse elbow	"	51.50	19.80	71.30
Sweep elbow	"	93.50	27.00	120.50
Partition	"	17.50	2.20	19.70
Straight tunnel	"	17.95	10.25	28.20
Elbow tunnel	"	26.90	12.90	39.80
Tee kit	"	28.20	14.85	43.05
Ceiling dropout	"	110.00	37.10	147.10
Coupling device	"	5.90	5.40	11.30
End cap	"	12.80	10.25	23.05
Lay-in wall duct, 18"	L.F.	28.40	2.95	31.35
Horizontal elbow	EA.	120.00	29.70	149.70
Edgeware elbow	"	68.00	37.10	105.10
Tee	"	120.00	40.70	160.70
Cross	"	120.00	46.40	166.40
Reverse elbow	"	69.00	27.00	96.00
Sweep elbow	"	120.00	29.70	149.70
Partition	"	17.95	3.70	21.65
Straight tunnel	"	21.80	14.85	36.65
Elbow tunnel	"	32.00	18.55	50.55
Tee kit	"	30.70	19.80	50.50
Ceiling dropout	"	110.00	40.70	150.70
Coupling device	"	8.85	7.40	16.25
Reducer coupling	"	27.40	14.85	42.25
Cabinet connector	"	33.30	74.00	107.30
End cap	"	17.95	14.85	32.80
16110.60 TRENCH DUCT				
Trench duct, with cover				
9"	L.F.	62.50	6.30	68.80
12"	"	70.00	7.40	77.40
18"	"	93.50	9.90	103.40

BASIC MATERIALS	UNIT	MAT.	INST.	TOTAL
16110.60 TRENCH DUCT				
24"	L.F.	120.00	12.90	132.90
30"	"	150.00	14.85	164.85
36"	"	180.00	21.20	201.20
Tees				
9"	EA.	230.00	64.50	294.50
12"	"	250.00	74.00	324.00
18"	"	330.00	82.50	412.50
24"	"	460.00	93.00	553.00
30"	"	620.00	110.00	730.00
36"	"	810.00	130.00	940.00
Vertical elbows				
9"	EA.	78.00	29.70	107.70
12"	"	82.00	40.70	122.70
18"	"	100.00	50.50	150.50
24"	"	110.00	62.00	172.00
30"	"	130.00	74.00	204.00
36"	"	140.00	93.00	233.00
Cabinet connectors				
9"	EA.	100.00	74.00	174.00
12"	"	110.00	78.00	188.00
18"	"	130.00	90.00	220.00
24"	"	160.00	93.00	253.00
30"	"	190.00	100.00	290.00
36"	"	210.00	110.00	320.00
End closers				
9"	EA.	24.20	22.85	47.05
12"	"	25.20	24.75	49.95
18"	"	37.70	29.70	67.40
24"	"	52.50	40.70	93.20
30"	"	60.50	47.90	108.40
36"	"	75.50	54.00	129.50
Horizontal elbows				
9"	EA.	230.00	56.00	286.00
12"	"	250.00	64.50	314.50
18"	"	330.00	78.00	408.00
24"	"	460.00	93.00	553.00
30"	"	620.00	110.00	730.00
36"	"	810.00	120.00	930.00
Crosses				
9"	EA.	360.00	74.00	434.00
12"	"	380.00	82.50	462.50
18"	"	460.00	93.00	553.00
24"	"	590.00	100.00	690.00
30"	"	690.00	120.00	810.00
36"	"	820.00	130.00	950.00
16110.80 WIREWAYS				
Wireway, hinge cover type				
2-1/2" x 2-1/2"				
1' section	EA.	8.40	5.70	14.10
2'	"	11.95	7.05	19.00

BASIC MATERIALS	UNIT	MAT.	INST.	TOTAL
16110.80 WIREWAYS				
3'	EA.	16.20	9.30	25.50
5'	"	26.80	14.15	40.95
10'	"	55.50	24.75	80.25
4" x 4"				
1'	EA.	8.15	9.30	17.45
2'	"	13.50	9.30	22.80
3'	"	20.00	11.40	31.40
4'	"	27.10	11.40	38.50
10'	"	61.00	29.70	90.70
6" x 6"				
1'	EA.	17.65	14.15	31.80
2'	"	22.05	14.15	36.20
3'	"	31.20	16.50	47.70
4'	"	40.70	16.50	57.20
5'	"	43.40	21.20	64.60
10'	"	100.00	33.00	133.00
8" x 8"				
1'	EA.	28.50	16.50	45.00
2'	"	43.40	16.50	59.90
3'	"	65.00	18.55	83.55
4'	"	78.50	18.55	97.05
5'	"	88.00	22.85	110.85
12" x 12"				
1'	EA.	39.40	22.85	62.25
2'	"	76.00	22.85	98.85
3'	"	110.00	27.00	137.00
4'	"	130.00	27.00	157.00
5'	"	160.00	33.00	193.00
Fittings				
2-1/2" x 2-1/2"				
Drop hanger	EA.	4.90	4.55	9.45
Bracket hanger	"	7.00	4.55	11.55
Panel adapter	"	8.40	18.55	26.95
End plate	"	2.75	4.55	7.30
U-connector	"	2.75	4.55	7.30
Tee	"	23.10	7.40	30.50
Cross	"	24.90	9.30	34.20
90 degree elbow	"	19.65	7.40	27.05
Sweep elbow	"	34.00	7.40	41.40
45 degree elbow	"	19.60	7.40	27.00
Lay-in adapter	"	8.90	5.40	14.30
4" x 4"				
Drop hanger	EA.	6.25	5.40	11.65
Bracket hanger	"	8.40	5.40	13.80
Panel adapter	"	9.85	22.85	32.70
End plate	"	3.50	5.40	8.90
U-connector	"	3.70	5.40	9.10
Tee	"	26.80	9.30	36.10
Cross	"	30.50	12.90	43.40
90 degree elbow	"	22.45	9.30	31.75
Sweep elbow	"	38.80	9.30	48.10
45 degree elbow	"	23.95	9.30	33.25

BASIC MATERIALS	UNIT	MAT.	INST.	TOTAL
16110.80 WIREWAYS				
Lay-in adapter	EA.	13.00	7.40	20.40
6" x 6"				
Drop hanger	EA.	11.25	5.40	16.65
Bracket hanger	"	12.00	5.40	17.40
Reducing bushing	"	24.90	5.40	30.30
Panel adapter	"	12.90	27.00	39.90
End plate	"	4.25	5.40	9.65
U-connector	"	4.75	5.40	10.15
Tee	"	30.50	9.30	39.80
Cross	"	37.90	12.90	50.80
90 degree elbow	"	24.90	9.30	34.20
Sweep elbow	"	56.50	9.30	65.80
45 degree elbow	"	24.90	9.30	34.20
Lay-in adapter	"	17.55	9.30	26.85
8" x 8"				
Drop hanger	EA.	14.75	5.40	20.15
Bracket hanger	"	12.70	5.40	18.10
Reducing bushing	"	29.50	5.40	34.90
Panel adapter	"	17.55	33.30	50.85
End plate	"	5.60	5.40	11.00
U-connector	"	7.00	5.40	12.40
Tee	"	56.50	14.85	71.35
Cross	"	60.00	16.50	76.50
90 degree elbow	"	40.60	14.85	55.45
Sweep elbow	"	36.10	14.85	50.95
45 degree elbow	"	40.60	14.85	55.45
Lay-in adapter	"	21.85	9.30	31.15
10" x 10"				
Drop hanger	EA.	14.85	5.40	20.25
Bracket hanger	"	15.10	5.40	20.50
Reducing bushing	"	42.00	5.40	47.40
Panel adapter	"	17.70	37.10	54.80
End plate	"	6.85	5.40	12.25
U-connector	"	7.10	5.40	12.50
Tee	"	54.50	16.50	71.00
Cross	"	80.50	18.55	99.05
90 degree elbow	"	45.80	16.50	62.30
Sweep elbow	"	44.50	16.50	61.00
45 degree elbow	"	46.00	16.50	62.50
Lay-in adapter	"	24.90	11.00	35.90
12" x 12"				
Drop hanger	EA.	41.60	5.40	47.00
Bracket hanger	"	28.40	5.40	33.80
Reducing bushing	"	42.50	5.40	47.90
Panel adapter	"	29.50	46.40	75.90
End plate	"	12.65	5.40	18.05
U-connector	"	16.80	22.85	39.65
Tee	"	110.00	19.80	129.80
Cross	"	110.00	22.85	132.85
90 degree elbow	"	74.00	22.85	96.85
Sweep elbow	"	100.00	22.85	122.85
45 degree elbow	"	110.00	22.85	132.85

BASIC MATERIALS	UNIT	MAT.	INST.	TOTAL
16110.80 WIREWAYS				
Lay-in adapter	EA.	27.30	19.80	47.10
Raintight wireway, 4" x 4"				
1' section	EA.	20.75	14.85	35.60
5'	"	49.00	14.85	63.85
10'	"	89.00	37.10	126.10
Fittings				
90 degree elbow	EA.	35.80	14.85	50.65
Tee	"	39.20	16.50	55.70
Cross	"	67.00	18.55	85.55
Panel adapter	"	16.70	27.00	43.70
End plate	"	8.65	5.40	14.05
Gusset bracket	"	6.35	5.40	11.75
6" x 6"				
1' section	EA.	30.00	18.55	48.55
5'	"	71.00	27.00	98.00
10'	"	130.00	56.00	186.00
Fittings				
90 degree elbow	EA.	45.00	18.55	63.55
Tee	"	55.00	18.55	73.55
Cross	"	78.00	27.00	105.00
Panel adapter	"	22.10	37.10	59.20
End plate	"	9.80	5.40	15.20
Gusset bracket	"	8.05	5.40	13.45
16120.41 ALUMINUM CONDUCTORS				
Type XHHW, stranded aluminum, 600v				
#8	L.F.	0.13	0.19	0.32
#6	"	0.19	0.22	0.41
#4	"	0.25	0.30	0.55
#2	"	0.30	0.33	0.63
1/0	"	0.49	0.41	0.90
2/0	"	0.58	0.45	1.03
3/0	"	0.71	0.52	1.23
4/0	"	0.85	0.56	1.41
300 MCM	"	1.20	0.74	1.94
350 MCM	"	1.30	0.85	2.15
400 MCM	"	1.55	1.05	2.60
500 MCM	"	1.85	1.25	3.10
600 MCM	"	2.15	1.50	3.65
700 MCM	"	2.25	1.75	4.00
750 MCM	"	3.10	1.90	5.00
THW, stranded				
#8	L.F.	0.11	0.19	0.30
#6	"	0.12	0.22	0.34
#4	"	0.16	0.30	0.46
#3	"	0.21	0.33	0.54
#1	"	0.30	0.37	0.67
1/0	"	0.34	0.41	0.75
2/0	"	0.39	0.44	0.83
3/0	"	0.49	0.44	0.93
4/0	"	0.56	0.56	1.12

BASIC MATERIALS	UNIT	MAT.	INST.	TOTAL
16120.41 ALUMINUM CONDUCTORS				
250 MCM	L.F.	0.66	0.67	1.33
300 MCM	"	0.87	0.74	1.61
350 MCM	"	0.95	0.85	1.80
400 MCM	"	1.05	1.05	2.10
500 MCM	"	1.25	1.25	2.50
600 MCM	"	1.50	1.50	3.00
700 MCM	"	1.70	1.75	3.45
750 MCM	"	1.80	1.90	3.70
XLP, stranded				
#6	L.F.	0.19	0.19	0.38
#4	"	0.25	0.30	0.55
#2	"	0.30	0.33	0.63
#1	"	0.41	0.37	0.78
1/0	"	0.49	0.41	0.90
2/0	"	0.59	0.45	1.04
3/0	"	0.71	0.52	1.23
4/0	"	0.85	0.56	1.41
250 MCM	"	1.05	0.60	1.65
300 MCM	"	1.20	0.74	1.94
350 MCM	"	1.30	0.85	2.15
400 MCM	"	1.55	1.05	2.60
500 MCM	"	1.85	1.25	3.10
600 MCM	"	2.15	1.50	3.65
700 MCM	"	2.25	1.75	4.00
750 MCM	"	3.10	1.90	5.00
1000 MCM	"	4.00	2.10	6.10
Bare stranded aluminum wire				
#4	L.F.	0.24	0.30	0.54
#2	"	0.25	0.33	0.58
1/0	"	0.30	0.41	0.71
2/0	"	0.35	0.45	0.80
3/0	"	0.38	0.52	0.90
4/0	"	0.41	0.56	0.97
Triplex XLP cable				
#4	L.F.	0.52	0.56	1.08
#2	"	0.65	0.74	1.39
1/0	"	1.05	1.10	2.15
4/0	"	1.90	1.80	3.70
Aluminum quadruplex XLP cable				
#4	L.F.	0.65	0.67	1.32
#2	"	0.86	0.85	1.71
1/0	"	1.45	1.20	2.65
2/0	"	1.70	1.55	3.25
4/0	"	2.50	2.35	4.85
Triplexed URD-XLP cable				
#6	L.F.	0.44	0.41	0.85
#4	"	0.62	0.52	1.14
#2	"	0.80	0.67	1.47
1/0	"	1.25	1.05	2.30
2/0	"	1.45	1.25	2.70
3/0	"	1.75	1.50	3.25
4/0	"	2.05	1.75	3.80

BASIC MATERIALS	UNIT	MAT.	INST.	TOTAL
16120.41 ALUMINUM CONDUCTORS				
250 MCM	L.F.	2.30	2.05	4.35
350 MCM	"	3.00	2.10	5.10
Type S.E.U. cable				
#8/3	L.F.	0.65	0.93	1.58
#6/3	"	0.65	1.05	1.70
#4/3	"	0.87	1.30	2.17
#2/3	"	1.20	1.40	2.60
#1/3	"	1.55	1.50	3.05
1/0-3	"	1.80	1.55	3.35
2/0-3	"	2.05	1.65	3.70
3/0-3	"	2.60	1.90	4.50
4/0-3	"	2.85	2.10	4.95
Type S.E.R. cable with ground				
#8/3	L.F.	0.80	1.05	1.85
#6/3	"	0.94	1.30	2.24
#4/3	"	1.05	1.40	2.45
#2/3	"	1.60	1.50	3.10
#1/3	"	2.00	1.65	3.65
1/0-3	"	2.40	1.85	4.25
2/0-3	"	2.80	2.05	4.85
3/0-3	"	3.30	2.20	5.50
4/0-3	"	3.95	2.45	6.40
#6/4	"	1.55	1.40	2.95
#4/4	"	1.80	1.65	3.45
#2/4	"	2.75	1.65	4.40
#1/4	"	3.40	1.85	5.25
1/0-4	"	4.05	1.90	5.95
2/0-4	"	4.70	2.10	6.80
3/0-4	"	5.60	2.35	7.95
4/0-4	"	6.85	2.85	9.70
16120.43 COPPER CONDUCTORS				
Copper conductors, type THW, solid				
#14	L.F.	0.06	0.15	0.21
#12	"	0.07	0.19	0.26
#10	"	0.08	0.22	0.30
Stranded				
#14	L.F.	0.08	0.15	0.23
#12	"	0.10	0.19	0.29
#10	"	0.12	0.22	0.34
#8	"	0.19	0.30	0.49
#6	"	0.25	0.33	0.58
#4	"	0.37	0.37	0.74
#3	"	0.53	0.37	0.90
#2	"	0.64	0.45	1.09
#1	"	0.82	0.52	1.34
1/0	"	1.00	0.59	1.59
2/0	"	1.20	0.74	1.94
3/0	"	1.45	0.93	2.38
4/0	"	1.80	1.05	2.85
250 MCM	"	2.25	1.10	3.35
300 MCM	"	2.65	1.25	3.90

BASIC MATERIALS		UNIT	MAT.	INST.	TOTAL
16120.43	**COPPER CONDUCTORS**				
350 MCM		L.F.	3.25	1.50	4.75
400 MCM		"	3.65	1.65	5.30
500 MCM		"	4.40	1.90	6.30
600 MCM		"	6.25	2.20	8.45
750 MCM		"	7.95	2.45	10.40
1000 MCM		"	8.80	2.85	11.65
THHN-THWN, solid					
#14		L.F.	0.06	0.15	0.21
#12		"	0.08	0.19	0.27
#10		"	0.10	0.22	0.32
Stranded					
#14		L.F.	0.08	0.15	0.23
#12		"	0.10	0.19	0.29
#10		"	0.12	0.22	0.34
#8		"	0.19	0.30	0.49
#6		"	0.25	0.33	0.58
#4		"	0.37	0.37	0.74
#2		"	0.64	0.45	1.09
#1		"	0.82	0.52	1.34
1/0		"	1.00	0.59	1.59
2/0		"	1.20	0.74	1.94
3/0		"	1.45	0.93	2.38
4/0		"	1.80	1.05	2.85
250 MCM		"	2.25	1.10	3.35
350 MCM		"	3.25	1.50	4.75
XHHW					
#14		L.F.	0.11	0.15	0.26
#10		"	0.18	0.22	0.40
#8		"	0.25	0.30	0.55
#6		"	0.27	0.33	0.60
#4		"	0.40	0.33	0.73
#2		"	0.64	0.41	1.05
#1		"	0.84	0.52	1.36
1/0		"	1.00	0.59	1.59
2/0		"	1.30	0.71	2.01
3/0		"	1.50	0.93	2.43
XLP, 600v					
#12		L.F.	0.12	0.19	0.31
#10		"	0.18	0.22	0.40
#8		"	0.25	0.30	0.55
#6		"	0.29	0.33	0.62
#4		"	0.43	0.37	0.80
#3		"	0.53	0.41	0.94
#2		"	0.66	0.45	1.11
#1		"	0.87	0.52	1.39
1/0		"	1.05	0.59	1.64
2/0		"	1.30	0.74	2.04
3/0		"	1.60	0.96	2.56
4/0		"	1.90	1.05	2.95
250 MCM		"	2.35	1.10	3.45
300 MCM		"	2.75	1.25	4.00
350 MCM		"	3.15	1.45	4.60

BASIC MATERIALS		UNIT	MAT.	INST.	TOTAL
16120.43	**COPPER CONDUCTORS**				
400 MCM		L.F.	3.75	1.65	5.40
500 MCM		"	4.40	1.90	6.30
600 MCM		"	6.35	2.20	8.55
750 MCM		"	7.65	2.45	10.10
1000 MCM		"	10.70	2.85	13.55
Bare solid wire					
#14		L.F.	0.07	0.15	0.22
#12		"	0.09	0.19	0.28
#10		"	0.11	0.22	0.33
#8		"	0.19	0.30	0.49
#6		"	0.25	0.33	0.58
#4		"	0.37	0.37	0.74
#2		"	0.60	0.45	1.05
Bare stranded wire					
#8		L.F.	0.19	0.30	0.49
#6		"	0.26	0.37	0.63
#4		"	0.37	0.37	0.74
#2		"	0.63	0.41	1.04
#1		"	0.81	0.52	1.33
1/0		"	1.00	0.67	1.67
2/0		"	1.15	0.74	1.89
3/0		"	1.45	0.93	2.38
4/0		"	1.80	1.05	2.85
250 MCM		"	2.20	1.10	3.30
300 MCM		"	2.60	1.25	3.85
350 MCM		"	3.00	1.50	4.50
400 MCM		"	3.55	1.65	5.20
500 MCM		"	4.30	1.90	6.20
Type "BX" solid armored cable					
#14/2		L.F.	0.34	0.93	1.27
#14/3		"	0.41	1.05	1.46
#14/4		"	0.57	1.15	1.72
#12/2		"	0.36	1.05	1.41
#12/3		"	0.53	1.15	1.68
#12/4		"	0.75	1.30	2.05
#10/2		"	0.62	1.15	1.77
#10/3		"	0.80	1.30	2.10
#10/4		"	1.25	1.50	2.75
#8/2		"	1.00	1.30	2.30
#8/3		"	1.30	1.50	2.80
Steel type, metal clad cable, solid, with ground					
#14/2		L.F.	0.36	0.67	1.03
#14/3		"	0.49	0.74	1.23
#14/4		"	0.68	0.85	1.53
#12/2		"	0.43	0.74	1.17
#12/3		"	0.60	0.93	1.53
#12/4		"	0.81	1.10	1.91
#10/2		"	0.72	0.85	1.57
#10/3		"	1.10	1.05	2.15
#10/4		"	1.45	1.25	2.70
Metal clad cable, stranded, with ground					
#8/2		L.F.	1.20	1.05	2.25

BASIC MATERIALS	UNIT	MAT.	INST.	TOTAL
16120.43 COPPER CONDUCTORS				
#8 / 3	L.F.	1.70	1.30	3.00
#8 / 4	"	2.20	1.55	3.75
#6 / 2	"	1.80	1.10	2.90
#6 / 3	"	2.15	1.40	3.55
#6 / 4	"	2.55	1.65	4.20
#4 / 2	"	2.30	1.50	3.80
#4 / 3	"	3.00	1.65	4.65
#4 / 4	"	3.55	2.05	5.60
#3 / 3	"	3.70	1.85	5.55
#3 / 4	"	4.35	2.20	6.55
#2 / 3	"	4.40	2.10	6.50
#2 / 4	"	5.35	2.45	7.80
#1 / 3	"	5.30	2.85	8.15
#4	"	1.70	1.30	3.00
16120.45 FLAT CONDUCTOR CABLE				
Flat conductor cable, with shield, 3 conductor				
#12 awg	L.F.	3.90	2.20	6.10
#10 awg	"	4.60	2.20	6.80
4 conductor				
#12 awg	L.F.	5.30	2.95	8.25
#10 awg	"	6.00	2.95	8.95
Transition boxes				
#12 awg	L.F.	6.60	3.30	9.90
#10 awg	"	7.45	3.30	10.75
Flat conductor cable communication, with shield				
10 conductor	L.F.	2.60	2.20	4.80
16 conductor	"	3.00	2.60	5.60
24 conductor	"	3.30	3.70	7.00
Power and communication heads, duplex receptacle	EA.	40.10	29.70	69.80
Double duplex receptacle	"	46.70	35.30	82.00
Telephone	"	26.70	29.70	56.40
Receptacle and telephone	"	67.00	35.30	102.30
Blank cover	"	6.65	5.40	12.05
Transition boxes				
Surface	EA.	120.00	27.00	147.00
Flush	"	72.50	37.10	109.60
Flat conductor cable fittings				
End caps	EA.	1.25	5.40	6.65
Insulators	"	13.95	11.00	24.95
Splice connectors	"	0.97	16.50	17.47
Tap connectors	"	1.05	16.50	17.55
Cable connectors	"	1.20	16.50	17.70
Terminal blocks	"	8.10	22.85	30.95
Tape	"			12.45
16120.47 SHEATHED CABLE				
Non-metallic sheathed cable				
Type NM cable with ground				
#14 / 2	L.F.	0.13	0.56	0.69
#12 / 2	"	0.19	0.59	0.78

BASIC MATERIALS	UNIT	MAT.	INST.	TOTAL
16120.47 SHEATHED CABLE				
#10/2	L.F.	0.30	0.66	0.96
#8/2	"	0.63	0.74	1.37
#6/2	"	0.94	0.93	1.87
#14/3	"	0.21	0.96	1.17
#12/3	"	0.30	0.99	1.29
#10/3	"	0.45	1.00	1.45
#8/3	"	0.97	1.00	1.97
#6/3	"	1.35	1.05	2.40
#4/3	"	2.00	1.20	3.20
#2/3	"	2.90	1.30	4.20
Type U.F. cable with ground				
#14/2	L.F.	0.17	0.59	0.76
#12/2	"	0.22	0.71	0.93
#10/2	"	0.30	0.74	1.04
#8/2	"	0.97	0.85	1.82
#6/2	"	1.15	1.00	2.15
#14/3	"	0.24	0.74	0.98
#12/3	"	0.32	0.81	1.13
#10/3	"	0.47	0.93	1.40
#8/3	"	1.30	1.05	2.35
#6/3	"	1.85	1.20	3.05
Type S.F.U. cable, 3 conductor				
#8	L.F.	0.85	1.05	1.90
#6	"	1.15	1.15	2.30
#3	"	2.20	1.50	3.70
#2	"	2.85	1.65	4.50
#1	"	4.15	1.85	6.00
#1/0	"	4.70	2.05	6.75
#2/0	"	5.55	2.35	7.90
#3/0	"	6.80	2.60	9.40
#4/0	"	7.65	2.85	10.50
Type SER cable, 4 conductor				
#6	L.F.	1.55	1.35	2.90
#4	"	2.40	1.45	3.85
#3	"	2.80	1.65	4.45
#2	"	3.65	1.80	5.45
#1	"	5.05	2.05	7.10
#1/0	"	6.65	2.35	9.00
#2/0	"	7.75	2.45	10.20
#3/0	"	8.80	2.85	11.65
#4/0	"	9.35	3.10	12.45
Flexible cord, type STO cord				
#18/2	L.F.	0.66	0.15	0.81
#18/3	"	0.78	0.19	0.97
#18/4	"	1.10	0.22	1.32
#16/2	"	0.76	0.15	0.91
#16/3	"	0.91	0.17	1.08
#16/4	"	1.20	0.19	1.39
#14/2	"	1.20	0.19	1.39
#14/3	"	1.50	0.23	1.73
#14/4	"	1.85	0.26	2.11
#12/2	"	1.60	0.22	1.82

BASIC MATERIALS	UNIT	MAT.	INST.	TOTAL
16120.47 SHEATHED CABLE				
#12/3	L.F.	1.95	0.25	2.20
#12/4	"	2.25	0.30	2.55
#10/2	"	2.05	0.26	2.31
#10/3	"	2.55	0.30	2.85
#10/4	"	2.85	0.33	3.18
#8/2	"	3.40	0.30	3.70
#8/3	"	3.80	0.33	4.13
#8/4	"	5.30	0.37	5.67
16130.10 FLOOR BOXES				
Adjustable floor boxes, steel	EA.	24.95	19.80	44.75
Cast bronze round	"	36.70	27.00	63.70
1 gang	"	42.60	29.70	72.30
2 gang	"	80.50	35.30	115.80
3 gang	"	110.00	37.10	147.10
Aluminum round	"	57.50	27.00	84.50
1 gang	"	48.50	29.70	78.20
2 gang	"	86.50	35.30	121.80
3 gang	"	120.00	37.10	157.10
Steel plate single recept	"	16.15	5.40	21.55
Duplex receptacle	"	15.40	6.75	22.15
Twist lock receptacle	"	16.35	6.75	23.10
Plug, 3/4"	"	20.55	5.40	25.95
1" plug	"	19.20	5.40	24.60
Carpet flange	"	24.95	5.40	30.35
Adjustable bronze plates for round cast boxes				
1/2" plug	EA.	22.00	5.40	27.40
3/4" plug	"	22.35	5.40	27.75
1" plug	"	22.70	5.40	28.10
1-1/4" plug	"	26.40	6.75	33.15
2" plug	"	27.20	7.40	34.60
Combination plug	"	22.70	7.40	30.10
Duplex receptacle plug	"	29.30	7.40	36.70
Adjustable aluminum plates for round cast boxes				
1/2" plug	EA.	24.95	5.40	30.35
3/4" plug	"	25.30	5.40	30.70
1" plug	"	26.00	5.40	31.40
1-1/4" plug	"	27.90	6.75	34.65
2" plug	"	28.20	7.40	35.60
Combination plug	"	25.30	7.40	32.70
Duplex receptacle plug	"	42.60	7.40	50.00
Adjustable bronze plates for gang type boxes				
1/2" plug	EA.	26.20	5.40	31.60
3/4" plug	"	26.60	5.40	32.00
1" plug	"	26.90	5.40	32.30
1-1/4" plug	"	28.10	6.75	34.85
2" plug	"	29.30	7.40	36.70
Carpet plate				
1 gang	EA.	23.20	5.40	28.60
2 gang	"	35.20	5.40	40.60
3 gang	"	47.00	7.40	54.40
Adjustable aluminum plates for gang type boxes				

BASIC MATERIALS	UNIT	MAT.	INST.	TOTAL
16130.10 — FLOOR BOXES				
1/2" plug	EA.	23.45	5.40	28.85
3/4" plug	"	24.00	5.40	29.40
1" plug	"	24.35	5.40	29.75
1-1/4" plug	"	27.90	6.75	34.65
2" plug	"	28.70	7.40	36.10
Duplex recept	"	23.60	7.40	31.00
Carpet plate				
1 gang	EA.	39.60	5.40	45.00
2 gang	"	63.00	5.40	68.40
3 gang	"	92.50	7.40	99.90
4 gang carpet plate	"	38.20	21.20	59.40
Telephone	"	34.70	18.55	53.25
Floor box nozzles, horizontal				
Duplex recept	EA.	110.00	19.80	129.80
Single recept	"	100.00	19.80	119.80
Double duplex recept	"	110.00	27.00	137.00
Vertical with duplex recept	"	110.00	22.85	132.85
Double duplex recept	"	110.00	27.00	137.00
Floor box bell nozzles split bell	"	19.05	9.30	28.35
One piece bell	"	60.00	9.30	69.30
Floor box standpipe				
1/2" x 3"	EA.	16.60	5.40	22.00
1/2" x 1"	"	16.35	5.40	21.75
Poke thru floor outlets				
2" floor	EA.	51.50	37.10	88.60
3" floor	"	56.00	44.30	100.30
4" floor	"	56.50	47.90	104.40
7" floor	"	60.00	56.00	116.00
9" floor	"	85.00	59.50	144.50
11" floor	"	88.00	67.50	155.50
13" floor	"	91.00	74.00	165.00
16130.40 — BOXES				
Round cast box, type SEH				
1/2"	EA.	14.40	12.90	27.30
3/4"	"	15.70	15.60	31.30
SEHC				
1/2"	EA.	18.10	12.90	31.00
3/4"	"	18.95	15.60	34.55
SEHL				
1/2"	EA.	22.30	12.90	35.20
3/4"	"	23.55	16.50	40.05
SEHT				
1/2"	EA.	25.40	15.60	41.00
3/4"	"	26.20	18.55	44.75
SEHX				
1/2"	EA.	27.50	18.55	46.05
3/4"	"	28.50	22.85	51.35
Blank cover	"	3.55	5.40	8.95
1/2", hub cover	"	2.70	5.40	8.10
Cover with gasket	"	3.00	6.60	9.60

BASIC MATERIALS	UNIT	MAT.	INST.	TOTAL
16130.40 BOXES				
Rectangle, type FS boxes				
1/2"	EA.	7.20	12.90	20.10
3/4"	"	7.60	14.85	22.45
1"	"	8.10	18.55	26.65
FSA				
1/2"	EA.	7.30	12.90	20.20
3/4"	"	8.25	14.85	23.10
FSC				
1/2"	EA.	8.00	12.90	20.90
3/4"	"	8.75	15.60	24.35
1"	"	10.90	18.55	29.45
FSL				
1/2"	EA.	8.80	12.90	21.70
3/4"	"	9.10	14.85	23.95
FSR				
1/2"	EA.	8.65	12.90	21.55
3/4"	"	11.00	14.85	25.85
FSS				
1/2"	EA.	7.70	12.90	20.60
3/4"	"	8.40	14.85	23.25
FSLA				
1/2"	EA.	8.80	12.90	21.70
3/4"	"	9.95	14.85	24.80
FSCA				
1/2"	EA.	11.70	12.90	24.60
3/4"	"	13.40	14.85	28.25
FSCC				
1/2"	EA.	9.40	14.85	24.25
3/4"	"	12.45	18.55	31.00
FSCT				
1/2"	EA.	9.80	14.85	24.65
3/4"	"	12.20	18.55	30.75
1"	"	15.40	21.20	36.60
FST				
1/2"	EA.	12.65	18.55	31.20
3/4"	"	14.25	21.20	35.45
FSX				
1/2"	EA.	12.90	22.85	35.75
3/4"	"	14.95	27.00	41.95
FSCD boxes				
1/2"	EA.	14.20	22.85	37.05
3/4"	"	15.05	27.00	42.05
Rectangle, type FS, 2 gang boxes				
1/2"	EA.	13.10	12.90	26.00
3/4"	"	14.05	14.85	28.90
1"	"	15.45	18.55	34.00
FSC, 2 gang boxes				
1/2"	EA.	14.25	12.90	27.15
3/4"	"	14.50	14.85	29.35
1"	"	18.10	18.55	36.65
FSS, 2 gang boxes				
3/4"	EA.	14.95	14.85	29.80

BASIC MATERIALS	UNIT	MAT.	INST.	TOTAL
16130.40 BOXES				
FS, tandem boxes				
1/2"	EA.	16.65	14.85	31.50
3/4"	"	18.65	16.50	35.15
FSC, tandem boxes				
1/2"	EA.	18.50	14.85	33.35
3/4"	"	19.60	16.50	36.10
FS, three gang boxes				
3/4"	EA.	20.10	16.50	36.60
1"	"	22.15	18.55	40.70
FSS, three gang boxes, 3/4"	"	25.90	18.55	44.45
Weatherproof cast aluminum boxes, 1 gang, 3 outlets				
1/2"	EA.	3.70	14.85	18.55
3/4"	"	4.05	18.55	22.60
2 gang, 3 outlets				
1/2"	EA.	7.10	18.55	25.65
3/4"	"	7.45	19.80	27.25
1 gang, 4 outlets				
1/2"	EA.	4.55	22.85	27.40
3/4"	"	4.90	27.00	31.90
2 gang, 4 outlets				
1/2"	EA.	7.75	22.85	30.60
3/4"	"	8.10	27.00	35.10
1 gang, 5 outlets				
1/2"	EA.	5.30	27.00	32.30
3/4"	"	6.40	29.70	36.10
2 gang, 5 outlets				
1/2"	EA.	9.50	27.00	36.50
3/4"	"	11.75	29.70	41.45
2 gang, 6 outlets				
1/2"	EA.	11.00	31.60	42.60
3/4"	"	11.75	33.30	45.05
2 gang, 7 outlets				
1/2"	EA.	11.25	37.10	48.35
3/4"	"	11.95	40.70	52.65
Weatherproof and type FS box covers, blank, 1 gang	"	1.60	5.40	7.00
Tumbler switch, 1 gang	"	1.95	5.40	7.35
1 gang, single recept	"	2.10	5.40	7.50
Duplex recept	"	1.95	5.40	7.35
Despard	"	1.90	5.40	7.30
Red pilot light	"	12.55	5.40	17.95
SW and				
Single recept	EA.	5.90	7.40	13.30
Duplex recept	"	4.85	7.40	12.25
2 gang				
Blank	EA.	2.15	6.75	8.90
Tumbler switch	"	3.00	6.75	9.75
Single recept	"	3.35	6.75	10.10
Duplex recept	"	3.60	6.75	10.35
3 gang				
Blank	EA.	4.05	7.40	11.45
Tumbler switch	"	5.05	7.40	12.45
4 gang				

BASIC MATERIALS	UNIT	MAT.	INST.	TOTAL
16130.40 — BOXES				
Tumbler switch	EA.	6.35	9.30	15.65
Explosion proof boxes type E				
1/2"	EA.	19.80	12.90	32.70
3/4"	"	19.05	14.85	33.90
1"	"	19.70	18.55	38.25
1-1/4"	"	34.10	21.20	55.30
1-1/2"	"	77.50	22.85	100.35
Type L.B.				
1/2"	EA.	18.90	14.85	33.75
3/4"	"	20.05	18.55	38.60
1"	"	20.90	21.20	42.10
1-1/4"	"	36.80	24.75	61.55
1-1/2"	"	73.50	27.00	100.50
2"	"	82.50	29.70	112.20
Type C				
1/2"	EA.	19.75	14.85	34.60
3/4"	"	21.20	18.55	39.75
1"	"	21.85	21.20	43.05
1-1/4"	"	36.80	24.75	61.55
1-1/2"	"	76.50	27.00	103.50
2"	"	86.50	29.70	116.20
Type CA				
1/2"	EA.	19.75	21.20	40.95
3/4"	"	22.45	27.00	49.45
Type L				
1/2"	EA.	20.65	14.85	35.50
3/4"	"	21.85	18.55	40.40
1"	"	22.85	21.20	44.05
1-1/4"	"	38.50	24.75	63.25
1-1/2"	"	75.50	27.00	102.50
2"	"	78.50	29.70	108.20
Type N				
1/2"	EA.	20.90	14.85	35.75
3/4"	"	21.80	18.55	40.35
1"	"	22.45	22.85	45.30
1-1/4"	"	39.10	24.75	63.85
Type T				
1/2"	EA.	20.05	19.80	39.85
3/4"	"	21.30	27.00	48.30
1"	"	22.10	31.60	53.70
1-1/4"	"	37.60	37.10	74.70
1-1/2"	"	77.00	43.00	120.00
2"	"	79.50	47.90	127.40
Type TA				
1/2"	EA.	23.10	27.00	50.10
3/4"	"	24.40	29.70	54.10
Type X				
1/2"	EA.	20.90	27.00	47.90
3/4"	"	22.40	31.60	54.00
1"	"	24.50	37.10	61.60
1-1/4"	"	44.20	43.00	87.20
1-1/2"	"	83.00	47.90	130.90

BASIC MATERIALS	UNIT	MAT.	INST.	TOTAL
16130.40 — BOXES				
2"	EA.	93.00	54.00	147.00
With union hubs				
1/2"	EA.	43.20	27.00	70.20
3/4"	"	44.90	29.70	74.60
Box covers				
Surface	EA.	9.25	7.40	16.65
Sealing	"	10.05	7.40	17.45
Dome	"	13.90	7.40	21.30
1/2" nipple	"	17.75	7.40	25.15
3/4" nipple	"	18.30	7.40	25.70
16130.45 — EXPLOSION PROOF FITTINGS				
Flexible couplings with female unions				
1/2" x 18"	EA.	80.00	7.40	87.40
3/4" x 18"	"	100.00	10.25	110.25
1" x 18"	"	170.00	12.90	182.90
1-1/4" x 18"	"	290.00	15.60	305.60
1-1/2" x 18"	"	360.00	18.55	378.55
2" x 18"	"	480.00	21.20	501.20
1/2" x 24"	"	93.50	9.30	102.80
3/4" x 24"	"	120.00	11.00	131.00
1" x 24"	"	220.00	14.85	234.85
1-1/4" x 24"	"	340.00	16.50	356.50
1-1/2" x 24"	"	410.00	21.20	431.20
2" x 24"	"	590.00	22.85	612.85
Female seal-offs				
1/2"	EA.	6.10	21.20	27.30
3/4"	"	7.20	24.75	31.95
1"	"	9.25	27.00	36.25
1-1/4"	"	11.20	31.60	42.80
1-1/2"	"	17.00	37.10	54.10
2"	"	22.10	43.00	65.10
2-1/2"	"	33.90	64.50	98.40
3"	"	44.90	80.00	124.90
4"	"	170.00	99.00	269.00
Conduit plugs				
1/2"	EA.	0.88	5.40	6.28
3/4"	"	1.00	5.40	6.40
1"	"	1.25	5.40	6.65
1-1/4"	"	1.55	9.30	10.85
1-1/2"	"	1.95	9.30	11.25
2"	"	3.40	11.00	14.40
2-1/2"	"	5.75	11.00	16.75
3"	"	8.20	12.90	21.10
4"	"	10.40	12.90	23.30
Sealing cement				
1 pound	EA.			5.50
5 pound	"			15.75
Fibre				
1 ounce	EA.			4.05
8 ounce	"			13.65
Male unions				

BASIC MATERIALS		UNIT	MAT.	INST.	TOTAL
16130.45	**EXPLOSION PROOF FITTINGS**				
1/2"		EA.	4.75	7.40	12.15
3/4"		"	6.75	9.00	15.75
1"		"	11.45	10.25	21.70
1-1/4"		"	21.00	11.00	32.00
1-1/2"		"	22.75	12.90	35.65
2"		"	29.40	15.60	45.00
2-1/2"		"	49.70	18.55	68.25
3"		"	65.00	27.00	92.00
4"		"	100.00	33.30	133.30
Female unions					
1/2"		EA.	4.35	7.40	11.75
3/4"		"	6.20	9.00	15.20
1"		"	10.85	10.25	21.10
1-1/4"		"	16.30	11.00	27.30
1-1/2"		"	20.90	12.90	33.80
2"		"	26.80	15.60	42.40
2-1/2"		"	40.50	18.55	59.05
3"		"	56.00	27.00	83.00
4"		"	87.50	33.30	120.80
Male elbows					
1/2"		EA.	5.80	9.30	15.10
3/4"		"	6.60	11.00	17.60
1"		"	9.60	12.90	22.50
1-1/4"		"	12.55	16.50	29.05
Female elbows					
1/2"		EA.	3.95	9.30	13.25
3/4"		"	4.35	11.00	15.35
1"		"	5.90	12.90	18.80
1-1/4"		"	9.60	16.50	26.10
Pulling elbows					
1/2"		EA.	27.70	12.90	40.60
3/4"		"	29.50	16.50	46.00
1"		"	74.50	18.55	93.05
1-1/4"		"	82.00	22.85	104.85
1-1/2"		"	110.00	27.00	137.00
2"		"	120.00	70.50	190.50
2-1/2"		"	260.00	93.00	353.00
3"		"	250.00	110.00	360.00
3-1/2"		"	470.00	130.00	600.00
4"		"	460.00	160.00	620.00
Male expansion couplings					
1/2"		EA.	11.05	9.30	20.35
3/4"		"	14.45	11.00	25.45
1"		"	27.00	16.50	43.50
Female expansion couplings					
1/2"		EA.	9.95	9.30	19.25
3/4"		"	15.60	11.00	26.60
1"		"	27.00	16.50	43.50
16130.60	**PULL AND JUNCTION BOXES**				
4" Octagon box		EA.	1.20	4.25	5.45

BASIC MATERIALS	UNIT	MAT.	INST.	TOTAL
16130.60 PULL AND JUNCTION BOXES				
Box extension	EA.	1.70	2.20	3.90
Plaster ring	"	0.91	2.20	3.11
Cover blank	"	0.38	2.20	2.58
Square box	"	1.40	4.25	5.65
Box extension	"	1.55	2.20	3.75
Plaster ring	"	0.71	2.20	2.91
Cover blank	"	0.44	2.20	2.64
4-11/16"				
Square box	EA.	2.55	4.25	6.80
Box extension	"	3.40	2.20	5.60
Plaster ring	"	1.95	2.20	4.15
Cover blank	"	0.77	2.20	2.97
Switch and device boxes				
2 gang	EA.	7.40	4.25	11.65
3 gang	"	9.10	4.25	13.35
4 gang	"	12.90	5.95	18.85
Device covers				
2 gang	EA.	3.60	2.20	5.80
3 gang	"	4.30	2.20	6.50
4 gang	"	6.20	2.20	8.40
Handy box	"	1.10	4.25	5.35
Extension	"	1.50	2.20	3.70
Switch cover	"	0.44	2.20	2.64
Switch box with knockout	"	1.25	5.40	6.65
Weatherproof cover, spring type	"	4.30	2.95	7.25
Cover plate, dryer receptacle 1 gang plastic	"	1.95	3.70	5.65
For 4" receptacle, 2 gang	"	3.40	3.70	7.10
Duplex receptacle cover plate, plastic	"	0.38	2.20	2.58
4", vertical bracket box, 1-1/2" with				
RMX clamps	EA.	1.80	5.40	7.20
BX clamps	"	2.70	5.40	8.10
4", octagon device cover				
1 switch	EA.	1.25	2.20	3.45
1 duplex recept	"	1.15	2.20	3.35
4", octagon swivel hanger box, 1/2" hub	"	4.20	2.20	6.40
3/4" hub	"	4.60	2.20	6.80
4" octagon adjustable bar hangers				
18-1/2"	EA.	1.40	1.85	3.25
26-1/2"	"	1.70	1.85	3.55
With clip				
18-1/2"	EA.	1.55	1.85	3.40
26-1/2"	"	1.50	1.85	3.35
4", square face bracket boxes, 1-1/2"				
RMX	EA.	2.95	5.40	8.35
BX	"	3.15	5.40	8.55
4" square to round plaster rings	"	1.15	2.20	3.35
2 gang device plaster rings	"	1.40	2.20	3.60
Surface covers				
1 gang switch	EA.	1.35	2.20	3.55
2 gang switch	"	1.40	2.20	3.60
1 single recept	"	1.35	2.20	3.55
1 20a twist lock recept	"	1.70	2.20	3.90

BASIC MATERIALS	UNIT	MAT.	INST.	TOTAL
16130.60 PULL AND JUNCTION BOXES				
1 30a twist lock recept	EA.	1.75	2.20	3.95
1 duplex recept	"	1.05	2.20	3.25
2 duplex recept	"	1.25	2.20	3.45
Switch and duplex recept	"	1.50	2.20	3.70
4-11/16" square to round plaster rings	"	3.15	2.20	5.35
2 gang device plaster rings	"	2.50	2.20	4.70
Surface covers				
1 gang switch	EA.	3.80	2.20	6.00
2 gang switch	"	4.10	2.20	6.30
1 single recept	"	3.80	2.20	6.00
1 20a twist lock recept	"	4.10	2.20	6.30
1 30a twist lock recept	"	4.30	2.20	6.50
1 duplex recept	"	3.80	2.20	6.00
2 duplex recept	"	3.90	2.20	6.10
Switch and duplex recept	"	3.95	2.20	6.15
4" plastic round boxes, ground straps				
Box only	EA.	0.87	5.40	6.27
Box w/clamps	"	1.10	7.40	8.50
Box w/16" bar	"	1.55	8.50	10.05
Box w/24" bar	"	1.65	9.30	10.95
4" plastic round box covers				
Blank cover	EA.	0.50	2.20	2.70
Plaster ring	"	0.56	2.20	2.76
4" plastic square boxes				
Box only	EA.	0.65	5.40	6.05
Box w/clamps	"	0.80	7.40	8.20
Box w/hanger	"	0.99	9.30	10.29
Box w/nails and clamp	"	1.40	9.30	10.70
4" plastic square box covers				
Blank cover	EA.	0.56	2.20	2.76
1 gang ring	"	0.53	2.20	2.73
2 gang ring	"	0.56	2.20	2.76
Round ring	"	0.65	2.20	2.85
16130.65 PULL BOXES AND CABINETS				
Galvanized pull boxes, screw cover				
4x4x4	EA.	3.90	7.05	10.95
4x6x4	"	4.85	7.05	11.90
6x6x4	"	5.80	7.05	12.85
6x8x4	"	8.40	7.05	15.45
8x8x4	"	9.85	9.30	19.15
8x10x4	"	11.40	9.00	20.40
8x12x4	"	12.75	9.30	22.05
Screw cover				
10x10x4	EA.	12.70	11.40	24.10
12x12x6	"	19.45	16.50	35.95
12x15x6	"	22.05	16.50	38.55
12x18x6	"	25.70	18.55	44.25
15x18x6	"	29.20	21.20	50.40
18x24x6	"	53.00	22.85	75.85
18x30x6	"	58.50	27.00	85.50

BASIC MATERIALS	UNIT	MAT.	INST.	TOTAL
16130.65 PULL BOXES AND CABINETS				
24x36x6	EA.	90.50	27.00	117.50
Cast iron junction box, unflanged				
6x6x4				
3/4" tap	EA.	63.50	18.55	82.05
1" tap	"	65.00	18.55	83.55
Two 1/2" taps	"	67.50	18.55	86.05
3/4" taps	"	68.00	18.55	86.55
6" adapter plate	"	28.50	12.90	41.40
6" exterior collar	"	46.40	12.90	59.30
Screw cover cabinet				
12x12x4	EA.	54.50	22.85	77.35
12x16x4	"	72.00	22.85	94.85
12x16x6	"	80.50	22.85	103.35
12x18x4	"	78.50	24.75	103.25
12x18x6	"	92.00	24.75	116.75
18x18x4	"	93.50	37.10	130.60
18x18x6	"	110.00	37.10	147.10
18x24x6	"	150.00	42.40	192.40
24x24x6	"	170.00	49.50	219.50
24x36x6	"	250.00	62.00	312.00
36x48x6	"	500.00	93.00	593.00
NEMA 3R, rain tight screw cover enclosures				
6x6x4	EA.	11.50	7.80	19.30
8x6x4	"	13.20	11.00	24.20
8x8x4	"	15.20	11.00	26.20
10x8x4	"	17.15	14.85	32.00
10x10x4	"	19.50	14.85	34.35
12x8x4	"	19.15	16.50	35.65
12x12x4	"	24.55	16.50	41.05
15x12x4	"	28.70	19.80	48.50
8x8x6	"	17.90	14.85	32.75
10x8x6	"	20.20	16.50	36.70
10x10x6	"	22.95	16.50	39.45
12x8x6	"	22.55	19.80	42.35
12x10x6	"	25.50	20.35	45.85
12x12x6	"	28.50	20.35	48.85
18x12x6	"	37.50	26.00	63.50
16130.80 RECEPTACLES				
Contractor grade duplex receptacles, 15 a 120v				
Duplex	EA.	1.35	7.40	8.75
125 volt, 20a, duplex, grounding type, standard grade	"	5.05	7.40	12.45
Ground fault interrupter type	"	39.30	11.00	50.30
250 volt, 20a, 2 pole, single receptacle, ground type	"	5.85	7.40	13.25
120/208v, 4 pole, single receptacle, twist lock				
20a	EA.	12.65	12.90	25.55
50a	"	24.10	12.90	37.00
125/250v, 3 pole, flush receptacle				
30a	EA.	8.00	11.00	19.00
50a	"	9.25	11.00	20.25
60a	"	37.40	12.90	50.30

BASIC MATERIALS	UNIT	MAT.	INST.	TOTAL
16130.80 RECEPTACLES				
277 v, 20a, 2 pole, grounding type, twist lock	EA.	9.25	7.40	16.65
Dryer receptacle, 250v, 30a/50a, 3 wire	"	15.50	11.00	26.50
Clock receptacle, 2 pole, grounding type	"	6.45	7.40	13.85
125v, 20a single recept. grounding type				
Standard grade	EA.	4.75	7.40	12.15
Specification	"	7.30	7.40	14.70
Hospital	"	7.55	7.40	14.95
Isolated ground orange	"	13.10	9.30	22.40
Duplex				
Specification grade	EA.	7.60	7.40	15.00
Hospital	"	9.80	7.40	17.20
Isolated ground orange	"	14.75	9.30	24.05
250v, 20a, duplex, 2 pole, grounding, spec. grade	"	11.25	7.40	18.65
Combination recepts, 20a, 125v and 250v, duplex	"	10.05	7.40	17.45
GFI hospital grade recepts, 20a, 125v, duplex	"	45.40	11.00	56.40
125/250v, 3 pole, 3 wire surface recepts				
30a	EA.	15.50	11.00	26.50
50a	"	17.20	11.00	28.20
60a	"	37.60	12.90	50.50
Cord set, 3 wire, 6' cord				
30a	EA.	16.75	11.00	27.75
50a	"	23.35	11.00	34.35
125/250v, 3 pole, 3 wire cap				
30a	EA.	13.70	14.85	28.55
50a	"	24.90	14.85	39.75
60a	"	32.10	16.50	48.60
16198.10 ELECTRIC MANHOLES				
Precast, handhole, 4' deep				
2'x2'	EA.	310.00	130.00	440.00
3'x3'	"	420.00	210.00	630.00
4'x4'	"	910.00	380.00	1,290
Power manhole, complete, precast, 8' deep				
4'x4'	EA.	1,370	520.00	1,890
6'x6'	"	1,810	740.00	2,550
8'x8'	"	2,090	780.00	2,870
6' deep, 9' x 12'	"	2,430	930.00	3,360
Cast in place, power manhole, 8' deep				
4'x4'	EA.	1,620	520.00	2,140
6'x6'	"	1,990	740.00	2,730
8'x8'	"	2,240	780.00	3,020
16199.10 UTILITY POLES & FITTINGS				
Wood pole, creosoted				
25'	EA.	140.00	87.50	227.50
30'	"	170.00	110.00	280.00
35'	"	360.00	130.00	490.00
40'	"	650.00	140.00	790.00
45'	"	730.00	260.00	990.00
50'	"	810.00	270.00	1,080
55'	"	890.00	280.00	1,170

BASIC MATERIALS	UNIT	MAT.	INST.	TOTAL
16199.10 UTILITY POLES & FITTINGS				
Treated, wood preservative, 6"x6"				
8'	EA.	12.95	18.55	31.50
10'	"	32.40	29.70	62.10
12'	"	36.30	33.00	69.30
14'	"	45.30	49.50	94.80
16'	"	65.00	59.50	124.50
18'	"	97.00	74.00	171.00
20'	"	130.00	74.00	204.00
Aluminum, brushed, no base				
8'	EA.	280.00	74.00	354.00
10'	"	320.00	99.00	419.00
15'	"	480.00	100.00	580.00
20'	"	750.00	120.00	870.00
25'	"	1,020	140.00	1,160
30'	"	1,550	160.00	1,710
35'	"	1,810	190.00	2,000
40'	"	2,460	230.00	2,690
Steel, no base				
10'	EA.	330.00	93.00	423.00
15'	"	390.00	110.00	500.00
20'	"	490.00	140.00	630.00
25'	"	650.00	170.00	820.00
30'	"	940.00	190.00	1,130
35'	"	1,420	230.00	1,650
Concrete, no base				
13'	EA.	520.00	200.00	720.00
16'	"	780.00	270.00	1,050
18'	"	940.00	330.00	1,270
25'	"	1,130	370.00	1,500
30'	"	1,470	450.00	1,920
35'	"	1,900	520.00	2,420
40'	"	2,200	590.00	2,790
45'	"	2,670	630.00	3,300
50'	"	3,370	670.00	4,040
55'	"	3,720	710.00	4,430
60'	"	4,270	740.00	5,010
Pole line hardware				
Wood crossarm				
4'	EA.	27.50	49.50	77.00
8'	"	55.00	62.00	117.00
10'	"	110.00	76.00	186.00
Angle steel brace				
1 piece	EA.	6.55	9.30	15.85
2 piece	"	14.00	12.90	26.90
Eye nut, 5/8"	"	3.25	1.85	5.10
Bolt (14-16"), 5/8"	"	8.60	7.40	16.00
Transformer, ground connection	"	3.80	9.30	13.10
Stirrup	"	11.65	11.40	23.05
Secondary lead support	"	13.05	14.85	27.90
Spool insulator	"	9.60	7.40	17.00
Guy grip, preformed				
7/16"	EA.	2.05	5.40	7.45

BASIC MATERIALS	UNIT	MAT.	INST.	TOTAL
16199.10 UTILITY POLES & FITTINGS				
1/2"	EA.	3.05	5.40	8.45
Hook	"	2.05	9.30	11.35
Strain insulator	"	16.75	13.50	30.25
Wire				
5/16"	L.F.	0.34	0.19	0.53
7/16"	"	0.67	0.22	0.89
1/2"	"	1.15	0.30	1.45
Soft drawn ground, copper, #8	"	0.27	0.30	0.57
Ground clamp	EA.	4.45	11.40	15.85
Perforated strapping for conduit, 1-1/2"	L.F.	1.70	5.40	7.10
Hot line clamp	EA.	13.05	29.70	42.75
Lightning arrester				
3kv	EA.	68.50	37.10	105.60
10kv	"	110.00	59.50	169.50
30kv	"	200.00	74.00	274.00
36kv	"	410.00	93.00	503.00
Fittings				
Plastic molding	L.F.	1.95	5.40	7.35
Molding staples	EA.	0.47	1.85	2.32
Ground wires staples	"	0.19	1.10	1.29
Copper butt plate	"	0.55	11.00	11.55
Anchor bond clamp	"	2.45	5.40	7.85
Guy wire				
1/4"	L.F.	0.26	1.10	1.36
3/8"	"	0.43	1.85	2.28
Guy grip				
1/4"	EA.	1.35	1.85	3.20
3/8"	"	2.40	1.85	4.25

POWER GENERATION	UNIT	MAT.	INST.	TOTAL
16210.10 GENERATORS				
Diesel generator, with auto transfer switch				
30kw	EA.	16,220	1,140	17,360
50kw	"	20,530	1,140	21,670
75kw	"	26,070	1,560	27,630
100kw	"	28,930	1,750	30,680
125kw	"	30,920	1,860	32,780
150kw	"	35,650	2,120	37,770
175kw	"	37,150	2,470	39,620
200kw	"	38,530	2,970	41,500
250kw	"	42,110	3,300	45,410
300kw	"	50,640	3,710	54,350
350kw	"	54,220	4,240	58,460

POWER GENERATION	UNIT	MAT.	INST.	TOTAL
16210.10 GENERATORS				
400kw	EA.	66,560	4,950	71,510
450kw	"	71,290	5,400	76,690
500kw	"	77,230	5,940	83,170
600kw	"	105,790	7,420	113,210
750kw	"	147,660	7,420	155,080
16230.10 CAPACITORS				
Three phase capacitors				
240v				
1.5 kvar	EA.	190.00	93.00	283.00
2.5 kvar	"	220.00	120.00	340.00
3.0 kvar	"	250.00	150.00	400.00
4 kvar	"	270.00	190.00	460.00
5 kvar	"	300.00	200.00	500.00
6 kvar	"	330.00	210.00	540.00
7.5 kvar	"	400.00	230.00	630.00
10 kvar	"	490.00	300.00	790.00
15 kvar	"	670.00	350.00	1,020
20 kvar	"	800.00	440.00	1,240
25 kvar	"	940.00	480.00	1,420
40 kvar	"	1,900	670.00	2,570
50 kvar	"	2,300	780.00	3,080
60 kvar	"	2,950	800.00	3,750
75 kvar	"	3,270	930.00	4,200
100 kvar	"	4,010	1,110	5,120
480v				
1.5 kvar	EA.	160.00	93.00	253.00
2.5 kvar	"	170.00	120.00	290.00
3 kvar	"	190.00	150.00	340.00
4 kvar	"	210.00	190.00	400.00
5 kvar	"	230.00	200.00	430.00
6 kvar	"	250.00	210.00	460.00
7.5 kvar	"	280.00	230.00	510.00
10 kvar	"	340.00	300.00	640.00
12.5 kvar	"	390.00	350.00	740.00
15 kvar	"	420.00	440.00	860.00
18 kvar	"	450.00	460.00	910.00
20 kvar	"	470.00	480.00	950.00
22.5 kvar	"	480.00	500.00	980.00
25 kvar	"	520.00	550.00	1,070
30 kvar	"	590.00	550.00	1,140
35 kvar	"	690.00	590.00	1,280
40 kvar	"	740.00	670.00	1,410
45 kvar	"	810.00	740.00	1,550
50 kvar	"	850.00	780.00	1,630
60 kvar	"	1,080	820.00	1,900
70 kvar	"	1,200	890.00	2,090
75 kvar	"	1,310	930.00	2,240
80 kvar	"	1,510	1,000	2,510
90 kvar	"	1,610	1,080	2,690
100 kvar	"	1,720	1,110	2,830

POWER GENERATION	UNIT	MAT.	INST.	TOTAL
16230.10 CAPACITORS				
125 kvar	EA.	2,310	1,230	3,540
150 kvar	"	2,650	1,370	4,020
16320.10 TRANSFORMERS				
Floor mounted, single phase, int. dry, 480v-120/240v				
3 kva	EA.	230.00	67.50	297.50
5 kva	"	320.00	110.00	430.00
7.5 kva	"	480.00	130.00	610.00
10 kva	"	540.00	140.00	680.00
15 kva	"	700.00	160.00	860.00
25 kva	"	890.00	280.00	1,170
37.5 kva	"	1,240	350.00	1,590
50 kva	"	1,400	380.00	1,780
75 kva	"	1,910	400.00	2,310
100 kva	"	2,540	430.00	2,970
Three phase, 480v-120/280v				
15 kva	EA.	890.00	220.00	1,110
30 kva	"	1,140	350.00	1,490
45 kva	"	1,530	400.00	1,930
75 kva	"	2,160	410.00	2,570
112.5 kva	"	2,920	470.00	3,390
150 kva	"	3,690	500.00	4,190
225 kva	"	5,020	570.00	5,590
Single phase, dry type, 2400v				
167 kva	EA.	6,930	830.00	7,760
250 kva	"	9,030	1,110	10,140
333 kva	"	11,090	1,390	12,480
5000v				
167 kva	EA.	7,430	830.00	8,260
250 kva	"	9,640	1,110	10,750
333 kva	"	11,800	1,390	13,190
8660v				
167 kva	EA.	7,810	1,020	8,830
250 kva	"	10,550	1,300	11,850
333 kva	"	12,650	2,520	15,170
1500v				
167 kva	EA.	8,800	1,020	9,820
250 kva	"	11,390	1,300	12,690
333 kva	"	13,610	1,580	15,190
Three phase, dry type transformer, 2400v				
225 kva	EA.	10,080	930.00	11,010
300 kva	"	11,980	1,020	13,000
500 kva	"	15,810	1,580	17,390
750 kva	"	20,770	1,950	22,720
5000v				
225.0 kva	EA.	10,170	930.00	11,100
300 kva	"	12,200	1,020	13,220
500 kva	"	15,890	1,580	17,470
750 kva	"	20,850	1,950	22,800
8660v				
225.0 kva	EA.	11,080	1,110	12,190
300 kva	"	13,910	1,210	15,120

POWER GENERATION	UNIT	MAT.	INST.	TOTAL
16320.10 **TRANSFORMERS**				
500 kva	EA.	17,960	1,770	19,730
750 kva	"	22,540	2,140	24,680
1500v				
225 kva	EA.	12,100	1,110	13,210
300 kva	"	15,120	1,210	16,330
500 kva	"	19,490	1,770	21,260
750 kva	"	24,530	2,140	26,670
Buck boost transformers				
.25 kva	EA.	62.50	37.10	99.60
.50 kva	"	82.00	46.40	128.40
.75 kva	"	110.00	56.00	166.00
1.00 kva	"	140.00	64.50	204.50
1.50 kva	"	160.00	74.00	234.00
2.00 kva	"	230.00	93.00	323.00
3.00 kva	"	320.00	110.00	430.00
16350.10 **CIRCUIT BREAKERS**				
Molded case, 240v, 15-60a, bolt-on				
1 pole	EA.	8.10	9.30	17.40
2 pole	"	17.95	12.90	30.85
70-100a, 2 pole	"	47.50	19.80	67.30
15-60a, 3 pole	"	58.50	14.85	73.35
70-100a, 3 pole	"	85.00	22.85	107.85
480v, 2 pole				
15-60a	EA.	130.00	11.00	141.00
70-100a	"	160.00	14.85	174.85
3 pole				
15-60a	EA.	160.00	14.85	174.85
70-100a	"	190.00	16.50	206.50
70-225a	"	430.00	22.85	452.85
Draw out air circuit breakers				
600a	EA.	7,450	590.00	8,040
800a	"	9,630	670.00	10,300
1600a	"	15,460	900.00	16,360
2000a	"	20,720	1,020	21,740
3000a	"	35,950	1,190	37,140
4000a	"	55,220	1,410	56,630
Load center circuit breakers, 240v				
1 pole, 10-60a	EA.	7.85	9.30	17.15
2 pole				
10-60a	EA.	17.95	14.85	32.80
70-100a	"	17.25	24.75	42.00
110-150a	"	110.00	27.00	137.00
3 pole				
10-60a	EA.	61.50	18.55	80.05
70-100a	"	92.50	27.00	119.50
Load center, G.F.I. breakers, 240v				
1 pole, 15-30a	EA.	61.50	11.00	72.50
2 pole, 15-30a	"	110.00	14.85	124.85
Key operated breakers, 240v, 1 pole, 10-30a	"	10.25	11.00	21.25
Tandem breakers, 240v				

POWER GENERATION

16350.10	CIRCUIT BREAKERS	UNIT	MAT.	INST.	TOTAL
1 pole, 15-30a		EA.	16.70	14.85	31.55
2 pole, 15-30a		"	32.10	19.80	51.90
Bolt-on, G.F.I. breakers, 240v, 1 pole, 15-30a		"	76.50	12.90	89.40
Enclosed breaker, 120v, 1 pole, 15-50a, NEMA 1		"	89.00	29.70	118.70
240v, 2 pole					
15-60a, NEMA 1		EA.	120.00	46.40	166.40
70-100a, NEMA 1		"	160.00	64.50	224.50
3 pole					
15-60a, NEMA 1		EA.	160.00	56.00	216.00
70-100a, NEMA 1		"	210.00	82.50	292.50
Enclosed circuit breakers					
120v, 1 pole, NEMA 3R, 15-50a		EA.	170.00	33.30	203.30
240v, 2 pole, NEMA 3R					
15-60a		EA.	210.00	46.40	256.40
70-100a		"	240.00	64.50	304.50
3 pole, NEMA 3R					
15-60a		EA.	230.00	56.00	286.00
70-100a		"	280.00	82.50	362.50
480v, NEMA 1					
1 pole, 15-50a		EA.	100.00	29.70	129.70
2 pole, 15-60a		"	180.00	46.40	226.40
70-100a		"	230.00	56.00	286.00
3 pole, NEMA 1					
15-60a		EA.	230.00	56.00	286.00
70-100a		"	260.00	82.50	342.50
480v, 1 pole, 15-50a, NEMA 3R		"	180.00	37.10	217.10
2 pole					
2 pole, 15-60a, NEMA 3R		EA.	260.00	46.40	306.40
70-100a, NEMA 3R		"	300.00	64.50	364.50
3 pole					
15-60a, NEMA 3R		EA.	300.00	56.00	356.00
70-100a, NEMA 3R		"	330.00	82.50	412.50
70-100a, NEMA 3R		"	210.00	46.40	256.40
70-100a, NEMA 1		"	250.00	64.50	314.50
3 pole					
15-60a, NEMA 1		EA.	260.00	64.50	324.50
70-100a, NEMA 1		"	300.00	82.50	382.50
Enclosed breakers, 600v, 2 phase, NEMA 3R					
15-60a		EA.	270.00	46.40	316.40
70-100a		"	330.00	64.50	394.50
3 phase, NEMA 3R					
15-60a		EA.	310.00	56.00	366.00
70-100a		"	360.00	82.50	442.50
600v, 3 phase, NEMA 1					
125a		EA.	640.00	82.50	722.50
150a		"	640.00	110.00	750.00
175a		"	640.00	110.00	750.00
200a		"	640.00	110.00	750.00
225a		"	640.00	110.00	750.00
250a		"	1,110	230.00	1,340
300a		"	1,110	230.00	1,340
350a		"	1,110	230.00	1,340

POWER GENERATION	UNIT	MAT.	INST.	TOTAL
16350.10 — CIRCUIT BREAKERS				
400a	EA.	1,110	230.00	1,340
500a	"	1,830	360.00	2,190
600a	"	1,830	360.00	2,190
700a	"	2,370	400.00	2,770
800a	"	2,370	400.00	2,770
900a	"	3,000	560.00	3,560
100a	"	3,000	560.00	3,560
1200a	"	6,720	690.00	7,410
1400a	"	6,720	690.00	7,410
1600a	"	6,720	890.00	7,610
1800a	"	7,060	1,110	8,170
2000a	"	7,060	1,110	8,170
600v, 3 phase, NEMA 3R				
125-225a	EA.	760.00	82.50	842.50
250-400a	"	1,420	210.00	1,630
500-600a	"	2,190	360.00	2,550
700-800a	"	2,880	410.00	3,290
900-1000a	"	3,360	560.00	3,920
1000-1200a	"	7,890	700.00	8,590
1400-1600a	"	8,010	890.00	8,900
1800-2000a	"	8,140	1,110	9,250
16360.10 — SAFETY SWITCHES				
Fused, 3 phase, 30 amp, 600v, heavy duty				
NEMA 1	EA.	120.00	42.40	162.40
NEMA 3r	"	200.00	42.40	242.40
NEMA 4	"	500.00	59.50	559.50
NEMA 12	"	190.00	64.50	254.50
60a				
NEMA 1	EA.	140.00	42.40	182.40
NEMA 3r	"	230.00	42.40	272.40
NEMA 4	"	570.00	59.50	629.50
NEMA 12	"	200.00	64.50	264.50
100a				
NEMA 1	EA.	260.00	64.50	324.50
NEMA 3r	"	370.00	64.50	434.50
NEMA 4	"	1,110	74.00	1,184
NEMA 12	"	310.00	93.00	403.00
200a				
NEMA 1	EA.	360.00	93.00	453.00
NEMA 3r	"	500.00	93.00	593.00
NEMA 4	"	1,560	100.00	1,660
NEMA 12	"	500.00	130.00	630.00
400a				
NEMA 1	EA.	990.00	200.00	1,190
NEMA 3r	"	1,180	200.00	1,380
NEMA 4	"	2,000	210.00	2,210
NEMA 12	"	1,090	260.00	1,350
600a				
NEMA 1	EA.	1,670	300.00	1,970
NEMA 3r	"	2,330	300.00	2,630

POWER GENERATION	UNIT	MAT.	INST.	TOTAL
16360.10 SAFETY SWITCHES				
NEMA 4	EA.	3,930	330.00	4,260
NEMA 12	"	1,840	460.00	2,300
Non-fused, 240-600v, heavy duty, 3 phase, 30 amp				
NEMA 1	EA.	61.00	42.40	103.40
NEMA 3r	"	110.00	42.40	152.40
NEMA 4	"	440.00	64.50	504.50
NEMA 12	"	130.00	64.50	194.50
60a				
NEMA1	EA.	110.00	42.40	152.40
NEMA 3r	"	190.00	42.40	232.40
NEMA 4	"	530.00	64.50	594.50
NEMA 12	"	170.00	64.50	234.50
100a				
NEMA 1	EA.	170.00	64.50	234.50
NEMA 3r	"	280.00	64.50	344.50
NEMA 4	"	1,080	93.00	1,173
NEMA 12	"	240.00	93.00	333.00
200a, NEMA 1	"	290.00	93.00	383.00
600a, NEMA 12	"	1,430	460.00	1,890
Bolt-on hubs				
3/4" - 1-1/2"	EA.	8.95	9.30	18.25
2"	"	16.25	11.00	27.25
2-1/2"	"	25.70	11.00	36.70
3"	"	47.50	12.90	60.40
3-1/2"	"	70.00	14.85	84.85
4"	"	90.00	14.85	104.85
Watertight hubs				
1/2"	EA.	7.20	9.30	16.50
3/4"	"	10.45	11.00	21.45
1"	"	11.00	14.85	25.85
1-1/4"	"	12.65	16.50	29.15
1-1/2"	"	18.60	17.45	36.05
2"	"	27.60	18.55	46.15
2-1/2"	"	32.90	19.80	52.70
3"	"	41.90	22.85	64.75
3-1/2"	"	66.00	29.70	95.70
4"	"	88.50	31.60	120.10
Non-fused, 600v, 3 pole, NEMA 7				
600a	EA.	530.00	82.50	612.50
100a	"	640.00	120.00	760.00
225a	"	1,430	150.00	1,580
NEMA 9				
60a	EA.	550.00	93.00	643.00
100a	"	650.00	120.00	770.00
225a	"	1,560	160.00	1,720
Fusible bolted pressure switches, 600v/3 pole, NEMA 1				
800a	EA.	3,240	590.00	3,830
1200a	"	3,930	820.00	4,750
1600a	"	4,240	930.00	5,170
2000a	"	4,370	1,110	5,480
2500a	"	4,990	1,300	6,290
3000a	"	6,740	1,670	8,410

POWER GENERATION	UNIT	MAT.	INST.	TOTAL
16360.10 SAFETY SWITCHES				
4000a	EA.	8,980	1,930	10,910
Non-fusible				
800a	EA.	3,120	540.00	3,660
1200a	"	3,490	740.00	4,230
1600a	"	3,740	850.00	4,590
2000a	"	3,970	1,040	5,010
2500a	"	4,550	1,300	5,850
3000a	"	6,490	1,670	8,160
4000a	"	8,730	1,930	10,660
Fusible load interrupter switches, 4.16 kv, NEMA 1				
200a	EA.	2,870	1,110	3,980
600a	"	4,500	2,600	7,100
Fusible load interrupter switch, 13.8 kv				
NEMA 1, 600a	EA.	4,990	3,710	8,700
NEMA 3R, 600a	"	5,990	3,710	9,700
4.16 kv, NEMA 3R				
200a	EA.	3,120	1,110	4,230
600a	"	5,610	2,600	8,210
Non-fused load interrupter switch, 4.16 kv, NEMA 1				
200a	EA.	2,490	1,110	3,600
600a	"	4,240	2,600	6,840
13.8 kv, NEMA 1, 600a	"	4,610	3,710	8,320
4.16 kv, NEMA 3R				
200a	EA.	2,620	1,110	3,730
600a	"	5,240	2,600	7,840
13.8 kv, NEMA 3R, 600a	"	5,610	3,710	9,320
Interrupter switch accessories, strip heater	"			340.00
Cable lugs	"			62.50
Key interlock	"			430.00
Auxiliary switch	"			240.00
Lightning arrester				
5 kva	EA.			1,030
15 kv	"			1,120
16365.10 FUSES				
Fuse, one-time, 250v				
30a	EA.	0.95	1.85	2.80
60a	"	1.20	1.85	3.05
100a	"	5.15	1.85	7.00
200a	"	12.95	1.85	14.80
400a	"	23.30	1.85	25.15
600a	"	33.70	1.85	35.55
600v				
30a	EA.	3.90	1.85	5.75
60a	"	5.15	1.85	7.00
100a	"	11.65	1.85	13.50
200a	"	22.00	1.85	23.85
400a	"	42.80	1.85	44.65
Fusetron, 600v				
200a	EA.	21.35	1.85	23.20
400a	"	42.80	1.85	44.65

POWER GENERATION	UNIT	MAT.	INST.	TOTAL
16365.10 **FUSES**				
Fuse, amp-trap, K1, 250v				
30a	EA.	6.35	1.85	8.20
60a	"	13.90	1.85	15.75
100a	"	32.10	1.85	33.95
200a	"	65.00	1.85	66.85
400a	"	130.00	1.85	131.85
600a	"	160.00	1.85	161.85
600v				
30a	EA.	13.65	1.85	15.50
60a	"	25.30	1.85	27.15
100a	"	52.00	1.85	53.85
200a	"	84.00	1.85	85.85
400a	"	160.00	1.85	161.85
K5, 250v				
30a	EA.	2.90	1.85	4.75
60a	"	5.30	1.85	7.15
100a	"	12.30	1.85	14.15
200a	"	23.95	1.85	25.80
400a	"	45.30	1.85	47.15
600a	"	67.50	1.85	69.35
600v				
30a	EA.	5.70	1.85	7.55
60a	"	10.95	1.85	12.80
100a	"	21.55	1.85	23.40
200a	"	42.10	1.85	43.95
400a	"	82.50	1.85	84.35
600a	"	120.00	1.85	121.85
J, 600v				
30a	EA.	10.50	1.85	12.35
60a	"	18.30	1.85	20.15
100a	"	41.80	1.85	43.65
200a	"	81.50	1.85	83.35
400a	"	150.00	1.85	151.85
L, 600v				
1200a	EA.	270.00	14.85	284.85
1600a	"	410.00	14.85	424.85
2000a	"	470.00	14.85	484.85
2500a	"	610.00	14.85	624.85
3000a	"	700.00	14.85	714.85
4000a	"	940.00	14.85	954.85
5000a	"	1,420	14.85	1,435
Fuse cl-ay 250v				
600a	EA.	260.00	11.00	271.00
1200a	"	260.00	11.00	271.00
1600a	"	320.00	11.00	331.00
2000a	"	420.00	11.00	431.00
600v				
1200a	EA.	250.00	11.00	261.00
1600a	"	320.00	11.00	331.00
2000a	"	420.00	11.00	431.00
Reducers, 600v				
60a-30a	EA.	5.15	5.40	10.55

POWER GENERATION	UNIT	MAT.	INST.	TOTAL
16365.10 FUSES				
100a-30a	EA.	6.45	5.40	11.85
100a-60a	"	18.40	5.40	23.80
200a-60a	"	33.90	9.30	43.20
200a-100a	"	12.95	9.30	22.25
400a-100a	"	50.00	12.90	62.90
400a-200a	"	53.00	12.90	65.90
600a-100a	"	88.50	14.85	103.35
600a-200a	"	86.00	14.85	100.85
600a-400a	"	88.50	14.85	103.35
16395.10 GROUNDING				
Ground rods, copper clad, 1/2" x				
6'	EA.	7.05	24.75	31.80
8'	"	9.15	27.00	36.15
10'	"	11.30	37.10	48.40
5/8" x				
5'	EA.	7.20	22.85	30.05
6'	"	8.50	27.00	35.50
8'	"	10.60	37.10	47.70
10'	"	13.75	46.40	60.15
3/4" x				
8'	EA.	17.10	27.00	44.10
10'	"	20.25	29.70	49.95
Ground rod clamp				
5/8"	EA.	2.25	4.55	6.80
3/4"	"	2.90	4.55	7.45
Coupling, on threaded rods, 3/4"	"	7.20	1.85	9.05
Ground receptacles	"	14.20	9.30	23.50
Bus bar, copper, 2" x 1/4"	L.F.	3.25	5.40	8.65
Copper braid, 1" x 1/8", for door ground	EA.	2.70	3.70	6.40
Brazed connection for				
#6 wire	EA.	9.15	18.55	27.70
#2 wire	"	11.65	29.70	41.35
#2/0 wire	"	13.75	37.10	50.85
#4/0 wire	"	20.60	42.40	63.00
Ground rod couplings				
1/2"	EA.	4.10	3.70	7.80
5/8"	"	5.30	3.70	9.00
Ground rod, driving stud				
1/2"	EA.	2.80	3.70	6.50
5/8"	"	2.90	3.70	6.60
3/4"	"	3.00	3.70	6.70
Ground rod clamps, #8-2 to				
1" pipe	EA.	3.05	7.40	10.45
2" pipe	"	4.20	9.30	13.50
3" pipe	"	19.90	11.00	30.90
5" pipe	"	30.10	12.90	43.00
6" pipe	"	45.80	16.50	62.30
#4-4/0 to				
1" pipe	EA.	8.50	7.40	15.90
2" pipe	"	13.85	9.30	23.15

POWER GENERATION

POWER GENERATION	UNIT	MAT.	INST.	TOTAL
16395.10 GROUNDING				
3" pipe	EA.	22.50	11.00	33.50
3" pipe	"	33.70	12.90	46.60
8 pipe	"	51.50	16.50	68.00
8 pipe	"	56.50	24.75	81.25
10 pipe	"	70.50	35.30	105.80
12 pipe	"	81.00	47.90	128.90

SERVICE AND DISTRIBUTION

SERVICE AND DISTRIBUTION	UNIT	MAT.	INST.	TOTAL
16425.10 SWITCHBOARDS				
Switchboard, 90" high, no main disconnect, 208/120v				
400a	EA.	1,500	290.00	1,790
600a	"	2,330	300.00	2,630
1000a	"	2,920	300.00	3,220
1200a	"	3,130	370.00	3,500
1600a	"	3,430	440.00	3,870
2000a	"	3,690	520.00	4,210
2500a	"	4,070	590.00	4,660
240/480v				
600a	EA.	2,670	300.00	2,970
800a	"	2,920	300.00	3,220
1600a	"	3,690	440.00	4,130
2000a	"	3,940	520.00	4,460
2500a	"	4,200	590.00	4,790
3000a	"	4,830	1,020	5,850
4000a	"	5,850	1,100	6,950
Main breaker sections, 600v				
1200a, GFI	EA.	10,550	630.00	11,180
1600a, GFI	"	12,460	720.00	13,180
2000a, GFI	"	13,230	740.00	13,970
2500a, GFI	"	14,620	930.00	15,550
3000a, GFI	"	16,410	1,110	17,520
4000a, GFI	"	19,330	1,300	20,630
Switchboard meter sections, 600v				
400a	EA.	1,910	300.00	2,210
600a	"	2,920	370.00	3,290
800a	"	3,560	410.00	3,970
1000a	"	4,320	500.00	4,820
2000a	"	5,340	590.00	5,930
2500a	"	5,980	740.00	6,720
3000a	"	6,230	930.00	7,160
4000a	"	6,990	1,110	8,100
Insulated case, draw out compartment, 208/120v				
800a	EA.	1,270	93.00	1,363

SERVICE AND DISTRIBUTION

16425.10 SWITCHBOARDS	UNIT	MAT.	INST.	TOTAL
1600a	EA.	1,520	110.00	1,630
2000a	"	1,910	130.00	2,040
2500a	"	2,290	130.00	2,420
3000a	"	3,810	150.00	3,960
4000a	"	5,340	180.00	5,520
Accessories for power trip breakers				
Shunt trip	EA.	610.00	18.55	628.55
Key interlock	"	290.00	82.50	372.50
Lifting and transport truck	"	1,910	170.00	2,080
Lifting device	"	190.00	49.50	239.50
Bus duct connection, 3 phase, 4 wire				
225a	EA.	270.00	110.00	380.00
400a	"	330.00	110.00	440.00
600a	"	430.00	120.00	550.00
800a	"	460.00	150.00	610.00
2500a	"	870.00	220.00	1,090
3000a	"	1,310	280.00	1,590
4000a	"	2,340	330.00	2,670
Provision for mounting current transformers				
800a & below primary	EA.	980.00	110.00	1,090
1000 to 1500a primary	"	1,220	110.00	1,330
2000 to 6000a primary	"	1,470	110.00	1,580
Provision for mounting potential transformers				
2000a max	EA.	3,500	140.00	3,640
Switchboard instruments				
Voltmeter	EA.	1,260	37.10	1,297
Ammeter, incoming line	"	1,260	37.10	1,297
Wattmeter	"	1,990	37.10	2,027
Varmeter	"	2,060	37.10	2,097
Power factor meter	"	2,380	37.10	2,417
Frequency meter	"	2,670	37.10	2,707
Recording voltmeter	"	5,670	74.00	5,744
Wattmeter	"	5,670	74.00	5,744
Power factor meter	"	6,610	74.00	6,684
Frequency meter	"	6,610	74.00	6,684
Instrument phase select switch	"	250.00	18.55	268.55
Enclosure, 90" high, 3 phase, 4 wire				
1000a	EA.	2,160	250.00	2,410
1200a	"	2,290	260.00	2,550
1600a	"	2,540	320.00	2,860
2000a	"	2,670	490.00	3,160
5500a	"	3,310	580.00	3,890
3000a	"	3,690	670.00	4,360
4000a	"	4,830	870.00	5,700
Circuit breakers, 600v, 100a, frame				
15-30a, 1 pole	EA.	70.00	11.00	81.00
15-60a, 2 pole	"	180.00	12.90	192.90
70-100a, 2 pole	"	220.00	14.85	234.85
15-60a, 3 pole	"	230.00	16.50	246.50
70-100a, 3 pole	"	270.00	18.55	288.55
Bolt on breakers, 600v, 225a frame, 110-225a				
2 pole	EA.	530.00	22.85	552.85

SERVICE AND DISTRIBUTION

16425.10	SWITCHBOARDS	UNIT	MAT.	INST.	TOTAL
3 pole		EA.	660.00	40.70	700.70
400a frame, 250-400a, 2 pole		"	960.00	46.40	1,006
800a frame					
450-600a, 2 pole		EA.	1,700	70.50	1,771
700-800a, 2 pole		"	1,980	93.00	2,073
450-600a, 3 pole		"	1,940	160.00	2,100
700-800a, 3 pole		"	2,510	160.00	2,670
Bolt on branch breakers, 600v					
1000-2000a, 2 pole		EA.	4,230	200.00	4,430
2500a, 2 pole		"	7,800	400.00	8,200
1000-2000a, 3 pole		"	5,420	300.00	5,720
2500a, 3 pole		"	9,520	410.00	9,930
3000a, 3 pole		"	18,120	740.00	18,860
Metal clad substation switch board, selector switch					
600a, 5kv		EA.	13,990	1,560	15,550
600a, 15kv		"	15,520	1,780	17,300
Fused switch, 600a					
5kv		EA.	10,170	1,300	11,470
15kv		"	14,110	1,300	15,410
1200a					
5kv		EA.	11,320	1,480	12,800
15kv		"	15,160	1,480	16,640
Oil cutout switch					
5 kv		EA.	3,940	560.00	4,500
15 kv		"	5,470	670.00	6,140
Liquid air terminal section		"	510.00	300.00	810.00
Dry air terminal section		"	1,020	320.00	1,340
Auxiliary compartment		"	7,250	1,110	8,360

16430.20	METERING	UNIT	MAT.	INST.	TOTAL
Outdoor wp meter sockets, 1 gang, 240v, 1 phase					
Includes sealing ring, 100a		EA.	24.45	56.00	80.45
150a		"	32.60	66.00	98.60
200a		"	40.80	74.00	114.80
Die cast hubs, 1-1/4"		"	3.80	11.85	15.65
1-1/2"		"	4.35	11.85	16.20
2"		"	5.25	11.85	17.10
Indoor meter center, main switch single phase, 240v					
400a		EA.	590.00	300.00	890.00
600a		"	1,150	410.00	1,560
800a		"	1,910	430.00	2,340
Main breaker					
400a		EA.	1,060	300.00	1,360
600a		"	1,600	410.00	2,010
800a		"	2,050	430.00	2,480
1000a		"	2,960	590.00	3,550
Terminal box		"	4,300	610.00	4,910
1600a		"	6,460	670.00	7,130
Terminal box					
800a		EA.	200.00	370.00	570.00
1600a		"	820.00	670.00	1,490
Main switch, three phase, 208v					

SERVICE AND DISTRIBUTION	UNIT	MAT.	INST.	TOTAL
16430.20 METERING				
400a	EA.	720.00	320.00	1,040
600a	"	1,300	440.00	1,740
800a	"	2,690	500.00	3,190
Main breaker				
400a	EA.	1,260	320.00	1,580
600a	"	1,980	440.00	2,420
800a	"	2,630	500.00	3,130
1000a	"	3,370	630.00	4,000
1200a	"	4,690	670.00	5,360
1600a	"	6,920	780.00	7,700
Terminal box				
800a	EA.	220.00	480.00	700.00
1600a	"	860.00	780.00	1,640
Indoor meter center				
2 meters	EA.	210.00	190.00	400.00
3 meters	"	260.00	230.00	490.00
4 meters	"	340.00	270.00	610.00
5 meters	"	430.00	300.00	730.00
6 meters	"	540.00	330.00	870.00
Plug on breakers, single phase, 208v				
60a	EA.	16.60	9.30	25.90
70a	"	33.20	9.30	42.50
80a	"	45.70	9.30	55.00
90a	"	47.90	9.30	57.20
100a	"	50.50	12.90	63.40
Indoor meter center, single phase, 125a breakers				
3 meters	EA.	280.00	230.00	510.00
4 meters	"	370.00	270.00	640.00
5 meters	"	460.00	300.00	760.00
6 meters	"	550.00	320.00	870.00
7 meters	"	640.00	370.00	1,010
8 meters	"	740.00	410.00	1,150
10 meters	"	920.00	440.00	1,360
150a breakers				
3 meters	EA.	1,330	230.00	1,560
4 meters	"	1,780	270.00	2,050
6 meters	"	2,670	300.00	2,970
7 meters	"	3,110	370.00	3,480
8 meters	"	3,560	410.00	3,970
200a breakers				
3 meters	EA.	1,390	230.00	1,620
4 meters	"	1,860	270.00	2,130
6 meters	"	2,790	300.00	3,090
7 meters	"	3,250	370.00	3,620
8 meters	"	3,720	410.00	4,130
Indoor meter center, three phase, 125a breakers				
3 meters	EA.	300.00	230.00	530.00
4 meters	"	400.00	270.00	670.00
5 meters	"	490.00	300.00	790.00
6 meters	"	590.00	330.00	920.00
7 meters	"	690.00	370.00	1,060
8 meters	"	790.00	410.00	1,200

SERVICE AND DISTRIBUTION

16430.20 METERING	UNIT	MAT.	INST.	TOTAL
10 meters	EA.	990.00	440.00	1,430
150a breakers				
3 meters	EA.	1,360	230.00	1,590
4 meters	"	1,820	270.00	2,090
6 meters	"	2,750	320.00	3,070
7 meters	"	3,180	410.00	3,590
8 meters	"	3,630	440.00	4,070
200a breakers				
3 meters	EA.	1,420	250.00	1,670
4 meters	"	1,900	270.00	2,170
6 meters	"	2,850	330.00	3,180
7 meters	"	3,320	410.00	3,730
8 meters	"	3,800	440.00	4,240
NEMA 3R, meter center, main switch, 1 phase, 240v				
400a	EA.	770.00	300.00	1,070
600a	"	1,460	370.00	1,830
800a	"	2,240	410.00	2,650
Main breaker				
400a	EA.	1,860	300.00	2,160
600a	"	2,390	370.00	2,760
800a	"	2,930	460.00	3,390
1000a	"	3,460	560.00	4,020
1200a	"	4,870	590.00	5,460
Terminal box				
225a	EA.	180.00	270.00	450.00
800a	"	270.00	430.00	700.00
1600a	"	890.00	670.00	1,560
NEMA 3R, three phase, 280v				
400a	EA.	870.00	320.00	1,190
600a	"	1,760	440.00	2,200
800a	"	3,260	480.00	3,740
Main breaker				
400a	EA.	2,160	320.00	2,480
600a	"	2,770	440.00	3,210
800a	"	3,510	480.00	3,990
1000a	"	3,940	630.00	4,570
1200a	"	5,400	670.00	6,070
Terminal box				
225a	EA.	200.00	300.00	500.00
800a	"	300.00	480.00	780.00
1600a	"	930.00	780.00	1,710
NEMA 3R meter center, single phase, 208v, 100a				
2 meters	EA.	220.00	190.00	410.00
3 meters	"	270.00	230.00	500.00
4 meters	"	400.00	270.00	670.00
5 meters	"	470.00	300.00	770.00
6 meters	"	750.00	330.00	1,080
4 meters	"	400.00	270.00	670.00
6 meters	"	590.00	300.00	890.00
7 meters	"	690.00	370.00	1,060
8 meters	"	790.00	410.00	1,200
125a, 3 meters	"	1,350	230.00	1,580

SERVICE AND DISTRIBUTION	UNIT	MAT.	INST.	TOTAL
16430.20 METERING				
4 meters	EA.	1,810	270.00	2,080
6 meters	"	2,710	310.00	3,020
7 meters	"	3,160	370.00	3,530
8 meters	"	3,620	410.00	4,030
150a, 3 meters	"	1,410	230.00	1,640
4 meters	"	1,890	270.00	2,160
6 meters	"	2,830	320.00	3,150
7 meters	"	3,310	370.00	3,680
8 meters	"	3,780	410.00	4,190
NEMA 3R center, 3 phase, 208v, 125a breakers				
3 meters	EA.	310.00	230.00	540.00
4 meters	"	420.00	270.00	690.00
6 meters	"	640.00	320.00	960.00
7 meters	"	740.00	370.00	1,110
8 meters	"	850.00	410.00	1,260
150a				
3 meters	EA.	1,410	250.00	1,660
4 meters	"	1,840	270.00	2,110
6 meters	"	2,770	330.00	3,100
7 meters	"	3,230	390.00	3,620
8 meters	"	3,690	430.00	4,120
200a				
3 meters	EA.	1,450	250.00	1,700
4 meters	"	1,930	270.00	2,200
6 meters	"	2,890	330.00	3,220
7 meters	"	3,390	410.00	3,800
8 meters	"	3,860	430.00	4,290
NEMA 3R, center plug-on breakers, 208v, 1 phase				
60a	EA.	16.60	9.30	25.90
70a	"	33.20	9.30	42.50
90a	"	47.90	9.30	57.20
100a	"	50.50	12.90	63.40
125a	"	97.00	14.85	111.85
16460.10 TRANSFORMERS				
Pad mounted, single phase, dry type, 480v-120/240v				
15 kva	EA.	1,000	300.00	1,300
25 kva	"	1,480	330.00	1,810
37.5 kva	"	2,190	370.00	2,560
50 kva	"	2,410	410.00	2,820
3 phase				
225 kva	EA.	5,140	930.00	6,070
300 kva	"	6,600	1,140	7,740
500 kva	"	10,390	1,410	11,800
750 kva	"	16,750	1,750	18,500
1000 kva	"	20,300	1,860	22,160
1500 kva	"	23,940	2,120	26,060
Substation transformers, outdoor, 5 kv - 208v				
112.5 kva	EA.	8,440	820.00	9,260
150 kva	"	9,170	890.00	10,060
225 kva	"	10,390	1,040	11,430

SERVICE AND DISTRIBUTION	UNIT	MAT.	INST.	TOTAL
16460.10 TRANSFORMERS				
300 kva	EA.	11,980	1,110	13,090
500 kva	"	15,900	1,670	17,570
750 kva	"	22,010	2,050	24,060
1000 kva	"	27,150	2,410	29,560
15 kv, 208v				
112 kva	EA.	10,570	1,040	11,610
150 kva	"	10,800	1,110	11,910
225 kva	"	11,160	1,300	12,460
300 kva	"	12,580	1,480	14,060
500 kva	"	17,690	1,860	19,550
750 kva	"	17,690	2,230	19,920
1000 kva	"	26,590	2,600	29,190
5kv, 480v				
112kva	EA.	8,070	820.00	8,890
150 kva	"	8,550	890.00	9,440
225 kva	"	9,620	1,040	10,660
300 kva	"	10,800	1,110	11,910
500 kva	"	14,480	1,670	16,150
750 kva	"	19,710	2,050	21,760
1000 kva	"	23,150	2,410	25,560
1500 kva	"	30,270	2,770	33,040
2000 kva	"	37,280	3,330	40,610
2500 kva	"	44,280	4,070	48,350
15 kv, 480v				
112.5 kva	EA.	10,330	1,040	11,370
150 kva	"	10,570	1,110	11,680
225 kva	"	10,680	1,300	11,980
300 kva	"	11,630	1,480	13,110
500 kva	"	16,500	1,860	18,360
750 kva	"	20,060	2,230	22,290
1000 kva	"	23,150	2,600	25,750
1500 kva	"	30,390	2,970	33,360
2000 kva	"	37,400	3,330	40,730
2500 kva	"	44,400	4,430	48,830
Pad mounted 3 phase, 15 kv outdoor				
50 kva	EA.	2,490	380.00	2,870
75 kva	"	3,500	440.00	3,940
112 kva	"	4,510	480.00	4,990
150 kva	"	6,410	540.00	6,950
225 kva	"	7,720	570.00	8,290
300 kva	"	9,500	630.00	10,130
500 kva	"	11,630	1,020	12,650
750 kva	"	14,840	1,350	16,190
1000 kva	"	17,570	1,650	19,220
1500 kva	"	22,440	1,980	24,420
Dry type, for power gear, 5 kv indoor				
75 kva	EA.	6,890	590.00	7,480
112.5 kva	"	7,720	690.00	8,410
150 kva	"	8,900	780.00	9,680
225 kva	"	11,040	870.00	11,910
300 kva	"	13,060	930.00	13,990
500 kva	"	17,250	1,020	18,270

SERVICE AND DISTRIBUTION	UNIT	MAT.	INST.	TOTAL
16460.10 TRANSFORMERS				
750 kva	EA.	22,680	1,350	24,030
16470.10 PANELBOARDS				
Indoor load center, 1 phase 240v main lug only				
30a - 2 spaces	EA.	10.50	74.00	84.50
100a - 8 spaces	"	35.50	90.00	125.50
150a - 16 spaces	"	86.00	110.00	196.00
200a - 24 spaces	"	110.00	130.00	240.00
200a - 42 spaces	"	210.00	150.00	360.00
Main circuit breaker				
100a - 8 spaces	EA.	86.00	90.00	176.00
100a - 16 spaces	"	110.00	100.00	210.00
150a - 16 spaces	"	180.00	110.00	290.00
150a - 24 spaces	"	210.00	120.00	330.00
200a - 24 spaces	"	210.00	130.00	340.00
200a - 42 spaces	"	290.00	130.00	420.00
3 phase, 480/277v, main lugs only, 120a, 30 circuits	"	580.00	130.00	710.00
277/480v, 4 wire, flush surface				
225a, 30 circuits	EA.	610.00	150.00	760.00
400a, 30 circuits	"	710.00	190.00	900.00
600a, 42 circuits	"	880.00	220.00	1,100
208/120v, main circuit breaker, 3 phase, 4 wire				
100a				
12 circuits	EA.	510.00	190.00	700.00
20 circuits	"	620.00	230.00	850.00
30 circuits	"	920.00	260.00	1,180
225a				
30 circuits	EA.	780.00	290.00	1,070
42 circuits	"	1,410	350.00	1,760
400a				
30 circuits	EA.	1,930	550.00	2,480
42 circuits	"	2,310	590.00	2,900
600a, 42 circuits	"	2,570	670.00	3,240
120/208v, flush, 3 ph., 4 wire, main only				
100a				
12 circuits	EA.	350.00	190.00	540.00
20 circuits	"	490.00	230.00	720.00
30 circuits	"	710.00	260.00	970.00
225a				
30 circuits	EA.	740.00	290.00	1,030
42 circuits	"	940.00	350.00	1,290
400a				
30 circuits	EA.	1,410	550.00	1,960
42 circuits	"	2,060	590.00	2,650
600a, 42 circuits	"	3,210	670.00	3,880
Panelboard accessories				
Grounding bus	EA.	14.10	12.90	27.00
Handle lock device	"	1.95	5.40	7.35
Factory assembled panel				
1 pole space	EA.	10.25	12.90	23.15

SERVICE AND DISTRIBUTION	UNIT	MAT.	INST.	TOTAL
16470.10 PANELBOARDS				
2 pole space	EA.	21.85	5.40	27.25
3 pole space	"	32.10	4.95	37.05
Panelboards 1 phase, 240/120v main circuit breaker				
Single phase, 3 wire, 120/240v flush				
100a, 20 circuits	EA.	420.00	130.00	550.00
225a, 30 circuits	"	830.00	150.00	980.00
240/120v, main lugs only				
100a				
8 circuits	EA.	250.00	110.00	360.00
12 circuits	"	270.00	110.00	380.00
20 circuits	"	280.00	110.00	390.00
225a				
24 circuits	EA.	300.00	130.00	430.00
30 circuits	"	340.00	140.00	480.00
42 circuits	"	380.00	140.00	520.00
Distribution panelboards, 3 ph, main breaker				
225a	EA.	950.00	590.00	1,540
400a	"	1,730	670.00	2,400
600a	"	2,700	820.00	3,520
800a	"	3,470	890.00	4,360
1000a	"	4,490	1,040	5,530
1200a	"	5,520	1,110	6,630
Single phase				
225a	EA.	830.00	520.00	1,350
400a	"	1,540	590.00	2,130
600a	"	2,180	740.00	2,920
800a	"	2,950	890.00	3,840
1000a	"	3,850	1,040	4,890
1200a	"	4,880	1,110	5,990
Fusible distribution panelboards, 3 phase, 600v				
100a	EA.	900.00	520.00	1,420
200a	"	1,030	590.00	1,620
400a	"	1,800	740.00	2,540
600a	"	2,300	890.00	3,190
800a	"	3,470	1,040	4,510
Single phase				
100a	EA.	770.00	440.00	1,210
200a	"	900.00	520.00	1,420
400a	"	1,930	670.00	2,600
600a	"	2,050	820.00	2,870
800a	"	2,820	960.00	3,780
Hospital panels, operating room				
3kv - 208v	EA.	2,820	230.00	3,050
3kv - 277v	"	2,950	230.00	3,180
5kv - 208v	"	2,950	230.00	3,180
5kv - 277v	"	3,210	230.00	3,440
Coronary care				
3kv - 208v	EA.	3,340	270.00	3,610
3kv - 277v	"	3,590	270.00	3,860
5kv - 208v	"	3,590	270.00	3,860
5kv - 277v	"	3,850	270.00	4,120
Intensive care				

SERVICE AND DISTRIBUTION	UNIT	MAT.	INST.	TOTAL
16470.10 PANELBOARDS				
3kv - 208v	EA.	3,720	300.00	4,020
3kv - 277v	"	3,850	300.00	4,150
5kv - 208v	"	3,850	300.00	4,150
5kv - 277v	"	4,110	300.00	4,410
15kv - 208v	"	7,320	440.00	7,760
15kv - 277v	"	7,580	440.00	8,020
25kv - 208v	"	7,830	590.00	8,420
25kv - 277v	"	8,090	590.00	8,680
Explosion proof, 240v, m.l.b. 20a, single phase				
6 breakers	EA.	2,050	410.00	2,460
8 breakers	"	2,310	440.00	2,750
10 breakers	"	2,700	460.00	3,160
12 breakers	"	3,080	490.00	3,570
14 breakers	"	3,720	520.00	4,240
16 breakers	"	4,490	520.00	5,010
18 breakers	"	4,880	580.00	5,460
20 breakers	"	5,140	600.00	5,740
22 breakers	"	5,260	630.00	5,890
24 breakers	"	5,520	660.00	6,180
16480.10 MOTOR CONTROLS				
Motor generator set, 3 phase, 480/277v, w/controls				
10kw	EA.	7,260	1,020	8,280
15kw	"	9,460	1,140	10,600
20kw	"	10,500	1,190	11,690
25kw	"	12,110	1,290	13,400
30kw	"	13,550	1,350	14,900
40kw	"	14,810	1,410	16,220
50kw	"	16,500	1,480	17,980
60kw	"	18,690	1,650	20,340
75kw	"	21,000	1,860	22,860
100kw	"	24,230	2,280	26,510
125kw	"	43,840	2,470	46,310
150kw	"	48,450	2,470	50,920
200kw	"	55,370	2,700	58,070
250kw	"	59,990	2,700	62,690
300kw	"	69,220	2,970	72,190
2 pole, 230 volt starter, w/NEMA-1				
1 hp, 9 amp, size 00	EA.	100.00	37.10	137.10
2 hp, 18amp, size 0	"	130.00	37.10	167.10
3 hp, 27amp, size 1	"	140.00	37.10	177.10
5 hp, 45amp, size 1p	"	190.00	37.10	227.10
7-1/2 hp, 45a, size 2	"	320.00	37.10	357.10
15 hp, 90a, size 3	"	480.00	37.10	517.10
2 pole, w/NEMA-4 enclosure				
2 hp, 18a, size 1	EA.	270.00	59.50	329.50
5 hp, 45amp, size 1p	"	350.00	59.50	409.50
7-1/2 hp, 45a, size 2	"	550.00	59.50	609.50
3 pole, 2 hp, 9a, 200-575v starter				
W/NEMA-1, size 00	EA.	130.00	49.50	179.50
W/NEMA-4 enclosure, size 00	"	270.00	64.50	334.50

SERVICE AND DISTRIBUTION	UNIT	MAT.	INST.	TOTAL
16480.10 MOTOR CONTROLS				
5hp, 18a				
W/NEMA-1 enclosure, size 0	EA.	180.00	49.50	229.50
W/NEMA-4 enclosure, size 0	"	300.00	64.50	364.50
7.5-10hp, 27a				
7.5-10hp 27a, w/NEMA-1 enclosure, size 1	EA.	180.00	49.50	229.50
W/NEMA-4 enclosure size 1	"	300.00	64.50	364.50
10-25hp, 45a				
W/NEMA-1 enclosure, size 2	EA.	320.00	49.50	369.50
W/NEMA-4 enclosure, size 2	"	560.00	64.50	624.50
25-50hp, 90a				
W/NEMA-1 enclosure, size 3	EA.	550.00	64.50	614.50
W/NEMA-4 enclosure, size 3	"	540.00	93.00	633.00
40-100hp, 135a				
W/NEMA-1 enclosure, size 4	EA.	1,190	93.00	1,283
W/NEMA-4 enclosure, size 4	"	1,830	130.00	1,960
75-200hp, 270a				
W/NEMA-1 enclosure, size 5	EA.	2,800	200.00	3,000
W/NEMA-4 enclosure, size 5	"	3,550	260.00	3,810
Magnetic starter accessories				
On-off-auto selector switch kit	EA.	36.20	11.85	48.05
With pilot light	"	66.00	12.90	78.90
Control center main lug only, 208v, 3 phase				
600a	EA.	1,120	440.00	1,560
1200a	"	2,490	590.00	3,080
Main circuit breakers, 208v, 3 phase				
400a	EA.	2,240	370.00	2,610
600a	"	2,490	520.00	3,010
800a	"	2,870	590.00	3,460
1000a	"	3,120	670.00	3,790
1200a	"	5,740	740.00	6,480
Non-reversing starters				
Size 1	EA.	440.00	27.00	467.00
Size 2	"	510.00	46.40	556.40
Size 3	"	570.00	56.00	626.00
Size 4	"	1,070	64.50	1,135
Reversing starters				
Size 1	EA.	570.00	27.00	597.00
Size 2	"	780.00	40.70	820.70
Fusible switch, non-revolving starters				
Size 1	EA.	390.00	27.00	417.00
Size 2	"	480.00	46.40	526.40
Size 3	"	620.00	56.00	676.00
Size 4	"	1,000	64.50	1,065
Reversing starters				
Size 1	EA.	410.00	27.00	437.00
Size 2	"	600.00	40.70	640.70
Two speed, non-reversing starter				
Size 1	EA.	570.00	27.00	597.00
Size 2	"	840.00	40.70	880.70
Magnetic starter, 600v, 2 pole, NEMA 3R				
Size 0, 2 hp	EA.	190.00	37.10	227.10
Size 1, 5hp	"	250.00	40.70	290.70

SERVICE AND DISTRIBUTION

	UNIT	MAT.	INST.	TOTAL
16480.10 **MOTOR CONTROLS**				
NEMA 3R				
Size 2, 7.5 hp	EA.	400.00	42.40	442.40
Size 3, 15 hp	"	880.00	44.30	924.30
NEMA 7				
Size 0, 2 hp	EA.	600.00	64.50	664.50
Size 1, 5 hp	"	680.00	74.00	754.00
Size 2, 7.5 hp	"	1,010	82.50	1,093
NEMA 12				
Size 0, 2 hp	EA.	190.00	56.00	246.00
Size 1, 5 hp	"	250.00	64.50	314.50
Size 2, 7.5 hp	"	400.00	74.00	474.00
Size 3, 15 hp	"	600.00	82.50	682.50
3 pole, NEMA 1				
Size 6	EA.	8,200	370.00	8,570
Size 7	"	11,050	440.00	11,490
Size 8	"	15,330	590.00	15,920
NEMA 4				
Size 6	EA.	10,340	520.00	10,860
Size 7	"	13,190	590.00	13,780
Size 8	"	17,470	740.00	18,210
NEMA 3R				
Size 0	EA.	200.00	46.40	246.40
Size 1	"	220.00	50.50	270.50
Size 2	"	430.00	67.50	497.50
Size 3	"	670.00	70.50	740.50
Size 4	"	1,660	100.00	1,760
NEMA 7				
Size 0	EA.	620.00	74.00	694.00
Size 1	"	650.00	80.00	730.00
Size 2	"	1,040	82.50	1,123
Size 3	"	1,560	100.00	1,660
Size 4	"	2,610	160.00	2,770
Size 5	"	6,060	360.00	6,420
Size 6	"	15,680	590.00	16,270
NEMA 12				
Size 00	EA.	110.00	64.50	174.50
Size 0	"	140.00	69.00	209.00
Size 1	"	160.00	70.50	230.50
Size 2	"	300.00	74.00	374.00
Size 3	"	470.00	93.00	563.00
Size 4	"	1,080	130.00	1,210
Size 5	"	2,610	270.00	2,880
Size 6	"	6,300	460.00	6,760
Size 7	"	9,030	520.00	9,550
Size 8	"	13,310	740.00	14,050
Reversing magnetic starters, 600v, 3 pole, NEMA 1				
Size 00	EA.	280.00	46.40	326.40
Size 0	"	330.00	47.90	377.90
Size 1	"	380.00	50.50	430.50
Size 2	"	750.00	56.00	806.00
Size 3	"	1,310	64.50	1,375
Size 4	"	2,970	74.00	3,044

SERVICE AND DISTRIBUTION	UNIT	MAT.	INST.	TOTAL
16480.10 MOTOR CONTROLS				
Size 5	EA.	6,420	200.00	6,620
Size 6	"	14,380	350.00	14,730
Size 7	"	19,600	410.00	20,010
Size 8	"	27,920	670.00	28,590
NEMA 4				
Size 0	EA.	530.00	64.50	594.50
Size 4	"	650.00	67.50	717.50
Size 2	"	1,160	70.50	1,231
Size 3	"	1,780	74.00	1,854
Size 4	"	4,040	93.00	4,133
Size 5	"	7,370	270.00	7,640
Size 6	"	16,520	460.00	16,980
Size 7	"	21,620	560.00	22,180
Size 8	"	29,940	740.00	30,680
NEMA 7				
Size 0	EA.	1,090	74.00	1,164
Size 1	"	1,140	82.50	1,223
Size 2	"	1,900	93.00	1,993
Size 3	"	3,090	110.00	3,200
NEMA 12				
Size 0	EA.	410.00	64.50	474.50
Size 1	"	460.00	70.50	530.50
Size 2	"	870.00	74.00	944.00
Size 3	"	1,540	82.50	1,623
Size 4	"	3,330	93.00	3,423
Size 5	"	7,370	270.00	7,640
Size 6	"	15,560	520.00	16,080
Size 7	"	20,790	590.00	21,380
Size 8	"	29,110	740.00	29,850
Electrically held lighting contactors, NEMA 1, 20a				
2 pole	EA.	130.00	37.10	167.10
3 pole	"	140.00	46.40	186.40
4 pole	"	180.00	56.00	236.00
6 pole	"	250.00	74.00	324.00
8 pole	"	330.00	93.00	423.00
10 pole	"	380.00	110.00	490.00
12 pole	"	440.00	130.00	570.00
30a				
2 pole	EA.	140.00	37.10	177.10
3 pole	"	150.00	46.40	196.40
4 pole	"	180.00	56.00	236.00
5 pole	"	240.00	64.50	304.50
60a				
2 pole	EA.	280.00	37.10	317.10
3 pole	"	300.00	46.40	346.40
4 pole	"	370.00	56.00	426.00
5 pole	"	530.00	64.50	594.50
100a				
2 pole	EA.	460.00	46.40	506.40
3 pole	"	490.00	64.50	554.50
4 pole	"	600.00	82.50	682.50
5 pole	"	870.00	100.00	970.00

SERVICE AND DISTRIBUTION	UNIT	MAT.	INST.	TOTAL
16480.10 — MOTOR CONTROLS				
200a				
2 pole	EA.	1,080	100.00	1,180
3 pole	"	1,150	110.00	1,260
4 pole	"	1,540	120.00	1,660
300a				
2 pole	EA.	2,290	160.00	2,450
3 pole	"	2,490	200.00	2,690
400a				
2 pole	EA.	5,990	160.00	6,150
3 pole	"	6,770	200.00	6,970
600a				
2 pole	EA.	7,370	250.00	7,620
3 pole	"	8,200	340.00	8,540
800a				
2 pole	EA.	8,790	300.00	9,090
3 pole	"	9,740	410.00	10,150
Mechanically held lighting contactors, NEMA 1, 20a				
2 pole	EA.	200.00	37.10	237.10
3 pole	"	210.00	46.40	256.40
4 pole	"	220.00	56.00	276.00
6 pole	"	370.00	74.00	444.00
8 pole	"	400.00	93.00	493.00
10 pole	"	450.00	110.00	560.00
30a				
2 pole	EA.	210.00	37.10	247.10
3 pole	"	220.00	46.40	266.40
4 pole	"	240.00	56.00	296.00
5 pole	"	300.00	64.50	364.50
60a				
2 pole	EA.	430.00	37.10	467.10
3 pole	"	440.00	46.40	486.40
4 pole	"	530.00	56.00	586.00
5 pole	"	690.00	64.50	754.50
100a				
2 pole	EA.	590.00	46.40	636.40
3 pole	"	630.00	64.50	694.50
4 pole	"	750.00	74.00	824.00
5 pole	"	1,010	93.00	1,103
200a				
2 pole	EA.	1,530	64.50	1,595
3 pole	"	1,730	93.00	1,823
4 pole	"	2,110	120.00	2,230
300a				
2 pole	EA.	2,730	160.00	2,890
3 pole	"	2,970	200.00	3,170
400a				
2 pole	EA.	6,530	160.00	6,690
3 pole	"	7,370	200.00	7,570
600a				
2 pole	EA.	7,840	250.00	8,090
3 pole	"	8,790	330.00	9,120
800a				

SERVICE AND DISTRIBUTION	UNIT	MAT.	INST.	TOTAL
16480.10 MOTOR CONTROLS				
2 pole	EA.	9,270	300.00	9,570
3 pole	"	10,340	420.00	10,760
AC relays, control type open, 15a, 600v				
2 pole	EA.	70.50	37.10	107.60
3 pole	"	85.00	46.40	131.40
4 pole	"	95.00	56.00	151.00
6 pole	"	140.00	74.00	214.00
8 pole	"	170.00	93.00	263.00
10 pole	"	240.00	110.00	350.00
12 pole	"	260.00	130.00	390.00
16490.10 SWITCHES				
Oil switches, medium voltage, bus components				
Switches, 277/120v, toggle device only	EA.	300.00	59.50	359.50
With oil 35kv, g&w gram 44, 4 way switch	"	9,900	300.00	10,200
Weatherproof enclosure				
3 way switch	EA.	12,230	370.00	12,600
4 way switch	"	12,810	410.00	13,220
Fused interrupter load, 35kv				
20A				
1 pole	EA.	13,980	590.00	14,570
2 pole	"	15,140	630.00	15,770
3 way	"	16,310	630.00	16,940
4 way	"	17,470	670.00	18,140
30a, 1 pole	"	13,980	590.00	14,570
3 way	"	16,310	630.00	16,940
4 way	"	17,470	670.00	18,140
Weatherproof switch, including box & cover, 20a				
1 pole	EA.	15,140	590.00	15,730
2 pole	"	16,310	630.00	16,940
3 way	"	17,470	670.00	18,140
4 way	"	18,640	670.00	19,310
3 way, oil switch, 15kv enclosure	"	12,810	440.00	13,250
Pedestal for 35kv double breaker switch	"	490.00	190.00	680.00
Bus terminal connector, 2	"	430.00	93.00	523.00
2 to 3	"	520.00	93.00	613.00
Support connector, 3	"	270.00	59.50	329.50
Tee connector, 2 to 3	"	400.00	74.00	474.00
Flexible bus stud connector	"	340.00	64.50	404.50
End cap 3	"	370.00	49.50	419.50
Weldment connection, 3	"	150.00	37.10	187.10
Plate switch, 1 gang	"	0.36	1.85	2.21
Start stop stations, manual motor starters	"	36.70	27.00	63.70
Lockout switch	"	6.10	9.30	15.40
Forward-reverse switch	"	42.80	27.00	69.80
On-off switch	"	42.80	27.00	69.80
Open-close switch	"	42.80	27.00	69.80
Forward-reverse-stop switch	"	48.90	37.10	86.00
Standard 3 button switch any standard legend	"	48.90	37.10	86.00
Standard 3 button with lockout	"	54.50	37.10	91.60
Manual motor starters, tog, 115/230v				
Size 1 gp	EA.	73.50	37.10	110.60

SERVICE AND DISTRIBUTION	UNIT	MAT.	INST.	TOTAL
16490.10 SWITCHES				
Size 2	EA.	97.50	37.10	134.60
Button				
Size 0	EA.	55.00	37.10	92.10
Size 1	"	78.50	37.10	115.60
Size 2	"	100.00	37.10	137.10
3-phase				
Size 0	EA.	90.50	49.50	140.00
Size 1	"	120.00	49.50	169.50
Time & float switches	"	180.00	59.50	239.50
Astronomical time switch, 40a, 240v	"	290.00	37.10	327.10
Timer switch 0-5 minute, with box	"	18.35	18.55	36.90
Single pole/single throw time, 277v, NEMA-1	"	61.50	27.00	88.50
Single toggle switch, 20a, 120v, with pilot	"	12.45	9.30	21.75
3-way toggle	"	34.30	11.00	45.30
Photo electric switches				
1000 watt				
105-135v	EA.	19.05	27.00	46.05
208-277v	"	25.70	27.00	52.70
3000 watt, 105-130v				
Double throw	EA.	77.00	37.10	114.10
Single throw	"	70.50	37.10	107.60
Double pole/single throw, 210-250v	"	89.50	49.50	139.00
Dimmer switch and switch plate				
600 w	EA.	19.25	11.40	30.65
1000 w	"	44.90	12.90	57.80
Dimmer switch incandescent				
1500w	EA.	100.00	26.00	126.00
2000w	"	150.00	27.70	177.70
Fluorescent				
12 lamps	EA.	44.90	18.55	63.45
20 lamps	"	77.00	20.45	97.45
30 lamps	"	130.00	22.30	152.30
40 lamps	"	150.00	26.00	176.00
Time clocks with skip, 40a, 120v				
SPST	EA.	51.50	27.70	79.20
SPDT	"	58.00	27.70	85.70
DPST	"	58.00	27.70	85.70
DPDT	"	89.50	37.10	126.60
SPST	"	95.00	37.10	132.10
Astronomic time clocks with skip, 40a, 120v				
DPST	EA.	89.50	27.70	117.20
SPST	"	140.00	37.10	177.10
SPDT	"	100.00	27.70	127.70
Raintight time clocks, 40a, 120v				
SPDT	EA.	64.00	37.10	101.10
DPST	"	77.00	37.10	114.10
Contractor grade wall switch 15a, 120v				
Single pole	EA.	1.50	5.95	7.45
Three way	"	2.60	7.40	10.00
Four way	"	12.80	9.90	22.70
Specification grade toggle switches, 20a, 120-277v				
Single pole	EA.	9.60	7.40	17.00

SERVICE AND DISTRIBUTION	UNIT	MAT.	INST.	TOTAL
16490.10 — SWITCHES				
Double pole	EA.	11.05	11.00	22.05
3 way	"	10.20	9.30	19.50
4 way	"	25.90	11.00	36.90
30a, 120-277v				
Single pole	EA.	14.40	7.40	21.80
Double pole	"	18.40	11.00	29.40
3 way	"	17.75	9.30	27.05
Specification grade key switches, 20a, 120-277v				
Single pole	EA.	14.75	7.40	22.15
Double pole	"	16.70	11.00	27.70
3 way	"	16.15	9.30	25.45
4 way	"	33.40	11.00	44.40
Red pilot light handle switches, 20a, 120-277v				
Single pole	EA.	12.90	7.40	20.30
Double pole	"	20.00	11.00	31.00
3 way	"	26.20	9.30	35.50
30a, 120-277v				
Single pole	EA.	18.60	7.40	26.00
Double pole	"	22.70	11.00	33.70
3 way	"	34.70	9.30	44.00
Momentary contact switches, 20a				
SPDT, ivory	EA.	17.45	9.30	26.75
SPDT, locking	"	23.10	11.00	34.10
Maintained contact switches				
SPDT ivory	EA.	35.30	9.30	44.60
DPDT ivory	"	35.70	9.30	45.00
SPDT locking	"	41.20	11.00	52.20
DPDT locking	"	41.70	12.90	54.60
Mercury switch, 3 way	"	8.70	9.30	18.00
Door switches, open on or off	"	15.40	18.55	33.95
Combination switch and pilot light, single pole	"	7.05	11.00	18.05
3 way	"	8.65	12.90	21.55
Combination switch and receptacle, single pole	"	10.05	11.00	21.05
3 way	"	10.15	11.00	21.15
Combination two switches, single pole / single pole	"	8.35	9.30	17.65
3 way	"	10.25	14.85	25.10
Switch plates, plastic ivory				
1 gang	EA.	0.36	2.95	3.31
2 gang	"	0.74	3.70	4.44
3 gang	"	1.10	4.45	5.55
4 gang	"	1.85	5.40	7.25
5 gang	"	4.20	5.95	10.15
6 gang	"	4.95	6.75	11.70
Stainless steel				
1 gang	EA.	1.25	2.95	4.20
2 gang	"	2.40	3.70	6.10
3 gang	"	4.20	4.55	8.75
4 gang	"	5.80	5.40	11.20
5 gang	"	7.95	5.95	13.90
6 gang	"	9.90	6.75	16.65
Brass				
1 gang	EA.	2.65	2.95	5.60

SERVICE AND DISTRIBUTION	UNIT	MAT.	INST.	TOTAL
16490.10 SWITCHES				
2 gang	EA.	5.30	3.70	9.00
3 gang	"	9.60	4.55	14.15
4 gang	"	12.95	5.40	18.35
5 gang	"	16.05	5.95	22.00
6 gang	"	19.15	6.75	25.90
16490.20 TRANSFER SWITCHES				
Automatic transfer switch 600v, 3 pole				
30a	EA.	1,430	130.00	1,560
60a	"	2,020	130.00	2,150
100a	"	2,970	180.00	3,150
150a	"	3,740	220.00	3,960
225a	"	4,510	300.00	4,810
260a	"	4,870	300.00	5,170
400a	"	6,530	370.00	6,900
600a	"	8,910	560.00	9,470
800a	"	10,930	670.00	11,600
1000a	"	15,680	780.00	16,460
1200a	"	17,820	850.00	18,670
1600a	"	21,980	930.00	22,910
2000a	"	23,520	1,100	24,620
2600a	"	41,590	1,560	43,150
Automatic transfer switches, 600v, 3 phase, 3000a	"	65,350	1,860	67,210
16490.80 SAFETY SWITCHES				
Safety switch, 600v, 3 pole, heavy duty, NEMA-1				
30a	EA.	120.00	37.10	157.10
60a	"	140.00	42.40	182.40
100a	"	280.00	59.50	339.50
200a	"	360.00	93.00	453.00
400a	"	1,040	200.00	1,240
600a	"	1,700	300.00	2,000
800a	"	3,040	390.00	3,430
1200a	"	3,770	530.00	4,300

LIGHTING	UNIT	MAT.	INST.	TOTAL
16510.05 INTERIOR LIGHTING				
Recessed fluorescent fixtures, 2'x2'				
2 lamp	EA.	52.00	27.00	79.00
4 lamp	"	69.50	27.00	96.50
2 lamp w/flange	"	65.00	37.10	102.10

LIGHTING	UNIT	MAT.	INST.	TOTAL
16510.05 INTERIOR LIGHTING				
4 lamp w/flange	EA.	82.00	37.10	119.10
1'x4'				
2 lamp	EA.	52.50	24.75	77.25
3 lamp	"	73.00	24.75	97.75
2 lamp w/flange	"	65.00	27.00	92.00
3 lamp w/flange	"	85.00	27.00	112.00
2'x4'				
2 lamp	EA.	65.00	27.00	92.00
3 lamp	"	76.00	27.00	103.00
4 lamp	"	74.50	27.00	101.50
2 lamp w/flange	"	77.50	37.10	114.60
3 lamp w/flange	"	88.50	37.10	125.60
4 lamp w/flange	"	87.00	37.10	124.10
4'x4'				
4 lamp	EA.	160.00	37.10	197.10
6 lamp	"	180.00	37.10	217.10
8 lamp	"	190.00	37.10	227.10
4 lamp w/flange	"	190.00	56.00	246.00
6 lamp w/flange	"	240.00	56.00	296.00
8 lamp, w/flange	"	220.00	56.00	276.00
Surface mounted incandescent fixtures				
40w	EA.	47.20	24.75	71.95
75w	"	51.00	24.75	75.75
100w	"	60.50	24.75	85.25
150w	"	71.50	24.75	96.25
Pendant				
40w	EA.	51.00	29.70	80.70
75w	"	56.50	29.70	86.20
100w	"	64.50	29.70	94.20
150w	"	74.00	29.70	103.70
Contractor grade recessed down lights				
100 watt housing only	EA.	18.85	37.10	55.95
150 watt housing only	"	22.65	37.10	59.75
100 watt trim	"	25.20	18.55	43.75
150 watt trim	"	31.50	18.55	50.05
Recessed incandescent fixtures				
40w	EA.	87.00	56.00	143.00
75w	"	91.00	56.00	147.00
100w	"	93.50	56.00	149.50
150w	"	110.00	56.00	166.00
Exit lights, 120v				
Recessed	EA.	67.50	46.40	113.90
Back mount	"	47.20	27.00	74.20
Universal mount	"	54.00	27.00	81.00
Emergency battery units, 6v-120v, 50 unit	"	100.00	56.00	156.00
With 1 head	"	120.00	56.00	176.00
With 2 heads	"	130.00	56.00	186.00
Mounting bucket	"	20.20	27.00	47.20
Light track single circuit				
2'	EA.	23.25	18.55	41.80
4'	"	43.40	18.55	61.95
8'	"	77.50	37.10	114.60

LIGHTING	UNIT	MAT.	INST.	TOTAL
16510.05 INTERIOR LIGHTING				
12'	EA.	120.00	56.00	176.00
Fittings and accessories				
Dead end	EA.	4.15	5.40	9.55
Starter kit	"	21.70	9.30	31.00
Conduit feed	"	26.30	5.40	31.70
Straight connector	"	14.60	5.40	20.00
Center feed	"	29.80	5.40	35.20
L-connector	"	23.25	5.40	28.65
T-connector	"	35.60	5.40	41.00
X-connector	"	46.50	7.40	53.90
Cord and plug	"	34.10	3.70	37.80
Rigid corner	"	44.90	5.40	50.30
Flex connector	"	69.50	5.40	74.90
2 way connector	"	100.00	7.40	107.40
Spacer clip	"	1.55	1.85	3.40
Grid box	"	8.50	5.40	13.90
T-bar clip	"	2.90	1.85	4.75
Utility hook	"	9.35	5.40	14.75
Fixtures, square				
R-20	EA.	62.00	5.40	67.40
R-30	"	64.50	5.40	69.90
40w flood	"	100.00	5.40	105.40
40w spot	"	110.00	5.40	115.40
100w flood	"	120.00	5.40	125.40
100w spot	"	120.00	5.40	125.40
Mini spot	"	85.50	5.40	90.90
Mini flood	"	88.00	5.40	93.40
Quartz, 500w	"	230.00	5.40	235.40
R-20 sphere	"	88.00	5.40	93.40
R-30 sphere	"	100.00	5.40	105.40
R-20 cylinder	"	48.00	5.40	53.40
R-30 cylinder	"	54.50	5.40	59.90
R-40 cylinder	"	77.50	5.40	82.90
R-30 wall wash	"	93.00	5.40	98.40
R-40 wall wash	"	120.00	5.40	125.40
Explosion proof, incan., surface mounted				
100w - 200w	EA.	220.00	64.50	284.50
300w	"	290.00	64.50	354.50
500w	"	410.00	64.50	474.50
With guard				
100w-200w	EA.	230.00	82.50	312.50
300w	"	300.00	82.50	382.50
500w	"	430.00	82.50	512.50
Reflectors for incan. light fixtures, dome	"	33.60	9.30	42.90
Angle	"	37.70	9.30	47.00
Highbay	"	81.00	11.00	92.00
Explosion proof fluor. fixtures, 800 ms.				
1 lamp	EA.	900.00	82.50	982.50
2 lamp	"	1,120	99.00	1,219
3 lamp	"	1,670	110.00	1,780
4 lamp	"	2,170	120.00	2,290
Explosion proof hp sodium fixtures				

LIGHTING	UNIT	MAT.	INST.	TOTAL
16510.05 INTERIOR LIGHTING				
50w-70w	EA.	610.00	82.50	692.50
100w	"	630.00	82.50	712.50
150w	"	650.00	93.00	743.00
200w	"	660.00	93.00	753.00
250w	"	690.00	93.00	783.00
310w	"	720.00	93.00	813.00
400w	"	1,080	93.00	1,173
With guard				
50w-70w	EA.	620.00	93.00	713.00
100w	"	640.00	93.00	733.00
150w	"	680.00	100.00	780.00
200w	"	690.00	100.00	790.00
250w	"	720.00	100.00	820.00
310w	"	750.00	100.00	850.00
400w	"	1,120	100.00	1,220
Explosion proof metal halide fixtures				
175w	EA.	530.00	93.00	623.00
250w	"	580.00	93.00	673.00
400w	"	850.00	93.00	943.00
With guard, 175w	"	540.00	100.00	640.00
250w	"	590.00	100.00	690.00
400w	"	890.00	100.00	990.00
Energy saving rapid start fluor. lamps				
F30cw	EA.	6.55	3.70	10.25
F40 cw	"	3.40	3.70	7.10
F40 cwx	"	5.85	3.70	9.55
F30 ww	"	7.90	3.70	11.60
F40 ww	"	4.25	3.70	7.95
F40 wwx	"	5.85	3.70	9.55
Slimline				
F48 cw	EA.	8.15	5.40	13.55
F96 cwx	"	12.60	5.40	18.00
F48 ww	"	9.65	5.40	15.05
F96 ww	"	9.20	5.40	14.60
F96 wwx	"	12.60	5.40	18.00
High output	"	8.20	5.40	13.60
F96 cwx	"	12.15	5.40	17.55
F96 cw	"	10.75	5.40	16.15
Power groove, F48 cw	"	19.65	5.40	25.05
Circle				
Fc6 cw	EA.	7.45	3.70	11.15
Fc8 cw	"	6.25	3.70	9.95
Fc12 cw	"	7.20	3.70	10.90
Fc16 cw	"	9.90	3.70	13.60
Fc6 ww	"	8.10	3.70	11.80
Fc8 ww	"	6.80	3.70	10.50
Fc12 ww	"	8.25	3.70	11.95
Fc16 ww	"	12.10	3.70	15.80
Incandescent lamps				
200w	EA.	3.40	3.70	7.10
300w	"	3.70	3.70	7.40
500w	"	8.30	3.70	12.00

LIGHTING	UNIT	MAT.	INST.	TOTAL
16510.05 — INTERIOR LIGHTING				
750w	EA.	20.30	5.40	25.70
1000w	"	22.15	7.40	29.55
1500w	"	32.40	7.40	39.80
Energy saving reflector floodlight lamps				
25w	EA.	7.85	3.70	11.55
30w	"	6.30	3.70	10.00
50w	"	7.00	3.70	10.70
75w	"	6.15	3.70	9.85
120w	"	7.35	3.70	11.05
150w	"	5.10	3.70	8.80
200w	"	10.65	3.70	14.35
300w	"	10.50	3.70	14.20
500w	"	24.45	5.40	29.85
750w	"	30.90	5.40	36.30
Reflector spotlight				
75w	EA.	5.55	3.70	9.25
100w	"	11.55	3.70	15.25
125w	"	15.25	3.70	18.95
150w	"	20.00	3.70	23.70
250w	"	14.75	3.70	18.45
300w	"	15.30	3.70	19.00
400w	"	15.60	5.40	21.00
500w	"	25.30	5.40	30.70
1000w	"	56.50	7.40	63.90
Medium par flood lamps				
75w	EA.	5.50	3.70	9.20
100w	"	6.10	3.70	9.80
150w	"	6.60	3.70	10.30
200w	"	25.30	3.70	29.00
300w	"	28.10	3.70	31.80
500w	"	55.00	5.40	60.40
Medium par spot lamps				
75w	EA.	6.85	3.70	10.55
120w	"	7.00	3.70	10.70
150w	"	7.35	3.70	11.05
Tubular quartz lamps				
100w	EA.	30.10	5.40	35.50
150w	"	18.20	5.40	23.60
200w	"	35.30	5.40	40.70
400w	"	32.60	5.40	38.00
500w	"	26.90	7.40	34.30
750w	"	33.30	7.40	40.70
1000w	"	47.10	9.30	56.40
1250w	"	49.80	9.30	59.10
1500w	"	40.40	9.30	49.70
Ballast replacements rapid start fluor				
1f-40-120v	EA.	14.85	27.00	41.85
1f-40-277v	"	24.95	27.00	51.95
1f-96-120v	"	77.00	27.00	104.00
1f-96-277v	"	93.00	27.00	120.00
2f-40-120v	"	21.15	27.00	48.15
2f-40-277v	"	24.95	27.00	51.95

LIGHTING	UNIT	MAT.	INST.	TOTAL
16510.05 INTERIOR LIGHTING				
2f-96-120v	EA.	66.00	27.00	93.00
2f-96-277v	"	68.50	27.00	95.50
Circline, 1fc6-1fc16	"	19.55	27.00	46.55
Very high output, 1500ma				
1f48-120v	EA.	110.00	27.00	137.00
1f48-277v	"	110.00	27.00	137.00
1f96-120v	"	100.00	27.00	127.00
1f96-277v	"	110.00	27.00	137.00
2f48-120v	"	100.00	27.00	127.00
2f48-277v	"	110.00	27.00	137.00
2f96-120v	"	110.00	27.00	137.00
2f96-277v	"	130.00	27.00	157.00
Mercury, multi tap				
475w	EA.	75.00	37.10	112.10
100w	"	70.50	37.10	107.60
175w	"	84.50	37.10	121.60
250w	"	100.00	37.10	137.10
400w	"	110.00	37.10	147.10
1000W	"	170.00	37.10	207.10
Metal halide, multi tap				
175w	EA.	100.00	37.10	137.10
250w	"	130.00	37.10	167.10
400w	"	160.00	37.10	197.10
1000w	"	240.00	37.10	277.10
1500w	"	310.00	37.10	347.10
High pressure sodium				
70w	EA.	140.00	37.10	177.10
100w	"	150.00	37.10	187.10
150w	"	170.00	37.10	207.10
250w	"	240.00	37.10	277.10
400w	"	270.00	37.10	307.10
1000w	"	370.00	37.10	407.10
Surface mounted fluorescent, wrap around lens				
1 lamp	EA.	46.70	29.70	76.40
2 lamps	"	50.50	33.00	83.50
4 lamps	"	80.00	37.10	117.10
Wall mounted fluorescent				
2-20w lamps	EA.	37.40	18.55	55.95
2-30w lamps	"	40.10	18.55	58.65
2-40w lamps	"	46.70	24.75	71.45
Indirect, with wood shielding, 2049w lamps				
4'	EA.	80.00	37.10	117.10
8'	"	110.00	59.50	169.50
Industrial fluorescent, 2 lamp				
4'	EA.	80.00	27.00	107.00
8'	"	130.00	49.50	179.50
Strip fluorescent				
4'				
1 lamp	EA.	26.70	24.75	51.45
2 lamps	"	33.30	24.75	58.05
8'				
1 lamp	EA.	53.50	27.00	80.50

LIGHTING	UNIT	MAT.	INST.	TOTAL
16510.05	**INTERIOR LIGHTING**			
2 lamps	EA.	60.00	33.00	93.00
Wire guard for strip fixture, 4' long	"	13.35	12.90	26.25
Strip fluorescent, 8' long, two 4' lamps	"	80.00	49.50	129.50
With four 4' lamps	"	110.00	59.50	169.50
Wet location fluorescent, plastic housing				
4' long				
1 lamp	EA.	93.50	37.10	130.60
2 lamps	"	100.00	49.50	149.50
8' long				
2 lamps	EA.	160.00	59.50	219.50
4 lamps	"	230.00	64.50	294.50
Parabolic troffer, 2'x2'				
With 2 "U" lamps	EA.	110.00	37.10	147.10
With 3 "U" lamps	"	130.00	42.40	172.40
2'x4'				
With 2 40w lamps	EA.	150.00	42.40	192.40
With 3 40w lamps	"	170.00	49.50	219.50
With 4 40w lamps	"	190.00	49.50	239.50
1'x4'				
With 1 T-12 lamp, 9 cell	EA.	110.00	27.00	137.00
With 2 T-12 lamps	"	130.00	33.00	163.00
With 1 T-12 lamp, 20 cell	"	130.00	27.00	157.00
With 2 T-12 lamps	"	140.00	33.00	173.00
Steel sided surface fluorescent, 2'x4'				
3 lamps	EA.	100.00	49.50	149.50
4 lamps	"	110.00	49.50	159.50
Outdoor sign fluor., 1 lamp, remote ballast				
4' long	EA.	1,910	220.00	2,130
6' long	"	2,290	300.00	2,590
Recess mounted, commercial, 2'x2', 13" high				
100w	EA.	640.00	150.00	790.00
250w	"	700.00	170.00	870.00
High pressure sodium, hi-bay open				
400w	EA.	350.00	64.50	414.50
1000w	"	600.00	90.00	690.00
Enclosed				
400w	EA.	540.00	90.00	630.00
1000w	"	800.00	110.00	910.00
Metal halide hi-bay, open				
400w	EA.	320.00	64.50	384.50
1000w	"	540.00	90.00	630.00
Enclosed				
400w	EA.	510.00	90.00	600.00
1000w	"	760.00	110.00	870.00
High pressure sodium, low bay, surface mounted				
100w	EA.	220.00	37.10	257.10
150w	"	240.00	42.40	282.40
250w	"	270.00	49.50	319.50
400w	"	350.00	59.50	409.50
Metal halide, low bay, pendant mounted				
175w	EA.	240.00	49.50	289.50
250w	"	320.00	59.50	379.50

LIGHTING	UNIT	MAT.	INST.	TOTAL
16510.05 INTERIOR LIGHTING				
400w	EA.	390.00	82.50	472.50
Indirect luminare, square, metal halide, freestanding				
175w	EA.	640.00	37.10	677.10
250w	"	650.00	37.10	687.10
400w	"	700.00	37.10	737.10
High pressure sodium				
150w	EA.	650.00	37.10	687.10
250w	"	680.00	37.10	717.10
400w	"	740.00	37.10	777.10
Round, metal halide				
175w	EA.	600.00	37.10	637.10
250w	"	640.00	37.10	677.10
400w	"	670.00	37.10	707.10
High pressure sodium				
150w	EA.	610.00	37.10	647.10
250w	"	660.00	37.10	697.10
400w	"	690.00	37.10	727.10
Wall mounted, metal halide				
175w	EA.	500.00	93.00	593.00
250w	"	530.00	93.00	623.00
400w	"	880.00	120.00	1,000
High pressure sodium				
150w	EA.	570.00	93.00	663.00
250w	"	700.00	93.00	793.00
400w	"	910.00	120.00	1,030
Wall pack lithonia, high pressure sodium				
35w	EA.	150.00	33.00	183.00
55w	"	170.00	37.10	207.10
150w	"	190.00	59.50	249.50
250w	"	240.00	64.50	304.50
Low pressure sodium				
35w	EA.	220.00	64.50	284.50
55w	"	330.00	74.00	404.00
Wall pack hubbell, high pressure sodium				
35w	EA.	160.00	33.00	193.00
150w	"	210.00	59.50	269.50
250w	"	270.00	64.50	334.50
Compact fluorescent				
2-7w	EA.	100.00	37.10	137.10
2-13w	"	110.00	49.50	159.50
1-18w	"	140.00	49.50	189.50
Handball & racquet ball court, 2'x2', metal halide				
250w	EA.	360.00	93.00	453.00
400w	"	440.00	100.00	540.00
High pressure sodium				
250w	EA.	390.00	93.00	483.00
400w	"	430.00	100.00	530.00
Bollard light, 42" w/found., high pressure sodium				
70w	EA.	600.00	95.50	695.50
100w	"	620.00	95.50	715.50
150w	"	620.00	95.50	715.50
Light fixture lamps				

LIGHTING	UNIT	MAT.	INST.	TOTAL
16510.05 INTERIOR LIGHTING				
Lamp				
20w med. bipin base, cool white, 24"	EA.	4.45	5.40	9.85
30w cool white, rapid start, 36"	"	5.75	5.40	11.15
40w cool white "U", 3"	"	11.05	5.40	16.45
40w cool white, rapid start, 48"	"	2.65	5.40	8.05
70w high pressure sodium, mogul base	"	57.00	7.40	64.40
75w slimline, 96"	"	7.35	7.40	14.75
100w				
Incandescent, 100a, inside frost	EA.	2.25	3.70	5.95
Mercury vapor, clear, mogul base	"	28.00	7.40	35.40
High pressure sodium, mogul base	"	61.00	7.40	68.40
150w				
Par 38 flood or spot, incandescent	EA.	6.85	3.70	10.55
High pressure sodium, 1/2 mogul base	"	63.50	7.40	70.90
175w				
Mercury vapor, clear, mogul base	EA.	48.30	7.40	55.70
Mercury halide, clear, mogul base	"	21.60	7.40	29.00
250w				
High pressure sodium, mogul base	EA.	67.50	7.40	74.90
Mercury vapor, clear, mogul base	"	38.10	7.40	45.50
Metal halide, clear, mogul base	"	61.00	7.40	68.40
High pressure sodium, mogul base	"	67.50	7.40	74.90
400w				
Mercury vapor, clear, mogul base	EA.	30.50	7.40	37.90
Metal halide, clear, mogul base	"	57.00	7.40	64.40
High pressure sodium, mogul base	"	72.50	7.40	79.90
1000w				
Mercury vapor, clear, mogul base	EA.	66.00	9.30	75.30
High pressure sodium, mogul base	"	170.00	9.30	179.30
16510.30 EXTERIOR LIGHTING				
Exterior light fixtures				
Rectangle, high pressure sodium				
70w	EA.	390.00	93.00	483.00
100w	"	400.00	95.50	495.50
150w	"	430.00	95.50	525.50
250w	"	580.00	100.00	680.00
400w	"	650.00	130.00	780.00
Flood, rectangular, high pressure sodium				
70w	EA.	400.00	93.00	493.00
100w	"	420.00	95.50	515.50
150w	"	460.00	95.50	555.50
400w	"	690.00	130.00	820.00
1000w	"	1,050	170.00	1,220
Round				
400w	EA.	720.00	130.00	850.00
1000w	"	1,120	170.00	1,290
Round, metal halide				
400w	EA.	790.00	130.00	920.00
1000w	"	1,170	170.00	1,340
Light fixture arms, cobra head, 6', high press. sodium				

LIGHTING	UNIT	MAT.	INST.	TOTAL
16510.30 EXTERIOR LIGHTING				
100w	EA.	430.00	74.00	504.00
150w	"	680.00	93.00	773.00
250w	"	710.00	93.00	803.00
400w	"	730.00	110.00	840.00
Flood, metal halide				
400w	EA.	750.00	130.00	880.00
1000w	"	1,140	170.00	1,310
1500w	"	1,540	220.00	1,760
Mercury vapor				
250w	EA.	440.00	100.00	540.00
400w	"	770.00	130.00	900.00
Incandescent				
300w	EA.	80.50	64.50	145.00
500w	"	130.00	74.00	204.00
1000w	"	150.00	120.00	270.00
16510.90 POWER LINE FILTERS				
Heavy duty power line filter, 240v				
100a	EA.	2,800	370.00	3,170
300a	"	9,300	590.00	9,890
600a	"	12,880	900.00	13,780
16600.20 CENTRAL INVERTER SYSTEMS				
Central inverter systems				
500va	EA.	6,170	110.00	6,280
1000va	"	6,740	150.00	6,890
1500va	"	7,880	200.00	8,080
2400va	"	10,170	250.00	10,420
3000va	"	10,740	320.00	11,060
4500va	"	16,340	370.00	16,710
6000va	"	20,560	410.00	20,970
7500va	"	23,760	520.00	24,280
10,000va	"	27,420	590.00	28,010
16,600va	"	47,520	850.00	48,370
25,000va	"	56,660	1,300	57,960
16610.30 UNINTERRUPTIBLE POWER				
Uninterruptible power systems, (U.P.S.)	EA.	11,140	300.00	11,440
5 kva	"	13,780	410.00	14,190
7.5 kva	"	16,800	590.00	17,390
10 kva	"	23,520	820.00	24,340
15 kva	"	26,810	850.00	27,660
20 kva	"	35,840	890.00	36,730
25 kva	"	44,690	930.00	45,620
30 kva	"	47,490	960.00	48,450
35 kva	"	50,330	1,000	51,330
40 kva	"	53,200	1,040	54,240
45 kva	"	55,980	1,080	57,060
50 kva	"	58,800	1,110	59,910
62.5 kva	"	67,650	1,190	68,840
75 kva	"	76,500	1,300	77,800

LIGHTING	UNIT	MAT.	INST.	TOTAL
16610.30 UNINTERRUPTIBLE POWER				
100 kva	EA.	98,560	1,340	99,900
150 kva	"	147,840	1,860	149,700
200 kva	"	197,120	2,050	199,170
300 kva	"	295,680	2,770	298,450
400 kva	"	394,240	3,330	397,570
500 kva	"	492,800	4,070	496,870
16670.10 LIGHTNING PROTECTION				
Lightning protection				
Copper point, nickel plated, 12'				
1/2" dia.	EA.	30.20	37.10	67.30
5/8" dia.	"	35.70	37.10	72.80

COMMUNICATIONS	UNIT	MAT.	INST.	TOTAL
16720.10 FIRE ALARM SYSTEMS				
Master fire alarm box, pedestal mounted	EA.	4,240	590.00	4,830
Master fire alarm box	"	2,180	220.00	2,400
Box light	"	120.00	18.55	138.55
Ground assembly for box	"	60.50	24.75	85.25
Bracket for pole type box	"	78.50	27.00	105.50
Pull station				
Waterproof	EA.	60.50	18.55	79.05
Manual	"	36.30	14.85	51.15
Horn, waterproof	"	54.50	37.10	91.60
Interior alarm	"	36.30	27.00	63.30
Coded transmitter, automatic	"	640.00	74.00	714.00
Control panel, 8 zone	"	1,090	300.00	1,390
Battery charger and cabinet	"	390.00	74.00	464.00
Batteries, nickel cadmium or lead calcium	"	150.00	190.00	340.00
CO2 pressure switch connection	"	60.50	27.00	87.50
Annunciator panels				
Fire detection annunciator, remote type, 8-zone	EA.	220.00	67.50	287.50
12-zone	"	280.00	74.00	354.00
16-zone	"	350.00	93.00	443.00
Fire alarm systems				
Bell	EA.	66.50	22.85	89.35
Weatherproof bell	"	82.50	24.75	107.25
Horn	"	38.80	27.00	65.80
Siren	"	390.00	74.00	464.00
Chime	"	48.50	22.85	71.35
Audio/visual	"	70.00	27.00	97.00
Strobe light	"	65.00	27.00	92.00
Smoke detector	"	100.00	24.75	124.75

COMMUNICATIONS	UNIT	MAT.	INST.	TOTAL
16720.10 FIRE ALARM SYSTEMS				
Heat detection	EA.	18.15	18.55	36.70
Thermal detector	"	16.95	18.55	35.50
Ionization detector	"	85.00	19.80	104.80
Duct detector	"	280.00	100.00	380.00
Test switch	"	48.50	18.55	67.05
Remote indicator	"	24.45	21.20	45.65
Door holder	"	110.00	27.00	137.00
Telephone jack	"	18.15	11.00	29.15
Fireman phone	"	260.00	37.10	297.10
Speaker	"	52.00	29.70	81.70
Remote fire alarm annunciator panel				
24 zone	EA.	1,330	250.00	1,580
48 zone	"	2,660	480.00	3,140
Control panel				
12 zone	EA.	910.00	110.00	1,020
16 zone	"	1,190	160.00	1,350
24 zone	"	1,820	250.00	2,070
48 zone	"	3,390	590.00	3,980
Power supply	"	210.00	56.00	266.00
Status command	"	5,750	190.00	5,940
Printer	"	2,240	56.00	2,296
Transponder	"	180.00	33.30	213.30
Transformer	"	130.00	24.75	154.75
Transceiver	"	180.00	27.00	207.00
Relays	"	72.50	18.55	91.05
Flow switch	"	240.00	74.00	314.00
Tamper switch	"	140.00	110.00	250.00
End of line resistor	"	10.30	12.90	23.20
Printed ckt. card	"	91.00	18.55	109.55
Central processing unit	"	11,860	230.00	12,090
UPS backup to c.p.u.	"	11,510	330.00	11,840
Smoke detector, fixed temp. & rate of rise comb.	"	180.00	59.50	239.50
16720.50 SECURITY SYSTEMS				
Sensors				
Balanced magnetic door switch, surface mounted	EA.	87.00	18.55	105.55
With remote test	"	120.00	37.10	157.10
Flush mounted	"	83.50	69.00	152.50
Mounted bracket	"	6.40	12.90	19.30
Mounted bracket spacer	"	5.75	12.90	18.65
Photoelectric sensor, for fence				
6 beam	EA.	9,400	100.00	9,500
9 beam	"	11,490	160.00	11,650
Photoelectric sensor, 12 volt dc				
500' range	EA.	280.00	59.50	339.50
800' range	"	310.00	74.00	384.00
Capacitance wire grid kit				
Surface	EA.	77.00	37.10	114.10
Duct	"	57.50	59.50	117.00
Tube grid kit	"	96.50	18.55	115.05
Vibration sensor, 30 max per zone	"	120.00	18.55	138.55
Audio sensor, 30 max per zone	"	120.00	18.55	138.55

COMMUNICATIONS	UNIT	MAT.	INST.	TOTAL
16720.50 SECURITY SYSTEMS				
Inertia sensor				
Outdoor	EA.	90.00	27.00	117.00
Indoor	"	57.50	18.55	76.05
Ultrasonic transmitter, 20 max per zone				
Omni-directional	EA.	64.00	59.50	123.50
Directional	"	70.50	49.50	120.00
Transceiver				
Omni-directional	EA.	70.50	37.10	107.60
Directional	"	77.00	37.10	114.10
Passive infra-red sensor, 20 max per zone	"	500.00	59.50	559.50
Access/secure control unit, balanced magnetic switch	"	310.00	59.50	369.50
Photoelectric sensor	"	510.00	59.50	569.50
Photoelectric fence sensor	"	520.00	59.50	579.50
Capacitance sensor	"	610.00	64.50	674.50
Audio and vibration sensor	"	540.00	59.50	599.50
Inertia sensor	"	720.00	59.50	779.50
Ultrasonic sensor	"	820.00	64.50	884.50
Infra-red sensor	"	510.00	74.00	584.00
Monitor panel, with access/secure tone, standard	"	340.00	64.50	404.50
High security	"	510.00	74.00	584.00
Emergency power indicator	"	210.00	18.55	228.55
Monitor rack with 115v power supply				
1 zone	EA.	290.00	37.10	327.10
10 zone	"	1,480	93.00	1,573
Monitor cabinet, wall mounted				
1 zone	EA.	450.00	37.10	487.10
5 zone	"	1,600	59.50	1,660
10 zone	"	710.00	64.50	774.50
20 zone	"	2,230	74.00	2,304
Floor mounted, 50 zone	"	2,320	150.00	2,470
Security system accessories				
Tamper assembly for monitor cabinet	EA.	57.50	16.50	74.00
Monitor panel blank	"	7.70	12.90	20.60
Audible alarm	"	65.50	18.55	84.05
Audible alarm control	"	270.00	12.90	282.90
Termination screw, terminal cabinet				
25 pair	EA.	190.00	59.50	249.50
50 pair	"	310.00	93.00	403.00
150 pair	"	510.00	190.00	700.00
Universal termination, cabinets & panel				
Remote test	EA.	44.90	64.50	109.40
No remote test	"	32.10	27.00	59.10
High security line supervision termination	"	230.00	37.10	267.10
Door cord for capacitance sensor, 12"	"	7.70	18.55	26.25
Insulation block kit for capacitance sensor	"	35.90	12.90	48.80
Termination block for capacitance sensor	"	7.80	12.90	20.70
Guard alert display	"	830.00	22.85	852.85
Uninterrupted power supply				
Plug-in 40kva transformer				
12 volt	EA.	32.10	12.90	45.00
18 volt	"	21.80	12.90	34.70
24 volt	"	15.40	12.90	28.30

COMMUNICATIONS	UNIT	MAT.	INST.	TOTAL
16720.50 SECURITY SYSTEMS				
Test relay	EA.	49.40	12.90	62.30
Coaxial cable, 50 ohm	L.F.	0.20	0.22	0.42
Door openers	EA.	57.50	18.55	76.05
Push buttons				
Standard	EA.	12.80	12.90	25.70
Weatherproof	"	19.25	16.50	35.75
Bells	"	48.80	27.00	75.80
Horns				
Standard	EA.	51.50	37.10	88.60
Weatherproof	"	96.50	46.40	142.90
Chimes	"	74.50	24.75	99.25
Flasher	"	54.00	22.85	76.85
Motion detectors	"	220.00	56.00	276.00
Intercom units	"	51.50	27.00	78.50
Remote annunciator	"	2,310	190.00	2,500
16730.20 CLOCK SYSTEMS				
Clock systems				
Single face	EA.	130.00	29.70	159.70
Double face	"	350.00	29.70	379.70
Skeleton	"	270.00	100.00	370.00
Master	"	2,820	190.00	3,010
Signal generator	"	2,440	150.00	2,590
Elapsed time indicator	"	490.00	29.70	519.70
Controller	"	100.00	19.80	119.80
Clock and speaker	"	210.00	40.70	250.70
Bell				
Standard	EA.	70.50	19.80	90.30
Weatherproof	"	90.00	29.70	119.70
Horn				
Standard	EA.	44.90	27.00	71.90
Weatherproof	"	58.00	35.30	93.30
Chime	"	51.50	19.80	71.30
Buzzer	"	19.25	19.80	39.05
Flasher	"	70.50	22.85	93.35
Control Board	"	320.00	130.00	450.00
Program unit	"	320.00	190.00	510.00
Block back box	"	19.25	18.55	37.80
Double clock back box	"	38.70	24.75	63.45
Wire guard	"	12.85	7.40	20.25
16740.10 TELEPHONE SYSTEMS				
Communication cable				
25 pair	L.F.	0.38	0.96	1.34
100 pair	"	1.95	1.05	3.00
150 pair	"	2.65	1.25	3.90
200 pair	"	3.55	1.50	5.05
300 pair	"	5.05	1.55	6.60
400 pair	"	6.50	1.65	8.15
Cable tap in manhole or junction box				

COMMUNICATIONS	UNIT	MAT.	INST.	TOTAL
16740.10 — TELEPHONE SYSTEMS				
25 pair cable	EA.	3.30	140.00	143.30
50 pair cable	"	6.60	280.00	286.60
75 pair cable	"	9.90	420.00	429.90
100 pair cable	"	13.20	560.00	573.20
150 pair cable	"	19.80	820.00	839.80
200 pair cable	"	26.40	1,100	1,126
300 pair cable	"	39.60	1,650	1,690
400 pair cable	"	53.00	2,280	2,333
Cable terminations, manhole or junction box				
25 pair cable	EA.	3.30	140.00	143.30
50 pair cable	"	6.60	280.00	286.60
100 pair cable	"	13.20	560.00	573.20
150 pair cable	"	19.80	820.00	839.80
200 pair cable	"	26.40	1,100	1,126
300 pair cable	"	39.60	1,650	1,690
400 pair cable	"	39.60	2,280	2,320
Telephones, standard				
1 button	EA.	69.00	110.00	179.00
2 button	"	86.00	130.00	216.00
6 button	"	130.00	200.00	330.00
12 button	"	260.00	280.00	540.00
18 button	"	300.00	330.00	630.00
Hazardous area				
Desk	EA.	1,200	270.00	1,470
Wall	"	560.00	190.00	750.00
Accessories				
Standard ground	EA.	19.80	59.50	79.30
Push button	"	19.80	59.50	79.30
Buzzer	"	21.15	59.50	80.65
Interface device	"	11.20	29.70	40.90
Long cord	"	11.90	29.70	41.60
Interior jack	"	7.25	14.85	22.10
Exterior jack	"	14.50	22.85	37.35
Hazardous area				
Selector switch	EA.	120.00	120.00	240.00
Bell	"	180.00	120.00	300.00
Horn	"	270.00	160.00	430.00
Horn relay	"	230.00	110.00	340.00
16740.30 — CALL SYSTEMS				
Call systems, single bed station	EA.	68.50	19.80	88.30
Double bed station	"	110.00	27.00	137.00
Call-in cord	"	41.20	7.40	48.60
Pull cord	"	62.00	7.40	69.40
Pillow speaker	"	41.20	10.25	51.45
Dome light	"	27.50	19.80	47.30
Zone light	"	30.20	19.80	50.00
Stake station	"	93.50	22.85	116.35
Duty station	"	68.50	18.55	87.05
Utility station	"	89.50	22.85	112.35
Nurses station	"	93.50	19.80	113.30
Surgical station	"	220.00	27.00	247.00

COMMUNICATIONS	UNIT	MAT.	INST.	TOTAL
16740.30 CALL SYSTEMS				
Master station	EA.	2,340	93.00	2,433
Control station	"	820.00	300.00	1,120
Annunciator	"	310.00	74.00	384.00
Power supply	"	230.00	57.00	287.00
Speakers	"	34.40	29.70	64.10
Foot switch	"	41.20	11.00	52.20
Code blue systems				
Bed station	EA.	230.00	27.00	257.00
Dome light	"	34.30	24.75	59.05
Zone light	"	62.00	27.00	89.00
Pull cord	"	68.50	9.30	77.80
Nurses station	"	96.00	19.80	115.80
Annunciator	"	310.00	74.00	384.00
Power supply	"	220.00	54.00	274.00
Nurse station indicator, alarm annunciators, flush				
4 circuit	EA.	1,790	150.00	1,940
6 circuit	"	2,340	300.00	2,640
12 circuit	"	3,980	450.00	4,430
Desktop				
4 circuit	EA.	2,200	130.00	2,330
6 circuit	"	2,750	130.00	2,880
12 circuit	"	4,260	190.00	4,450
16750.20 SIGNALING SYSTEMS				
Signaling systems				
4" bell	EA.	69.50	22.30	91.80
6" bell	"	84.00	24.15	108.15
10" bell	"	99.00	27.70	126.70
Buzzer				
Size 0	EA.	18.50	16.50	35.00
Size 1	"	19.75	16.50	36.25
Size 2	"	21.00	18.55	39.55
Size 3	"	22.20	19.80	42.00
Horn	"	61.50	22.85	84.35
Chime	"	99.00	19.80	118.80
Push button				
Standard	EA.	21.00	14.85	35.85
Weatherproof	"	31.80	18.55	50.35
Door opener				
Mortise	EA.	22.20	18.55	40.75
Rim	"	31.20	14.85	46.05
Transformer	"	11.20	16.50	27.70
Contractor grade doorbell chime kit				
Chime	EA.	21.35	37.10	58.45
Doorbutton	"	2.95	11.85	14.80
Transformer	"	9.45	18.55	28.00
16770.30 SOUND SYSTEMS				
Power amplifiers	EA.	610.00	130.00	740.00
Pre-amplifiers	"	480.00	100.00	580.00
Tuner	"	310.00	54.00	364.00
Horn				

COMMUNICATIONS		UNIT	MAT.	INST.	TOTAL
16770.30	**SOUND SYSTEMS**				
Equilizer		EA.	780.00	59.50	839.50
Mixer		"	330.00	82.50	412.50
Tape recorder		"	1,020	69.00	1,089
Microphone		"	89.50	37.10	126.60
Cassette Player		"	520.00	80.00	600.00
Record player		"	45.30	70.50	115.80
Equipment rack		"	58.00	47.90	105.90
Speaker					
Ceiling		EA.	110.00	50.50	160.50
Wall		"	320.00	150.00	470.00
Paging		"	120.00	29.70	149.70
Column		"	170.00	19.80	189.80
Single		"	38.80	22.85	61.65
Double		"	120.00	160.00	280.00
Volume control		"	38.80	19.80	58.60
Plug-in		"	120.00	29.70	149.70
Desk		"	97.00	14.85	111.85
Outlet		"	19.35	14.85	34.20
Stand		"	38.80	11.00	49.80
Console		"	9,320	300.00	9,620
Power supply		"	170.00	47.90	217.90
16780.10	**ANTENNAS AND TOWERS**				
Guy cable, alumaweld					
1x3, 7/32"		L.F.	0.27	1.85	2.12
1x3, 1/4"		"	0.32	1.85	2.17
1x3, 25/64"		"	0.46	2.20	2.66
1x19, 1/2"		"	1.15	2.60	3.75
1x7, 35/64"		"	1.35	2.95	4.30
1x19, 13/16"		"	1.45	3.70	5.15
Preformed alumaweld end grip					
1/4" cable		EA.	2.05	3.70	5.75
3/8" cable		"	2.75	3.70	6.45
1/2" cable		"	3.45	5.40	8.85
9/16" cable		"	4.30	7.40	11.70
5/8" cable		"	5.25	9.30	14.55
Fiberglass guy rod, white epoxy coated					
1/4" dia.		L.F.	1.35	5.40	6.75
3/8" dia		"	2.05	5.40	7.45
1/2" dia		"	2.75	7.40	10.15
5/8" dia		"	3.45	9.30	12.75
Preformed glass grip end grip, guy rod					
1/4" dia.		EA.	7.60	5.40	13.00
3/8" dia.		"	8.80	7.40	16.20
1/2" dia.		"	10.80	9.30	20.10
5/8" dia.		"	12.10	9.30	21.40
Spelter socket end grip, 1/4" dia. guy rod					
Standard strength		EA.	22.50	18.55	41.05
High performance		"	27.70	18.55	46.25
3/8" dia. guy rod					
Standard strength		EA.	21.90	12.90	34.80

COMMUNICATIONS	UNIT	MAT.	INST.	TOTAL
16780.10 — ANTENNAS AND TOWERS				
High performance	EA.	27.70	18.55	46.25
Timber pole, Douglas Fir				
80-85 ft	EA.	1,680	720.00	2,400
90-95 ft	"	2,020	820.00	2,840
Southern yellow pine				
35-45 ft	EA.	1,010	410.00	1,420
50-55 ft	"	1,340	520.00	1,860
16780.50 — TELEVISION SYSTEMS				
TV outlet, self terminating, w/ cover plate	EA.	7.20	11.40	18.60
Thru splitter	"	15.70	59.50	75.20
End of line	"	13.05	49.50	62.55
In line splitter multitap				
4 way	EA.	26.20	67.50	93.70
2 way	"	19.60	63.00	82.60
Equipment cabinet	"	65.50	59.50	125.00
Antenna				
Broad band uhf	EA.	130.00	130.00	260.00
Lightning arrester	"	35.30	27.00	62.30
TV cable	L.F.	0.29	0.19	0.48
Coaxial cable rg	"	0.20	0.19	0.39
Cable drill, with replacement tip	EA.	6.55	18.55	25.10
Cable blocks for in-line taps	"	13.05	27.00	40.05
In-line taps ptu-series 36 tv system	"	15.70	42.40	58.10
Control receptacles	"	10.25	16.65	26.90
Coupler	"	19.60	90.00	109.60
Head end equipment	"	2,480	250.00	2,730
TV camera	"	1,340	62.00	1,402
TV power bracket	"	120.00	29.70	149.70
TV monitor	"	1,030	54.00	1,084
Video recorder	"	1,990	78.00	2,068
Console	"	4,120	320.00	4,440
Selector switch	"	640.00	51.00	691.00
TV controller	"	310.00	52.00	362.00

RESISTANCE HEATING	UNIT	MAT.	INST.	TOTAL
16850.10 — ELECTRIC HEATING				
Baseboard heater				
2', 375w	EA.	38.50	37.10	75.60
3', 500w	"	47.50	37.10	84.60
4', 750w	"	59.00	42.40	101.40

RESISTANCE HEATING	UNIT	MAT.	INST.	TOTAL
16850.10 — ELECTRIC HEATING				
5', 935w	EA.	78.00	49.50	127.50
6', 1125w	"	87.50	59.50	147.00
7', 1310w	"	110.00	67.50	177.50
8', 1500w	"	120.00	74.00	194.00
9', 1680w	"	130.00	82.50	212.50
10', 1875w	"	140.00	85.00	225.00
Unit heater wall mounted				
1500w	EA.	180.00	59.50	239.50
2500w	"	190.00	67.50	257.50
4000w	"	260.00	85.00	345.00
Thermostat				
Integral	EA.	29.50	18.55	48.05
Line voltage	"	29.50	18.55	48.05
Electric heater connection	"	1.30	9.30	10.60
Fittings				
Inside corner	EA.	19.25	14.85	34.10
Outside corner	"	20.55	14.85	35.40
Receptacle section	"	21.80	14.85	36.65
Blank section	"	27.00	14.85	41.85
Unit heater, wall mounted				
750w	EA.	110.00	59.50	169.50
1500w	"	130.00	62.00	192.00
2000w	"	180.00	64.50	244.50
3000w	"	220.00	74.00	294.00
Infrared heaters				
600w	EA.	140.00	37.10	177.10
2000w	"	170.00	44.30	214.30
3000w	"	220.00	74.00	294.00
4000w	"	270.00	93.00	363.00
Controller	"	51.50	24.75	76.25
Wall bracket	"	100.00	27.00	127.00
Radiant ceiling heater panels				
500w	EA.	130.00	37.10	167.10
750w	"	150.00	37.10	187.10
Unit heaters, suspended, single phase				
3.0 kw	EA.	210.00	100.00	310.00
5.0 kw	"	240.00	100.00	340.00
7.5 kw	"	360.00	120.00	480.00
10.0 kw	"	400.00	140.00	540.00
Three phase				
5 kw	EA.	340.00	100.00	440.00
7.5 kw	"	370.00	120.00	490.00
10 kw	"	450.00	140.00	590.00
15 kw	"	730.00	160.00	890.00
20 kw	"	930.00	200.00	1,130
25 kw	"	1,090	240.00	1,330
30 kw	"	1,280	300.00	1,580
35 kw	"	1,620	300.00	1,920
Unit heater thermostat	"	34.50	19.80	54.30
Mounting bracket	"	35.40	27.00	62.40
Relay	"	44.90	22.85	67.75
Duct heaters, three phase				

RESISTANCE HEATING	UNIT	MAT.	INST.	TOTAL
16850.10 ELECTRIC HEATING				
10 kw	EA.	400.00	140.00	540.00
15 kw	"	470.00	140.00	610.00
17.5 kw	"	500.00	150.00	650.00
20 kw	"	530.00	230.00	760.00

CONTROLS	UNIT	MAT.	INST.	TOTAL
16910.40 CONTROL CABLE				
Control cable, 600v, #14 THWN, PVC jacket				
2 wire	L.F.	0.19	0.30	0.49
4 wire	"	0.28	0.37	0.65
6 wire	"	0.45	4.85	5.30
8 wire	"	0.53	5.40	5.93
10 wire	"	0.62	5.95	6.57
12 wire	"	1.05	6.75	7.80
14 wire	"	1.05	7.80	8.85
16 wire	"	1.20	8.25	9.45
18 wire	"	1.30	9.00	10.30
20 wire	"	1.45	9.30	10.75
22 wire	"	1.55	10.60	12.15
Audio cables, shielded, #24 gauge				
3 conductor	L.F.	0.15	0.15	0.30
4 conductor	"	0.18	0.22	0.40
5 conductor	"	0.20	0.26	0.46
6 conductor	"	0.24	0.33	0.57
7 conductor	"	0.26	0.41	0.67
8 conductor	"	0.29	0.45	0.74
9 conductor	"	0.31	0.52	0.83
10 conductor	"	0.35	0.56	0.91
15 conductor	"	0.62	0.67	1.29
20 conductor	"	0.83	0.85	1.68
25 conductor	"	1.05	1.00	2.05
30 conductor	"	1.25	1.10	2.35
40 conductor	"	1.60	1.35	2.95
50 conductor	"	2.00	1.55	3.55
#22 gauge				
3 conductor	L.F.	0.13	0.15	0.28
4 conductor	"	0.30	0.22	0.52
#20 gauge				
3 conductor	L.F.	0.17	0.15	0.32
10 conductor	"	0.55	0.56	1.11
15 conductor	"	0.71	0.67	1.38
#18 gauge				
3 conductor	L.F.	0.20	0.15	0.35

CONTROLS	UNIT	MAT.	INST.	TOTAL
16910.40 CONTROL CABLE				
4 conductor	L.F.	0.26	0.22	0.48
Microphone cables, #24 gauge				
2 conductor	L.F.	0.20	0.15	0.35
3 conductor	"	0.24	0.19	0.43
#20 gauge				
1 conductor	L.F.	0.22	0.15	0.37
2 conductor	"	0.36	0.15	0.51
2 conductor	"	0.43	0.15	0.58
3 conductor	"	0.50	0.22	0.72
4 conductor	"	0.72	0.26	0.98
5 conductor	"	0.88	0.33	1.21
7 conductor	"	0.97	0.41	1.38
8 conductor	"	1.10	0.45	1.55
Computer cables shielded, #24 gauge				
1 pair	L.F.	0.15	0.15	0.30
2 pair	"	0.20	0.15	0.35
3 pair	"	0.24	0.22	0.46
4 pair	"	0.27	0.26	0.53
5 pair	"	0.35	0.33	0.68
6 pair	"	0.41	0.41	0.82
7 pair	"	0.44	0.45	0.89
8 pair	"	0.50	0.52	1.02
50 pair	"	3.00	1.45	4.45
Coaxial cables				
RG 6/u	L.F.	0.15	0.22	0.37
RG 6a/u	"	0.31	0.22	0.53
RG 8/u	"	0.36	0.22	0.58
RG 8a/u	"	0.45	0.22	0.67
RG 9/u	"	2.00	0.22	2.22
RG 11/u	"	0.37	0.22	0.59
RG 58/u	"	0.15	0.22	0.37
RG 59/u	"	0.20	0.22	0.42
RG 62/u	"	0.17	0.22	0.39
RG 174/u	"	0.15	0.22	0.37
RG 213/u	"	0.46	0.22	0.68
MATV and CCTV camera cables				
1 conductor	L.F.	0.20	0.15	0.35
2 conductor	"	0.26	0.19	0.45
4 conductor	"	0.63	0.22	0.85
7 conductor	"	0.85	0.33	1.18
12 conductor	"	1.30	0.56	1.86
13 conductor	"	1.40	0.59	1.99
14 conductor	"	1.45	0.67	2.12
28 conductor	"	3.45	1.00	4.45
Fire alarm cables, #22 gauge				
6 conductor	L.F.	0.20	0.37	0.57
9 conductor	"	0.26	0.56	0.82
12 conductor	"	0.30	0.59	0.89
#18 gauge				
2 conductor	L.F.	0.20	0.19	0.39
4 conductor	"	0.26	0.26	0.52
#16 gauge				

CONTROLS	UNIT	MAT.	INST.	TOTAL
16910.40 CONTROL CABLE				
2 conductor	L.F.	0.20	0.26	0.46
4 conductor	"	0.28	0.30	0.58
#14 gauge				
2 conductor	L.F.	0.29	0.30	0.59
#12 gauge				
2 conductor	L.F.	0.37	0.37	0.74
Plastic jacketed thermostat cable				
2 conductor	L.F.	0.15	0.15	0.30
3 conductor	"	0.21	0.19	0.40
4 conductor	"	0.26	0.22	0.48
5 conductor	"	0.29	0.30	0.59
6 conductor	"	0.36	0.33	0.69
7 conductor	"	0.39	0.45	0.84
8 conductor	"	0.45	0.48	0.93

Man-Hour Tables

The man-hour productivities used to develop the labor costs are listed in the following section of this book. These productivities represent typical installation labor for thousands of construction items. The data takes into account all activities involved in normal construction under commonly experienced working conditions. As with the Costbook pages, these items are listed according to the CSI MASTERFORMAT. In order to best use the information in this book, please review this sample page and read the "Features in this Book" section.

CSI MASTERFORMAT Division

CSI Broadscope Category

CSI Mediumscope Category (First 5 Digits)

Detailed Description — Complete descriptions of items may include information listed above a particular line. Review of the whole category is recommended for a complete description.

Unit of Measurement — Each item is defined in terms of the common estimating unit. Quantities listed are defined as man-hour per unit.

Man-Hours — Man-hour quantities represent typical installation times and take into account all activities involved in normal construction under commonly experienced working conditions.

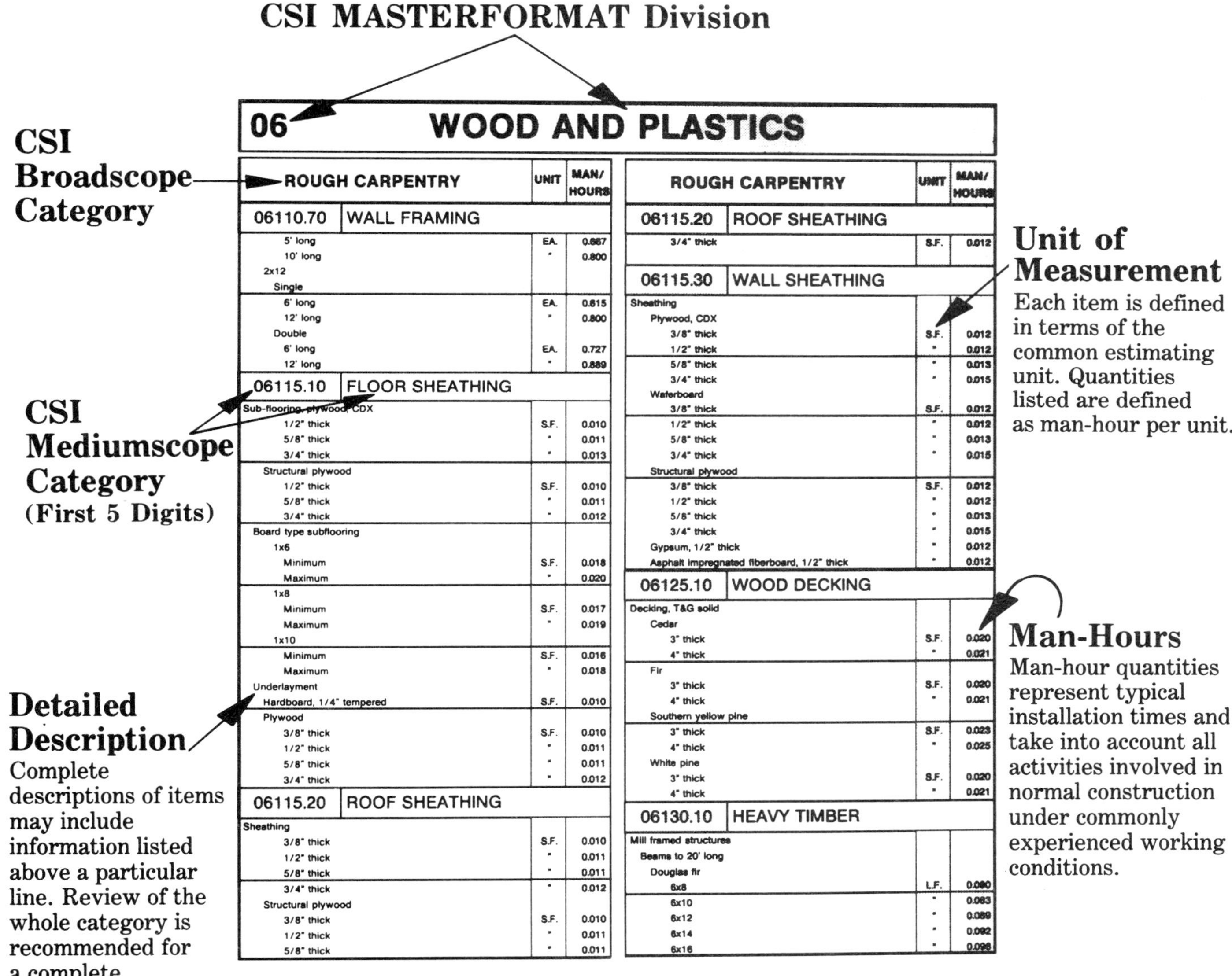

06 — WOOD AND PLASTICS

ROUGH CARPENTRY		UNIT	MAN/ HOURS
06110.70	**WALL FRAMING**		
	5' long	EA.	0.667
	10' long	"	0.800
2x12			
Single			
	6' long	EA.	0.615
	12' long	"	0.800
Double			
	6' long	EA.	0.727
	12' long	"	0.889
06115.10	**FLOOR SHEATHING**		
Sub-flooring, plywood, CDX			
	1/2" thick	S.F.	0.010
	5/8" thick	"	0.011
	3/4" thick	"	0.013
Structural plywood			
	1/2" thick	S.F.	0.010
	5/8" thick	"	0.011
	3/4" thick	"	0.012
Board type subflooring			
1x6			
	Minimum	S.F.	0.018
	Maximum	"	0.020
1x8			
	Minimum	S.F.	0.017
	Maximum	"	0.019
1x10			
	Minimum	S.F.	0.016
	Maximum	"	0.018
Underlayment			
Hardboard, 1/4" tempered		S.F.	0.010
Plywood			
	3/8" thick	S.F.	0.010
	1/2" thick	"	0.011
	5/8" thick	"	0.011
	3/4" thick	"	0.012
06115.20	**ROOF SHEATHING**		
Sheathing			
	3/8" thick	S.F.	0.010
	1/2" thick	"	0.011
	5/8" thick	"	0.011
	3/4" thick	"	0.012
Structural plywood			
	3/8" thick	S.F.	0.010
	1/2" thick	"	0.011
	5/8" thick	"	0.011

ROUGH CARPENTRY		UNIT	MAN/ HOURS
06115.20	**ROOF SHEATHING**		
	3/4" thick	S.F.	0.012
06115.30	**WALL SHEATHING**		
Sheathing			
Plywood, CDX			
	3/8" thick	S.F.	0.012
	1/2" thick	"	0.012
	5/8" thick	"	0.013
	3/4" thick	"	0.015
Waferboard			
	3/8" thick	S.F.	0.012
	1/2" thick	"	0.012
	5/8" thick	"	0.013
	3/4" thick	"	0.015
Structural plywood			
	3/8" thick	S.F.	0.012
	1/2" thick	"	0.012
	5/8" thick	"	0.013
	3/4" thick	"	0.015
	Gypsum, 1/2" thick	"	0.012
	Asphalt impregnated fiberboard, 1/2" thick	"	0.012
06125.10	**WOOD DECKING**		
Decking, T&G solid			
Cedar			
	3" thick	S.F.	0.020
	4" thick	"	0.021
Fir			
	3" thick	S.F.	0.020
	4" thick	"	0.021
Southern yellow pine			
	3" thick	S.F.	0.023
	4" thick	"	0.025
White pine			
	3" thick	S.F.	0.020
	4" thick	"	0.021
06130.10	**HEAVY TIMBER**		
Mill framed structures			
Beams to 20' long			
Douglas fir			
	6x8	L.F.	0.080
	6x10	"	0.083
	6x12	"	0.089
	6x14	"	0.092
	6x16	"	0.096

DEMOLITION

	UNIT	MAN/HOURS
02060.10 BUILDING DEMOLITION		
Cut-outs		
Concrete, elevated slabs, mesh reinforcing		
Under 5 cf	C.F.	0.800
Over 5 cf	"	0.667
Bar reinforcing		
Under 5 cf	C.F.	1.333
Over 5 cf	"	1.000
Rubbish handling		
Load in dumpster or truck		
Minimum	C.F.	0.018
Maximum	"	0.027
For use of elevators, add		
Minimum	C.F.	0.004
Maximum	"	0.008
Rubbish hauling		
Hand loaded on trucks, 2 mile trip	C.Y.	0.320
Machine loaded on trucks, 2 mile trip	"	0.240
02075.80 CORE DRILLING		
Concrete		
6" thick		
3" dia.	EA.	0.571
4" dia.	"	0.667
6" dia.	"	0.800
8" dia.	"	1.333
8" thick		
3" dia.	EA.	0.800
4" dia.	"	1.000
6" dia.	"	1.143
8" dia.	"	1.600
10" thick		
3" dia.	EA.	1.000
4" dia.	"	1.143
6" dia.	"	1.333
8" dia.	"	2.000
12" thick		
3" dia.	EA.	1.333
4" dia.	"	1.600
6" dia.	"	2.000
8" dia.	"	2.667

HAZARDOUS WASTE

	UNIT	MAN/HOURS
02080.10 ASBESTOS REMOVAL		
Enclosure using wood studs & poly, install & remove	S.F.	0.020
02080.12 DUCT INSULATION REMOVAL		
Remove duct insulation, duct size		
6" x 12"	L.F.	0.044
x 18"	"	0.062
x 24"	"	0.089
8" x 12"	"	0.067
x 18"	"	0.073
x 24"	"	0.100
12" x 12"	"	0.067
x 18"	"	0.089
x 24"	"	0.114
02080.15 PIPE INSULATION REMOVAL		
Removal, asbestos insulation		
2" thick, pipe		
1" to 3" dia.	L.F.	0.067
4" to 6" dia.	"	0.076
3" thick		
7" to 8" dia.	L.F.	0.080
9" to 10" dia.	"	0.084
11" to 12" dia.	"	0.089
13" to 14" dia.	"	0.094
15" to 18" dia.	"	0.100

SITE DEMOLITION

	UNIT	MAN/HOURS
02105.43 GAS PIPING		
Remove welded steel pipe, not including excavation		
4" dia.	L.F.	0.150
5" dia.	"	0.240
6" dia.	"	0.300
8" dia.	"	0.480
10" dia.	"	0.600
02105.45 SANITARY PIPING		
Remove sewer pipe, not including excavation		
4" dia.	L.F.	0.096
6" dia.	"	0.109
8" dia.	"	0.120
10" dia.	"	0.126
12" dia.	"	0.133
15" dia.	"	0.141
18" dia.	"	0.160

SITE DEMOLITION		UNIT	MAN/HOURS
02105.45	**SANITARY PIPING**		
24" dia.		L.F.	0.200
30" dia.		"	0.240
36" dia.		"	0.300
02105.48	**WATER PIPING**		
Remove water pipe, not including excavation			
4" dia.		L.F.	0.109
6" dia.		"	0.114
8" dia.		"	0.126
10" dia.		"	0.133
12" dia.		"	0.141
14" dia.		"	0.150
16" dia.		"	0.160
18" dia.		"	0.171
20" dia.		"	0.185
Remove valves			
6"		EA.	1.200
10"		"	1.333
14"		"	1.500
18"		"	2.000
02105.60	**UNDERGROUND TANKS**		
Remove underground storage tank, and backfill			
50 to 250 gals		EA.	8.000
600 gals		"	8.000
1000 gals		"	12.000
4000 gals		"	19.200
5000 gals		"	19.200
10,000 gals		"	32.000
12,000 gals		"	40.000
15,000 gals		"	48.000
20,000 gals		"	60.000
02105.66	**SEPTIC TANKS**		
Remove septic tank			
1000 gals		EA.	2.000
2000 gals		"	2.400
5000 gals		"	3.000
15,000 gals		"	24.000
25,000 gals		"	32.000
40,000 gals		"	48.000
02162.10	**TRENCH SHEETING**		
Closed timber, including pull and salvage, excavation			
8' deep		S.F.	0.064
20' deep		"	0.091

EARTHWORK		UNIT	MAN/HOURS
02210.10	**HAULING MATERIAL**		
Haul material by 10 cy dump truck, round trip distance			
1 mile		C.Y.	0.044
2 mile		"	0.053
5 mile		"	0.073
10 mile		"	0.080
20 mile		"	0.089
30 mile		"	0.107
Site grading, cut & fill, sandy clay, 200' haul, 75 hp dozer		"	0.032
Spread topsoil by equipment on site		"	0.036
Site grading (cut and fill to 6") less than 1 acre			
75 hp dozer		C.Y.	0.053
1.5 cy backhoe/loader		"	0.080
02210.30	**BULK EXCAVATION**		
Hydraulic excavator			
1 cy capacity			
Light material		C.Y.	0.040
Medium material		"	0.048
Wet material		"	0.060
Blasted rock		"	0.069
Wheel mounted front-end loader			
7/8 cy capacity			
Light material		C.Y.	0.020
Medium material		"	0.023
Wet material		"	0.027
Blasted rock		"	0.032
Track mounted front-end loader			
1-1/2 cy capacity			
Light material		C.Y.	0.013
Medium material		"	0.015
Wet material		"	0.016
Blasted rock		"	0.018
02220.40	**BUILDING EXCAVATION**		
Structural excavation, unclassified earth			
3/8 cy backhoe		C.Y.	0.107
3/4 cy backhoe		"	0.080
1 cy backhoe		"	0.067
Foundation backfill and compaction by machine		"	0.160
02220.50	**UTILITY EXCAVATION**		
Trencher, sandy clay, 8" wide trench			
18" deep		L.F.	0.018
24" deep		"	0.020
36" deep		"	0.023
Trench backfill, 95% compaction			
Tamp by hand		C.Y.	0.500
Vibratory compaction		"	0.400
Trench backfilling, with borrow sand, place & compact		"	0.400
02220.60	**TRENCHING**		
Trenching and continuous footing excavation			
By gradall			

EARTHWORK

02220.60 TRENCHING

	UNIT	MAN/HOURS
1 cy capacity		
Light soil	C.Y.	0.023
Medium soil	"	0.025
Heavy/wet soil	"	0.027
Loose rock	"	0.029
Blasted rock	"	0.031
By hydraulic excavator		
1/2 cy capacity		
Light soil	C.Y.	0.027
Medium soil	"	0.029
Heavy/wet soil	"	0.032
Loose rock	"	0.036
Blasted rock	"	0.040
Hand excavation		
Bulk, wheeled 100'		
Normal soil	C.Y.	0.889
Sand or gravel	"	0.800
Medium clay	"	1.143
Heavy clay	"	1.600
Loose rock	"	2.000
Trenches, up to 2' deep		
Normal soil	C.Y.	1.000
Sand or gravel	"	0.889
Medium clay	"	1.333
Heavy clay	"	2.000
Loose rock	"	2.667
Trenches, to 6' deep		
Normal soil	C.Y.	1.143
Sand or gravel	"	1.000
Medium clay	"	1.600
Heavy clay	"	2.667
Loose rock	"	4.000
Backfill trenches		
With compaction		
By hand	C.Y.	0.667
By 60 hp tracked dozer	"	0.020
By 200 hp tracked dozer	"	0.009
By small front-end loader	"	0.023
Backfill trenches, sand bedding, no compaction		
By hand	C.Y.	0.667
By small front-end loader	"	0.023

02220.90 HAND EXCAVATION

	UNIT	MAN/HOURS
Excavation		
To 2' deep		
Normal soil	C.Y.	0.889
Sand and gravel	"	0.800
Medium clay	"	1.000
Heavy clay	"	1.143
Loose rock	"	1.333
To 6' deep		
Normal soil	C.Y.	1.143
Sand and gravel	"	1.000

EARTHWORK

02220.90 HAND EXCAVATION

	UNIT	MAN/HOURS
Medium clay	C.Y.	1.333
Heavy clay	"	1.600
Loose rock	"	2.000
Compaction of backfill around structures or in trench		
By hand with air tamper	C.Y.	0.571
By hand with vibrating plate tamper	"	0.533
1 ton roller	"	0.400
Miscellaneous hand labor		
Trim slopes, sides of excavation	S.F.	0.001
Trim bottom of excavation	"	0.002
Excavation around obstructions and services	C.Y.	2.667

02240.05 SOIL STABILIZATION

	UNIT	MAN/HOURS
Straw bale secured with rebar	L.F.	0.027
Filter barrier, 18" high filter fabric	"	0.080
Sediment fence, 36" fabric with 6" mesh	"	0.100

UTILITIES

02605.30 MANHOLES

	UNIT	MAN/HOURS
Precast sections, 48" dia.		
Base section	EA.	2.000
1'0" riser	"	1.600
1'4" riser	"	1.714
2'8" riser	"	1.846
4'0" riser	"	2.000
2'8" cone top	"	2.400
Precast manholes, 48" dia.		
4' deep	EA.	4.800
6' deep	"	6.000
7' deep	"	6.857
8' deep	"	8.000
10' deep	"	9.600
Cast-in-place, 48" dia., with frame and cover		
5' deep	EA.	12.000
6' deep	"	13.714
8' deep	"	16.000
10' deep	"	19.200
Brick manholes, 48" dia. with cover, 8" thick		
4' deep	EA.	8.000
6' deep	"	8.889
8' deep	"	10.000
10' deep	"	11.429
12' deep	"	13.333
14' deep	"	16.000

UTILITIES	UNIT	MAN/ HOURS
02605.30 MANHOLES		
Inverts for manholes		
Single channel	EA.	3.200
Triple channel	"	4.000
Frames and covers, 24" diameter		
300 lb	EA.	0.800
400 lb	"	0.889
500 lb	"	1.143
Watertight, 350 lb	"	2.667
For heavy equipment, 1200 lb	"	4.000
Steps for manholes		
7" x 9"	EA.	0.160
8" x 9"	"	0.178
Curb inlet, 4' throat, cast in place		
12"-30" pipe	EA.	12.000
36"-48" pipe	"	13.714
Raise exist frame and cover, repaving	"	4.800
02610.10 CAST IRON FLANGED PIPE		
Cast iron flanged sections		
4" pipe, with one bolt set		
3' section	EA.	0.218
4' section	"	0.240
5' section	"	0.267
6' section	"	0.300
8' section	"	0.343
10' section	"	0.480
12' section	"	0.800
15' section	"	1.200
18' section	"	1.600
6" pipe, with one bolt set		
3' section	EA.	0.240
4' section	"	0.282
5' section	"	0.320
6' section	"	0.369
8' section	"	0.533
10' section	"	0.600
12' section	"	0.800
15' section	"	1.200
18' section	"	1.714
8" pipe, with one bolt set		
3' section	EA.	0.300
4' section	"	0.343
5' section	"	0.400
6' section	"	0.480
8' section	"	0.686
10' section	"	0.800
12' section	"	1.200
15' section	"	1.600
18' section	"	2.000
10" pipe, with one bolt set		
3' section	EA.	0.308
4' section	"	0.353
5' section	"	0.414
6' section	"	0.500

UTILITIES	UNIT	MAN/ HOURS
02610.10 CAST IRON FLANGED PIPE		
8' section	EA.	0.727
10' section	"	0.857
12' section	"	1.333
15' section	"	1.714
18' section	"	2.400
12" pipe, with one bolt set		
3' section	EA.	0.333
4' section	"	0.387
5' section	"	0.462
6' section	"	0.545
8' section	"	0.800
10' section	"	0.923
12' section	"	1.500
15' section	"	2.000
18' section	"	2.667
02610.11 CAST IRON FITTINGS		
Mechanical joint, with 2 bolt kits		
90 deg bend		
4"	EA.	0.533
6"	"	0.615
8"	"	0.800
10"	"	1.143
12"	"	1.600
14"	"	2.000
16"	"	2.667
45 deg bend		
4"	EA.	0.533
6"	"	0.615
8"	"	0.800
10"	"	1.143
12"	"	1.600
14"	"	2.000
16"	"	2.667
Tee, with 3 bolt kits		
4" x 4"	EA.	0.800
6" x 6"	"	1.000
8" x 8"	"	1.333
10" x 10"	"	2.000
12" x 12"	"	2.667
Wye, with 3 bolt kits		
6" x 6"	EA.	1.000
8" x 8"	"	1.333
10" x 10"	"	2.000
12" x 12"	"	2.667
Reducer, with 2 bolt kits		
6" x 4"	EA.	1.000
8" x 6"	"	1.333
10" x 8"	"	2.000
12" x 10"	"	2.667
Flanged, 90 deg bend, 125 lb.		
4"	EA.	0.667
6"	"	0.800

Left Column

UTILITIES	UNIT	MAN/HOURS
02610.11 CAST IRON FITTINGS		
8"	EA.	1.000
10"	"	1.333
12"	"	2.000
14"	"	2.667
16"	"	2.667
Tee		
4"	EA.	1.000
6"	"	1.143
8"	"	1.333
10"	"	1.600
12"	"	2.000
14"	"	2.667
16"	"	4.000
02610.13 GATE VALVES		
Gate valve, (AWWA) mechanical joint, with adjustable box		
4" valve	EA.	0.800
6" valve	"	0.960
8" valve	"	1.200
10" valve	"	1.412
12" valve	"	1.714
14" valve	"	2.000
16" valve	"	2.182
18" valve	"	2.400
Flanged, with box, post indicator (AWWA)		
4" valve	EA.	0.960
6" valve	"	1.091
8" valve	"	1.333
10" valve	"	1.600
12" valve	"	2.000
14" valve	"	2.400
16" valve	"	3.000
02610.15 WATER METERS		
Water meter, displacement type		
1"	EA.	0.800
1-1/2"	"	0.889
2"	"	1.000
02610.17 CORPORATION STOPS		
Stop for flared copper service pipe		
3/4"	EA.	0.400
1"	"	0.444
1-1/4"	"	0.533
1-1/2"	"	0.667
2"	"	0.800
02610.40 DUCTILE IRON PIPE		
Ductile iron pipe, cement lined, slip-on joints		
4"	L.F.	0.067
6"	"	0.071
8"	"	0.075

Right Column

UTILITIES	UNIT	MAN/HOURS
02610.40 DUCTILE IRON PIPE		
10"	L.F.	0.080
12"	"	0.096
14"	"	0.120
16"	"	0.133
18"	"	0.150
20"	"	0.171
Mechanical joint pipe		
4"	L.F.	0.092
6"	"	0.100
8"	"	0.109
10"	"	0.120
12"	"	0.160
14"	"	0.185
16"	"	0.218
18"	"	0.240
20"	"	0.267
Fittings, mechanical joint		
90 degree elbow		
4"	EA.	0.533
6"	"	0.615
8"	"	0.800
10"	"	1.143
12"	"	1.600
14"	"	2.000
16"	"	2.667
18"	"	3.200
20"	"	4.000
45 degree elbow		
4"	EA.	0.533
6"	"	0.615
8"	"	0.800
10"	"	1.143
12"	"	1.600
14"	"	2.000
16"	"	2.667
18"	"	4.000
20"	"	4.000
Tee		
4"x3"	EA.	1.000
4"x4"	"	1.000
6"x3"	"	1.143
6"x4"	"	1.143
6"x6"	"	1.143
8"x4"	"	1.333
8"x6"	"	1.333
8"x8"	"	1.333
10"x4"	"	1.600
10"x6"	"	1.600
10"x8"	"	1.600
10"x10"	"	1.600
12"x4"	"	2.000
12"x6"	"	2.000
12"x8"	"	2.000
12"x10"	"	2.000

UTILITIES	UNIT	MAN/HOURS
02610.40 DUCTILE IRON PIPE		
12"x12"	EA.	2.133
14"x4"	"	2.286
14"x6"	"	2.286
14"x8"	"	2.286
14"x10"	"	2.286
14"x12"	"	2.462
14"x14"	"	2.462
16"x4"	"	2.667
16"x6"	"	2.667
16"x8"	"	2.667
16"x10"	"	2.667
16"x12"	"	2.667
16"x14"	"	2.667
16"x16"	"	2.667
18"x6"	"	2.909
18"x8"	"	2.909
18"x10"	"	2.909
18"x12"	"	2.909
18"x14"	"	2.909
18"x16"	"	2.909
18"x18"	"	2.909
20"x6"	"	3.200
20"x8"	"	3.200
20"x10"	"	3.200
20"x12"	"	3.200
20"x14"	"	3.200
20"x16"	"	3.200
20"x18"	"	3.200
20"x20"	"	3.200
Cross		
4"x3"	EA.	1.333
4"x4"	"	1.333
6"x3"	"	1.600
6"x4"	"	1.600
6"x6"	"	1.600
8"x4"	"	1.778
8"x6"	"	1.778
8"x8"	"	1.778
10"x4"	"	2.000
10"x6"	"	2.000
10"x8"	"	2.000
10"x10"	"	2.000
12"x4"	"	2.286
12"x6"	"	2.286
12"x8"	"	2.286
12"x10"	"	2.462
12"x12"	"	2.462
14"x4"	"	2.667
14"x6"	"	2.667
14"x8"	"	2.667
14"x10"	"	2.667
14"x12"	"	2.909
14"x14"	"	2.909
16"x4"	"	3.200

UTILITIES	UNIT	MAN/HOURS
02610.40 DUCTILE IRON PIPE		
16"x6"	EA.	3.200
16"x8"	"	3.200
16"x10"	"	3.200
16"x12"	"	3.200
16"x14"	"	3.200
16"x16"	"	3.200
18"x6"	"	3.556
18"x8"	"	3.556
18"x10"	"	3.556
18"x12"	"	3.556
18"x14"	"	3.556
18"x16"	"	3.556
18"x18"	"	3.556
20"x6"	"	3.810
20"x8"	"	3.810
20"x10"	"	3.810
20"x12"	"	3.810
20"x14"	"	3.810
20"x16"	"	3.810
20"x18"	"	4.000
20"x20"	"	4.000
02610.60 PLASTIC PIPE		
PVC, class 150 pipe		
4" dia.	L.F.	0.060
6" dia.	"	0.065
8" dia.	"	0.069
10" dia.	"	0.075
12" dia.	"	0.080
Schedule 40 pipe		
1-1/2" dia.	L.F.	0.047
2" dia.	"	0.050
2-1/2" dia.	"	0.053
3" dia.	"	0.057
4" dia.	"	0.067
6" dia.	"	0.080
90 degree elbows		
1"	EA.	0.133
1-1/2"	"	0.133
2"	"	0.145
2-1/2"	"	0.160
3"	"	0.178
4"	"	0.200
6"	"	0.267
45 degree elbows		
1"	EA.	0.133
1-1/2"	"	0.133
2"	"	0.145
2-1/2"	"	0.160
3"	"	0.178
4"	"	0.200
6"	"	0.267
Tees		

UTILITIES	UNIT	MAN/ HOURS
02610.60 **PLASTIC PIPE**		
1"	EA.	0.160
1-1/2"	"	0.160
2"	"	0.178
2-1/2"	"	0.200
3"	"	0.229
4"	"	0.267
6"	"	0.320
Couplings		
1"	EA.	0.133
1-1/2"	"	0.133
2"	"	0.145
2-1/2"	"	0.160
3"	"	0.178
4"	"	0.200
6"	"	0.267
Drainage pipe		
PVC schedule 80		
1" dia.	L.F.	0.047
1-1/2" dia.	"	0.047
ABS, 2" dia.	"	0.050
2-1/2" dia.	"	0.053
3" dia.	"	0.057
4" dia.	"	0.067
6" dia.	"	0.080
8" dia.	"	0.063
10" dia.	"	0.075
12" dia.	"	0.080
90 degree elbows		
1"	EA.	0.133
1-1/2"	"	0.133
2"	"	0.145
2-1/2"	"	0.160
3"	"	0.178
4"	"	0.200
6"	"	0.267
45 degree elbows		
1"	EA.	0.133
1-1/2"	"	0.133
2"	"	0.145
2-1/2"	"	0.160
3"	"	0.178
4"	"	0.200
6"	"	0.267
Tees		
1"	EA.	0.160
1-1/2"	"	0.160
2"	"	0.178
2-1/2"	"	0.200
3"	"	0.229
4"	"	0.267
6"	"	0.320
Couplings		
1"	EA.	0.133
1-1/2"	"	0.133

UTILITIES	UNIT	MAN/ HOURS
02610.60 **PLASTIC PIPE**		
2"	EA.	0.145
2-1/2"	"	0.160
3"	"	0.178
4"	"	0.200
6"	"	0.267
Pressure pipe		
PVC, class 200 pipe		
3/4"	L.F.	0.040
1"	"	0.042
1-1/4"	"	0.044
1-1/2"	"	0.047
2"	"	0.050
2-1/2"	"	0.053
3"	"	0.057
4"	"	0.067
6"	"	0.080
8"	"	0.069
90 degree elbows		
3/4"	EA.	0.133
1"	"	0.133
1-1/4"	"	0.133
1-1/2"	"	0.133
2"	"	0.145
2-1/2"	"	0.160
3"	"	0.178
4"	"	0.200
6"	"	0.267
8"	"	0.400
45 degree elbows		
3/4"	EA.	0.133
1"	"	0.133
1-1/4"	"	0.133
1-1/2"	"	0.133
2"	"	0.145
2-1/2"	"	0.160
3"	"	0.178
4"	"	0.200
6"	"	0.267
8"	"	0.400
Tees		
3/4"	EA.	0.160
1"	"	0.160
1-1/4"	"	0.160
1-1/2"	"	0.160
2"	"	0.178
2-1/2"	"	0.200
3"	"	0.229
4"	"	0.267
6"	"	0.320
8"	"	0.444
Couplings		
3/4"	EA.	0.133
1"	"	0.133
1-1/4"	"	0.133

UTILITIES

		UNIT	MAN/HOURS
02610.60	**PLASTIC PIPE**		
1-1/2"		EA.	0.133
2"		"	0.145
2-1/2"		"	0.160
3"		"	0.178
4"		"	0.178
6"		"	0.200
8"		"	0.267
02610.90	**VITRIFIED CLAY PIPE**		
Vitrified clay pipe, extra strength			
6" dia.		L.F.	0.109
8" dia.		"	0.114
10" dia.		"	0.120
12" dia.		"	0.160
15" dia.		"	0.240
18" dia.		"	0.267
24" dia.		"	0.343
30" dia.		"	0.480
36" dia.		"	0.686
02630.10	**TAPPING SADDLES & SLEEVES**		
Tapping saddle, tap size to 2"			
4" saddle		EA.	0.400
6" saddle		"	0.500
8" saddle		"	0.667
10" saddle		"	0.800
12" saddle		"	1.143
14" saddle		"	1.600
Tapping sleeve			
4x4		EA.	0.533
6x4		"	0.615
6x6		"	0.615
8x4		"	0.800
8x6		"	0.800
10x4		"	0.960
10x6		"	0.960
10x8		"	0.960
10x10		"	1.000
12x4		"	1.000
12x6		"	1.091
12x8		"	1.200
12x10		"	1.333
12x12		"	1.500
Tapping valve, mechanical joint			
4" valve		EA.	3.000
6" valve		"	4.000
8" valve		"	6.000
10" valve		"	8.000
12" valve		"	12.000
Tap hole in pipe			
4" hole		EA.	1.000
6" hole		"	1.600
8" hole		"	2.667

UTILITIES

		UNIT	MAN/HOURS
02630.10	**TAPPING SADDLES & SLEEVES**		
10" hole		EA.	3.200
12" hole		"	4.000
02640.15	**VALVE BOXES**		
Valve box, adjustable, for valves up to 20"			
3' deep		EA.	0.267
4' deep		"	0.320
5' deep		"	0.400
02640.19	**THRUST BLOCKS**		
Thrust block, 3000# concrete			
1/4 c.y.		EA.	1.333
1/2 c.y.		"	1.600
3/4 c.y.		"	2.667
1 c.y.		"	5.333
02645.10	**FIRE HYDRANTS**		
Standard, 3 way post, 6" mechanical joint			
2' deep		EA.	8.000
4' deep		"	9.600
6' deep		"	12.000
8' deep		"	13.714
02665.10	**CHILLED WATER SYSTEMS**		
Chilled water pipe, 2" thick insulation, w/casing			
Align and tack weld on sleepers			
1-1/2" dia.		L.F.	0.022
3" dia.		"	0.034
4" dia.		"	0.048
6" dia.		"	0.060
8" dia.		"	0.069
10" dia.		"	0.080
12" dia.		"	0.096
14" dia.		"	0.104
16" dia.		"	0.120
Align and tack weld on trench bottom			
18" dia.		L.F.	0.133
20" dia.		"	0.150
Preinsulated fittings			
Align and tack weld on sleepers			
Elbows			
1-1/2"		EA.	0.500
3"		"	0.800
4"		"	1.000
6"		"	1.333
8"		"	1.600
Tees			
1-1/2"		EA.	0.533
3"		"	0.889
4"		"	1.143
6"		"	1.600
8"		"	2.000

UTILITIES	UNIT	MAN/HOURS
02665.10 CHILLED WATER SYSTEMS		
Reducers		
3"	EA.	0.667
4"	"	0.800
6"	"	1.000
8"	"	1.333
Anchors, not including concrete		
4"	EA.	1.000
6"	"	1.000
Align and tack weld on trench bottom		
Elbows		
10"	EA.	1.500
12"	"	1.714
14"	"	1.846
16"	"	2.000
18"	"	2.182
20"	"	2.400
Tees		
10"	EA.	1.500
12"	"	1.714
14"	"	1.846
16"	"	2.000
18"	"	2.182
20"	"	2.400
Reducers		
10"	EA.	1.000
12"	"	1.091
14"	"	1.200
16"	"	1.333
18"	"	1.500
20"	"	1.714
Anchors, not including concrete		
10"	EA.	1.000
12"	"	1.091
14"	"	1.200
16"	"	1.333
18"	"	1.500
20"	"	1.714
02670.10 WELLS		
Domestic water, drilled and cased		
4" dia.	L.F.	1.600
6" dia.	"	1.846
8" dia.	"	2.400
02685.10 GAS DISTRIBUTION		
Gas distribution lines		
Polyethylene, 60 psi coils		
1-1/4" dia.	L.F.	0.053
1-1/2" dia.	"	0.057
2" dia.	"	0.067
3" dia.	"	0.080
30' pipe lengths		
3" dia.	L.F.	0.089

UTILITIES	UNIT	MAN/HOURS
02685.10 GAS DISTRIBUTION		
4" dia.	L.F.	0.100
6" dia.	"	0.133
8" dia.	"	0.160
Steel, schedule 40, plain end		
1" dia.	L.F.	0.067
2" dia.	"	0.073
3" dia.	"	0.080
4" dia.	"	0.160
5" dia.	"	0.171
6" dia.	"	0.200
8" dia.	"	0.218
Natural gas meters, direct digital reading, threaded		
250 cfh @ 5 lbs	EA.	1.600
425 cfh @ 10 lbs	"	1.600
800 cfh @ 20 lbs	"	2.000
1000 cfh @ 25 lbs	"	2.000
1,400 cfh @ 100 lbs	"	2.667
2,300 cfh @ 100 lbs	"	4.000
5,000 cfh @ 100 lbs	"	8.000
Gas pressure regulators		
Threaded		
3/4"	EA.	1.000
1"	"	1.333
1-1/4"	"	1.333
1-1/2"	"	1.333
2"	"	1.600
Flanged		
3"	EA.	2.000
4"	"	2.667
02690.10 STORAGE TANKS		
Oil storage tank, underground		
Steel		
500 gals	EA.	3.000
1,000 gals	"	4.000
4,000 gals	"	8.000
5,000 gals	"	12.000
10,000 gals	"	24.000
Fiberglass, double wall		
550 gals	EA.	4.000
1,000 gals	"	4.000
2,000 gals	"	6.000
4,000 gals	"	12.000
6,000 gals	"	16.000
8,000 gals	"	24.000
10,000 gals	"	30.000
12,000 gals	"	40.000
15,000 gals	"	53.333
20,000 gals	"	60.000
Above ground		
Steel		
275 gals	EA.	2.400
500 gals	"	4.000

UTILITIES	UNIT	MAN/HOURS
02690.10 STORAGE TANKS		
1,000 gals	EA.	4.800
1,500 gals	"	6.000
2,000 gals	"	8.000
5,000 gals	"	12.000
Fill cap	"	0.800
Vent cap	"	0.800
Level indicator	"	0.800
02695.80 STEAM METERS		
In-line turbine, direct reading, 300 lb, flanged		
2"	EA.	1.000
3"	"	1.333
4"	"	1.600
Threaded, 2"		
5" line	EA.	8.000
6" line	"	8.000
8" line	"	8.000
10" line	"	8.000
12" line	"	8.000
14" line	"	8.000
16" line	"	8.000

SEWERAGE AND DRAINAGE	UNIT	MAN/HOURS
02720.10 CATCH BASINS		
Standard concrete catch basin		
Cast in place, 3'8" x 3'8", 6" thick wall		
2' deep	EA.	6.000
3' deep	"	6.000
4' deep	"	8.000
5' deep	"	8.000
6' deep	"	9.600
4'x4', 8" thick wall, cast in place		
2' deep	EA.	6.000
3' deep	"	6.000
4' deep	"	8.000
5' deep	"	8.000
6' deep	"	9.600
Frames and covers, cast iron		
Round		
24" dia.	EA.	2.000
26" dia.	"	2.000
28" dia.	"	2.000
Rectangular		
23"x23"	EA.	2.000
27"x20"	"	2.000

SEWERAGE AND DRAINAGE	UNIT	MAN/HOURS
02720.10 CATCH BASINS		
24"x24"	EA.	2.000
26"x26"	"	2.000
Curb inlet frames and covers		
27"x27"	EA.	2.000
24"x36"	"	2.000
24"x25"	"	2.000
24"x22"	"	2.000
20"x22"	"	2.000
Airfield catch basin frame and grating, galvanized		
2'x4'	EA.	2.000
2'x2'	"	2.000
02720.40 STORM DRAINAGE		
Concrete pipe		
Plain, bell and spigot joint, class II		
6" pipe	L.F.	0.109
8" pipe	"	0.120
10" pipe	"	0.126
12" pipe	"	0.133
15" pipe	"	0.141
18" pipe	"	0.150
21" pipe	"	0.160
24" pipe	"	0.171
Reinforced, class III, tongue and groove joint		
12" pipe	L.F.	0.133
15" pipe	"	0.141
18" pipe	"	0.150
21" pipe	"	0.160
24" pipe	"	0.171
27" pipe	"	0.185
30" pipe	"	0.200
36" pipe	"	0.218
42" pipe	"	0.240
48" pipe	"	0.267
54" pipe	"	0.300
60" pipe	"	0.343
66" pipe	"	0.400
72" pipe	"	0.480
Flared end-section, concrete		
12" pipe	L.F.	0.133
15" pipe	"	0.141
18" pipe	"	0.150
24" pipe	"	0.171
30" pipe	"	0.200
36" pipe	"	0.218
42" pipe	"	0.240
48" pipe	"	0.267
54" pipe	"	0.300
Corrugated metal pipe, coated, paved invert		
16 ga.		
8" pipe	L.F.	0.080
10" pipe	"	0.083
12" pipe	"	0.086

SEWERAGE AND DRAINAGE

02720.40 STORM DRAINAGE

	UNIT	MAN/HOURS
15" pipe	L.F.	0.092
18" pipe	"	0.100
21" pipe	"	0.109
24" pipe	"	0.120
30" pipe	"	0.133
36" pipe	"	0.150
12 ga., 48" pipe	"	0.171
10 ga.		
60" pipe	L.F.	0.200
72" pipe	"	0.240
Galvanized or aluminum, plain		
16 ga.		
8" pipe	L.F.	0.080
10" pipe	"	0.083
12" pipe	"	0.086
15" pipe	"	0.092
18" pipe	"	0.100
24" pipe	"	0.120
30" pipe	"	0.133
36" pipe	"	0.150
12 ga., 48" pipe	"	0.171
10 ga., 60" pipe	"	0.200
Galvanized or aluminum, coated oval arch		
16 ga.		
17" x 13"	L.F.	0.109
21" x 15"	"	0.120
14 ga.		
28" x 20"	L.F.	0.133
35" x 24"	"	0.171
12 ga.		
42" x 29"	L.F.	0.200
57" x 38"	"	0.240
64" x 43"	"	0.253
Oval arch culverts, plain		
16 ga.		
17" x 13"	L.F.	0.109
21" x 15"	"	0.120
14 ga.		
28" x 20"	L.F.	0.133
35" x 24"	"	0.171
12 ga.		
57" x 38"	L.F.	0.200
64" x 43"	"	0.240
71" x 47"	"	0.253
Nestable corrugated metal pipe		
16 ga.		
10" pipe	L.F.	0.083
12" pipe	"	0.086
15" pipe	"	0.092
18" pipe	"	0.100
24" pipe	"	0.120
30" pipe	"	0.133
14 ga., 36" pipe	"	0.150
Headwalls, cast in place, 30 deg wingwall		

SEWERAGE AND DRAINAGE

02720.40 STORM DRAINAGE

	UNIT	MAN/HOURS
12" pipe	EA.	2.000
15" pipe	"	2.000
18" pipe	"	2.286
24" pipe	"	2.286
30" pipe	"	2.667
36" pipe	"	4.000
42" pipe	"	4.000
48" pipe	"	5.333
54" pipe	"	6.667
60" pipe	"	8.000
4" cleanout for storm drain		
4" pipe	EA.	1.000
6" pipe	"	1.000
8" pipe	"	1.000
Connect new drain line		
To existing manhole	EA.	2.667
To new manhole	"	1.600

02730.10 SANITARY SEWERS

	UNIT	MAN/HOURS
Clay		
6" pipe	L.F.	0.080
8" pipe	"	0.086
10" pipe	"	0.092
12" pipe	"	0.100
PVC		
4" pipe	L.F.	0.060
6" pipe	"	0.063
8" pipe	"	0.067
10" pipe	"	0.071
12" pipe	"	0.075
Cleanout		
4" pipe	EA.	1.000
6" pipe	"	1.000
8" pipe	"	1.000
Connect new sewer line		
To existing manhole	EA.	2.667
To new manhole	"	1.600

02740.10 DRAINAGE FIELDS

	UNIT	MAN/HOURS
Perforated PVC pipe, for drain field		
4" pipe	L.F.	0.053
6" pipe	"	0.057

02740.50 SEPTIC TANKS

	UNIT	MAN/HOURS
Septic tank, precast concrete		
1000 gals	EA.	4.000
2000 gals	"	6.000
5000 gals	"	12.000
25,000 gals	"	48.000
40,000 gals	"	80.000
Leaching pit, precast concrete, 72" diameter		
3' deep	EA.	3.000

SEWERAGE AND DRAINAGE	UNIT	MAN/HOURS
02740.50 SEPTIC TANKS		
6' deep	EA.	3.429
8' deep	"	4.000
02760.10 PIPELINE RESTORATION		
Relining existing water main		
6" dia.	L.F.	0.240
8" dia.	"	0.253
10" dia.	"	0.267
12" dia.	"	0.282
14" dia.	"	0.300
16" dia.	"	0.320
18" dia.	"	0.343
20" dia.	"	0.369
24" dia.	"	0.400
36" dia.	"	0.480
48" dia.	"	0.533
72" dia.	"	0.600
Replacing in line gate valves		
6" valve	EA.	3.200
8" valve	"	4.000
10" valve	"	4.800
12" valve	"	6.000
16" valve	"	6.857
18" valve	"	8.000
20" valve	"	9.600
24" valve	"	12.000
36" valve	"	16.000

POWER & COMMUNICATIONS	UNIT	MAN/HOURS
02780.20 HIGH VOLTAGE CABLE		
High voltage XLP copper cable, shielded, 5000v		
#6 awg	L.F.	0.013
#4 awg	"	0.016
#2 awg	"	0.019
#1 awg	"	0.021
#1/0 awg	"	0.024
#2/0 awg	"	0.029
#3/0 awg	"	0.034
#4/0 awg	"	0.036
#250 awg	"	0.043
#300 awg	"	0.048
#350 awg	"	0.053
#500 awg	"	0.073
#750 awg	"	0.080
Ungrounded, 15,000v		

POWER & COMMUNICATIONS	UNIT	MAN/HOURS
02780.20 HIGH VOLTAGE CABLE		
#1 awg	L.F.	0.031
#1/0 awg	"	0.034
#2/0 awg	"	0.036
#3/0 awg	"	0.040
#4/0 awg	"	0.046
#250 awg	"	0.048
#300 awg	"	0.053
#350 awg	"	0.062
#500 awg	"	0.080
#750 awg	"	0.098
#1000 awg	"	0.123
Aluminum cable, shielded, 5000v		
#6 awg	L.F.	0.011
#4 awg	"	0.013
#2 awg	"	0.015
#1 awg	"	0.017
#1/0 awg	"	0.019
#2/0 awg	"	0.020
#3/0 awg	"	0.021
#4/0 awg	"	0.024
#250 awg	"	0.026
#300 awg	"	0.031
#350 awg	"	0.034
#500 awg	"	0.036
#750 awg	"	0.044
#1000 awg	"	0.050
Ungrounded, 15,000v		
#1 awg	L.F.	0.021
#1/0 awg	"	0.025
#2/0 awg	"	0.027
#3/0 awg	"	0.028
#4/0 awg	"	0.029
#250 awg	"	0.031
#300 awg	"	0.032
#350 awg	"	0.036
#500 awg	"	0.043
#750 awg	"	0.052
#1000 awg	"	0.064
Indoor terminations, 5000v		
#6 - #4	EA.	0.157
#2 - #2/0	"	0.157
#3/0 - #250	"	0.157
#300 - #750	"	2.759
#1000	"	3.810
In-line splice, 5000v		
#6 - #4/0	EA.	3.810
#250 - #500	"	10.000
#750 - #1000	"	13.008
T-splice, 5000v		
#2 - #4/0	EA.	11.994
#250 - #500	"	20.000
#750 - #1000	"	25.000
Indoor terminations, 15,000v		
#2 - #2/0	EA.	3.478

POWER & COMMUNICATIONS

02780.20   HIGH VOLTAGE CABLE

	UNIT	MAN/HOURS
#3/0 - #500	EA.	5.333
#750 - #1000	"	6.154
In-line splice, 15,000v		
#2 - #4/0	EA.	8.999
#250 - #500	"	11.994
#750 - #1000	"	18.018
T-splice, 15,000v		
#4	EA.	18.018
#250 - #500	"	29.963
#750 - #1000	"	44.944
Compression lugs, 15,000v		
#4	EA.	0.400
#2	"	0.533
#1	"	0.533
#1/0	"	0.667
#2/0	"	0.667
#3/0	"	0.851
#4/0	"	0.851
#250	"	0.952
#300	"	0.952
#350	"	1.159
#500	"	1.250
#750	"	1.509
#1000	"	1.905
Compression splices, 15,000v		
#4	EA.	0.667
#2	"	0.727
#1	"	0.899
#1/0	"	1.000
#2/0	"	1.159
#3/0	"	1.250
#4/0	"	1.404
#250	"	1.509
#350	"	1.739
#500	"	2.000
#750	"	2.500

02780.40   SUPPORTS & CONNECTORS

	UNIT	MAN/HOURS
Cable supports for conduit		
1-1/2"	EA.	0.348
2"	"	0.348
2-1/2"	"	0.400
3"	"	0.400
3-1/2"	"	0.500
4"	"	0.500
5"	"	0.667
6"	"	0.727
Split bolt connectors		
#10	EA.	0.200
#8	"	0.200
#6	"	0.200
#4	"	0.400
#3	"	0.400
#2	"	0.400

POWER & COMMUNICATIONS

02780.40   SUPPORTS & CONNECTORS

	UNIT	MAN/HOURS
#1/0	EA.	0.667
#2/0	"	0.667
#3/0	"	0.667
#4/0	"	0.667
#250	"	1.000
#350	"	1.000
#500	"	1.000
#750	"	1.509
#1000	"	1.509
Single barrel lugs		
#6	EA.	0.250
#1/0	"	0.500
#250	"	0.667
#350	"	0.667
#500	"	0.667
#600	"	0.899
#800	"	0.899
#1000	"	0.899
Double barrel lugs		
#1/0	EA.	0.899
#250	"	1.290
#350	"	1.290
#600	"	1.905
#800	"	1.905
#1000	"	1.905
Three barrel lugs		
#2/0	EA.	1.290
#250	"	1.905
#350	"	1.905
#600	"	2.667
#800	"	2.667
#1000	"	2.667
Four barrel lugs		
#250	EA.	2.759
#350	"	2.759
#600	"	3.478
#800	"	3.478
Compression conductor adapters		
#6	EA.	0.296
#4	"	0.348
#2	"	0.444
#1	"	0.444
#1/0	"	0.533
#250	"	0.800
#350	"	0.851
#500	"	1.096
#750	"	1.143
Terminal blocks, 2 screw		
3 circuit	EA.	0.200
6 circuit	"	0.200
8 circuit	"	0.200
10 circuit	"	0.296
12 circuit	"	0.296
18 circuit	"	0.296

POWER & COMMUNICATIONS	UNIT	MAN/HOURS
02780.40 SUPPORTS & CONNECTORS		
24 circuit	EA.	0.348
36 circuit	"	0.348
Compression splice		
#8 awg	EA.	0.381
#6 awg	"	0.276
#4 awg	"	0.276
#2 awg	"	0.533
#1 awg	"	0.533
#1/0 awg	"	0.533
#2/0 awg	"	0.851
#3/0 awg	"	0.851
#4/0 awg	"	0.851
#250 awg	"	1.356
#300 awg	"	1.356
#350 awg	"	1.404
#400 awg	"	1.404
#500 awg	"	1.509
#600 awg	"	1.509
#750 awg	"	1.739
#1000 awg	"	1.739

SITE IMPROVEMENTS	UNIT	MAN/HOURS

FORMWORK

03110.20 ELEVATED SLAB FORMWORK

	UNIT	MAN/HOURS
Elevated slab formwork		
Slab, with drop panels		
1 use	S.F.	0.064
5 uses	"	0.055
Floor slab, hung from steel beams		
1 use	S.F.	0.062
5 uses	"	0.053
Floor slab, with pans or domes		
1 use	S.F.	0.073
5 uses	"	0.062
Equipment curbs, 12" high		
1 use	L.F.	0.080
5 uses	"	0.067

03110.25 EQUIPMENT PAD FORMWORK

	UNIT	MAN/HOURS
Equipment pad, job built		
1 use	S.F.	0.100
2 uses	"	0.094
3 uses	"	0.089
4 uses	"	0.084
5 uses	"	0.080

03110.35 FOOTING FORMWORK

	UNIT	MAN/HOURS
Wall footings, job built, continuous		
1 use	S.F.	0.080
5 uses	"	0.067
Column footings, spread		
1 use	S.F.	0.100
5 uses	"	0.080

03110.55 SLAB/MAT FORMWORK

	UNIT	MAN/HOURS
Edge forms		
6" high		
1 use	L.F.	0.073
5 uses	"	0.062
12" high		
1 use	L.F.	0.080
5 uses	"	0.067
Formwork for openings		
1 use	S.F.	0.160
5 uses	"	0.114

03110.65 WALL FORMWORK

	UNIT	MAN/HOURS
Wall forms, exterior, job built		
Up to 8' high wall		
1 use	S.F.	0.080
5 uses	"	0.067
Over 8' high wall		
1 use	S.F.	0.100
5 uses	"	0.080
Column pier and pilaster		
1 use	S.F.	0.160
5 uses	"	0.114

FORMWORK

03110.65 WALL FORMWORK

	UNIT	MAN/HOURS
Interior wall forms		
Up to 8' high		
1 use	S.F.	0.073
5 uses	"	0.062
Over 8' high		
1 use	S.F.	0.089
5 uses	"	0.073

REINFORCEMENT

03210.20 ELEVATED SLAB REINFORCING

	UNIT	MAN/HOURS
Elevated slab		
#3 - #4	TON	10.000
#5 - #6	"	8.889
Galvanized		
#3 - #4	TON	10.000
#5 - #6	"	8.889

03210.25 EQUIP. PAD REINFORCING

	UNIT	MAN/HOURS
Equipment pad		
#3 - #4	TON	16.000
#5 - #6	"	14.545
#7 - #8	"	13.333
#9 - #10	"	12.308
#11 - #12	"	11.429

03210.35 FOOTING REINFORCING

	UNIT	MAN/HOURS
Footings		
Grade 50		
#3 - #4	TON	13.333
#5 - #6	"	11.429
Grade 60		
#3 - #4	TON	13.333
#5 - #6	"	11.429

03210.55 SLAB/MAT REINFORCING

	UNIT	MAN/HOURS
Bars, slabs		
#3 - #4	TON	13.333
#5 - #6	"	11.429
#7 - #8	"	10.000
Galvanized		
#3 - #4	TON	13.333
#5 - #6	"	11.429

REINFORCEMENT

03210.65 — WALL REINFORCING

	UNIT	MAN/HOURS
Walls		
#3 - #4	TON	11.429
#5 - #6	"	10.000
Galvanized		
#3 - #4	TON	11.429
#5 - #6	"	10.000
Masonry wall (horizontal)		
#3 - #4	TON	32.000
#5 - #6	"	26.667
Galvanized		
#3 - #4	TON	32.000
#5 - #6	"	26.667
Masonry wall (vertical)		
#3 - #4	TON	40.000
#5 - #6	"	32.000
Galvanized		
#3 - #4	TON	40.000
#5 - #6	"	32.000

PLACING CONCRETE

03380.20 — ELEVATED SLAB CONCRETE

	UNIT	MAN/HOURS
Elevated slab		
2500# or 3000# concrete		
By crane	C.Y.	0.480
By pump	"	0.369
By hand buggy	"	0.800
5000# concrete		
By crane	C.Y.	0.480
By pump	"	0.369
By hand buggy	"	0.800

03380.25 — EQUIPMENT PAD CONCRETE

	UNIT	MAN/HOURS
Equipment pad		
2500# or 3000# concrete		
By chute	C.Y.	0.267
By pump	"	0.686
By crane	"	0.800
3500# or 4000# concrete		
By chute	C.Y.	0.267
By pump	"	0.686
By crane	"	0.800
5000# concrete		
By chute	C.Y.	0.267
By pump	"	0.686
By crane	"	0.800

PLACING CONCRETE

03380.35 — FOOTING CONCRETE

	UNIT	MAN/HOURS
Continuous footing		
2500# or 3000# concrete		
By chute	C.Y.	0.267
By pump	"	0.600
By crane	"	0.686
5000# concrete		
By chute	C.Y.	0.267
By pump	"	0.600
By crane	"	0.686
Spread footing		
2500# or 3000# concrete		
Under 5 cy		
By chute	C.Y.	0.267
By pump	"	0.640
By crane	"	0.738
5000# concrete		
Under 5 c.y.		
By chute	C.Y.	0.267
By pump	"	0.640
By crane	"	0.738

03380.55 — SLAB/MAT CONCRETE

	UNIT	MAN/HOURS
Slab on grade		
2500# or 3000# concrete		
By chute	C.Y.	0.200
By crane	"	0.400
By pump	"	0.343
By hand buggy	"	0.533
5000# concrete		
By chute	C.Y.	0.200
By crane	"	0.400
By pump	"	0.343
By hand buggy	"	0.533

03380.65 — WALL CONCRETE

	UNIT	MAN/HOURS
Walls		
2500# or 3000# concrete		
To 4'		
By chute	C.Y.	0.229
By crane	"	0.800
By pump	"	0.738
To 8'		
By crane	C.Y.	0.873
By pump	"	0.800
Filled block (CMU)		
3000# concrete, by pump		
4" wide	S.F.	0.034
6" wide	"	0.040
8" wide	"	0.048
10" wide	"	0.056
12" wide	"	0.069
Pilasters, 3000# concrete	C.F.	0.960
Wall cavity, 2" thick, 3000# concrete	S.F.	0.032

MORTAR AND GROUT

	UNIT	MAN/HOURS
04100.10 MASONRY GROUT		
Grout, non shrink, non-metallic, trowelable	C.F.	0.016
Grout-filled individual CMU cells		
4" wide	L.F.	0.012
8" wide	"	0.012
04150.50 MASONRY FLASHING		
Through-wall flashing		
5 oz. coated copper	S.F.	0.067
0.030" elastomeric	"	0.053

UNIT MASONRY

	UNIT	MAN/HOURS
04210.10 BRICK MASONRY		
Standard size brick, running bond		
Face brick, red (6.4/sf)		
Veneer	S.F.	0.133
Cavity wall	"	0.114
9" solid wall	"	0.229
Chimney, standard brick, including flue		
16" x 16"	L.F.	0.800
16" x 20"	"	0.800
16" x 24"	"	0.800
20" x 20"	"	1.000
20" x 24"	"	1.000
20" x 32"	"	1.143
04220.10 CONCRETE MASONRY UNITS		
Hollow, load bearing		
4"	S.F.	0.059
8"	"	0.067
Solid, load bearing		
4"	S.F.	0.059
8"	"	0.067
Back-up block, 8" x 16"		
4"	S.F.	0.047
8"	"	0.053
Steel angles and plates		
Minimum	Lb.	0.011
Maximum	"	0.020
Various size angle lintels		
1/4" stock		
3" x 3"	L.F.	0.050
3" x 3-1/2"	"	0.050
3/8" stock		
3" x 4"	L.F.	0.050
4" x 4"	"	0.050

UNIT MASONRY

	UNIT	MAN/HOURS
04220.10 CONCRETE MASONRY UNITS		
1/2" stock		
6" x 4"	L.F.	0.050
04295.10 PARGING/MASONRY PLASTER		
Parging		
1/2" thick	S.F.	0.053
3/4" thick	"	0.067

MASONRY RESTORATION

	UNIT	MAN/HOURS
04520.10 RESTORATION AND CLEANING		
Masonry cleaning		
Washing brick		
Smooth surface	S.F.	0.013
Rough surface	"	0.018
Pointing masonry		
Brick	S.F.	0.032
Concrete block	"	0.023
Brick removal and replacement		
Minimum	EA.	0.100
Average	"	0.133
Maximum	"	0.400
04550.10 REFRACTORIES		
Flue liners		
Rectangular		
8" x 12"	L.F.	0.133
12" x 12"	"	0.145
12" x 18"	"	0.160
16" x 16"	"	0.178
18" x 18"	"	0.190
20" x 20"	"	0.200
24" x 24"	"	0.229
Round		
18" dia.	L.F.	0.190
24" dia.	"	0.229

METAL FASTENING	UNIT	MAN/HOURS
05050.10 STRUCTURAL WELDING		
Welding		
Single pass		
1/8"	L.F.	0.040
3/16"	"	0.053
1/4"	"	0.067
Miscellaneous steel shapes		
Plain	Lb.	0.002
Galvanized	"	0.003
Plates		
Plain	Lb.	0.002
Galvanized	"	0.003
05050.95 METAL LINTELS		
Lintels, steel		
Plain	Lb.	0.020
Galvanized	"	0.020

COLD FORMED FRAMING	UNIT	MAN/HOURS
05410.10 METAL FRAMING		
Furring channel, galvanized		
Beams and columns, 3/4"		
12" o.c.	S.F.	0.080
16" o.c.	"	0.073
Walls, 3/4"		
12" o.c.	S.F.	0.040
16" o.c.	"	0.033
Stud, load bearing		
16" o.c.		
16 ga.		
2-1/2"	S.F.	0.036
3-5/8"	"	0.036
24" o.c.		
16 ga.		
2-1/2"	S.F.	0.031
3-5/8"	"	0.031

FASTENERS AND ADHESIVES

FASTENERS AND ADHESIVES	UNIT	MAN/HOURS
06050.10 **ACCESSORIES**		
Anchors		
Bolts, threaded two ends, with nuts and washers		
1/2" dia.		
4" long	E.A.	0.050
7-1/2" long	"	0.050
3/4" dia.		
7-1/2" long	E.A.	0.050
15" long	"	0.050
Bolts, carriage		
1/4 x 4	E.A.	0.080
5/16 x 6	"	0.084
3/8 x 6	"	0.084
1/2 x 6	"	0.084
Joist and beam hangers		
18 ga.		
2 x 4	E.A.	0.080
2 x 6	"	0.080
2 x 8	"	0.080
2 x 10	"	0.089
2 x 12	"	0.100
Strap ties, 14 ga., 1-3/8" wide		
12" long	E.A.	0.067
18" long	"	0.073
24" long	"	0.080
36" long	"	0.089

ROUGH CARPENTRY	UNIT	MAN/HOURS
06110.10 **BLOCKING**		
Steel construction		
Walls		
2x4	L.F.	0.053
2x6	"	0.062
2x8	"	0.067
2x10	"	0.073
2x12	"	0.080
Ceilings		
2x4	L.F.	0.062
2x6	"	0.073
2x8	"	0.080
2x10	"	0.089
2x12	"	0.100
Wood construction		
Walls		
2x4	L.F.	0.044
2x6	"	0.050

ROUGH CARPENTRY	UNIT	MAN/HOURS
06110.10 **BLOCKING**		
2x8	L.F.	0.053
2x10	"	0.057
2x12	"	0.062
Ceilings		
2x4	L.F.	0.050
2x6	"	0.057
2x8	"	0.062
2x10	"	0.067
2x12	"	0.073
06110.20 **CEILING FRAMING**		
Ceiling joists		
12" o.c.		
2x4	S.F.	0.019
2x8	"	0.021
16" o.c.		
2x4	S.F.	0.015
2x8	"	0.017
Sister joists for ceilings		
2x4	L.F.	0.057
2x6	"	0.067
2x8	"	0.080
2x10	"	0.100
2x12	"	0.133
06110.30 **FLOOR FRAMING**		
Floor joists		
12" o.c.		
2x6	S.F.	0.016
2x8	"	0.016
2x10	"	0.017
2x12	"	0.017
16" o.c.		
2x6	S.F.	0.013
2x8	"	0.014
2x10	"	0.014
2x12	"	0.014
Sister joists for floors		
2x4	L.F.	0.050
2x6	"	0.057
2x8	"	0.067
2x10	"	0.080
2x12	"	0.100
3x6	"	0.080
3x8	"	0.089
3x10	"	0.100
3x12	"	0.114
4x6	"	0.080
4x8	"	0.089
4x10	"	0.100
4x12	"	0.114

ROUGH CARPENTRY	UNIT	MAN/HOURS
06110.40 FURRING		
Furring, wood strips		
Walls		
On masonry or concrete walls		
1x2 furring		
12" o.c.	S.F.	0.025
16" o.c.	"	0.023
1x3 furring		
12" o.c.	S.F.	0.025
16" o.c.	"	0.023
On wood walls		
1x2 furring		
12" o.c.	S.F.	0.018
16" o.c.	"	0.016
1x3 furring		
12" o.c.	S.F.	0.018
16" o.c.	"	0.016
Ceilings		
On masonry or concrete ceilings		
1x2 furring		
12" o.c.	S.F.	0.044
16" o.c.	"	0.040
1x3 furring		
12" o.c.	S.F.	0.044
16" o.c.	"	0.040
On wood ceilings		
1x2 furring		
12" o.c.	S.F.	0.030
16" o.c.	"	0.027
1x3		
12" o.c.	S.F.	0.030
16" o.c.	"	0.027
06110.50 ROOF FRAMING		
Roof framing		
Rafters, gable end		
0-2 pitch (flat to 2-in-12)		
12" o.c.		
2x4	S.F.	0.017
2x6	"	0.017
2x8	"	0.018
16" o.c.		
2x6	S.F.	0.014
2x8	"	0.015
Sister rafters		
2x4	L.F.	0.057
2x6	"	0.067
2x8	"	0.080
2x10	"	0.100
2x12	"	0.133
06110.60 SLEEPERS		
Sleepers, over concrete		
12" o.c.		

ROUGH CARPENTRY	UNIT	MAN/HOURS
06110.60 SLEEPERS		
1x2	S.F.	0.018
1x3	"	0.019
2x4	"	0.022
2x6	"	0.024
16" o.c.		
1x2	S.F.	0.016
1x3	"	0.016
2x4	"	0.019
2x6	"	0.020
06110.65 SOFFITS		
Soffit framing		
2x3	L.F.	0.057
2x4	"	0.062
2x6	"	0.067
2x8	"	0.073
06110.70 WALL FRAMING		
Framing wall, studs		
12" o.c.		
2x3	S.F.	0.015
2x4	"	0.015
16" o.c.		
2x3	S.F.	0.013
2x4	"	0.013
24" o.c.		
2x3	S.F.	0.011
2x4	"	0.011
Plates, top or bottom		
2x3	L.F.	0.024
2x4	"	0.025
Headers, door or window		
2x6		
Single		
3' long	EA.	0.400
6' long	"	0.500
Double		
3' long	EA.	0.444
6' long	"	0.571
2x8		
Single		
4' long	EA.	0.500
8' long	"	0.615
Double		
4' long	EA.	0.571
8' long	"	0.727
2x10		
Single		
5' long	EA.	0.615
10' long	"	0.800
Double		
5' long	EA.	0.667

ROUGH CARPENTRY		UNIT	MAN/ HOURS
06110.70	**WALL FRAMING**		
10' long		EA.	0.800
06115.10	**FLOOR SHEATHING**		
Sub-flooring, plywood, CDX			
1/2" thick		S.F.	0.010
5/8" thick		"	0.011
3/4" thick		"	0.013
Structural plywood			
1/2" thick		S.F.	0.010
Board type subflooring			
1x6			
Minimum		S.F.	0.018
Maximum		"	0.020
1x8			
Minimum		S.F.	0.017
Maximum		"	0.019
1x10			
Minimum		S.F.	0.016
Maximum		"	0.018
Underlayment			
Hardboard, 1/4" tempered		S.F.	0.010
Plywood, CDX			
3/8" thick		S.F.	0.010
1/2" thick		"	0.011
5/8" thick		"	0.011
3/4" thick		"	0.012
06115.20	**ROOF SHEATHING**		
Sheathing			
Plywood, CDX			
3/8" thick		S.F.	0.010
1/2" thick		"	0.011
5/8" thick		"	0.011
06115.30	**WALL SHEATHING**		
Sheathing			
Plywood, CDX			
3/8" thick		S.F.	0.012
1/2" thick		"	0.012
5/8" thick		"	0.013
3/4" thick		"	0.015
06220.10	**MILLWORK**		
Countertop, laminated plastic			
25" x 7/8" thick			
Minimum		L.F.	0.200
Average		"	0.267
Maximum		"	0.320
25" x 1-1/4" thick			
Minimum		L.F.	0.267

ROUGH CARPENTRY		UNIT	MAN/ HOURS
06220.10	**MILLWORK**		
Average		L.F.	0.320
Maximum		"	0.400
Add for cutouts		EA.	0.500
Backsplash, 4" high, 7/8" thick		L.F.	0.160
Base cabinets, 34-1/2" high, 24" deep, hardwood, no tops			
Minimum		L.F.	0.320
Average		"	0.400
Maximum		"	0.533
Wall cabinets			
Minimum		L.F.	0.267
Average		"	0.320
Maximum		"	0.400

MOISTURE PROTECTION	UNIT	MAN/HOURS
07150.10 **DAMPROOFING**		
Silicone dampproofing, sprayed on		
Concrete surface		
1 coat	S.F.	0.004
2 coats	"	0.006
Concrete block		
1 coat	S.F.	0.005
2 coats	"	0.007
Brick		
1 coat	S.F.	0.006
2 coats	"	0.008
07160.10 **BITUMINOUS DAMPPROOFING**		
Building paper, asphalt felt		
15 lb	S.F.	0.032
30 lb	"	0.033
07190.10 **VAPOR BARRIERS**		
Vapor barrier, polyethylene		
2 mil	S.F.	0.004
6 mil	"	0.004
8 mil	"	0.004
10 mil	"	0.004

INSULATION	UNIT	MAN/HOURS
07210.10 **BATT INSULATION**		
Ceiling, fiberglass, unfaced		
3-1/2" thick, R11	S.F.	0.009
6" thick, R19	"	0.011
9" thick, R30	"	0.012
Suspended ceiling, unfaced		
3-1/2" thick, R11	S.F.	0.009
6" thick, R19	"	0.010
9" thick, R30	"	0.011
Crawl space, unfaced		
3-1/2" thick, R11	S.F.	0.012
6" thick, R19	"	0.013
9" thick, R30	"	0.015
Wall, fiberglass		
Paper backed		
2" thick, R7	S.F.	0.008
3" thick, R8	"	0.009
4" thick, R11	"	0.009
6" thick, R19	"	0.010
Foil backed, 1 side		

INSULATION	UNIT	MAN/HOURS
07210.10 **BATT INSULATION**		
2" thick, R7	S.F.	0.008
3" thick, R11	"	0.009
4" thick, R14	"	0.009
6" thick, R21	"	0.010
Foil backed, 2 sides		
2" thick, R7	S.F.	0.009
3" thick, R11	"	0.010
4" thick, R14	"	0.011
6" thick, R21	"	0.011
Unfaced		
2" thick, R7	S.F.	0.008
3" thick, R9	"	0.009
4" thick, R11	"	0.009
6" thick, R19	"	0.010
Mineral wool batts		
Paper backed		
2" thick, R6	S.F.	0.008
4" thick, R12	"	0.009
6" thick, R19	"	0.010
Fasteners, self adhering, attached to ceiling deck		
2-1/2" long	EA.	0.013
4-1/2" long	"	0.015
Capped, self-locking washers for fastening insulation	"	0.008
07210.20 **BOARD INSULATION**		
Insulation, rigid		
Fiberglass, roof		
0.75" thick, R2.78	S.F.	0.007
1.06" thick, R4.17	"	0.008
1.31" thick, R5.26	"	0.008
1.63" thick, R6.67	"	0.008
2.25" thick, R8.33	"	0.009
Perlite board, roof		
1.00" thick, R2.78	S.F.	0.007
1.50" thick, R4.17	"	0.007
2.00" thick, R5.92	"	0.007
2.50" thick, R6.67	"	0.008
3.00" thick, R8.33	"	0.008
4.00" thick, R10.00	"	0.008
5.25" thick, R14.29	"	0.009
07210.60 **LOOSE FILL INSULATION**		
Blown-in type		
Fiberglass		
5" thick, R11	S.F.	0.007
6" thick, R13	"	0.008
9" thick, R19	"	0.011
Rockwool, attic application		
6" thick, R13	S.F.	0.008
8" thick, R19	"	0.010
10" thick, R22	"	0.012
12" thick, R26	"	0.013

INSULATION

	UNIT	MAN/HOURS
07210.60 LOOSE FILL INSULATION		
15" thick, R30	S.F.	0.016
Poured type		
Fiberglass		
1" thick, R4	S.F.	0.005
2" thick, R8	"	0.006
3" thick, R12	"	0.007
4" thick, R16	"	0.008
Mineral wool		
1" thick, R3	S.F.	0.005
2" thick, R6	"	0.006
3" thick, R9	"	0.007
4" thick, R12	"	0.008
Vermiculite or perlite		
2" thick, R4.8	S.F.	0.006
3" thick, R7.2	"	0.007
4" thick, R9.6	"	0.008
Masonry, poured vermiculite or perlite		
4" block	S.F.	0.004
6" block	"	0.005
8" block	"	0.006
10" block	"	0.006
12" block	"	0.007
07210.70 SPRAYED INSULATION		
Foam, sprayed on		
Polystyrene		
1" thick, R4	S.F.	0.008
2" thick, R8	"	0.011
Urethane		
1" thick, R7.7	S.F.	0.008
2" thick, R15.4	"	0.011
07250.10 FIREPROOFING		
Sprayed on		
1" thick		
On beams	S.F.	0.018
On columns	"	0.016
On decks		
Flat surface	S.F.	0.008
Fluted surface	"	0.010
1-1/2" thick		
On beams	S.F.	0.023
On columns	"	0.020
On decks		
Flat surface	S.F.	0.010
Fluted surface	"	0.013

SHINGLES AND TILES

	UNIT	MAN/HOURS
07310.10 ASPHALT SHINGLES		
Standard asphalt shingles, strip shingles		
210 lb/square	SQ.	0.800
235 lb/square	"	0.889
240 lb/square	"	1.000
260 lb/square	"	1.143
300 lb/square	"	1.333
385 lb/square	"	1.600
Roll roofing, mineral surface		
90 lb	SQ.	0.571
110 lb	"	0.667
140 lb	"	0.800
07310.30 METAL SHINGLES		
Aluminum, .020" thick		
Plain	SQ.	1.600
Colors	"	1.600
Steel, galvanized		
26 ga.		
Plain	SQ.	1.600
Colors	"	1.600
Porcelain enamel, 22 ga.		
Minimum	SQ.	2.000
Average	"	2.000
Maximum	"	2.000
Replacement shingles		
Small jobs	EA.	0.267
Large jobs	S.F.	0.133
07310.70 WOOD SHINGLES		
Wood shingles, on roofs		
White cedar, #1 shingles		
4" exposure	SQ.	2.667
5" exposure	"	2.000
On walls		
White cedar, #1 shingles		
4" exposure	SQ.	4.000
5" exposure	"	3.200
07310.80 WOOD SHAKES		
Shakes, hand split, 24" red cedar, on roofs		
5" exposure	SQ.	4.000
9" exposure	"	2.667

MEMBRANE ROOFING	UNIT	MAN/HOURS
07510.10 BUILT-UP ASPHALT ROOFING		
Built-up roofing, asphalt felt, including gravel		
2 ply	SQ.	2.000
3 ply	"	2.667
4 ply	"	3.200
07530.10 SINGLE-PLY ROOFING		
Elastic sheet roofing		
Neoprene, 1/16" thick	S.F.	0.010
EPDM rubber		
45 mil	S.F.	0.010
60 mil	"	0.010
PVC		
45 mil	S.F.	0.010
60 mil	"	0.010
Flashing		
Pipe flashing, 90 mil thick		
1" pipe	EA.	0.200
2" pipe	"	0.200
3" pipe	"	0.211
4" pipe	"	0.211
5" pipe	"	0.222
6" pipe	"	0.222
8" pipe	"	0.235
10" pipe	"	0.267
12" pipe	"	0.267
Neoprene flashing, 60 mil thick strip		
6" wide	L.F.	0.067
12" wide	"	0.100
18" wide	"	0.133
24" wide	"	0.200
Adhesives		
Mastic sealer, applied at joints only		
1/4" bead	L.F.	0.004
Ballast, 3/4" through 1-1/2" dia. river gravel, 100lb/sf	S.F.	0.800
Walkway for membrane roofs, 1/2" thick	"	0.027

FLASHING AND SHEET METAL	UNIT	MAN/HOURS
07610.10 METAL ROOFING		
Sheet metal roofing, copper, 16 oz, batten seam	SQ.	5.333
Standing seam	"	5.000
Aluminum roofing, natural finish		
Corrugated, on steel frame		
.0175" thick	SQ.	2.286
.032" thick	"	2.286
Corrugated galvanized steel roofing, on steel frame		

FLASHING AND SHEET METAL	UNIT	MAN/HOURS
07610.10 METAL ROOFING		
28 ga.	SQ.	2.286
22 ga.	"	2.286
07620.10 FLASHING AND TRIM		
Counter flashing		
Aluminum, .032"	S.F.	0.080
Stainless steel, .015"	"	0.080
16 oz.	"	0.080
Base flashing		
Aluminum, .040"	S.F.	0.067
Stainless steel, .018"	"	0.067
16 oz.	"	0.067
24 oz.	"	0.067
Scupper outlets		
10" x 10" x 4"	EA.	0.200
22" x 4" x 4"	"	0.200
8" x 8" x 5"	"	0.200
Drainage boots, roof, cast iron		
2 x 3	L.F.	0.100
3 x 4	"	0.100
4 x 5	"	0.107
4 x 6	"	0.107
5 x 7	"	0.114
Pitch pocket, copper, 16 oz.		
4 x 4	EA.	0.200
6 x 6	"	0.200
8 x 8	"	0.200
8 x 10	"	0.200
8 x 12	"	0.200
Reglets, copper 10 oz.	L.F.	0.053
Stainless steel, .020"	"	0.053
07620.20 GUTTERS AND DOWNSPOUTS		
Copper gutter and downspout		
Downspouts, 16 oz. copper		
Round		
3" dia.	L.F.	0.053
4" dia.	"	0.053
Rectangular, corrugated		
2" x 3"	L.F.	0.050
3" x 4"	"	0.050
Rectangular, flat surface		
2" x 3"	L.F.	0.053
3" x 4"	"	0.053
Lead-coated copper downspouts		
Round		
3" dia.	L.F.	0.050
4" dia.	"	0.057
Rectangular, corrugated		
2" x 3"	L.F.	0.053
3" x 4"	"	0.053
Rectangular, plain		
2" x 3"	L.F.	0.053

FLASHING AND SHEET METAL

07620.20 GUTTERS AND DOWNSPOUTS

	UNIT	MAN/HOURS
3" x 4"	L.F.	0.053
Gutters, 16 oz. copper		
Half round		
4" wide	L.F.	0.080
5" wide	"	0.089
Type K		
4" wide	L.F.	0.080
5" wide	"	0.089
Lead-coated copper gutters		
Half round		
4" wide	L.F.	0.080
6" wide	"	0.089
Type K		
4" wide	L.F.	0.080
5" wide	"	0.089
Aluminum gutter and downspout		
Downspouts		
2" x 3"	L.F.	0.053
3" x 4"	"	0.057
4" x 5"	"	0.062
Round		
3" dia.	L.F.	0.053
4" dia.	"	0.057
Gutters, stock units		
4" wide	L.F.	0.084
5" wide	"	0.089
Galvanized steel gutter and downspout		
Downspouts, round corrugated		
3" dia.	L.F.	0.053
4" dia.	"	0.053
5" dia.	"	0.057
6" dia.	"	0.057
Rectangular		
2" x 3"	L.F.	0.053
3" x 4"	"	0.050
4" x 4"	"	0.050
Gutters, stock units		
5" wide		
Plain	L.F.	0.089
Painted	"	0.089
6" wide		
Plain	L.F.	0.094
Painted	"	0.094

07700.10 MANUFACTURED SPECIALTIES

	UNIT	MAN/HOURS
Ceiling access doors		
Swing up model, metal frame		
Steel door		
2'6" x 2'6"	EA.	0.800
2'6" x 3'0"	"	0.800
Aluminum door		
2'6" x 2'6"	EA.	0.800
2'6" x 3'0"	"	0.800

FLASHING AND SHEET METAL

07700.10 MANUFACTURED SPECIALTIES

	UNIT	MAN/HOURS
Swing down model, metal frame		
Steel door		
2'6" x 2'6"	EA.	0.800
2'6" x 3'0"	"	0.800
Aluminum door		
2'6" x 2'6"	EA.	0.800
2'6" x 3'0"	"	0.800
Gravity ventilators, with curb, base, damper and screen		
Stationary siphon		
6" dia.	EA.	0.533
12" dia.	"	0.533
24" dia.	"	0.800
36" dia.	"	0.800
Wind driven spinner		
6" dia.	EA.	0.533
12" dia.	"	0.533
24" dia.	"	0.800
36" dia.	"	0.800
Stationary mushroom		
16" dia.	EA.	0.800
30" dia.	"	1.000
36" dia.	"	1.333
42" dia.	"	1.600

JOINT SEALERS

07920.10 CAULKING

	UNIT	MAN/HOURS
Caulk exterior, two component		
1/4 x 1/2	L.F.	0.040
3/8 x 1/2	"	0.044
1/2 x 1/2	"	0.050
Caulk interior, single component		
1/4 x 1/2	L.F.	0.038
3/8 x 1/2	"	0.042
1/2 x 1/2	"	0.047

SPECIALTIES

SPECIALTIES	UNIT	MAN/HOURS
10165.10 TOILET PARTITIONS		
Toilet partition, plastic laminate		
Ceiling mounted	EA.	2.667
Floor mounted	"	2.000
Metal		
Ceiling mounted	EA.	2.667
Floor mounted	"	2.000
Wheel chair partition, plastic laminate		
Ceiling mounted	EA.	2.667
Floor mounted	"	2.000
Painted metal		
Ceiling mounted	EA.	2.667
Floor mounted	"	2.000
Urinal screen, plastic laminate		
Wall hung	EA.	1.000
Floor mounted	"	1.000
Porcelain enameled steel, floor mounted	"	1.000
Painted metal, floor mounted	"	1.000
Stainless steel, floor mounted	"	1.000
Metal toilet partitions		
Toilet partitions, front door and side divider, floor mounted		
Porcelain enameled steel	EA.	2.000
Painted steel	"	2.000
Stainless steel	"	2.000
10185.10 SHOWER STALLS		
Shower receptors		
Precast, terrazzo		
32" x 32"	EA.	0.667
32" x 48"	"	0.800
Concrete		
32" x 32"	EA.	0.667
48" x 48"	"	0.889
Shower door, trim and hardware		
Economy, 24" wide, chrome frame, tempered glass	EA.	0.800
Porcelain enameled steel, flush	"	0.800
Baked enameled steel, flush	"	0.800
Aluminum frame, tempered glass, 48" wide, sliding	"	1.000
Folding	"	1.000
Aluminum frame and tempered glass, molded plastic		
Complete with receptor and door		
32" x 32"	EA.	2.000
36" x 36"	"	2.000
40" x 40"	"	2.286
Shower compartment, precast concrete receptor		
Single entry type		
Porcelain enameled steel	EA.	8.000
Baked enameled steel	"	8.000
Stainless steel	"	8.000
Double entry type		
Porcelain enameled steel	EA.	10.000
Baked enameled steel	"	10.000
Stainless steel	"	10.000

SPECIALTIES	UNIT	MAN/HOURS
10270.40 ACCESS & PEDESTAL FLOOR		
Panels, no covering, 2'x2'		
Plain	S.F.	0.010
Perforated	"	0.400
Pedestals		
For 6" to 12" clearance	EA.	0.080
Stringers		
2'	L.F.	0.038
6'	"	0.027
Accessories		
Ramp assembly	S.F.	0.032
Elevated floor assembly	"	0.030
Handrail	L.F.	0.400
Fascia plate	"	0.200
RF shielding components, floor liner		
Hot rolled steel sheet		
14 ga.	S.F.	0.020
11 ga.	"	0.062
10290.10 PEST CONTROL		
Termite control		
Under slab spraying		
Minimum	S.F.	0.002
Average	"	0.004
Maximum	"	0.008
10500.10 LOCKERS		
Locker bench, floor mounted, laminated maple		
4'	EA.	0.667
6'	"	0.667
Wardrobe locker, single tier type		
12" x 15" x 72"	EA.	0.800
18" x 15" x 72"	"	0.842
12" x 18" x 72"	"	0.889
18" x 18" x 72"	"	0.941
10520.10 FIRE PROTECTION		
Portable fire extinguishers		
Water pump tank type		
2.5 gal.		
Red enameled galvanized	EA.	0.533
Red enameled copper	"	0.533
Polished copper	"	0.533
Carbon dioxide type, red enamel steel		
Squeeze grip with hose and horn		
2.5 lb	EA.	0.533
5 lb	"	0.615
10 lb	"	0.800
15 lb	"	1.000
20 lb	"	1.000
Wheeled type		
125 lb	EA.	1.600
250 lb	"	1.600

SPECIALTIES

SPECIALTIES	UNIT	MAN/HOURS
10520.10 FIRE PROTECTION		
500 lb	EA.	1.600
Dry chemical, pressurized type		
Red enameled steel		
2.5 lb	EA.	0.533
5 lb	"	0.615
10 lb	"	0.800
20 lb	"	1.000
30 lb	"	1.000
Chrome plated steel, 2.5 lb	"	0.533
Other type extinguishers		
2.5 gal, stainless steel, pressurized water tanks	EA.	0.533
Soda and acid type	"	0.533
Cartridge operated, water type	"	0.533
Loaded stream, water type	"	0.533
Foam type	"	0.533
40 gal, wheeled foam type	"	1.600
Fire extinguisher cabinets		
Enameled steel		
8" x 12" x 27"	EA.	1.600
8" x 16" x 38"	"	1.600
Aluminum		
8" x 12" x 27"	EA.	1.600
8" x 16" x 38"	"	1.600
8" x 12" x 27"	"	1.600
Stainless steel		
8" x 16" x 38"	EA.	1.600
10670.10 SHELVING		
Metal storage shelving, baked enamel		
7 shelf unit, 72" or 84" high		
10" shelf	L.F.	0.800
18" shelf	"	0.941
4 shelf unit, 40" high		
10" shelf	L.F.	0.667
18" shelf	"	0.842
3 shelf unit, 32" high		
10" shelf	L.F.	0.400
18" shelf	"	0.471
Single shelf unit, attached to masonry		
10" shelf	L.F.	0.133
18" shelf	"	0.163
10750.10 TELEPHONE ENCLOSURES		
Telephone enclosure, wall mounted, shelf, 28" x 30" x 15"	EA.	2.000
Directory shelf, stainless steel, 3 binders	"	1.333
10800.10 BATH ACCESSORIES		
Ash receiver, wall mounted, aluminum	EA.	0.400
Grab bar, 1-1/2" dia., stainless steel, wall mounted		
24" long	EA.	0.400
36" long	"	0.421
42" long	"	0.444
48" long	"	0.471

SPECIALTIES	UNIT	MAN/HOURS
10800.10 BATH ACCESSORIES		
52" long	EA.	0.500
1" dia., stainless steel		
12" long	EA.	0.348
18" long	"	0.364
24" long	"	0.400
30" long	"	0.421
36" long	"	0.444
48" long	"	0.471
Hand dryer, surface mounted, 110 volt	"	1.000
Medicine cabinet, 16 x 22, baked enamel, steel, lighted	"	0.320
With mirror, lighted	"	0.533
Mirror, 1/4" plate glass, up to 10 sf	S.F.	0.080
Mirror, stainless steel frame		
18"x24"	EA.	0.267
18"x32"	"	0.320
18"x36"	"	0.400
24"x30"	"	0.400
24"x36"	"	0.444
24"x48"	"	0.667
24"x60"	"	0.800
30"x30"	"	0.800
30"x72"	"	1.000
48"x72"	"	1.333
With shelf, 18"x24"	"	0.320
Sanitary napkin dispenser, stainless steel, wall mounted	"	0.533
Shower rod, 1" diameter		
Chrome finish over brass	EA.	0.400
Stainless steel	"	0.400
Soap dish, stainless steel, wall mounted	"	0.533
Toilet tissue dispenser, stainless, wall mounted		
Single roll	EA.	0.200
Double roll	"	0.229
Towel dispenser, stainless steel		
Flush mounted	EA.	0.444
Surface mounted	"	0.400
Combination towel dispenser and waste receptacle	"	0.533
Towel bar, stainless steel		
18" long	EA.	0.320
24" long	"	0.364
30" long	"	0.400
36" long	"	0.444
Toothbrush and tumbler holder	"	0.267
Waste receptacle, stainless steel, wall mounted	"	0.667

ARCHITECTURAL EQUIPMENT

11010.10 MAINTENANCE EQUIPMENT

	UNIT	MAN/HOURS
Vacuum cleaning system		
3 valves		
1.5 hp	EA.	8.889
2.5 hp	"	11.429
5 valves	"	16.000
7 valves	"	20.000

11060.10 THEATER EQUIPMENT

	UNIT	MAN/HOURS
Roll out stage, steel frame, wood floor		
Manual	S.F.	0.050
Electric	"	0.080
Portable stages		
8" high	S.F.	0.040
18" high	"	0.044
36" high	"	0.047
48" high	"	0.050
Band risers		
Minimum	S.F.	0.040
Maximum	"	0.040
Chairs for risers		
Minimum	EA.	0.036
Maximum	"	0.036

11080.10 POLICE EQUIPMENT

	UNIT	MAN/HOURS
Firing range equipment, rifle		
3 position	EA.	26.667
4 position	"	40.000
5 position	"	44.444
6 position	"	47.059

11090.10 CHECKROOM EQUIPMENT

	UNIT	MAN/HOURS
Motorized checkroom equipment		
No shelf system, 6'4" height		
7'6" length	EA.	8.000
14'6" length	"	8.000
28' length	"	8.000
One shelf, 6'8" height		
7'6" length	EA.	8.000
14'6" length	"	8.000
28' length	"	8.000
Two shelves, 7'5" height		
7'6" length	EA.	8.000
14'6" length	"	8.000
28' length	"	8.000
Three shelves, 8' height		
7'6" length	EA.	16.000
14'6" length	"	16.000
28' length	"	16.000
Four shelves, 8'7" height		
7'6" length	EA.	16.000
14'6" length	"	16.000
28' length	"	16.000
200 lb		

ARCHITECTURAL EQUIPMENT

11110.10 LAUNDRY EQUIPMENT

	UNIT	MAN/HOURS
High capacity, heavy duty		
Washer extractors		
135 lb		
Standard	EA.	6.667
Pass through	"	6.667
200 lb		
Standard	EA.	6.667
Pass through	"	6.667
110 lb dryer	"	6.667
Hand operated presser	"	8.889
Mushroom press	"	8.889
Spreader feeders		
2 station	EA.	8.889
4 station	"	16.000
Delivery carts		
12 bushel	EA.	0.100
Low capacity		
Pressers		
Air operated	EA.	3.200
Hand operated	"	3.200
Extractor, low capacity	"	3.200
Ironer, 48"	"	1.600
Coin washers		
10 lb capacity	EA.	1.600
20 lb capacity	"	1.600
Coin dryer	"	1.000
Coin dry cleaner, 20 lb	"	3.200

11161.10 LOADING DOCK EQUIPMENT

	UNIT	MAN/HOURS
Dock leveler, 10 ton capacity		
6' x 8'	EA.	8.000
7' x 8'	"	8.000

11170.10 WASTE HANDLING

	UNIT	MAN/HOURS
Incinerator, electric		
100 lb/hr		
Minimum	EA.	8.000
Maximum	"	8.000
400 lb/hr		
Minimum	EA.	16.000
Maximum	"	16.000
1000 lb/hr		
Minimum	EA.	24.242
Maximum	"	24.242
Incinerator, medical-waste		
25 lb/hr, 2-7 x 4-0	EA.	16.000
50 lb/hr, 2-11 x 4-11	"	16.000
75 lb/hr, 3-8 x 5-0	"	32.000
100 lb/hr, 3-8 x 6-0	"	32.000
Industrial compactor		
1 cy	EA.	8.889
3 cy	"	11.429

ARCHITECTURAL EQUIPMENT	UNIT	MAN/ HOURS
11170.10 WASTE HANDLING		
5 cy	EA.	16.000
11400.10 FOOD SERVICE EQUIPMENT		
Unit kitchens		
30" compact kitchen		
Refrigerator, with range, sink	EA.	4.000
Sink only	"	2.667
Range only	"	2.000
Cabinet for upper wall section	"	1.143
Stainless shield, for rear wall	"	0.320
Side wall	"	0.320
42" compact kitchen		
Refrigerator with range, sink	EA.	4.444
Sink only	"	4.000
Cabinet for upper wall section	"	1.333
Stainless shield, for rear wall	"	0.333
Side wall	"	0.333
54" compact kitchen		
Refrigerator, oven, range, sink	EA.	5.714
Cabinet for upper wall section	"	1.600
Stainless shield, for		
Rear wall	EA.	0.364
Side wall	"	0.364
60" compact kitchen		
Refrigerator, oven, range, sink	EA.	5.714
Cabinet for upper wall section	"	1.600
Stainless shield, for		
Rear wall	EA.	0.364
Side wall	"	0.364
72" compact kitchen		
Refrigerator, oven, range, sink	EA.	6.667
Cabinet for upper wall section	"	1.600
Stainless shield for		
Rear wall	EA.	0.400
Side wall	"	0.400
Bake oven		
Single deck		
Minimum	EA.	1.000
Maximum	"	2.000
Double deck		
Minimum	EA.	1.333
Maximum	"	2.000
Triple deck		
Minimum	EA.	1.333
Maximum	"	2.667
Convection type oven, electric, 40" x 45" x 57"		
Minimum	EA.	1.000
Maximum	"	2.000
Broiler, without oven, 69" x 26" x 39"		
Minimum	EA.	1.000
Maximum	"	1.333
Coffee urns, 10 gallons		

ARCHITECTURAL EQUIPMENT	UNIT	MAN/ HOURS
11400.10 FOOD SERVICE EQUIPMENT		
Minimum	EA.	2.667
Maximum	"	4.000
Fryer, with submerger		
Single		
Minimum	EA.	1.600
Maximum	"	2.667
Double		
Minimum	EA.	2.000
Maximum	"	2.667
Griddle, counter		
3' long		
Minimum	EA.	1.333
Maximum	"	1.600
5' long		
Minimum	EA.	2.000
Maximum	"	2.667
Kettles, steam, jacketed		
20 gallons		
Minimum	EA.	2.000
Maximum	"	4.000
40 gallons		
Minimum	EA.	2.000
Maximum	"	4.000
60 gallons		
Minimum	EA.	2.000
Maximum	"	4.000
Range		
Heavy duty, single oven, open top		
Minimum	EA.	1.000
Maximum	"	2.667
Fry top		
Minimum	EA.	1.000
Maximum	"	2.667
Steamers, electric		
27 kw		
Minimum	EA.	2.000
Maximum	"	2.667
18 kw		
Minimum	EA.	2.000
Maximum	"	2.667
Dishwasher, rack type		
Single tank, 190 racks/hr	EA.	4.000
Double tank		
234 racks/hr	EA.	4.444
265 racks/hr	"	5.333
Dishwasher, automatic 100 meals/hr	"	2.667
Disposals		
100 gal/hr	EA.	2.667
120 gal/hr	"	2.759
250 gal/hr	"	2.857
Exhaust hood for dishwasher, gutter 4 sides, s-steel		
4'x4'x2'	EA.	2.963
4'x7'x2'	"	3.200
Food preparation machines		

ARCHITECTURAL EQUIPMENT	UNIT	MAN/HOURS
11400.10 FOOD SERVICE EQUIPMENT		
Vertical cutter mixers		
25 quart	EA.	2.667
40 quart	"	2.667
80 quart	"	4.000
130 quart	"	6.667
Choppers		
5 lb	EA.	2.000
16 lb	"	2.667
40 lb	"	4.000
Mixers, floor models		
20 quart	EA.	1.000
60 quart	"	1.000
80 quart	"	1.143
140 quart	"	1.600
Ice cube maker		
50 lb per day		
Minimum	EA.	8.000
Maximum	"	8.000
500 lb per day		
Minimum	EA.	13.333
Maximum	"	13.333
Ice flakers		
300 lb per day	EA.	8.000
600 lb per day	"	13.333
1000 lb per day	"	17.778
2000 lb per day	"	20.000
Refrigerated cases		
Dairy products		
Multi deck type	L.F.	0.533
Delicatessen case, service deli		
Single deck	L.F.	4.000
Multi deck	"	5.000
Meat case		
Single deck	L.F.	4.706
Multi deck	"	5.000
Produce case		
Single deck	L.F.	4.706
Multi deck	"	5.000
Bottle coolers		
6' long		
Minimum	EA.	16.000
Maximum	"	16.000
10' long		
Minimum	EA.	26.667
Maximum	"	26.667
Frozen food cases		
Chest type	L.F.	4.706
Reach-in, glass door	"	5.000
Island case, single	"	4.706
Multi deck	"	5.000
Ice storage bins		
500 lb capacity	EA.	11.429
1000 lb capacity	"	22.857

ARCHITECTURAL EQUIPMENT	UNIT	MAN/HOURS
11450.10 RESIDENTIAL EQUIPMENT		
Compactor, 4 to 1 compaction	"	2.000
Dishwasher, built-in		
2 cycles	EA.	4.000
4 or more cycles	"	4.000
Disposal		
Garbage disposer	EA.	2.667
Heaters, electric, built-in		
Ceiling type	EA.	2.667
Wall type		
Minimum	EA.	2.000
Maximum	"	2.667
Hood for range, 2-speed, vented		
30" wide	EA.	2.667
42" wide	"	2.667
Ice maker, automatic		
30 lb per day	EA.	1.143
50 lb per day	"	4.000
Ranges electric		
Built-in, 30", 1 oven	EA.	2.667
2 oven	"	2.667
Counter top, 4 burner, standard	"	2.000
With grill	"	2.000
Free standing, 21", 1 oven	"	2.667
30", 1 oven	"	1.600
2 oven	"	1.600
Water softener		
30 grains per gallon	EA.	2.667
70 grains per gallon	"	4.000
11470.10 DARKROOM EQUIPMENT		
Dryers		
36" x 25" x 68"	EA.	4.000
48" x 25" x 68"	"	4.000
Processors, film		
Black and white	EA.	4.000
Color negatives	"	4.000
Prints	"	4.000
Transparencies	"	4.000
Sinks with cabinet and/or stand		
5" sink with stand		
24" x 48"	EA.	2.000
32" x 64"	"	2.667
38" x 52"	"	2.667
42" x 132"	"	4.000
48" x 52"	"	4.000
5" sink with cabinet		
24" x 48"	EA.	2.000
32" x 64"	"	2.667
38" x 52"	"	2.667
42" x 132"	"	4.000
48" x 52"	"	4.000
10" sink with stand		
24" x 48"	EA.	2.000
32" x 64"	"	2.667

ARCHITECTURAL EQUIPMENT

11470.10 DARKROOM EQUIPMENT

Item	UNIT	MAN/HOURS
38" x 52"	EA.	2.667
10" sink with cabinet		
24" x 48"	EA.	2.000
38" x 52"	"	2.667

11500.10 INDUSTRIAL EQUIPMENT

Item	UNIT	MAN/HOURS
Vehicular paint spray booth, solid back, 14'4" x 9'6"		
24' deep	EA.	8.000
26'6" deep	"	8.000
28'6" deep	"	8.000
Drive through, 14'9" x 9'6"		
24' deep	EA.	8.000
26'6" deep	"	8.000
28'6" deep	"	8.000
Water wash, paint spray booth		
5' x 11'2" x 10'8"	EA.	8.000
6' x 11'2" x 10'8"	"	8.000
8' x 11'2" x 10'8"	"	8.000
10' x 11'2" x 11'2"	"	8.000
12' x 12'2" x 11'2"	"	8.000
14' x 12'2" x 11'2"	"	8.000
16' x 12'2" x 11'2"	"	8.000
20' x 12'2" x 11'2"	"	8.000
Dry type spray booth, with paint arrestors		
5'4" x 7'2" x 6'8"	EA.	8.000
6'4" x 7'2" x 6'8"	"	8.000
8'4" x 7'2" x 9'2"	"	8.000
10'4" x 7'2" x 9'2"	"	8.000
12'4" x 7'6" x 9'2"	"	8.000
14'4" x 7'6" x 9'8"	"	8.000
16'4" x 7'7" x 9'8"	"	8.000
20'4" x 7'7" x 10'8"	"	8.000
Air compressor, electric		
1 hp		
115 volt	EA.	5.333
5 hp		
115 volt	EA.	8.000
230 volt	"	8.000
Hydraulic lifts		
8,000 lb capacity	EA.	20.000
11,000 lb capacity	"	32.000
24,000 lb capacity	"	53.333
Power tools		
Band saws		
10"	EA.	0.667
14"	"	0.800
Motorized shaper	"	0.615
Motorized lathe	"	0.667
Bench saws		
9" saw	EA.	0.533
10" saw	"	0.571
12" saw	"	0.667
Electric grinders		
1/3 hp	EA.	0.320

ARCHITECTURAL EQUIPMENT

11500.10 INDUSTRIAL EQUIPMENT

Item	UNIT	MAN/HOURS
1/2 hp	EA.	0.348
3/4 hp	"	0.348

11600.10 LABORATORY EQUIPMENT

Item	UNIT	MAN/HOURS
Cabinets, base		
Minimum	L.F.	0.667
Maximum	"	0.667
Full storage, 7' high		
Minimum	L.F.	0.667
Maximum	"	0.667
Wall		
Minimum	L.F.	0.800
Maximum	"	0.800
Counter tops		
Minimum	S.F.	0.100
Average	"	0.114
Maximum	"	0.133
Tables		
Open underneath	S.F.	0.400
Doors underneath	"	0.500
Medical laboratory equipment		
Analyzer		
Chloride	EA.	0.400
Blood	"	0.667
Bath, water, utility, countertop unit	"	0.800
Hot plate, lab, countertop	"	0.727
Stirrer	"	0.727
Incubator, anaerobic, 23x23x36"	"	4.000
Dry heat bath	"	1.333
Incinerator, for sterilizing	"	0.080
Meter, serum protein	"	0.100
Ph analog, general purpose	"	0.114
Refrigerator, blood bank, undercounter type 153 litres	"	1.333
5.4 cf, undercounter type	"	1.333
Refrigerator/freezer, 4.4 cf, undercounter type	"	1.333
Sealer, impulse, free standing, 20x12x4"	"	0.267
Timer, electric, 1-60 minutes, bench or wall mounted	"	0.444
Glassware washer - dryer, undercounter	"	10.000
Balance, torsion suspension, tabletop, 4.5 lb capacity	"	0.444
Binocular microscope, with in base illuminator	"	0.308
Centrifuge, table model, 19x16x13"	"	0.320
Clinical model, with four place head	"	0.178

11700.10 MEDICAL EQUIPMENT

Item	UNIT	MAN/HOURS
Hospital equipment, lights		
Examination, portable	EA.	0.667
Meters		
Air flow meter	EA.	0.444
Oxygen flow meters	"	0.333
Physical therapy		
Chair, hydrotherapy	EA.	0.133
Diathermy, shortwave, portable, on casters	"	0.320
Exercise bicycle, floor standing, 35" x 15"	"	0.267
Hydrocollator, 4 pack, portable, 129 x 90 x 160"	"	0.114

ARCHITECTURAL EQUIPMENT

11700.10   MEDICAL EQUIPMENT

	UNIT	MAN/HOURS
Lamp, infrared, mobile with variable heat control	EA.	0.615
Ultra violet, base mounted	"	0.615
Stimulator, galvanic-faradic, hand held	"	0.053
Ultrasound, muscle stimulator, portable, 13x13x8"	"	0.067
Whirlpool, 85 gallon	"	4.000
65 gallon capacity	"	4.000
Radiology		
Radiographic table, motor driven tilting table	EA.	80.000
Fluoroscope image/tv system	"	160
Processor for washing and drying radiographs		
Water filter unit, 30" x 48-1/2" x 37-1/2"	EA.	13.333
Steam sterilizers		
For heat and moisture stable materials	EA.	0.800
For fast drying after sterilization	"	1.000
Compact unit	"	1.000
Semi-automatic	"	4.000
Floor loading		
Single door	EA.	6.667
Double door	"	8.000
Utensil washer, sanitizer	"	6.154
Automatic washer/sterilizer	"	16.000
16 x 16 x 26", including generator & accessories	"	26.667
Steam generator, elec., 10 kw to 180 kw	"	16.000
Surgical scrub		
Minimum	EA.	2.667
Maximum	"	2.667
Gas sterilizers		
Automatic, free standing, 21x19x29"	EA.	8.000
Surgical lights, ceiling mounted		
Minimum	EA.	13.333
Maximum	"	16.000
Water stills		
4 liters/hr	EA.	2.667
8 liters/hr	"	2.667
19 liters/hr	"	6.667
X-ray equipment		
Mobile unit		
Minimum	EA.	4.000
Maximum	"	8.000
Minimum	"	8.000
Maximum	"	8.000
Incubators		
15 cf	EA.	4.000
29 cf	"	6.667
Infant transport, portable	"	4.211
Headwall		
Aluminum, with back frame and console	EA.	4.000
Hospital ground detection system		
Power ground module	EA.	2.286
Ground slave module	"	1.739
Master ground module	"	1.509
Remote indicator	"	1.600
X-ray indicator	"	1.739
Micro ammeter	"	2.000

ARCHITECTURAL EQUIPMENT

11700.10   MEDICAL EQUIPMENT

	UNIT	MAN/HOURS
Supervisory module	EA.	1.739
Ground cords	"	0.296
Hospital isolation monitors, 5 ma		
120v	EA.	3.478
208v	"	3.478
240v	"	3.478
Digital clock-timers separate display	"	1.600
One display	"	1.600
Remote control	"	1.250
Battery pack	"	1.250
Surgical chronometer clock and 3 timers	"	2.500
Auxilary control	"	1.159

11700.20   DENTAL EQUIPMENT

	UNIT	MAN/HOURS
Dental care equipment		
Drill console with accessories	EA.	13.333
Amalgamator	"	0.400
Lathe	"	0.267
Finish polisher	"	0.533
Model trimmer	"	0.364
Motor, wall mounted	"	0.364
Cleaner, ultrasonic	"	0.800
Curing unit, bench mounted	"	1.333
Oral evacuation system, dual pump	"	1.000
Sterilizer, table top, self contained	"	0.444
Dental lights		
Light, floor or ceiling mounted	EA.	4.000
X-ray unit		
Portable	EA.	2.000
Wall mounted with remote control	"	6.667
Illuminator, single panel	"	11.429
X-ray film processor	"	6.667
Shield, portable x-ray, lead lined	"	0.533

INTERIOR	UNIT	MAN/ HOURS
12302.10 CASEWORK		
Kitchen base cabinet, prefinished, 24" deep, 35" high		
12"wide	EA.	0.800
18" wide	"	0.800
24" wide	"	0.889
27" wide	"	0.889
36" wide	"	1.000
48" wide	"	1.000
Corner cabinet, 36" wide	"	1.000
Wall cabinet, 12" deep, 12" high		
30" wide	EA.	0.800
36" wide	"	0.800
15" high		
30" wide	EA.	0.889
36" wide	"	0.889
24" high		
30" wide	EA.	0.889
36" wide	"	0.889
30" high		
12" wide	EA.	1.000
18" wide	"	1.000
24" wide	"	1.000
27" wide	"	1.000
30" wide	"	1.143
36" wide	"	1.143
Corner cabinet, 30" high		
24" wide	EA.	1.333
30" wide	"	1.333
36" wide	"	1.333
Wardrobe	"	2.000
Vanity with top, laminated plastic		
24" wide	EA.	2.000
30" wide	"	2.000
36" wide	"	2.667
48" wide	"	3.200
12390.10 COUNTER TOPS		
Stainless steel, counter top, with backsplash	S.F.	0.200
Acid-proof, kemrock surface	"	0.133

CONSTRUCTION	UNIT	MAN/HOURS
13152.10 **SWIMMING POOL EQUIPMENT**		
Diving boards		
14' long		
Aluminum	EA.	4.444
Fiberglass	"	4.444
Ladders, heavy duty		
2 steps		
Minimum	EA.	1.600
Maximum	"	1.600
4 steps		
Minimum	EA.	2.000
Maximum	"	2.000
Lifeguard chair		
Minimum	EA.	8.000
Maximum	"	8.000
Lights, underwater		
12 volt, with transformer	EA.	2.000
110 volt		
Minimum	EA.	2.000
Maximum	"	2.000
Ground fault interrupter for 110 volt, each light	"	0.667
Pool covers		
Reinforced polyethylene	S.F.	0.062
Vinyl water tube		
Minimum	S.F.	0.062
Maximum	"	0.062
Slides with water tube		
Minimum	EA.	6.667
Maximum	"	6.667

ELEVATORS

ELEVATORS	UNIT	MAN/HOURS
14210.10 ELEVATORS		
Passenger elevators, electric, geared		
Based on a shaft of 6 stops and 6 openings		
50 fpm, 2000 lb	EA.	24.000
100 fpm, 2000 lb	"	26.667
150 fpm		
2000 lb	EA.	30.000
3000 lb	"	34.286
4000 lb	"	40.000
Based on a shaft of 8 stops and 8 openings		
300 fpm		
3000 lb	EA.	48.000
3500 lb	"	48.000
4000 lb	"	53.333
5000 lb	"	57.143
Freight elevators, electric		
Based on a shaft of 6 stops and 6 openings		
50 fpm		
3500 lb	EA.	26.667
4000 lb	"	26.667
5000 lb	"	30.000
100 fpm		
3500 lb	EA.	30.000
4000 lb	"	30.000
5000 lb	"	34.286
14300.10 ESCALATORS		
Escalators		
32" wide, floor to floor		
12' high	EA.	40.000
15' high	"	48.000
18' high	"	60.000
22' high	"	80.000
25' high	"	96.000

LIFTS	UNIT	MAN/HOURS
14410.10 PERSONNEL LIFTS		
Residential stair climber, per story	EA.	6.667

BLANK LINE

MATERIAL HANDLING	UNIT	MAN/HOURS
14580.10 PNEUMATIC SYSTEMS		
Pneumatic message tube system		
Average, 20 station job		
3" round system	EA.	72.727
4" round system	"	80.000
6" round system	"	88.889
4" x 7" oval system	"	160
Trash and linen tube system		
10 stations	EA.	120
15 stations	"	160
20 stations	"	185
30 stations	"	218

HOISTS AND CRANES	UNIT	MAN/HOURS
14600.10 INDUSTRIAL HOISTS		
Industrial hoists, electric, light to medium duty		
500 lb	EA.	4.000
1000 lb	"	4.211
2000 lb	"	4.444
3000 lb	"	4.706
4000 lb	"	5.000
5000 lb	"	5.333
6000 lb	"	5.517
7500 lb	"	5.714
10,000 lb	"	5.926
15,000 lb	"	6.154
20,000 lb	"	6.667
25,000 lb	"	7.273
30,000 lb	"	8.000
Heavy duty		
500 lb	EA.	4.000
1000 lb	"	4.211
2000 lb	"	4.444
3000 lb	"	4.706
4000 lb	"	5.000
5000 lb	"	5.333
6000 lb	"	5.517
7500 lb	"	5.714
10,000 lb	"	5.926
15,000 lb	"	6.154
20,000 lb	"	6.667
25,000 lb	"	7.273
30,000 lb	"	8.000

Left Column

BASIC MATERIALS	UNIT	MAN/HOURS
15100.10 SPECIALTIES		
Wall penetration		
Concrete wall, 6" thick		
2" dia.	EA.	0.267
4" dia.	"	0.400
8" dia.	"	0.571
12" thick		
2" dia.	EA.	0.364
4" dia.	"	0.571
8" dia.	"	0.889
Non-destructive testing, piping systems		
X-ray of welds		
3" dia. pipe	EA.	0.800
4" dia. pipe	"	0.800
6" dia. pipe	"	0.800
8" dia. pipe	"	1.000
10" dia. pipe	"	1.000
Liquid penetration of welds		
2" dia. pipe	EA.	0.500
3" dia. pipe	"	0.500
4" dia. pipe	"	0.500
6" dia. pipe	"	0.500
8" dia. pipe	"	0.500
10" dia. pipe	"	0.500
15120.10 BACKFLOW PREVENTERS		
Backflow preventer, flanged, cast iron, with valves		
3" pipe	EA.	4.000
4" pipe	"	4.444
6" pipe	"	6.667
8" pipe	"	8.000
Threaded		
3/4" pipe	EA.	0.500
2" pipe	"	0.800
Reduced pressure assembly, bronze, threaded		
3/4"	EA.	0.500
1"	"	0.571
1-1/4"	"	0.667
1-1/2"	"	0.800
15140.10 PIPE HANGERS, HEAVY		
Hangers		
1/2" pipe, clevis pipe hanger		
Black steel	EA.	0.267
Galvanized	"	0.267
U bolt	"	0.080
3/4" pipe, clevis pipe hanger		
Black steel	EA.	0.267
Galvanized	"	0.267
U bolt	"	0.080
1" pipe, clevis pipe hanger		
Black steel	EA.	0.267
Galvanized	"	0.267

Right Column

BASIC MATERIALS	UNIT	MAN/HOURS
15140.10 PIPE HANGERS, HEAVY		
U bolt	EA.	0.080
1-1/4" pipe, clevis pipe hanger		
Black steel	EA.	0.267
Galvanized	"	0.267
U bolt	"	0.080
1-1/2" pipe, clevis pipe hanger		
Black steel	EA.	0.267
Galvanized	"	0.267
U bolt	"	0.080
2" pipe, clevis pipe hanger		
Black steel	EA.	0.267
Galvanized	"	0.267
Adjustable pipe roll stand	"	1.000
U bolt	"	0.080
Adjustable roller hanger	"	0.267
2-1/2" pipe, clevis pipe hanger		
Black steel	EA.	0.267
Galvanized	"	0.267
Adjustable pipe roll stand	"	0.267
Adjustable roller hanger	"	0.267
3" pipe, clevis pipe hanger		
Black steel	EA.	0.267
Galvanized	"	0.267
Adjustable pipe roll stand	"	0.267
U bolt	"	0.080
Adjustable roller hanger	"	0.267
3-1/2" pipe, clevis pipe hanger		
Black steel	EA.	0.267
Galvanized	"	0.267
Adjustable pipe roll stand	"	0.267
U bolt	"	0.080
Adjustable roller hanger	"	0.267
4" pipe, clevis pipe hanger		
Black steel	EA.	0.267
Galvanized	"	0.267
Adjustable pipe roll stand	"	0.267
U bolt	"	0.080
Adjustable roller hanger	"	0.267
5" pipe, clevis pipe hanger		
Black steel	EA.	0.320
Galvanized	"	0.320
Adjustable pipe roll stand	"	0.320
U bolt	"	0.080
Adjustable roller hanger	"	0.320
6" pipe, clevis pipe hanger		
Black steel	EA.	0.320
Galvanized	"	0.320
Adjustable pipe roll stand	"	0.320
U bolt	"	0.100
Adjustable roller hanger	"	0.320
8" pipe, clevis pipe hanger		
Black steel	EA.	0.320
Galvanized	"	0.320
Adjustable pipe roll stand	"	0.320

BASIC MATERIALS	UNIT	MAN/HOURS
15140.10 PIPE HANGERS, HEAVY		
U bolt	EA.	0.100
Adjustable roller hanger	"	0.320
10" clevis pipe hanger		
Black steel	EA.	0.320
Galvanized	"	0.320
Adjustable pipe roll stand	"	0.320
Adjustable roller hanger	"	0.320
12" pipe, clevis pipe hanger		
Black steel	EA.	0.320
Galvanized	"	0.320
Adjustable pipe roll stand	"	0.320
Adjustable roller hanger	"	0.320
14" pipe, clevis pipe hanger		
Black steel	EA.	0.364
Galvanized	"	0.364
Threaded rod, galvanized		
C-clamp, steel, with lock nut		
3/8"	EA.	0.100
1/2"	"	0.100
5/8"	"	0.100
3/4"	"	0.100
7/8"	"	0.100
Angle support, medium, welded steel		
12"x18"	EA.	0.800
18"x24"	"	0.800
24"x30"	"	0.800
Heavy, welded steel		
12"x18"	EA.	0.800
18"x24"	"	0.800
24"x30"	"	0.800
15140.11 PIPE HANGERS, LIGHT		
A band, black iron		
1/2"	EA.	0.057
1"	"	0.059
1-1/4"	"	0.062
1-1/2"	"	0.067
2"	"	0.073
2-1/2"	"	0.080
3"	"	0.089
4"	"	0.100
5"	"	0.107
6"	"	0.114
8"	"	0.133
Copper		
1/2"	EA.	0.057
3/4"	"	0.059
1"	"	0.059
1-1/4"	"	0.062
1-1/2"	"	0.067
2"	"	0.073
2-1/2"	"	0.080
3"	"	0.089
4"	"	0.100

BASIC MATERIALS	UNIT	MAN/HOURS
15140.11 PIPE HANGERS, LIGHT		
Black riser friction hangers		
3/4"	EA.	0.067
1"	"	0.070
1-1/4"	"	0.073
1-1/2"	"	0.076
2"	"	0.080
2-1/2"	"	0.089
3"	"	0.100
4"	"	0.114
5"	"	0.123
6"	"	0.133
8"	"	0.145
10"	"	0.160
Short pattern black riser clamps		
1-1/2"	EA.	0.073
2"	"	0.076
3"	"	0.080
4"	"	0.089
Copper riser friction hanger		
1/2"	EA.	0.062
3/4"	"	0.064
1"	"	0.067
1-1/4"	"	0.070
1-1/2"	"	0.073
2"	"	0.076
2-1/2"	"	0.080
3"	"	0.080
4"	"	0.089
Auto grip hangers, galvanized		
1/2"	EA.	0.057
3/4"	"	0.062
1"	"	0.064
1-1/4"	"	0.067
1-1/2"	"	0.070
2"	"	0.073
2-1/2"	"	0.076
3"	"	0.080
4"	"	0.089
Copper		
1/2"	EA.	0.057
3/4"	"	0.062
1"	"	0.064
1-1/4"	"	0.067
1-1/2"	"	0.070
2"	"	0.073
2-1/2"	"	0.076
3"	"	0.080
4"	"	0.089
Split rings (F&M), galvanized		
3/8"	EA.	0.057
1/2"	"	0.062
3/4"	"	0.064
1"	"	0.067
1-1/4"	"	0.070

BASIC MATERIALS	UNIT	MAN/HOURS		BASIC MATERIALS	UNIT	MAN/HOURS
15140.11 PIPE HANGERS, LIGHT				**15140.11** PIPE HANGERS, LIGHT		
1-1/2"	EA.	0.073		1-1/2"	EA.	0.040
2"	"	0.076		2"	"	0.040
2-1/2"	"	0.080		3"	"	0.042
3"	"	0.084		4"	"	0.042
4"	"	0.089		PVC coated hangers, galvanized, 28 ga.		
Copper				1-1/2" x 12"	EA.	0.053
1/4"	EA.	0.057		2" x 12"	"	0.057
3/8"	"	0.057		3" x 12"	"	0.062
1/2"	"	0.062		4" x 12"	"	0.067
1"	"	0.067		Copper, 30 ga.		
1-1/4"	"	0.070		1-1/2" x 12"	EA.	0.053
1-1/2"	"	0.073		2" x 12"	"	0.057
2"	"	0.076		3" x 12"	"	0.062
2-1/2"	"	0.080		4" x 12"	"	0.067
3"	"	0.084		2" x 24"	"	0.062
4"	"	0.089		3" x 24"	"	0.067
F&M plates, galvanized				4" x 24"	"	0.073
3/8"	EA.	0.057		Milford hangers		
1/2"	"	0.059		1/2" x 6"	EA.	0.057
Copper				1/2" x 12"	"	0.062
3/8"	EA.	0.057		3/4" x 6"	"	0.059
1/2"	"	0.059		3/4" x 12"	"	0.064
2 hole clips, galvanized				1" x 6"	"	0.062
3/4"	EA.	0.053		1" x 12"	"	0.064
1"	"	0.055		1-1/4" x 6"	"	0.064
1-1/4"	"	0.057		1-1/4" x 12"	"	0.067
1-1/2"	"	0.059		1-1/2" x 6"	"	0.067
2"	"	0.062		1-1/2" x 12"	"	0.070
2-1/2"	"	0.064		2" x 6"	"	0.070
3"	"	0.067		2" x 12"	"	0.073
4"	"	0.073		Wire hook hangers		
Perforated strap				Black wire, 1/2" x		
3/4"				4"	EA.	0.040
Galvanized, 20 ga.	L.F.	0.040		6"	"	0.042
Copper, 22 ga.	"	0.040		8"	"	0.044
Threaded rod-couplings				10"	"	0.044
1/4"	EA.	0.050		12"	"	0.047
3/4"	"	0.053		3/4" x		
1/2"	EA.	0.057		4"	EA.	0.042
5/8"	"	0.062		6"	"	0.044
Reducing rod coupling, 1/2" x 3/8"	"	0.057		8"	"	0.047
C-clamps				10"	"	0.050
3/4"	EA.	0.080		12"	"	0.053
Top beam clamp				1" x		
3/8"	EA.	0.067		4"	EA.	0.044
1/2"	"	0.073		6"	"	0.047
Side beam connector				8"	"	0.050
3/8"	EA.	0.067		10"	"	0.053
1/2"	"	0.073		12"	"	0.057
J-Hooks				1-1/4" x		
1/2"	EA.	0.036		4"	EA.	0.047
3/4"	"	0.036		6"	"	0.050
1"	"	0.038		8"	"	0.053
1-1/4"	"	0.039		10"	"	0.057

BASIC MATERIALS

15140.11   PIPE HANGERS, LIGHT

Description	UNIT	MAN/HOURS
12"	EA.	0.062
1-1/2" x		
6"	EA.	0.053
8"	"	0.057
10"	"	0.062
12"	"	0.067
2" x		
6"	EA.	0.057
8"	"	0.062
10"	"	0.067
12"	"	0.073
Copper wire hooks		
1/2" x		
4"	EA.	0.040
6"	"	0.042
8"	"	0.044
10"	"	0.047
12"	"	0.050
3/4" x		
4"	EA.	0.042
6"	"	0.044
8"	"	0.047
10"	"	0.050
12"	"	0.053
1" x		
4"	EA.	0.044
6"	"	0.047
8"	"	0.050
10"	"	0.053
12"	"	0.057
1-1/4" x		
6"	EA.	0.047
8"	"	0.050
10"	"	0.053
12"	"	0.057
1-1/2" x		
6"	EA.	0.053
8"	"	0.057
10"	"	0.062
12"	"	0.067
2" x		
6"	EA.	0.057
8"	"	0.062
10"	"	0.067
12"	"	0.073

15175.60   EXPANSION TANKS

Description	UNIT	MAN/HOURS
Expansion tank, 125 psi, galvanized steel		
20 gallon	EA.	1.000
30 gallon	"	1.333
65 gallon	"	2.000
80 gallon	"	2.286

BASIC MATERIALS

15240.10   VIBRATION CONTROL

Description	UNIT	MAN/HOURS
Vibration isolator, in-line, stainless connector, screwed		
1/2"	EA.	0.444
3/4"	"	0.471
1"	"	0.500
1-1/4"	"	0.533
1-1/2"	"	0.571
2"	"	0.615
2-1/2"	"	0.667
3"	"	0.727
4"	"	0.800
6"	"	0.889
Flanged		
8"	EA.	1.000
10"	"	1.143
12"	"	1.333

INSULATION

15260.10   FIBERGLASS PIPE INSULATION

Description	UNIT	MAN/HOURS
Fiberglass insulation on 1/2" pipe		
1" thick	L.F.	0.027
1-1/2" thick	"	0.033
3/4" pipe		
1" thick	L.F.	0.027
1-1/2" thick	"	0.033
1" pipe		
1" thick	L.F.	0.027
1-1/2" thick	"	0.033
2" thick	"	0.040
1-1/4" dia. pipe		
1" thick	L.F.	0.033
1-1/2" thick	"	0.036
2" thick	"	0.040
1-1/2" pipe		
1" thick	L.F.	0.033
1-1/2" thick	"	0.036
2" thick	"	0.038
2-1/2" thick	"	0.040
3-1/2" thick	"	0.042
2" pipe		
1" thick	L.F.	0.033
1-1/2" thick	"	0.036
2" thick	"	0.040
2-1/2" thick	"	0.044
3-1/2" thick	"	0.050
2-1/2" pipe		

INSULATION	UNIT	MAN/HOURS
15260.10 FIBERGLASS PIPE INSULATION		
1" thick	L.F.	0.033
1-1/2" thick	"	0.036
2" thick	"	0.040
2-1/2" thick	"	0.044
3" thick	"	0.050
3-1/2" thick	"	0.057
3" pipe		
1" thick	L.F.	0.038
1-1/2" thick	"	0.040
2" thick	"	0.044
2-1/2" thick	"	0.050
3" thick	"	0.057
3-1/2" thick	"	0.067
4" pipe		
1" thick	L.F.	0.038
1-1/2" thick	"	0.040
2" thick	"	0.044
2-1/2" thick	"	0.050
3" thick	"	0.057
3-1/2" thick	"	0.067
5" pipe		
1" thick	L.F.	0.038
2" thick	"	0.040
3" thick	"	0.047
4" thick	"	0.062
6" pipe		
1" thick	L.F.	0.042
2" thick	"	0.044
4" thick	"	0.053
6" thick	"	0.073
8" pipe		
2" thick	L.F.	0.042
3" thick	"	0.044
4" thick	"	0.053
6" thick	"	0.073
10" pipe		
2" thick	L.F.	0.042
3" thick	"	0.044
4" thick	"	0.053
6" thick	"	0.073
12" pipe		
2" thick	L.F.	0.042
3" thick	"	0.044
4" thick	"	0.053
6" thick	"	0.073
15260.20 CALCIUM SILICATE		
Calcium silicate insulation, 6" pipe		
2" thick	L.F.	0.057
2-1/2" thick	"	0.062
3" thick	"	0.067
4" thick	"	0.073
6" thick	"	0.080

INSULATION	UNIT	MAN/HOURS
15260.20 CALCIUM SILICATE		
8" pipe		
2" thick	L.F.	0.062
2-1/2" thick	"	0.067
3" thick	"	0.073
4" thick	"	0.080
6" thick	"	0.089
10" pipe		
2" thick	L.F.	0.062
2-1/2" thick	"	0.067
3" thick	"	0.073
4" thick	"	0.080
6" thick	"	0.089
12" pipe		
2" thick	L.F.	0.062
2-1/2" thick	"	0.067
3" thick	"	0.073
4" thick	"	0.080
6" thick	"	0.089
15260.60 EXTERIOR PIPE INSULATION		
Fiberglass insulation, aluminum jacket		
1/2" pipe		
1" thick	L.F.	0.062
1-1/2" thick	"	0.067
3/4" pipe		
1" thick	L.F.	0.062
1-1/2" thick	"	0.067
1" pipe		
1" thick	L.F.	0.062
1-1/2" thick	"	0.067
3-1/2" thick	"	0.089
1-1/4" pipe		
1" thick	L.F.	0.073
1-1/2" thick	"	0.076
3-1/2" thick	"	0.089
1-1/2" pipe		
1" thick	L.F.	0.073
1-1/2" thick	"	0.076
2" thick	"	0.080
2-1/2" thick	"	0.084
3-1/2" thick	"	0.089
2" pipe		
1" thick	L.F.	0.073
1-1/2" thick	"	0.076
2" thick	"	0.080
2-1/2" thick	"	0.084
3-1/2" thick	"	0.089
2-1/2" pipe		
1" thick	L.F.	0.073
1-1/2" thick	"	0.076
2" thick	"	0.080
2-1/2" thick	"	0.084
3" thick	"	0.089
3-1/2" thick	"	0.094

INSULATION	UNIT	MAN/HOURS
15260.60 EXTERIOR PIPE INSULATION		
3" pipe		
1" thick	L.F.	0.080
1-1/2" thick	"	0.084
2" thick	"	0.089
2-1/2" thick	"	0.094
3" thick	"	0.100
3-1/2" thick	"	0.107
4" pipe		
1" thick	L.F.	0.080
1-1/2" thick	"	0.084
2" thick	"	0.089
2-1/2" thick	"	0.094
3" thick	"	0.100
3-1/2" thick	"	0.107
5" pipe		
1" thick	L.F.	0.080
2" thick	"	0.084
2-1/2" thick	"	0.089
3" thick	"	0.094
3-1/2" thick	"	0.100
6" pipe		
1" thick	L.F.	0.089
2" thick	"	0.094
2-1/2" thick	"	0.100
3" thick	"	0.107
3-1/2" thick	"	0.114
4" thick	"	0.123
6" thick	"	0.133
8" pipe		
2" thick	L.F.	0.089
3" thick	"	0.094
3-1/2" thick	"	0.100
4" thick	"	0.107
6" thick	"	0.114
10" pipe		
2" thick	L.F.	0.089
3" thick	"	0.094
3-1/2" thick	"	0.100
4" thick	"	0.107
6" thick	"	0.114
12" pipe		
2" thick	L.F.	0.089
3" thick	"	0.094
3-1/2" thick	"	0.100
4" thick	"	0.107
6" thick	"	0.114
Calcium silicate with aluminum jacket, 6" pipe		
2" thick	L.F.	0.100
2-1/2" thick	"	0.107
3" thick	"	0.114
4" thick	"	0.133
8" pipe		
2" thick	L.F.	0.100
2-1/2" thick	"	0.107

INSULATION	UNIT	MAN/HOURS
15260.60 EXTERIOR PIPE INSULATION		
3" thick	L.F.	0.114
4" thick	"	0.123
6" thick	"	0.133
10" pipe		
2" thick	L.F.	0.114
2-1/2" thick	"	0.123
3" thick	"	0.133
4" thick	"	0.145
6" thick	"	0.160
12" pipe		
2" thick	L.F.	0.114
2-1/2" thick	"	0.123
3" thick	"	0.133
4" thick	"	0.145
6" thick	"	0.160
15260.90 PIPE INSULATION FITTINGS		
Insulation protection saddle		
1" thick covering		
1/2" pipe	EA.	0.320
3/4" pipe	"	0.320
1" pipe	"	0.320
1-1/4" pipe	"	0.320
1-1/2" pipe	"	0.320
2" pipe	"	0.320
2-1/2" pipe	"	0.320
3" pipe	"	0.364
4" pipe	"	0.400
6" pipe	"	0.500
1-1/2" thick covering		
3/4" pipe	EA.	0.320
1" pipe	"	0.320
1-1/4" pipe	"	0.320
1-1/2" pipe	"	0.320
2" pipe	"	0.320
2-1/2" pipe	"	0.320
3" pipe	"	0.320
4" pipe	"	0.400
6" pipe	"	0.500
8" pipe	"	0.667
10" pipe	"	0.667
12" pipe	"	0.667
2" thick covering		
3/4" pipe	EA.	0.320
1" pipe	"	0.320
1-1/4" pipe	"	0.320
1-1/2" pipe	"	0.320
2" pipe	"	0.320
2-1/2" pipe	"	0.320
3" pipe	"	0.320
4" pipe	"	0.400
6" pipe	"	0.500
8" pipe	"	0.667

INSULATION

		UNIT	MAN/HOURS
15260.90	**PIPE INSULATION FITTINGS**		
	10" pipe	EA.	0.667
	12" pipe	"	0.667
	3" thick covering		
	2" pipe	EA.	0.320
	2-1/2" pipe	"	0.320
	3" pipe	"	0.320
	4" pipe	"	0.400
	6" pipe	"	0.500
	8" pipe	"	0.667
	10" pipe	"	0.667
	12" pipe	"	0.667
15280.10	**EQUIPMENT INSULATION**		
Equipment insulation, 2" thick, cellular glass		S.F.	0.050
Urethane, rigid, field applied jacket, plastered finish		"	0.100
Fiberglass, rigid, with vapor barrier		"	0.044
15290.10	**DUCTWORK INSULATION**		
Fiberglass duct insulation, plain blanket			
1-1/2" thick		S.F.	0.010
2" thick		"	0.013
With vapor barrier			
1-1/2" thick		S.F.	0.010
2" thick		"	0.013
Rigid with vapor barrier			
2" thick		S.F.	0.027
3" thick		"	0.032
4" thick		"	0.040
6" thick		"	0.053
Weatherproof, polystyrene, 3" thick, w/vapor barrier		"	0.080
Urethane board with vapor barrier		"	0.100

FIRE PROTECTION

		UNIT	MAN/HOURS
15330.10	**WET SPRINKLER SYSTEM**		
Sprinkler head, 212 deg, brass, exposed piping		EA.	0.320
Chrome, concealed piping		"	0.444
Water motor alarm		"	1.333
Fire department inlet connection		"	1.600
Wall plate for fire dept connection		"	0.667
Swing check valve flanged iron body, 4"		"	2.667
Check valve, 6"		"	4.000
Wet pipe valve, flange to groove, 4"		"	0.889
Flange to flange			

FIRE PROTECTION

		UNIT	MAN/HOURS
15330.10	**WET SPRINKLER SYSTEM**		
	6"	EA.	1.333
	8"	"	2.667
Alarm valve, flange to flange, (wet valve)			
	4"	EA.	0.889
	8"	"	6.667
Inspector's test connection		"	0.667
Wall hydrant, polished brass, 2-1/2" x 2-1/2", single		"	0.571
2-way		"	0.571
3-way		"	0.571
Wet valve trim, includes retard chamber & gauges, 4"-6"		"	0.667
Retard pressure switch for wet systems		"	1.600
Air maintenance device		"	0.667
Wall hydrant non-freeze, 8" thick wall, vacuum breaker		"	0.400
12" thick wall		"	0.400
15330.50	**DRY SPRINKLER SYSTEM**		
Dry pipe valve, flange to flange			
4"		EA.	1.600
6"		"	2.000
Trim, 4" and 6", includes gauges		"	0.667
Field testing and flushing		"	6.667
Disinfection		"	6.667
Pressure switch double circuit, open/close contacts		"	2.000
Low air			
Supervisory unit		EA.	1.333
Pressure switch		"	0.667
15330.70	**CO2 SYSTEM**		
CO2 system, high pressure, 75# cylinder with			
Valve assemblies		EA.	1.600
Storage rack		"	1.143
Manifold		"	5.714
Flexible loops		"	0.100
Beam scale for cylinders		"	1.333
Mechanically control head		"	0.533
Electrically control head		"	0.533
Stop valves		"	0.800
Check valves		"	1.000
Activation station		"	0.800
Nozzles		"	0.667
Hose reel with 75' of 3/4" hose		"	4.000
Main/reserve transfer switch		"	1.333
Pressure switch		"	0.800
Heat responsive device		"	1.333
Battery and charger		"	4.000
Low pressure			
Battery and charger		EA.	4.000
Pressure switch		"	0.889
Nozzles		"	0.727
Master selector valve		"	1.333
Selector valve		"	1.333
Low pressure hose reel with 75' of 3/4" hose		"	4.000
Tank fill lines		"	1.000
Activation stations		"	0.667

FIRE PROTECTION	UNIT	MAN/HOURS
15330.70 **CO2 SYSTEM**		
Electro manual pilot panels	EA.	1.000
Halon fire protection system, per computer room		

PLUMBING	UNIT	MAN/HOURS
15410.05 **C.I. PIPE, ABOVE GROUND**		
No hub pipe		
1-1/2" pipe	L.F.	0.057
2" pipe	"	0.067
3" pipe	"	0.080
4" pipe	"	0.133
6" pipe	"	0.160
8" pipe	"	0.267
10" pipe	"	0.320
No hub fittings, 1-1/2" pipe		
1/4 bend	EA.	0.267
1/8 bend	"	0.267
Sanitary tee	"	0.400
Sanitary cross	"	0.400
Wye	"	0.400
Tapped tee	"	0.267
P-trap	"	0.267
Tapped cross	"	0.267
2" pipe		
1/4 bend	EA.	0.320
1/8 bend	"	0.320
Sanitary tee	"	0.533
Sanitary cross	"	0.533
Wye	"	0.667
Double wye	"	0.667
2x1-1/2" wye & 1/8 bend	"	0.500
Double wye & 1/8 bend	"	0.667
Test tee less 2" plug	"	0.320
Tapped tee		
2"x2"	EA.	0.320
2"x1-1/2"	"	0.320
P-trap		
2"x2"	EA.	0.320
Tapped cross		
2"x1-1/2"	EA.	0.320
3" pipe		
1/4 bend	EA.	0.400
1/8 bend	"	0.400
Sanitary tee	"	0.500
3"x2" sanitary tee	"	0.500
3"x1-1/2" sanitary tee	"	0.500
Sanitary cross	"	0.667

PLUMBING	UNIT	MAN/HOURS
15410.05 **C.I. PIPE, ABOVE GROUND**		
3x2" sanitary cross	EA.	0.667
Wye	"	0.667
3x2" wye	"	0.667
Double wye	"	0.667
3x2" double wye	"	0.667
3x2" wye & 1/8 bend	"	0.571
3x1-1/2" wye & 1/8 bend	"	0.571
Double wye & 1/8 bend	"	0.667
3x2" double wye & 1/8 bend	"	0.667
3x2" reducer	"	0.364
Test tee, less 3" plug	"	0.400
3x3" tapped tee	"	0.400
3x2" tapped tee	"	0.400
3x1-1/2" tapped tee	"	0.400
P-trap	"	0.400
3x2" tapped cross	"	0.400
3x1-1/2" tapped cross	"	0.400
Closet flange, 3-1/2" deep	"	0.200
4" pipe		
1/4 bend	EA.	0.400
1/8 bend	"	0.400
Sanitary tee	"	0.667
4x3" sanitary tee	"	0.667
4x2" sanitary tee	"	0.667
Sanitary cross	"	0.800
4x3" sanitary cross	"	0.800
4x2" sanitary cross	"	0.800
Wye	"	0.667
4x3" wye	"	0.667
4x2" wye	"	0.667
Double wye	"	0.800
4x3" double wye	"	0.800
4x2" double wye	"	0.800
Wye & 1/8 bend	"	0.667
4x3" wye & 1/8 bend	"	0.667
4x2" wye & 1/8 bend	"	0.667
Double wye & 1/8 bend	"	0.800
4x3" double wye & 1/8 bend	"	0.800
4x2" double wye & 1/8 bend	"	0.800
4x3" reducer	"	0.400
4x2" reducer	"	0.400
Test tee, less 4" plug	"	0.400
4x2" tapped tee	"	0.400
4x1-1/2" tapped tee	"	0.400
P-trap	"	0.400
4x2" tapped cross	"	0.400
4x1-1/2" tapped cross	"	0.400
Closet flange		
3" deep	EA.	0.400
8" deep	"	0.400
6" pipe		
1/4 bend	EA.	0.667
1/8 bend	"	0.667
Sanitary tee	"	0.800

15410.05 — C.I. PIPE, ABOVE GROUND

PLUMBING	UNIT	MAN/HOURS
6x4" sanitary tee	EA.	0.800
Wye	"	0.800
6x4" wye	"	0.800
6x3" wye	"	0.800
6x2" wye	"	0.800
Double wye	"	1.000
6x4" double wye	"	1.000
Wye & 1/8 bend	"	0.800
6x4" wye & 1/8 bend	"	0.800
6x3" wye & 1/8 bend	"	0.800
6x2" wye & 1/8 bend	"	0.800
6x4" reducer	"	0.444
6x4" reducer	"	0.400
6x3" reducer	"	0.444
Test tee		
Less 6" plug	EA.	0.500
Plug	"	
P-trap	"	0.500
8" pipe		
1/4 bend	EA.	0.667
1/8 bend	"	0.667
Sanitary tee	"	1.000
8x6" sanitary tee	"	1.000
Wye	"	0.800
8x6" wye	"	0.667
8x4" wye	"	0.667
Double wye	"	0.667
Wye & 1/8 bend	"	0.667
8x6" wye & 1/8 bend	"	0.667
8x4" wye & 1/8 bend	"	0.667
8x6" reducer	"	0.667
8x4" reducer	"	0.667
8x3" reducer	"	0.667
8x2" reducer	"	0.444
Test tee		
Less 8" plug	EA.	0.667
Plug	"	
10" pipe		
1/4 bend	EA.	0.667
1/8 bend	"	0.667
Wye	"	1.333
10x8" wye	"	1.333
10x6" wye	"	1.333
10x4" wye	"	1.333
10x8" reducer	"	0.667
10x6" reducer	"	0.667
10x4" reducer	"	0.667

15410.06 — C.I. PIPE, BELOW GROUND

PLUMBING	UNIT	MAN/HOURS
No hub pipe		
1-1/2" pipe	L.F.	0.040
2" pipe	"	0.044
3" pipe	"	0.050

15410.06 — C.I. PIPE, BELOW GROUND

PLUMBING	UNIT	MAN/HOURS
4" pipe	L.F.	0.067
6" pipe	"	0.073
8" pipe	"	0.089
10" pipe	"	0.100
Fittings, 1-1/2"		
1/4 bend	EA.	0.229
1/8 bend	"	0.229
Wye	"	0.320
Wye & 1/8 bend	"	0.229
P-trap	"	0.229
2"		
1/4 bend	EA.	0.267
1/8 bend	"	0.267
Double wye	"	0.500
Wye & 1/8 bend	"	0.400
Double wye & 1/8 bend	"	0.500
P-trap	"	0.267
3"		
1/4 bend	EA.	0.320
1/8 bend	"	0.320
Wye	"	0.500
3x2" wye	"	0.500
Wye & 1/8 bend	"	0.500
Double wye & 1/8 bend	"	0.500
3x2" double wye & 1/8 bend	"	0.500
3x2" reducer	"	0.320
P-trap	"	0.320
4"		
1/4 bend	EA.	0.320
1/8 bend	"	0.320
Wye	"	0.500
4x3" wye	"	0.500
4x2" wye	"	0.500
Double wye	"	0.667
4x3" double wye	"	0.667
4x2" double wye	"	0.667
Wye & 1/8 bend	"	0.500
4x3" wye & 1/8 bend	"	0.500
4x2" wye & 1/8 bend	"	0.500
Double wye & 1/8 bend	"	0.667
4x3" double wye & 1/8 bend	"	0.667
4x2" double wye & 1/8 bend	"	0.667
4x3" reducer	"	0.320
4x2" reducer	"	0.320
6"		
1/4 bend	EA.	0.500
1/8 bend	"	0.500
Wye & 1/8 bend	"	0.667
6x4" wye & 1/8 bend	"	0.667
6x3" wye & 1/8 bend	"	0.667
6x2" wye & 1/8 bend	"	0.667
6x3" reducer	"	0.364
P-trap	"	0.400
8"		

15410.06   C.I. PIPE, BELOW GROUND

PLUMBING	UNIT	MAN/HOURS
1/4 bend	EA.	0.500
1/8 bend	"	0.500
Wye	"	0.667
8x6" wye	"	0.500
8x4" wye	"	0.500
8x6" wye & 1/8 bend	"	0.500
8x4" reducer	"	0.400
8x3" reducer	"	0.400
8x2" reducer	"	0.400
10"		
1/4 bend	EA.	0.500
1/8 bend	"	0.500
Wye	"	1.000
10x8" wye	"	1.000
10x6" wye	"	1.000
10x4" wye	"	1.000
10x8" reducer	"	0.500
10x6" reducer	"	0.500

15410.08   EXTRA HEAVY SOIL PIPE

PLUMBING	UNIT	MAN/HOURS
Extra heavy soil pipe, single hub		
2" x 5'	EA.	0.160
3" x 5'	"	0.170
4" x 5'	"	0.186
6" x 5'	"	0.200
Double hub		
2" x 5'	EA.	0.200
4" x 5'	"	0.211
5" x 5'	"	0.222
6" x 5'	"	0.229
Single hub		
3" x 10'	EA.	0.133
4" x 10'	"	0.138
6" x 10'	"	0.145
8" x 10'	"	0.154
"Mini", single hub		
2" x 42"	EA.	0.160
3" x 42"	"	0.170
4" x 42"	"	0.186
6" x 42"	"	0.200
8" x 42"	"	0.211
Fittings, 1/4" bend		
2"	EA.	0.267
3"	"	0.320
4"	"	0.320
1/8 bend		
2"	EA.	0.267
3"	"	0.320
4"	"	0.320
5"	"	0.400
6"	"	0.500
8"	"	0.500
Long sweep		
2"	EA.	0.267

15410.08   EXTRA HEAVY SOIL PIPE

PLUMBING	UNIT	MAN/HOURS
3"	EA.	0.320
4"	"	0.320
Straight T		
4" x 2"	EA.	0.500
4" x 3"	"	0.500
4"	"	0.571
Sanitary T		
3" x 2"	EA.	0.500
3"	"	0.500
4" x 2"	"	0.571
4" x 3"	"	0.571
4"	"	0.615
Wye		
4" x 2"	EA.	0.571
4" x 3"	"	0.571
4"	"	0.615
Combination Y and 1/8 bend, 4"	"	0.615
Double wye, 4"	"	0.615
Tapped sanitary T, 4" x 2"	"	0.667
P trap		
2"	EA.	0.320
4"	"	0.533
Dandy		
4", with 3" brass plug	EA.	0.444
6", 4" brass plug	"	0.500

15410.09   SERVICE WEIGHT PIPE

PLUMBING	UNIT	MAN/HOURS
Service weight pipe, single hub		
2" x 5'	EA.	0.160
3" x 5'	"	0.170
4" x 5'	"	0.178
5" x 5'	"	0.190
6" x 5'	"	0.200
8" x 5'	"	0.211
10" x 5'	"	0.222
12" x 5'	"	0.250
Double hub		
2" x 5'	EA.	0.200
3" x 5'	"	0.216
4" x 5'	"	0.229
5" x 5'	"	0.250
6" x 5'	"	0.267
10" x 5'	"	0.286
12' x 5'	"	0.308
Single hub		
2" x 10'	EA.	0.200
3" x 10'	"	0.216
4" x 10'	"	0.229
5" x 10'	"	0.250
6" x 10'	"	0.267
8" x 10'	"	0.286
10" x 10'	"	0.308
12" x 10'	"	0.333

15410.09 SERVICE WEIGHT PIPE

PLUMBING	UNIT	MAN/HOURS
Shorty		
2" x 42"	EA.	0.160
3" x 42"	"	0.170
4" x 42"	"	0.178
5" x 42"	"	0.190
6" x 42"	"	0.200
8" x 42"	"	0.211
10" x 42"	"	0.222
Soil plug		
2"	EA.	0.267
3"	"	0.286
4"	"	0.308
5"	"	0.320
6"	"	0.333
1/5 bend		
3"	EA.	0.320
4"	"	0.364
1/6 bend		
2"	EA.	0.267
3"	"	0.320
4"	"	0.364
1/8 bend		
2"	EA.	0.267
3"	"	0.320
4"	"	0.364
5"	"	0.381
6"	"	0.400
8"	"	0.500
10"	"	0.571
12"	"	0.667
1/16 bend		
2"	EA.	0.267
3"	"	0.320
4"	"	0.364
5"	"	0.381
6"	"	0.400
1/4 bend		
2"	EA.	0.267
3"	"	0.320
4"	"	0.364
5"	"	0.381
6"	"	0.400
8"	"	0.500
10"	"	0.571
12"	"	0.667
1/4 bend, long		
2" x 12"	EA.	0.286
4" x 18"	"	0.400
4" x 12"	"	0.400
Sweep		
2"	EA.	0.267
3"	"	0.320
4"	"	0.364
5"	"	0.381

15410.09 SERVICE WEIGHT PIPE

PLUMBING	UNIT	MAN/HOURS
6"	EA.	0.400
8"	"	0.500
Reducing long sweep		
3" x 2"	EA.	0.320
4" x 3"	"	0.400
Straight T		
2"	EA.	0.500
3"	"	0.533
4" x 2"	"	0.571
4" x 3"	"	0.615
4"	"	0.667
Sanitary T		
2"	EA.	0.500
3" x 2"	"	0.533
3"	"	0.571
4" x 2"	"	0.615
4" x 3"	"	0.667
4"	"	0.667
5"	"	0.727
6"	"	0.727
Wye		
2"	EA.	0.400
3" x 2"	"	0.444
3"	"	0.444
4" x 2"	"	0.471
4" x 3"	"	0.471
4"	"	0.471
5" x 2"	"	0.500
5" x 3"	"	0.500
5" x 4"	"	0.500
5"	"	0.500
6" x 2"	"	0.500
6" x 3"	"	0.500
6" x 4"	"	0.533
6" x 5"	"	0.533
6"	"	0.571
8" x 4"	"	0.571
8" x 6"	"	0.615
8"	"	0.667
10" x 4"	"	0.667
10" x 6"	"	0.727
10" x 8"	"	0.800
Service wye		
10"	EA.	1.000
12" x 4"	"	1.000
12" x 6"	"	1.000
12" x 8"	"	1.143
12"	"	1.333
Combination wye and 1/8 bend		
2"	EA.	0.533
3" x 2"	"	0.571
3"	"	0.571
4" x 2"	"	0.615
4" x 3"	"	0.615

PLUMBING	UNIT	MAN/HOURS
15410.09 SERVICE WEIGHT PIPE		
4"	EA.	0.667
5" x 4"	"	0.667
6" x 4"	"	0.800
8"	"	1.000
Straight cross, 4"	"	0.667
Sanitary cross		
2"	EA.	0.571
3"	"	0.615
3" x 2"	"	0.615
4"	"	0.667
4" x 3"	"	0.667
4" x 2"	"	0.727
Double wye		
2"	EA.	0.571
3" x 2"	"	0.615
3"	"	0.615
4" x 2"	"	0.667
4" x 3"	"	0.667
4"	"	0.667
5"	"	0.800
6" x 4"	"	1.000
Combination double wye and 1/8 bend		
2"	EA.	0.571
3" x 2"	"	0.615
3"	"	0.615
4" x 3"	"	0.667
4"	"	0.667
Tapped sanitary T		
2" x 1-1/2"	EA.	0.571
2" x 2"	"	0.571
3" x 1-1/2"	"	0.615
3" x 2"	"	0.615
4" x 1-1/2"	"	0.667
4" x 2"	"	0.667
Tapped straight T		
3" x 1-1/2"	EA.	0.615
3" x 2"	"	0.615
4" x 2"	"	0.667
4" x 1-1/2"	"	0.667
Tapped Y		
4" x 2"	EA.	0.667
Tapped sanitary cross		
2" x 1-1/2"	EA.	0.571
2" x 2"	"	0.615
3" x 1-1/2"	"	0.615
3" x 2"	"	0.615
3" x 3"	"	0.615
4" x 1-1/2"	"	0.667
4" x 2"	"	0.667
Cleanout, dandy, with brass plug		
2", 1-1/2" plug	EA.	0.571
3", 2" plug	"	0.615
4", 3" plug	"	0.667
5", 4" plug	"	0.727

PLUMBING	UNIT	MAN/HOURS
15410.09 SERVICE WEIGHT PIPE		
6", 4" plug	EA.	0.800
8", 6" plug	"	1.000
Reducer, 3" x 2"	"	0.571
4" x		
2"	EA.	0.615
3"	"	0.615
5" x		
2"	EA.	0.667
3"	"	0.667
4"	"	0.667
6" x		
2"	EA.	0.800
3"	"	0.800
4"	"	0.800
5"	"	0.800
8" x		
4"	EA.	1.000
6"	"	1.000
10" x		
4"	EA.	1.143
6"	"	1.143
8"	"	1.143
12" x		
4"	EA.	1.333
6"	"	1.333
8"	"	1.333
10"	"	1.333
P trap		
2"	EA.	0.571
3"	"	0.615
4"	"	0.667
5"	"	0.800
6"	"	1.000
15410.10 COPPER PIPE		
Type "K" copper		
1/2"	L.F.	0.025
3/4"	"	0.027
1"	"	0.029
1-1/4"	"	0.031
1-1/2"	"	0.033
2"	"	0.036
2-1/2"	"	0.040
3"	"	0.042
4"	"	0.044
DWV, copper		
1-1/4"	L.F.	0.033
1-1/2"	"	0.036
2"	"	0.040
3"	"	0.044
4"	"	0.050
6"	"	0.057
Refrigeration tubing, copper, sealed		

PLUMBING

15410.10 COPPER PIPE

	UNIT	MAN/HOURS
1/8"	L.F.	0.032
3/16"	"	0.033
1/4"	"	0.035
5/16"	"	0.036
3/8"	"	0.038
1/2"	"	0.040
7/8"	"	0.046
1-1/8"	"	0.053
1-3/8"	"	0.062
Type "L" copper		
1/4"	L.F.	0.024
3/8"	"	0.024
1/2"	"	0.025
3/4"	"	0.027
1"	"	0.029
1-1/4"	"	0.031
1-1/2"	"	0.033
2"	"	0.036
2-1/2"	"	0.040
3"	"	0.042
3-1/2"	"	0.043
4"	"	0.044
Type "M" copper		
1/2"	L.F.	0.025
3/4"	"	0.027
1"	"	0.029
1-1/4"	"	0.031
2"	"	0.036
2-1/2"	"	0.040
3"	"	0.042
4"	"	0.044

15410.11 COPPER FITTINGS

	UNIT	MAN/HOURS
Coupling, with stop		
1/4"	EA.	0.267
3/8"	"	0.320
1/2"	"	0.348
5/8"	"	0.400
3/4"	"	0.444
1"	"	0.471
3"	"	0.800
4"	"	1.000
Reducing coupling		
1/4" x 1/8"	EA.	0.320
3/8" x 1/4"	"	0.348
1/2" x		
3/8"	EA.	0.400
1/4"	"	0.400
1/8"	"	0.400
3/4" x		
3/8"	EA.	0.444
1/2"	"	0.444
1" x		

PLUMBING

15410.11 COPPER FITTINGS

	UNIT	MAN/HOURS
3/8"	EA.	0.500
1" x 1/2"	"	0.500
1" x 3/4"	"	0.500
1-1/4" x		
1/2"	EA.	0.533
3/4"	"	0.533
1"	"	0.533
1-1/2" x		
1/2"	EA.	0.571
3/4"	"	0.571
1"	"	0.571
1-1/4"	"	0.571
2" x		
1/2"	EA.	0.667
3/4"	"	0.667
1"	"	0.667
1-1/4"	"	0.667
1-1/2"	"	0.667
2-1/2" x		
1"	EA.	0.800
1-1/4"	"	0.800
1-1/2"	"	0.800
2"	"	0.800
3" x		
1-1/2"	EA.	1.000
2"	"	1.000
2-1/2"	"	1.000
4" x		
2"	EA.	1.143
2-1/2"	"	1.143
3"	"	1.143
Slip coupling		
1/4"	EA.	0.267
1/2"	"	0.320
3/4"	"	0.400
1"	"	0.444
1-1/4"	"	0.500
1-1/2"	"	0.533
2"	"	0.667
2-1/2"	"	0.667
3"	"	0.800
4"	"	1.000
Coupling with drain		
1/2"	EA.	0.400
3/4"	"	0.444
1"	"	0.500
Reducer		
3/8" x 1/4"	EA.	0.320
1/2" x 3/8"	"	0.320
3/4" x		
1/4"	EA.	0.364
3/8"	"	0.364
1/2"	"	0.364
1" x		

PLUMBING	UNIT	MAN/HOURS
15410.11 COPPER FITTINGS		
1/2"	EA.	0.400
3/4"	"	0.400
1-1/4" x		
1/2"	EA.	0.444
3/4"	"	0.444
1"	"	0.444
1-1/2" x		
1/2"	EA.	0.500
3/4"	"	0.500
1"	"	0.500
1-1/4"	"	0.500
2" x		
1/2"	EA.	0.571
3/4"	"	0.571
1"	"	0.571
1-1/4"	"	0.571
1-1/2"	"	0.571
2-1/2" x		
1"	EA.	0.667
1-1/4"	"	0.667
1-1/2"	"	0.667
2"	"	0.667
3" x		
1-1/4"	EA.	0.800
1-1/2"	"	0.800
2"	"	0.800
2-1/2"	"	0.800
4" x		
2"	EA.	1.000
3"	"	1.000
Female adapters		
1/4"	EA.	0.320
3/8"	"	0.364
1/2"	"	0.400
3/4"	"	0.444
1"	"	0.444
1-1/4"	"	0.500
1-1/2"	"	0.500
2"	"	0.533
2-1/2"	"	0.571
3"	"	0.667
4"	"	0.800
Increasing female adapters		
1/8" x		
3/8"	EA.	0.320
1/2"	"	0.320
1/4" x 1/2"	"	0.348
3/8" x 1/2"	"	0.364
1/2" X		
3/4"	EA.	0.400
1"	"	0.400
3/4" X		
1"	EA.	0.444
1-1/4"	"	0.444

PLUMBING	UNIT	MAN/HOURS
15410.11 COPPER FITTINGS		
1" x		
1-1/4"	EA.	0.444
1-1/2"	"	0.444
1-1/4" x		
1-1/2"	EA.	0.500
2"	"	0.500
1-1/2" x 2"	"	0.533
Reducing female adapters		
3/8" x 1/4"	EA.	0.364
1/2" x		
1/4"	EA.	0.400
3/8"	"	0.400
3/4" x 1/2"	"	0.444
1" x		
1/2"	EA.	0.444
3/4"	"	0.444
1-1/4" x		
1/2"	EA.	0.500
3/4"	"	0.500
1"	"	0.500
1-1/2" x		
1"	EA.	0.533
1-1/4"	"	0.533
2" x		
1"	EA.	0.571
1-1/4"	"	0.571
1-1/2"	"	0.571
Female fitting adapters		
1/2"	EA.	0.400
3/4"	"	0.400
3/4" x 1/2"	"	0.421
1"	"	0.444
1-1/4"	"	0.471
1-1/2"	"	0.500
2"	"	0.533
Male adapters		
1/4"	EA.	0.364
3/8"	"	0.364
3"	"	0.667
4"	"	0.800
Increasing male adapters		
3/8" x 1/2"	EA.	0.364
1/2" x		
3/4"	EA.	0.400
1"	"	0.400
3/4" x		
1"	EA.	0.421
1-1/4"	"	0.421
1" x 1-1/4"	"	0.444
1-1/2" x		
3/4"	EA.	0.471
1"	"	0.471
1-1/4"	"	0.471
2" x		

PLUMBING	UNIT	MAN/HOURS
15410.11 COPPER FITTINGS		
1"	EA.	0.500
1-1/4"	"	0.500
1-1/2"	"	0.500
2" x 2-1/2"	"	0.533
Reducing male adapters		
1/2" x		
1/4"	EA.	0.400
3/8"	"	0.400
3/4" x 1/2"	"	0.421
1" x		
1/2"	EA.	0.444
3/4"	"	0.444
1-1/4" x		
3/4"	EA.	0.471
1"	"	0.471
1-1/2" x		
3/4"	EA.	0.500
1"	"	0.500
1-1/4'	"	0.500
2" x		
3/4"	EA.	0.533
1"	"	0.533
1-1/4"	"	0.533
1-1/2"	"	0.533
2-1/2" x		
2"	EA.	0.615
Fitting x male adapters		
1/2"	EA.	0.400
3/4"	"	0.421
1"	"	0.444
1-1/4"	"	0.471
1-1/2"	"	0.500
2"	"	0.533
90 ells		
1/8"	EA.	0.320
1/4"	"	0.320
3/8"	"	0.364
1/2"	"	0.400
3/4"	"	0.421
1"	"	0.444
1-1/4"	"	0.471
1-1/2"	"	0.500
2"	"	0.533
2-1/2"	"	0.615
3"	"	0.667
4"	"	0.800
Reducing 90 ell		
3/8" x 1/4"	EA.	0.364
1/2" x		
1/4"	EA.	0.400
3/8"	"	0.400
3/4" x 1/2"	"	0.421
1" x		
1/2"	EA.	0.444

PLUMBING	UNIT	MAN/HOURS
15410.11 COPPER FITTINGS		
3/4"	EA.	0.444
1-1/4" x		
1/2"	EA.	0.471
3/4"	"	0.471
1"	"	0.471
1-1/2" x		
1/2"	EA.	0.500
3/4"	"	0.500
1"	"	0.500
x 1-1/4"	"	0.500
2" x		
3/4"	EA.	0.533
x 1"	"	0.533
x 1-1/4"	"	0.533
x 1-1/2"	"	0.533
2-1/2" x		
1-1/2"	EA.	0.615
x 2"	"	0.615
3" x		
2"	EA.	0.667
x 2-1/2"	"	0.667
Street ells, copper		
1/4"	EA.	0.320
3/8"	"	0.364
1/2"	"	0.400
3/4"	"	0.421
1"	"	0.444
1-1/4"	"	0.471
1-1/2"	"	0.500
2"	"	0.533
2-1/2"	"	0.615
3"	"	0.667
4"	"	0.800
Female, 90 ell		
1/2"	EA.	0.400
3/4"	"	0.421
1"	"	0.444
1-1/4"	"	0.471
1-1/2"	"	0.500
2"	"	0.533
Female increasing, 90 ell		
3/8" x 1/2"	EA.	0.364
1/2" x		
3/4"	EA.	0.400
1"	"	0.400
3/4" x 1"	"	0.421
1" x 1-1/4"	"	0.444
1-1/4" x 1-1/2"	"	0.471
Female reducing, 90 ell		
1/2" x 3/8"	EA.	0.400
3/4" x 1/2"	"	0.421
1" x		
1/2"	EA.	0.444
x 3/4"	"	0.444

PLUMBING		UNIT	MAN/HOURS
15410.11	COPPER FITTINGS		
1-1/4" x			
3/4"		EA.	0.471
x 1"		"	0.471
x 1/2"		"	0.471
1-1/2" x			
1"		EA.	0.500
x 1-1/4"		"	0.500
2" x 1-1/2"		"	0.533
Male, 90 ell			
1/4"		EA.	0.320
3/8"		"	0.364
1/2"		"	0.400
3/4"		"	0.421
1"		"	0.444
1-1/4"		"	0.471
1-1/2"		"	0.500
2"		"	0.533
Male, increasing 90 ell			
1/2" x			
3/4"		EA.	0.400
1"		"	0.400
3/4" x 1"		"	0.421
1" x 1-1/4"		"	0.444
1-1/4" x 1-1/2"		"	0.471
Male, reducing 90 ell			
1/2" x 3/8"		EA.	0.400
3/4" x 1/2"		"	0.421
1" x			
1/2"		EA.	0.444
3/4"		"	0.444
1-1/4" x 1"		"	0.471
Drop ear ells			
1/2"		EA.	0.400
Female drop ear ells			
1/2"		EA.	0.400
1/2" x 3/8"		"	0.400
3/4"		"	0.421
Female flanged sink ell			
1/2"		EA.	0.400
45 ells			
1/4"		EA.	0.320
3/8"		"	0.364
3"		"	0.667
4"		"	0.800
45 street ell			
1/4"		EA.	0.320
3/8"		"	0.364
1/2"		"	0.400
3/4"		"	0.421
1"		"	0.444
1-1/2"		"	0.500
2"		"	0.533
2-1/2"		"	0.615
3"		"	0.667

PLUMBING		UNIT	MAN/HOURS
15410.11	COPPER FITTINGS		
4"		EA.	0.800
Tee			
1/8"		EA.	0.320
1/4"		"	0.320
3/8"		"	0.364
3"		"	0.667
4"		"	0.800
Caps			
1/4"		EA.	0.320
3/8"		"	0.364
3"		"	0.667
4"		"	0.800
Test caps			
1/2"		EA.	0.400
3/4"		"	0.421
1"		"	0.444
1-1/4"		"	0.471
1-1/2"		"	0.500
2"		"	0.533
3"		"	0.667
Flush bushing			
1/4" x 1/8"		EA.	0.320
1/2" x			
1/4"		EA.	0.400
3/8"		"	0.400
3/4" x			
3/8"		EA.	0.421
1/2"		"	0.421
1" x			
1/2"		EA.	0.444
3/4"		"	0.444
1-1/4" x			
1/2"		EA.	0.471
3/4"		"	0.471
x 1"		"	0.471
1-1/2" x			
1/2"		EA.	0.500
1-1/4"		"	0.500
2" x			
1"		EA.	0.533
1-1/4"		"	0.533
1-1/2"		"	0.533
Female flush bushing			
1/2" x			
1/2" x 1/8"		EA.	0.400
1/4"		"	0.400
Union			
1/4"		EA.	0.320
3/8"		"	0.364
3"		"	0.667
Female			
1/2"		EA.	0.400
3/4"		"	0.421
Male			

PLUMBING		UNIT	MAN/HOURS
15410.11	**COPPER FITTINGS**		
1/2"		EA.	0.400
3/4"		"	0.421
1"		"	0.444
45 degree wye			
1/2"		EA.	0.400
3/4"		"	0.421
1"		"	0.444
1" x 3/4" x 3/4"		"	0.444
1-1/4"		"	0.471
1-1/4" x 1" x 1"		"	0.471
Twin ells			
1" x 3/4" x 3/4"		EA.	0.444
1" x 1" x 1"		"	0.444
1-1/4" x 1" x 1"		"	0.471
Companion flanges, 125#			
2" x 6"		EA.	0.533
2-1/2"		"	0.615
3" x 7-1/2"		"	0.667
4" x 9"		"	0.800
90 union ells, male			
1/2"		EA.	0.400
3/4"		"	0.421
1"		"	0.444
DWV fittings, coupling with stop			
1-1/4"		EA.	0.471
1-1/2"		"	0.500
1-1/2" x 1-1/4"		"	0.500
2"		"	0.533
2" x 1-1/4"		"	0.533
2" x 1-1/2"		"	0.533
3"		"	0.667
3" x 1-1/2"		"	0.667
3" x 2"		"	0.667
4"		"	0.800
Slip coupling			
1-1/2"		EA.	0.500
2"		"	0.533
3"		"	0.667
90 ells			
1-1/2"		EA.	0.500
1-1/2" x 1-1/4"		"	0.500
2"		"	0.533
2" x 1-1/2"		"	0.533
3"		"	0.667
4"		"	0.800
Street, 90 elbows			
1-1/2"		EA.	0.500
2"		"	0.533
3"		"	0.667
4"		"	0.800
Female, 90 elbows			
1-1/2"		EA.	0.500
2"		"	0.533
Male, 90 elbows			

PLUMBING		UNIT	MAN/HOURS
15410.11	**COPPER FITTINGS**		
1-1/2"		EA.	0.500
2"		"	0.533
90 with side inlet			
3" x 3" x 1"		EA.	0.667
3" x 3" x 1-1/2"		"	0.667
3" x 3" x 2"		"	0.667
45 ells			
1-1/4"		EA.	0.471
1-1/2"		"	0.500
2"		"	0.533
3"		"	0.667
4"		"	0.800
Street, 45 ell			
1-1/2"		EA.	0.500
2"		"	0.533
3"		"	0.667
60 ell			
1-1/2"		EA.	0.500
2"		"	0.533
3"		"	0.667
22-1/2 ell			
1-1/2"		EA.	0.500
2"		"	0.533
3"		"	0.667
11-1/4 ell			
1-1/2"		EA.	0.500
2"		"	0.533
3"		"	0.667
Wye			
1-1/4"		EA.	0.471
1-1/2"		"	0.500
2"		"	0.533
2" x 1-1/2" x 1-1/2"		"	0.533
2" x 1-1/2" x 2"		"	0.533
2" x 1-1/2" x 2"		"	0.533
3"		"	0.667
3" x 3" x 1-1/2"		"	0.667
3" x 3" x 2"		"	0.667
4"		"	0.800
4" x 4" x 2"		"	0.800
4" x 4" x 3"		"	0.800
Sanitary tee			
1-1/4"		EA.	0.471
1-1/2"		"	0.500
2"		"	0.533
2" x 1-1/2" x 1-1/2"		"	0.533
2" x 1-1/2" x 2"		"	0.533
2" x 2" x 1-1/2"		"	0.533
3"		"	0.667
3" x 3" x 1-1/2"		"	0.667
3" x 3" x 2"		"	0.667
4"		"	0.800
4" x 4" x 3"		"	0.800
Female sanitary tee			

PLUMBING	UNIT	MAN/ HOURS
15410.11 COPPER FITTINGS		
1-1/2"	EA.	0.500
Long turn tee		
1-1/2"	EA.	0.500
2"	"	0.533
3" x 1-1/2"	"	0.667
Double wye		
1-1/2"	EA.	0.500
2"	"	0.533
2" x 2" x 1-1/2" x 1-1/2"	"	0.533
3"	"	0.667
3" x 3" x 1-1/2" x 1-1/2"	"	0.667
3" x 3" x 2" x 2"	"	0.667
4" x 4" x 1-1/2" x 1-1/2"	"	0.800
Double sanitary tee		
1-1/2"	EA.	0.500
2"	"	0.533
2" x 2" x 1-1/2"	"	0.533
3"	"	0.667
3" x 3" x 1-1/2" x 1-1/2"	"	0.667
3" x 3" x 2" x 2"	"	0.667
4" x 4" x 1-1/2" x 1-1/2"	"	0.800
Long		
2" x 1-1/2"	EA.	0.533
Twin elbow		
1-1/2"	EA.	0.500
2"	"	0.533
2" x 1-1/2" x 1-1/2"	"	0.533
Spigot adapter, manoff		
1-1/2" x 2"	EA.	0.500
1-1/2" x 3"	"	0.500
2"	"	0.533
2" x 3"	"	0.533
2" x 4"	"	0.533
3"	"	0.667
3" x 4"	"	0.667
4"	"	0.800
No-hub adapters		
1-1/2" x 2"	EA.	0.500
2"	"	0.533
2" x 3"	"	0.533
3"	"	0.667
3" x 4"	"	0.667
4"	"	0.800
Fitting reducers		
1-1/2" x 1-1/4"	EA.	0.500
2" x 1-1/2"	"	0.533
3" x 1-1/2"	"	0.667
3" x 2"	"	0.667
Slip joint (Desanco)		
1-1/4"	EA.	0.471
1-1/2"	"	0.500
1-1/2" x 1-1/4"	"	0.500
Street x slip joint (Desanco)		
1-1/2"	EA.	0.500

PLUMBING	UNIT	MAN/ HOURS
15410.11 COPPER FITTINGS		
1-1/2" x 1-1/4"	EA.	0.500
Flush bushing		
1-1/2" x 1-1/4"	EA.	0.500
2" x 1-1/2"	"	0.533
3" x 1-1/2"	"	0.667
3" x 2"	"	0.667
Male hex trap bushing		
1-1/4" x 1-1/2"	EA.	0.471
1-1/2"	"	0.500
1-1/2" x 2"	"	0.500
2"	"	0.533
Round trap bushing		
1-1/2"	EA.	0.500
2"	"	0.533
Female adapter		
1-1/4"	EA.	0.471
1-1/2"	"	0.500
1-1/2" x 2"	"	0.500
2"	"	0.533
2" x 1-1/2"	"	0.533
3"	"	0.667
Fitting x female adapter		
1-1/2"	EA.	0.500
2"	"	0.533
Male adapters		
1-1/4"	EA.	0.471
1-1/4" x 1-1/2"	"	0.471
1-1/2"	"	0.500
1-1/2" x 2"	"	0.500
2"	"	0.533
2" x 1-1/2"	"	0.533
3"	"	0.667
Male x slip joint adapters		
1-1/2" x 1-1/4"	EA.	0.500
Dandy cleanout		
1-1/2"	EA.	0.500
2"	"	0.533
3"	"	0.667
End cleanout, flush pattern		
1-1/2" x 1"	EA.	0.500
2" x 1-1/2"	"	0.533
3" x 2-1/2"	"	0.667
Copper caps		
1-1/2"	EA.	0.500
2"	"	0.533
Closet flanges		
3"	EA.	0.667
4"	"	0.800
Drum traps, with cleanout		
1-1/2" x 3" x 6"	EA.	0.500
P-trap, swivel, with cleanout		
1-1/2"	EA.	0.500
P-trap, solder union		
1-1/2"	EA.	0.500

PLUMBING	UNIT	MAN/HOURS
15410.11 COPPER FITTINGS		
2"	EA.	0.533
With cleanout		
1-1/2"	EA.	0.500
2"	"	0.533
2" x 1-1/2"	"	0.533
Swivel joint, with cleanout		
1-1/2" x 1-1/4"	EA.	0.500
1-1/2"	"	0.500
2" x 1-1/2"	"	0.533
Estabrook TY, with inlets		
3", with 1-1/2" inlet	EA.	0.667
Fine thread adapters		
1/2"	EA.	0.400
1/2" x 1/2" IPS	"	0.400
1/2" x 3/4" IPS	"	0.400
1/2" x male	"	0.400
1/2" x female	"	0.400
15410.14 BRASS I.P.S. FITTINGS		
Fittings, iron pipe size, 45 deg ell		
1/8"	EA.	0.320
1/4"	"	0.320
3/8"	"	0.364
1/2"	"	0.400
3/4"	"	0.421
1"	"	0.444
1-1/4"	"	0.471
1-1/2"	"	0.500
2"	"	0.533
90 deg ell		
1/8"	EA.	0.320
1/4"	"	0.320
3/8"	"	0.364
1/2"	"	0.400
3/4"	"	0.421
1"	"	0.444
1-1/4"	"	0.471
1-1/2"	"	0.500
2"	"	0.533
2-1/2"	"	0.615
90 deg ell, reducing		
1/4" x 1/8"	EA.	0.320
3/8" x 1/8"	"	0.364
3/8" x 1/4"	"	0.364
1/2" x 1/4"	"	0.400
1/2" x 3/8"	"	0.400
3/4" x 1/2"	"	0.421
1" x 3/8"	"	0.444
1" x 1/2"	"	0.444
1" x 3/4"	"	0.444
1-1/4" x 3/4"	"	0.471
1-1/4" x 1"	"	0.471
1-1/2" x 1/2"	"	0.500

PLUMBING	UNIT	MAN/HOURS
15410.14 BRASS I.P.S. FITTINGS		
1-1/2" x 3/4"	EA.	0.500
1-1/2" x 1"	"	0.500
2" x 1"	"	0.533
2" x 1-1/4"	"	0.533
2" x 1-1/2"	"	0.533
Street ell, 45 deg		
1/2"	EA.	0.400
3/4"	"	0.421
2"	"	0.533
90 deg		
1/8"	EA.	0.320
1/4"	"	0.320
3/8"	"	0.364
1/2"	"	0.400
3/4"	"	0.421
1"	"	0.444
1-1/4"	"	0.471
1-1/2"	"	0.500
2"	"	0.533
Tee, 1/8"	"	0.320
1/4"	"	0.320
3/8"	"	0.364
1/2"	"	0.400
3/4"	"	0.421
1"	"	0.444
1-1/4"	"	0.471
1-1/2"	"	0.500
2"	"	0.533
2-1/2"	"	0.615
Tee, reducing, 3/8" x		
1/4"	EA.	0.364
1/2"	"	0.364
1/2" x		
1/4"	EA.	0.400
3/8"	"	0.400
3/4"	"	0.400
3/4" x		
1/4"	EA.	0.421
1/2"	"	0.421
1"	"	0.421
1" x		
1/2"	EA.	0.444
3/4"	"	0.444
1-1/4" x		
1/2"	EA.	0.471
1"	"	0.471
1-1/2" x		
1/2"	EA.	0.500
3/4"	"	0.500
1"	"	0.500
1-1/2"	"	0.500
2" x		
1/2"	EA.	0.533
3/4"	"	0.533

PLUMBING	UNIT	MAN/HOURS
15410.14 BRASS I.P.S. FITTINGS		
1"	EA.	0.533
1-1/4"	"	0.533
1-1/2"	"	0.533
2-1/2" x 2"	"	0.615
Tee, reducing		
1/2" x 3/8" x 1/2"	EA.	0.400
3/4" x 1/2" x 1/2"	"	0.421
3/4" x 1/2" x 3/4"	"	0.421
1" x 1/2" x 1/2"	"	0.444
1" x 1/2" x 3/4"	"	0.444
1" x 3/4" x 1/2"	"	0.444
1" x 3/4" x 3/4"	"	0.444
1-1/4" x 1/2" x 1-1/4"	"	0.471
1-1/4" x 1" x 1"	"	0.471
1-1/4" x 1-1/4" x 3/4"	"	0.471
1-1/2" x 1-1/4" x 1-1/4"	"	0.500
Union		
1/8"	EA.	0.320
1/4"	"	0.320
3/8"	"	0.364
1/2"	"	0.400
3/4"	"	0.421
1"	"	0.444
1-1/4"	"	0.471
1-1/2"	"	0.500
2"	"	0.533
Brass face bushing		
3/8" x 1/4"	EA.	0.364
1/2" x 3/8"	"	0.400
3/4" x 1/2"	"	0.421
1" x 3/4"	"	0.444
1-1/4" x 1"	"	0.471
Hex bushing, 1/4" x 1/8"	"	0.320
1/2" x		
1/4"	EA.	0.400
3/8"	"	0.400
5/8" x		
1/8"	EA.	0.400
1/4"	"	0.400
3/4" x		
1/8"	EA.	0.421
1/4"	"	0.421
3/8"	"	0.421
1/2"	"	0.421
1" x		
1/4"	EA.	0.444
3/8"	"	0.444
1/2"	"	0.444
3/4"	"	0.444
1-1/4" x		
3/8"	EA.	0.471
1/2"	"	0.471
3/4"	"	0.471
1"	"	0.471

PLUMBING	UNIT	MAN/HOURS
15410.14 BRASS I.P.S. FITTINGS		
1-1/2" x		
1/4"	EA.	0.500
1/2"	"	0.500
3/4"	"	0.500
1"	"	0.500
1-1/4"	"	0.500
2" x		
1/2"	EA.	0.533
3/4"	"	0.533
1"	"	0.533
1-1/4"	"	0.533
1-1/2"	"	0.533
2-1/2" x		
1"	EA.	0.615
2"	"	0.615
3" x		
1-1/2"	EA.	0.667
2"	"	0.667
2-1/2"	"	0.667
4" x		
2"	EA.	0.800
3"	"	0.800
Caps		
1/8"	EA.	0.320
1/4"	"	0.320
3/8"	"	0.364
1/2"	"	0.400
3/4"	"	0.421
1"	"	0.444
1-1/4"	"	0.471
1-1/2"	"	0.500
2"	"	0.533
Couplings		
1/8"	EA.	0.320
1/4"	"	0.320
3/8"	"	0.364
1/2"	"	0.400
3/4"	"	0.421
1"	"	0.444
1-1/4"	"	0.471
1-1/2"	"	0.500
2"	"	0.533
2-1/2"	"	0.615
Couplings, reducing, 1/4" x 1/8"	"	0.320
3/8" x		
1/8"	EA.	0.364
1/4"	"	0.364
1/2" x		
1/8"	EA.	0.400
1/4"	"	0.400
3/8"	"	0.400
3/4" x		
1/4"	EA.	0.421
3/8"	"	0.421

PLUMBING — 15410.14 BRASS I.P.S. FITTINGS

PLUMBING	UNIT	MAN/HOURS
1/2"	EA.	0.421
1/2"	"	0.421
3/4"	"	0.421
1-1/4" x		
1/2"	EA.	0.471
3/4"	"	0.471
1"	"	0.471
1-1/2" x		
3/4"	EA.	0.500
1"	"	0.500
1-1/4"	"	0.500
2" x		
3/4"	EA.	0.533
1"	"	0.533
1-1/4"	"	0.533
1-1/2"	"	0.533
2-1/2" x 1-1/2"	"	0.615
Square head plug, solid		
1/8"	EA.	0.320
1/4"	"	0.320
3/8"	"	0.364
1/2"	"	0.400
3/4"	"	0.421
Cored		
1/2"	EA.	0.400
3/4"	"	0.421
1"	"	0.444
1-1/4"	"	0.471
1-1/2"	"	0.500
2"	"	0.533
3"	"	0.667
4"	"	0.800
Countersunk		
1/2"	EA.	0.400
3/4"	"	0.421
1-1/2"	"	0.500
2"	"	0.533
Locknut		
3/4"	EA.	0.421
1"	"	0.444
1-1/4"	"	0.471
2"	"	0.533
Close standard red nipple, 1/8"	"	0.320
1/8" x		
1-1/2"	EA.	0.320
2"	"	0.320
2-1/2"	"	0.320
3"	"	0.320
3-1/2"	"	0.320
4"	"	0.320
4-1/2"	"	0.320
5"	"	0.320
5-1/2"	"	0.320
6"	"	0.320

PLUMBING — 15410.14 BRASS I.P.S. FITTINGS

PLUMBING	UNIT	MAN/HOURS
1/4" x close	EA.	0.320
1/4" x		
1-1/2"	EA.	0.320
2"	"	0.320
2-1/2"	"	0.320
3"	"	0.320
3-1/2"	"	0.320
4"	"	0.320
4-1/2"	"	0.320
5"	"	0.320
5-1/2"	"	0.320
6"	"	0.320
3/8" x close	"	0.364
3/8" x		
1-1/2"	EA.	0.364
2"	"	0.364
2-1/2"	"	0.364
3"	"	0.364
3-1/2"	"	0.364
4"	"	0.364
4-1/2"	"	0.364
5"	"	0.364
5-1/2"	"	0.364
6"	"	0.364
1/2" x close	"	0.400
1/2" x		
1-1/2"	EA.	0.400
2"	"	0.400
2-1/2"	"	0.400
3"	"	0.400
3-1/2"	"	0.400
4"	"	0.400
4-1/2"	"	0.400
5"	"	0.400
5-1/2"	"	0.400
6"	"	0.400
7-1/2"	"	0.400
8"	"	0.400
3/4" x close	"	0.421
3/4" x		
1-1/2"	EA.	0.421
2"	"	0.421
2-1/2"	"	0.421
3"	"	0.421
3-1/2"	"	0.421
4"	"	0.421
4-1/2"	"	0.421
5"	"	0.421
5-1/2"	"	0.421
6"	"	0.421
1" x close	"	0.444
1" x		
2"	EA.	0.444
2-1/2"	"	0.444

PLUMBING	UNIT	MAN/HOURS
15410.14 BRASS I.P.S. FITTINGS		
3"	EA.	0.444
3-1/2"	"	0.444
4"	"	0.444
4-1/2"	"	0.444
5"	"	0.444
5-1/2"	"	0.444
6"	"	0.444
1-1/4" x close	"	0.471
1-1/4" x		
2"	EA.	0.471
2-1/2"	"	0.471
3"	"	0.471
3-1/2"	"	0.471
4"	"	0.471
4-1/2"	"	0.471
5"	"	0.471
5-1/2"	"	0.471
6"	"	0.471
1-1/2" x close	"	0.500
1-1/2" x		
2"	EA.	0.500
2-1/2"	"	0.500
3"	"	0.500
3-1/2"	"	0.500
4-1/2"	"	0.500
5"	"	0.500
5-1/2"	"	0.500
6"	"	0.500
2" x close	"	0.533
2"		
2-1/2"	EA.	0.533
3"	"	0.533
3-1/2"	"	0.533
4"	"	0.533
4-1/2"	"	0.533
5"	"	0.533
5-1/2"	"	0.533
6"	"	0.533
2-1/2" x close	"	0.615
2-1/2" x		
3"	EA.	0.615
3-1/2"	"	0.615
4"	"	0.615
2-1/2" x 5"	"	0.615
3"		
Close	EA.	0.667
15410.15 BRASS FITTINGS		
Compression fittings, union		
3/8"	EA.	0.133
1/2"	"	0.133
5/8"	"	0.133
Union elbow		
3/8"	EA.	0.133

PLUMBING	UNIT	MAN/HOURS
15410.15 BRASS FITTINGS		
1/2"	EA.	0.133
5/8"	"	0.133
Union tee		
3/8"	EA.	0.133
1/2"	"	0.133
5/8"	"	0.133
Male connector		
3/8"	EA.	0.133
1/2"	"	0.133
5/8"	"	0.133
Female connector		
3/8"	EA.	0.133
1/2"	"	0.133
5/8"	"	0.133
Brass flare fittings, union		
3/8"	EA.	0.129
1/2"	"	0.129
5/8"	"	0.129
90 deg elbow union		
3/8"	EA.	0.129
1/2"	"	0.129
5/8"	"	0.129
Three way tee		
3/8"	EA.	0.216
1/2"	"	0.216
5/8"	"	0.216
Cross		
3/8"	EA.	0.286
1/2"	"	0.286
5/8"	"	0.286
Male connector, half union		
3/8"	EA.	0.129
1/2"	"	0.129
5/8"	"	0.129
Female connector, half union		
3/8"	EA.	0.129
1/2"	"	0.129
5/8"	"	0.129
Long forged nut		
3/8"	EA.	0.129
1/2"	"	0.129
5/8"	"	0.129
Short forged nut		
3/8"	EA.	0.129
1/2"	"	0.129
5/8"	"	0.129
Compression elbow		
1/4"	EA.	0.129
5/16"	"	0.129
3/4"	"	0.129
Sleeve		
1/8"	EA.	0.160
1/4"	"	0.160
5/16"	"	0.160

PLUMBING		UNIT	MAN/HOURS
15410.15	**BRASS FITTINGS**		
	3/8"	EA.	0.160
	1/2"	"	0.160
	5/8"	"	0.160
	3/4"	"	0.160
	7/8"	"	0.160
Tee			
	1/4"	EA.	0.229
	5/16"	"	0.229
Male tee			
	5/16" x 1/8"	EA.	0.229
Female union			
	1/8" x 1/8"	EA.	0.200
	1/4" x 3/8"	"	0.200
	3/8" x 1/4"	"	0.200
	3/8" x 1/2"	"	0.200
	5/8" x 1/2"	"	0.229
Male union, 1/4"			
	1/4" x 1/4"	EA.	0.200
	3/8"	"	0.200
	1/2"	"	0.200
5/16" x			
	1/8"	EA.	0.200
	1/4"	"	0.200
	3/8"	"	0.200
3/8" x			
	1/8"	EA.	0.200
	1/4"	"	0.200
	1/2"	"	0.200
	3/4"	"	0.200
1/2" x			
	1/4"	EA.	0.229
	3/8"	"	0.229
5/8" x			
	3/8"	EA.	0.229
	1/2"	"	0.229
	3/4"	"	0.229
	1/2"	"	0.229
7/8" x			
	1/2"	EA.	0.229
	3/4"	"	0.229
Female elbow, 1/4" x 1/4"		"	0.229
5/16" x			
	1/8"	EA.	0.229
	1/4"	"	0.229
3/8" x			
	3/8"	EA.	0.229
	1/2"	"	0.229
Male elbow, 1/8" x 1/8"		"	0.229
	3/16" x 1/4"	"	0.229
1/4" x			
	1/8"	EA.	0.229
	1/4"	"	0.229
	3/8"	"	0.229
	5/16" x		

PLUMBING		UNIT	MAN/HOURS
15410.15	**BRASS FITTINGS**		
	1/8"	EA.	0.229
	1/4"	"	0.229
	3/8"	"	0.229
	3/8" x		
	1/8"	EA.	0.229
	1/4"	"	0.229
	3/8"	"	0.229
	1/2"	"	0.229
1/2" x			
	1/4"	EA.	0.267
	3/8"	"	0.267
	1/2"	"	0.267
5/8" x			
	3/8"	EA.	0.267
	1/2"	"	0.267
	3/4"	"	0.267
3/4" x			
	1/2"	EA.	0.267
	3/4"	"	0.267
7/8" x			
	3/4"	EA.	0.267
Union			
	1/8"	EA.	0.229
	3/16"	"	0.229
	1/4"	"	0.229
	5/16"	"	0.229
	3/8"	"	0.229
	3/4"	"	0.229
	7/8"	"	0.229
Reducing union			
	3/8" x 1/4"	EA.	0.267
	5/8" x		
	3/8"	EA.	0.267
	1/2"	"	0.267
15410.17	**CHROME PLATED FITTINGS**		
Fittings			
90 ell			
	3/8"	EA.	0.200
	1/2"	"	0.200
45 ell			
	3/8"	EA.	0.200
	1/2"	"	0.200
Tee			
	3/8"	EA.	0.267
	1/2"	"	0.267
Coupling			
	3/8"	EA.	0.200
	1/2"	"	0.200
Union			
	3/8"	EA.	0.200
	1/2"	"	0.200
Tee			
	1/2" x 3/8" x 3/8"	EA.	0.267

PLUMBING

15410.17   CHROME PLATED FITTINGS

	UNIT	MAN/HOURS
1/2" x 3/8" x 1/2"	EA.	0.267

15410.18   GLASS PIPE

	UNIT	MAN/HOURS
Glass pipe		
1-1/2" dia.	L.F.	0.160
2" dia.	"	0.178
3" dia.	"	0.200
4" dia.	"	0.229
6" dia.	"	0.267

15410.30   PVC/CPVC PIPE

	UNIT	MAN/HOURS
PVC schedule 40		
1/2" pipe	L.F.	0.033
3/4" pipe	"	0.036
1" pipe	"	0.040
1-1/4" pipe	"	0.044
1-1/2" pipe	"	0.050
2" pipe	"	0.057
2-1/2" pipe	"	0.067
3" pipe	"	0.080
4" pipe	"	0.100
6" pipe	"	0.200
8" pipe	"	0.267
Fittings, 1/2"		
90 deg ell	EA.	0.100
45 deg ell	"	0.100
Tee	"	0.114
Reducing insert	"	0.133
Threaded	"	0.100
Male adapter	"	0.133
Female adapter	"	0.100
Union	"	0.160
Cap	"	0.133
Flange	"	0.160
3/4"		
90 deg elbow	EA.	0.133
45 deg elbow	"	0.133
Tee	"	0.160
Reducing insert	"	0.114
Threaded	"	0.133
1"		
90 deg elbow	EA.	0.160
45 deg elbow	"	0.160
Tee	"	0.178
Reducing insert	"	0.160
Threaded	"	0.178
Male adapter	"	0.200
Female adapter	"	0.200
Union	"	0.267
Cap	"	0.160
Flange	"	0.267
1-1/4"		

PLUMBING

15410.30   PVC/CPVC PIPE

	UNIT	MAN/HOURS
90 deg elbow	EA.	0.229
45 deg elbow	"	0.229
Tee	"	0.267
Reducing insert	"	0.267
Threaded	"	0.267
Male adapter	"	0.267
Female adapter	"	0.267
Union	"	0.320
Cap	"	0.267
Flange	"	0.320
1-1/2"		
90 deg elbow	EA.	0.229
45 deg elbow	"	0.229
Tee	"	0.267
Reducing insert	"	0.267
Threaded	"	0.267
Male adapter	"	0.267
Female adapter	"	0.267
Union	"	0.400
Cap	"	0.267
Flange	"	0.400
2"		
90 deg elbow	EA.	0.267
45 deg elbow	"	0.267
Tee	"	0.320
Reducing insert	"	0.320
Threaded	"	0.320
Male adapter	"	0.320
Female adapter	"	0.320
Union	"	0.500
Cap	"	0.320
Flange	"	0.500
2-1/2"		
90 deg elbow	EA.	0.500
45 deg elbow	"	0.500
Tee	"	0.533
Reducing insert	"	0.533
Threaded	"	0.533
Male adapter	"	0.533
Female adapter	"	0.533
Union	"	0.667
Cap	"	0.500
Flange	"	0.667
3"		
90 deg elbow	EA.	0.667
45 deg elbow	"	0.667
Tee	"	0.727
Reducing insert	"	0.667
Threaded	"	0.667
Male adapter	"	0.667
Female adapter	"	0.667
Union	"	0.800
Cap	"	0.667
Flange	"	0.800

PLUMBING	UNIT	MAN/HOURS
15410.30 PVC/CPVC PIPE		
4"		
90 deg elbow	EA.	0.800
45 deg elbow	"	0.800
Tee	"	0.889
Reducing insert	"	0.800
Threaded	"	0.800
Male adapter	"	0.800
Female adapter	"	0.800
Union	"	1.000
Cap	"	0.800
Flange	"	1.000
PVC schedule 80 pipe		
1-1/2" pipe	L.F.	0.050
2" pipe	"	0.057
3" pipe	"	0.080
4" pipe	"	0.100
Fittings, 1-1/2"		
90 deg elbow	EA.	0.267
45 deg elbow	"	0.267
Tee	"	0.400
Reducing insert	"	0.267
Threaded	"	0.267
Male adapter	"	0.267
Female adapter	"	0.267
Union	"	0.400
Cap	"	0.267
Flange	"	0.400
2"		
90 deg elbow	EA.	0.320
45 deg elbow	"	0.320
Tee	"	0.500
Reducing insert	"	0.320
Threaded	"	0.320
Male adapter	"	0.320
Female adapter	"	0.320
2-1/2"		
90 deg elbow	EA.	0.500
45 deg elbow	"	0.500
Tee	"	0.667
Reducing insert	"	0.500
Threaded	"	0.500
Male adapter	"	0.500
Female adapter	"	0.500
Union	"	0.667
Cap	"	0.500
Flange	"	0.667
3"		
90 deg elbow	EA.	0.667
45 deg elbow	"	0.667
Tee	"	0.800
Reducing insert	"	0.667
Threaded	"	0.667
Male adapter	"	0.667
Female adapter	"	0.667

PLUMBING	UNIT	MAN/HOURS
15410.30 PVC/CPVC PIPE		
Union	EA.	0.800
Cap	"	0.667
Flange	"	0.800
4"		
90 deg elbow	EA.	0.800
45 deg elbow	"	0.800
Tee	"	1.000
Reducing insert	"	0.800
Threaded	"	0.800
Male adapter	"	0.800
Union	"	1.000
Cap	"	0.800
Flange	"	1.000
CPVC schedule 40		
1/2" pipe	L.F.	0.033
3/4" pipe	"	0.036
1" pipe	"	0.040
1-1/4" pipe	"	0.044
1-1/2" pipe	"	0.050
2" pipe	"	0.057
Fittings, CPVC, schedule 80		
1/2", 90 deg ell	EA.	0.080
Tee	"	0.133
3/4", 90 deg ell	"	0.080
Tee	"	0.133
1", 90 deg ell	"	0.089
Tee	"	0.145
1-1/4", 90 deg ell	"	0.089
Tee	"	0.145
1-1/2", 90 deg ell	"	0.160
Tee	"	0.200
2", 90 deg ell	"	0.160
Tee	"	0.200
Polypropylene, acid resistant, DWV pipe		
Schedule 40		
1-1/2" pipe	L.F.	0.057
2" pipe	"	0.067
3" pipe	"	0.080
4" pipe	"	0.100
6" pipe	"	0.200
Fittings		
1-1/2"		
1/4 bend	EA.	0.200
1/8 bend	"	0.200
Sanitary tee	"	0.400
Cleanout with plug	"	0.400
Wye	"	0.400
Combination wye & 1/8 bend	"	0.400
P-trap	"	0.500
Hub adapter	"	0.200
Coupling	"	0.200
2"		
1/4 bend	EA.	0.229
1/8 bend	"	0.229

PLUMBING	UNIT	MAN/HOURS
15410.30 **PVC/CPVC PIPE**		
Sanitary tee	EA.	0.471
Cleanout with plug	"	0.471
Wye	"	0.471
Combination wye & 1/8 bend	"	0.471
P-trap	"	0.667
Hub adapter	"	0.267
Mechanical joint adapter	"	0.267
Coupling	"	0.267
3"		
1/4 bend	EA.	0.267
1/8 bend	"	0.267
Sanitary tee	"	0.533
Cleanout with plug	"	0.533
Wye	"	0.533
Combination wye & 1/8 bend	"	0.533
P-trap	"	0.800
Hub adapter	"	0.267
Mechanical joint adapter	"	0.267
Coupling	"	0.267
4"		
1/4 bend	EA.	0.400
1/8 bend	"	0.400
Sanitary tee	"	0.800
Cleanout with plug	"	0.800
Wye	"	0.800
Combination wye & 1/8 bend	"	0.800
P-trap	"	1.333
Hub adapter	"	0.400
Mechanical joint adapter	"	0.400
Coupling	"	0.400
6"		
1/4 bend	EA.	0.667
1/8 bend	"	0.667
Sanitary tee	"	1.333
Cleanout with plug	"	1.333
Wye	"	1.333
Hub adapter	"	0.667
Coupling	"	0.667
Polyethylene pipe and fittings		
SDR-21		
3" pipe	L.F.	0.100
4" pipe	"	0.133
6" pipe	"	0.200
8" pipe	"	0.229
10" pipe	"	0.267
12" pipe	"	0.320
14" pipe	"	0.400
16" pipe	"	0.500
18" pipe	"	0.615
20" pipe	"	0.800
22" pipe	"	0.889
24" pipe	"	1.000
Fittings, 3"		
90 deg elbow	EA.	0.400

PLUMBING	UNIT	MAN/HOURS
15410.30 **PVC/CPVC PIPE**		
45 deg elbow	EA.	0.400
Tee	"	0.667
45 deg wye	"	0.667
Reducer	"	0.500
Flange assembly	"	0.400
4"		
90 deg elbow	EA.	0.500
45 deg elbow	"	0.500
Tee	"	0.800
45 deg wye	"	0.800
Reducer	"	0.667
Flange assembly	"	0.500
8"		
90 deg elbow	EA.	1.000
45 deg elbow	"	1.000
Tee	"	1.600
45 deg wye	"	1.600
Reducer	"	1.333
Flange assembly	"	1.000
10"		
90 deg elbow	EA.	1.333
45 deg elbow	"	1.333
Tee	"	2.000
45 deg wye	"	2.000
Reducer	"	1.600
Flange assembly	"	1.333
12"		
90 deg elbow	EA.	1.600
45 deg elbow	"	1.600
Tee	"	2.667
45 deg wye	"	2.667
Reducer	"	2.000
Flange assembly	"	1.600
14"		
90 deg elbow	EA.	2.000
45 deg elbow	"	2.000
Tee	"	3.200
45 deg wye	"	3.200
Reducer	"	2.667
Flange assembly	"	2.000
16"		
90 deg elbow	EA.	2.000
45 deg elbow	"	2.000
Tee	"	3.200
45 deg wye	"	3.200
Reducer	"	2.667
Flange assembly	"	2.000
18"		
90 deg elbow	EA.	2.667
45 deg elbow	"	2.667
Tee	"	4.000
45 deg wye	"	4.000
Reducer	"	2.667
Flange assembly	"	2.667

PLUMBING

15410.30 PVC/CPVC PIPE

	UNIT	MAN/HOURS
20"		
90 deg elbow	EA.	2.667
45 deg elbow	"	2.667

15410.33 ABS DWV PIPE

	UNIT	MAN/HOURS
Schedule 40 ABS		
1-1/2" pipe	L.F.	0.040
2" pipe	"	0.044
3" pipe	"	0.057
4" pipe	"	0.080
6" pipe	"	0.100
Fittings		
1/8 bend		
1-1/2"	EA.	0.160
2"	"	0.200
3"	"	0.267
4"	"	0.320
6"	"	0.400
Tee, sanitary		
1-1/2"	EA.	0.267
2"	"	0.320
3"	"	0.400
4"	"	0.500
6"	"	0.667
Tee, sanitary reducing		
2 x 1-1/2 x 1-1/2	EA.	0.320
2 x 1-1/2 x 2	"	0.333
2 x 2 x 1-1/2	"	0.364
3 x 3 x 1-1/2	"	0.400
3 x 3 x 2	"	0.444
4 x 4 x 1-1/2	"	0.500
4 x 4 x 2	"	0.571
4 x 4 x 3	"	0.615
6 x 6 x 4	"	0.667
Wye		
1-1/2"	EA.	0.229
2"	"	0.320
3"	"	0.400
4"	"	0.500
6"	"	0.667
Reducer		
2 x 1-1/2	EA.	0.200
3 x 1-1/2	"	0.267
3 x 2	"	0.267
4 x 2	"	0.320
4 x 3	"	0.320
6 x 4	"	0.400
P-trap		
1-1/2"	EA.	0.267
2"	"	0.296
3"	"	0.348
4"	"	0.400
6"	"	0.500

PLUMBING

15410.33 ABS DWV PIPE

	UNIT	MAN/HOURS
Double sanitary, tee		
1-1/2"	EA.	0.320
2"	"	0.400
3"	"	0.500
4"	"	0.667
Long sweep, 1/4 bend		
1-1/2"	EA.	0.160
2"	"	0.200
3"	"	0.267
4"	"	0.400
Wye, standard		
1-1/2"	EA.	0.267
2"	"	0.320
3"	"	0.400
4"	"	0.500
Wye, reducing		
2 x 1-1/2 x 1-1/2	EA.	0.267
2 x 2 x 1-1/2	"	0.320
4 x 4 x 2	"	0.500
4 x 4 x 3	"	0.533
Double wye		
1-1/2"	EA.	0.320
2"	"	0.400
3"	"	0.500
4"	"	0.667
2 x 2 x 1-1/2 x 1-1/2	"	0.400
3 x 3 x 2 x 2	"	0.500
4 x 4 x 3 x 3	"	0.667
Combination wye and 1/8 bend		
1-1/2"	EA.	0.267
2"	"	0.320
3"	"	0.400
4"	"	0.500
2 x 2 x 1-1/2	"	0.320
3 x 3 x 1-1/2	"	0.400
3 x 3 x 2	"	0.400
4 x 4 x 2	"	0.500
4 x 4 x 3	"	0.500

15410.35 PLASTIC PIPE

	UNIT	MAN/HOURS
Fiberglass reinforced pipe		
2" pipe	L.F.	0.062
3" pipe	"	0.067
4" pipe	"	0.073
6" pipe	"	0.080
8" pipe	"	0.133
10" pipe	"	0.160
12" pipe	"	0.200
Fittings		
90 deg elbow, flanged		
2"	EA.	0.800
3"	"	0.889
4"	"	1.000
6"	"	1.333

Left Column

PLUMBING	UNIT	MAN/HOURS
15410.35 PLASTIC PIPE		
8"	EA.	1.600
10"	"	2.000
12"	"	2.667
45 deg elbow, flanged		
2"	EA.	0.667
3"	"	0.800
4"	"	1.000
6"	"	1.333
8"	"	1.600
10"	"	2.000
12"	"	2.667
Tee, flanged		
2"	EA.	1.000
3"	"	1.143
4"	"	1.333
6"	"	1.600
8"	"	2.000
10"	"	2.667
12"	"	4.000
Wye, flanged		
2"	EA.	1.000
3"	"	1.143
4"	"	1.333
6"	"	1.600
8"	"	2.000
10"	"	2.667
12"	"	4.000
Concentric reducer, flanged		
2"	EA.	0.667
4"	"	0.800
6"	"	1.143
8"	"	1.600
10"	"	2.000
12"	"	2.667
Adapter, bell x male or female		
2"	EA.	0.667
3"	"	0.727
4"	"	0.800
6"	"	1.143
8"	"	1.600
10"	"	2.000
12"	"	2.667
Nipples		
2" x 6"	EA.	0.080
2" x 12"	"	0.100
3" x 8"	"	0.123
3" x 12"	"	0.133
4" x 8"	"	0.133
4" x 12"	"	0.160
6" x 12"	"	0.200
8" x 18"	"	0.200
8" x 24"	"	0.229
10" x 18"	"	0.267
10" x 24"	"	0.320

Right Column

PLUMBING	UNIT	MAN/HOURS
15410.35 PLASTIC PIPE		
12" x 18"	EA.	0.364
12" x 24"	"	0.400
Sleeve coupling		
2"	EA.	0.667
3"	"	0.800
4"	"	1.143
6"	"	1.600
8"	"	2.000
10"	"	2.667
Flanges		
2"	EA.	0.667
3"	"	0.800
4"	"	1.143
6"	"	1.600
8"	"	2.000
10"	"	2.667
12"	"	2.667
15410.70 STAINLESS STEEL PIPE		
Stainless steel, schedule 40, threaded		
1/2" pipe	L.F.	0.114
3/4" pipe	"	0.118
1" pipe	"	0.123
1-1/2" pipe	"	0.133
2" pipe	"	0.145
2-1/2" pipe	"	0.160
3" pipe	"	0.178
4" pipe	"	0.200
Fittings, 1/2"		
90 deg ell	EA.	1.000
45 deg ell	"	1.000
Tee	"	1.333
Cap	"	0.500
Reducer	"	0.667
Union	"	1.000
Flange	"	1.000
3/4"		
90 deg ell	EA.	1.000
45 deg ell	"	1.000
Tee	"	1.333
Cap	"	0.500
Reducer	"	0.667
Union	"	1.000
Flange	"	1.000
1"		
90 deg ell	EA.	1.000
45 deg ell	"	1.000
Tee	"	1.333
Cap	"	0.500
Reducer	"	1.000
Union	"	1.000
Flange	"	1.000
1-1/4"		

PLUMBING	UNIT	MAN/HOURS
15410.70 STAINLESS STEEL PIPE		
90 deg ell	EA.	1.000
45 deg ell	"	1.000
Tee	"	1.333
Cap	"	0.500
Union	"	1.000
Flange	"	1.000
1-1/2"		
90 deg ell	EA.	1.333
45 deg ell	"	1.333
Tee	"	1.600
Cap	"	0.667
Reducer	"	0.800
Union	"	1.143
Flange	"	1.143
2"		
90 deg ell	EA.	1.600
45 deg ell	"	1.600
Tee	"	2.667
Cap	"	0.800
Reducer	"	1.000
Union	"	1.600
Flange	"	1.600
Type 304, sch 10 pipe		
1" pipe	L.F.	0.100
1-1/4" pipe	"	0.133
1-1/2" pipe	"	0.145
2" pipe	"	0.178
2-1/2" pipe	"	0.200
3" pipe	"	0.229
4" pipe	"	0.267
6" pipe	"	0.320
Fittings, 1"		
90 deg elbow	EA.	1.600
45 deg elbow	"	1.600
Tee	"	2.667
Reducer	"	1.600
Flange	"	1.000
Cap	"	1.000
1-1/4"		
90 deg elbow	EA.	1.600
45 deg elbow	"	1.600
Tee	"	2.667
Reducer	"	1.600
Flange	"	1.000
Cap	"	1.000
1-1/2"		
90 deg elbow	EA.	1.333
45 deg elbow	"	1.333
Tee	"	2.667
Reducer	"	2.000
Flange	"	1.333
Cap	"	1.333
2"		
90 deg elbow	EA.	1.600

PLUMBING	UNIT	MAN/HOURS
15410.70 STAINLESS STEEL PIPE		
45 deg elbow	EA.	1.600
Tee	"	4.000
Reducer	"	2.667
Flange	"	2.000
Cap	"	2.000
2-1/2"		
90 deg elbow	EA.	1.600
45 deg elbow	"	1.600
Tee	"	4.000
Reducer	"	2.667
Flange	"	2.000
Cap	"	2.000
3"		
90 deg elbow	EA.	2.000
45 deg elbow	"	2.000
Tee	"	5.333
Reducer	"	4.000
Flange	"	2.667
Cap	"	2.667
4"		
90 deg elbow	EA.	2.000
45 deg elbow	"	2.000
Reducer	"	4.000
Flange	"	2.667
Cap	"	2.667
6"		
90 deg elbow	EA.	2.667
45 deg elbow	"	2.667
Tee	"	11.429
Reducer	"	4.000
Flange	"	4.000
Cap	"	4.000
Type 304 tubing		
.035 wall		
1/4"	L.F.	0.044
3/8"	"	0.050
1/2"	"	0.057
5/8"	"	0.067
3/4"	"	0.080
7/8"	"	0.089
1"	"	0.100
.049 wall		
1/4"	L.F.	0.047
3/8"	"	0.053
1/2"	"	0.062
5/8"	"	0.073
3/4"	"	0.089
7/8"	"	0.100
1"	"	0.114
.065 wall		
1/4"	L.F.	0.053
3/8"	"	0.067
1/2"	"	0.073
5/8"	"	0.089

PLUMBING	UNIT	MAN/HOURS
15410.70 STAINLESS STEEL PIPE		
3/4"	L.F.	0.114
7/8"	"	0.133
1"	"	0.160
Type 316 tubing		
.035 wall		
1/4"	L.F.	0.044
3/8"	"	0.050
1/2"	"	0.057
5/8"	"	0.067
3/4"	"	0.080
7/8"	"	0.089
1"	"	0.100
.049 wall		
1/4"	L.F.	0.053
3/8"	"	0.067
1/2"	"	0.073
5/8"	"	0.089
3/4"	"	0.114
7/8"	"	0.133
1"	"	0.160
.065 wall		
1/4"	L.F.	0.053
3/8"	"	0.067
1/2"	"	0.073
5/8"	"	0.089
3/4"	"	0.114
7/8"	"	0.133
1"	"	0.160
Fittings, 1/4"		
90 deg elbow	EA.	0.160
Union tee	"	0.267
Union	"	0.267
Male connector	"	0.200
3/8"		
90 deg elbow	EA.	0.200
Union tee	"	0.308
Union	"	0.308
Male connector	"	0.200
1/2"		
90 deg elbow	EA.	0.211
Union tee	"	0.333
Union	"	0.333
Male connector	"	0.200
5/8"		
90 deg elbow	EA.	0.267
Union tee	"	0.400
Union	"	0.400
Male connector	"	0.267
3/4"		
90 deg elbow	EA.	0.267
Union tee	"	0.400
Union	"	0.400
Male connector	"	0.267
7/8"		

PLUMBING	UNIT	MAN/HOURS
15410.70 STAINLESS STEEL PIPE		
90 deg elbow	EA.	0.286
Union tee	"	0.444
Union	"	0.444
Male connector	"	0.286
1"		
90 deg elbow	EA.	0.364
Union tee	"	0.500
Union	"	0.500
Male connector	"	0.400
Type 316 valves		
Gate valves		
1/4"	EA.	0.267
3/8"	"	0.320
1/2"	"	0.348
3/4"	"	0.400
1"	"	0.533
Globe valves		
1/4"	EA.	0.267
3/8"	"	0.320
1/2"	"	0.348
3/4"	"	0.400
1"	"	0.533
Check valves		
1/4"	EA.	0.267
3/8"	"	0.320
1/2"	"	0.348
3/4"	"	0.400
1"	"	0.533
15410.80 STEEL PIPE		
Black steel, extra heavy pipe, threaded		
1/2" pipe	L.F.	0.032
3/4" pipe	"	0.032
1" pipe	"	0.040
1-1/2" pipe	"	0.044
2-1/2" pipe	"	0.100
3" pipe	"	0.133
4" pipe	"	0.160
5" pipe	"	0.200
6" pipe	"	0.200
8" pipe	"	0.267
10" pipe	"	0.320
12" pipe	"	0.400
Fittings, malleable iron, threaded, 1/2" pipe		
90 deg ell	EA.	0.267
45 deg ell	"	0.267
Tee	"	0.400
Reducing tee	"	0.400
Cap	"	0.160
Coupling	"	0.320
Union	"	0.267
Nipple, 4" long	"	0.267
3/4" pipe		
90 deg ell	EA.	0.267

15410.80   STEEL PIPE

PLUMBING	UNIT	MAN/HOURS
45 deg ell	EA.	0.400
Tee	"	0.400
Reducing tee	"	0.267
Cap	"	0.160
Coupling	"	0.267
Union	"	0.267
Nipple, 4" long	"	0.267
1" pipe		
90 deg ell	EA.	0.320
45 deg ell	"	0.320
Tee	"	0.444
Reducing tee	"	0.444
Cap	"	0.160
Coupling	"	0.320
Union	"	0.320
Nipple, 4" long	"	0.320
1-1/2" pipe		
90 deg ell	EA.	0.400
45 deg ell	"	0.400
Tee	"	0.571
Reducing tee	"	0.571
Cap	"	0.200
Coupling	"	0.400
Union	"	0.400
Nipple, 4" long	"	0.400
2-1/2" pipe		
90 deg ell	EA.	1.000
45 deg ell	"	1.000
Tee	"	1.333
Reducing tee	"	1.333
Cap	"	0.500
Coupling	"	1.333
Union	"	1.333
Nipple, 4" long	"	1.333
3" pipe		
90 deg ell	EA.	1.333
45 deg ell	"	1.333
Tee	"	2.000
Reducing tee	"	2.000
Cap	"	0.667
Coupling	"	1.333
Union	"	1.333
Nipple, 4" long	"	1.333
4" pipe		
90 deg ell	EA.	1.600
45 deg ell	"	1.600
Tee	"	2.667
Reducing tee	"	2.667
Cap	"	2.667
Coupling	"	0.800
Union	"	2.667
Nipple, 4" long	"	2.667
6" pipe		
90 deg ell	EA.	1.600

15410.80   STEEL PIPE

PLUMBING	UNIT	MAN/HOURS
45 deg ell	EA.	1.600
Tee	"	2.667
Reducing tee	"	2.667
Cap	"	0.800
8" pipe		
90 deg ell	EA.	3.200
45 deg ell	"	3.200
Tee	"	5.000
Reducing tee	"	4.211
Cap	"	1.600
10" pipe		
90 deg ell	EA.	4.000
45 deg ell	"	4.000
Tee	"	5.000
Reducing tee	"	2.000
Cap	"	2.000
12" pipe		
90 deg ell	EA.	5.000
45 deg ell	"	5.000
Tee	"	6.667
Reducing tee	"	6.667
Cap	"	2.667
Butt welded, 1/2" pipe		
90 deg ell	EA.	0.267
45 deg ell	"	0.267
Tee	"	0.400
3/4" pipe		
90 deg ell	EA.	0.267
45 deg. ell	"	0.267
Tee	"	0.400
1" pipe		
90 deg ell	EA.	0.320
45 deg ell	"	0.320
Tee	"	0.444
1-1/2" pipe		
90 deg ell	EA.	0.400
45 deg. ell	"	0.400
Tee	"	0.571
Reducing tee	"	0.571
Cap	"	0.320
2-1/2" pipe		
90 deg. ell	EA.	0.800
45 deg. ell	"	0.800
Tee	"	1.143
Reducing tee	"	1.143
Cap	"	0.400
3" pipe		
90 deg ell	EA.	1.000
45 deg. ell	"	1.000
Tee	"	1.333
Reducing tee	"	1.333
Cap	"	0.667
4" pipe		
90 deg ell	EA.	1.333

PLUMBING	UNIT	MAN/HOURS
15410.80 STEEL PIPE		
45 deg. ell	EA.	1.333
Tee	"	2.000
Reducing tee	"	2.000
Cap	"	0.667
6" pipe		
90 deg. ell	EA.	1.600
45 deg. ell	"	1.600
Tee	"	2.667
Reducing tee	"	2.667
Cap	"	0.800
8" pipe		
90 deg. ell	EA.	2.667
45 deg. ell	"	2.667
Tee	"	4.000
Reducing tee	"	4.000
Cap	"	1.600
10" pipe		
90 deg ell	EA.	2.667
45 deg. ell	"	2.667
Tee	"	4.000
Reducing tee	"	4.000
Cap	"	2.000
12" pipe		
90 deg. ell	EA.	3.200
45 deg. ell	"	3.200
Tee	"	5.714
Reducing tee	"	5.714
Cap	"	2.000
90 deg. ell	"	0.267
45 deg. ell	"	0.267
Tee	"	0.400
Reducing tee	"	0.400
3/4" pipe		
90 deg. ell	EA.	0.267
45 deg. ell	"	0.267
Tee	"	0.400
Reducing tee	"	0.400
1" pipe		
90 deg. ell	EA.	0.320
45 deg. ell	"	0.320
Tee	"	0.444
Reducing tee	"	0.444
1-1/2" pipe		
90 deg. ell	EA.	0.400
45 deg. ell	"	0.400
Tee	"	0.571
Reducing tee	"	0.571
2-1/2" pipe		
90 deg. ell	EA.	0.800
45 deg. ell	"	0.800
Tee	"	1.143
Reducing tee	"	1.143
3" pipe		
90 deg. ell	EA.	1.000

PLUMBING	UNIT	MAN/HOURS
15410.80 STEEL PIPE		
45 deg. ell	EA.	1.000
Tee	"	1.600
Reducing tee	"	1.600
4" pipe		
90 deg. ell	EA.	1.333
45 deg. ell	"	1.333
Tee	"	2.000
Reducing tee	"	2.000
6" pipe		
90 deg. ell	EA.	1.333
45 deg. ell	"	1.333
Tee	"	2.000
Reducing tee	"	2.000
8" pipe		
90 deg. ell	EA.	2.667
45 deg. ell	"	2.667
Tee	"	4.000
Reducing tee	"	4.000
15410.82 GALVANIZED STEEL PIPE		
Galvanized pipe		
1/2" pipe	L.F.	0.080
3/4" pipe	"	0.100
1" pipe	"	0.114
1-1/4" pipe	"	0.133
1-1/2" pipe	"	0.160
2" pipe	"	0.200
2-1/2" pipe	"	0.267
3" pipe	"	0.286
4" pipe	"	0.333
6" pipe	"	0.667
90 degree ell, 150 lb malleable iron, galvanized		
1/2"	EA.	0.160
3/4"	"	0.200
1"	"	0.211
1-1/4"	"	0.235
1-1/2"	"	0.267
2"	"	0.320
2-1/2"	EA.	0.500
3"	"	0.615
4"	"	0.667
5"	"	0.800
6"	"	0.800
45 degree ell, 150 lb m.i., galv.		
1/2"	EA.	0.160
3/4"	"	0.200
1"	"	0.211
1-1/4"	"	0.235
1-1/2"	"	0.267
2"	"	0.320
2-1/2"	"	0.500
3"	"	0.615
4"	"	0.800
5"	"	0.800

PLUMBING	UNIT	MAN/HOURS
15410.82 GALVANIZED STEEL PIPE		
6"	EA.	1.000
Tees, straight, 150 lb m.i., galv.		
1/2"	EA.	0.200
3/4"	"	0.229
1"	"	0.267
1-1/4"	"	0.320
1-1/2"	"	0.400
2"	"	0.500
2-1/2"	"	0.667
3"	"	0.800
4"	"	1.000
5"	"	1.143
6"	"	1.333
Tees, reducing, out, 150 lb m.i., galv.		
1/2"	EA.	0.200
3/4"	"	0.229
1"	"	0.267
1-1/4"	"	0.320
1-1/2"	"	0.400
2"	"	0.500
2-1/2"	"	0.667
3"	"	0.800
4"	"	1.000
5"	"	1.143
6"	"	1.333
Couplings, straight, 150 lb m.i., galv.		
1/2"	EA.	0.160
3/4"	"	0.178
1"	"	0.200
1-1/4"	"	0.229
1-1/2"	"	0.267
2"	"	0.320
2-1/2"	"	0.500
3"	"	0.667
4"	"	0.727
5"	"	0.800
6"	"	0.800
Couplings, reducing, 150 lb m.i., galv		
1/2"	EA.	0.160
3/4"	"	0.178
1"	"	0.200
1-1/4"	"	0.229
1-1/2"	"	0.267
2"	"	0.320
2-1/2"	"	0.500
3"	"	0.667
4"	"	0.727
5"	"	0.800
6"	"	0.800
Caps, 150 lb m.i., galv.		
1/2"	EA.	0.080
3/4"	"	0.084
1"	"	0.089
1-1/4"	"	0.094

PLUMBING	UNIT	MAN/HOURS
15410.82 GALVANIZED STEEL PIPE		
1-1/2"	EA.	0.100
2"	"	0.114
2-1/2"	"	0.145
3"	"	0.200
4"	"	0.250
5"	"	0.308
6"	"	0.400
Unions, 150 lb m.i., galv.		
1/2"	EA.	0.200
3/4"	"	0.229
1"	"	0.267
1-1/4"	"	0.320
1-1/2"	"	0.400
2"	"	0.444
2-1/2"	"	0.533
3"	"	0.667
Nipples, galvanized steel, 4" long		
1/2"	EA.	0.100
3/4"	"	0.107
1"	"	0.114
1-1/4"	"	0.123
1-1/2"	"	0.133
2"	"	0.145
2-1/2"	"	0.160
3"	"	0.200
4"	"	0.267
90 degree reducing ell, 150 lb m.i., galv.		
3/4" x 1/2"	EA.	0.160
1" x 3/4"	"	0.178
1-1/4" x 1"	"	0.200
1-1/4" x 3/4"	"	0.229
1-1/4" x 1/2"	"	0.267
1-1/2" x 1-1/4"	"	0.267
1-1/2" x 1"	"	0.267
1-1/2" x 3/4"	"	0.267
2" x 1-1/2"	"	0.348
2" x 1-1/4"	"	0.348
2" x 1"	"	0.348
2" x 3/4"	"	0.348
2-1/2" x 2"	"	0.500
2-1/2" x 1-1/2"	"	0.500
3" x 2-1/2"	"	0.667
3" x 2"	"	0.667
4" x 3"	"	0.727
Square head plug (C.I.)		
1/2"	EA.	0.089
3/4"	"	0.100
1"	"	0.107
1-1/4"	"	0.114
1-1/2"	"	0.123
2"	"	0.133
2-1/2"	"	0.178
3"	"	0.200
4"	"	0.267

PLUMBING		UNIT	MAN/HOURS
15410.82	**GALVANIZED STEEL PIPE**		
5"		EA.	0.320
6"		"	0.400
Screwed flanges, galv.			
1"		EA.	0.400
1-1/4"		"	0.444
1-1/2"		"	0.500
2"		"	0.500
2-1/2"		"	0.533
3"		"	0.727
4"		"	1.000
5"		"	1.333
6"		"	1.333
15430.23	**CLEANOUTS**		
Cleanout, wall			
2"		EA.	0.533
3"		"	0.533
4"		"	0.667
6"		"	0.800
8"		"	1.000
Floor			
2"		EA.	0.667
3"		"	0.667
4"		"	0.800
6"		"	1.000
8"		"	1.143
15430.24	**GREASE TRAPS**		
Grease traps, cast iron, 3" pipe			
35 gpm, 70 lb capacity		EA.	8.000
50 gpm, 100 lb capacity		"	10.000
15430.25	**HOSE BIBBS**		
Hose bibb			
1/2"		EA.	0.267
3/4"		"	0.267
15430.60	**VALVES**		
Gate valve, 125 lb, bronze, soldered			
1/2"		EA.	0.200
3/4"		"	0.200
1"		"	0.267
1-1/2"		"	0.320
2"		"	0.400
2-1/2"		"	0.500
Threaded			
1/4", 125 lb		EA.	0.320
1/2"			
125 lb		EA.	0.320
150 lb		"	0.320
300 lb		"	0.320
3/4"			

PLUMBING		UNIT	MAN/HOURS
15430.60	**VALVES**		
125 lb		EA.	0.320
150 lb		"	0.320
300 lb		"	0.320
1"			
125 lb		EA.	0.320
150 lb		"	0.320
300 lb		"	0.400
1-1/2"			
125 lb		EA.	0.400
150 lb		"	0.400
300 lb		"	0.444
2"			
125 lb		EA.	0.571
150 lb		"	0.571
300 lb		"	0.667
Cast iron, flanged			
2", 150 lb		EA.	0.667
2-1/2"			
125 lb		EA.	0.667
150 lb		"	0.667
250 lb		"	0.667
3"			
125 lb		EA.	0.800
150 lb		"	0.800
250 lb		"	0.800
4"			
125 lb		EA.	1.143
150 lb		"	1.143
250 lb		"	1.143
6"			
125 lb		EA.	1.600
250 lb		"	1.600
8"			
125 lb		EA.	2.000
250 lb		"	2.000
OS&Y, flanged			
2"			
125 lb		EA.	0.667
250 lb		"	0.667
2-1/2"			
125 lb		EA.	0.667
250 lb		"	0.800
3"			
125 lb		EA.	0.800
250 lb		"	0.800
4"			
125 lb		EA.	1.333
250 lb		"	1.333
6"			
125 lb		EA.	1.600
250 lb		"	1.600
Check valve, bronze, soldered, 125 lb			
1/2"		EA.	0.200
3/4"		"	0.200

PLUMBING

15430.60   VALVES

	UNIT	MAN/HOURS
1"	EA.	0.267
1-1/4"	"	0.320
1-1/2"	"	0.320
2"	"	0.400
Threaded		
1/2"		
125 lb	EA.	0.267
150 lb	"	0.267
200 lb	"	0.267
3/4"		
125 lb	EA.	0.320
150 lb	"	0.320
200 lb	"	0.320
1"		
125 lb	EA.	0.400
150 lb	"	0.400
200 lb	"	0.400
Flow check valve, cast iron, threaded		
1"	EA.	0.320
1-1/4"	"	0.400
1-1/2"		
125 lb	EA.	0.400
150 lb	"	0.400
200 lb	"	0.444
2"		
125 lb	EA.	0.444
150 lb	"	0.444
200 lb	"	0.500
2-1/2"		
125 lb	EA.	0.667
250 lb	"	0.800
3"		
125 lb	EA.	0.800
250 lb	"	1.000
4"		
125 lb	EA.	1.143
250 lb	"	1.333
6"		
125 lb	EA.	1.600
250 lb	"	1.600
Vertical check valve, bronze, 125 lb, threaded		
1/2"	EA.	0.320
3/4"	"	0.364
1"	"	0.400
1-1/4"	"	0.444
1-1/2"	"	0.500
2"	"	0.571
Cast iron, flanged		
2-1/2"	EA.	0.800
3"	"	1.000
4"	"	1.333
6	"	1.600
8"	"	2.000
10"	"	2.667

PLUMBING

15430.60   VALVES

	UNIT	MAN/HOURS
12"	EA.	3.200
Globe valve, bronze, soldered, 125 lb		
1/2"	EA.	0.229
3/4"	"	0.250
1"	"	0.267
1-1/4"	"	0.286
1-1/2"	"	0.333
2"	"	0.400
Threaded		
1/2"		
125 lb	EA.	0.267
150 lb	"	0.267
300 lb	"	0.267
3/4"		
125 lb	EA.	0.320
150 lb	"	0.320
300 lb	"	0.320
1"		
125 lb	EA.	0.400
150 lb	"	0.400
300 lb	"	0.400
1-1/4"		
125 lb	EA.	0.400
150 lb	"	0.400
300 lb	"	0.400
1-1/2"		
125 lb	EA.	0.444
150 lb	"	0.444
300 lb	"	0.444
2"		
125 lb	EA.	0.533
150 lb	"	0.533
300 lb	"	0.533
Cast iron flanged		
2-1/2"		
125 lb	EA.	0.800
250 lb	"	0.800
3"		
125 lb	EA.	1.000
250 lb	"	1.000
4"		
125 lb	EA.	1.333
250 lb	"	1.333
6"		
125 lb	EA.	1.600
250 lb	"	1.600
8"		
125 lb	EA.	2.000
250 lb	"	2.000
Butterfly valve, cast iron, wafer type		
2"		
150 lb	EA.	0.571
200 lb	"	0.667
2-1/2"		

15430.60   VALVES

PLUMBING	UNIT	MAN/HOURS
150 lb	EA.	0.667
200 lb	"	0.727
3"		
150 lb	EA.	0.800
200 lb	"	0.889
4"		
150 lb	EA.	1.143
200 lb	"	1.333
6"		
150 lb	EA.	1.600
200 lb	"	1.600
8"		
150 lb	EA.	1.778
200 lb	"	2.000
10"		
150 lb	EA.	2.000
200 lb	"	2.667
Ball valve, bronze, 250 lb, threaded		
1/2"	EA.	0.320
3/4"	"	0.320
1"	"	0.400
1-1/4"	"	0.444
1-1/2"	"	0.500
2"	"	0.571
Angle valve, bronze, 150 lb, threaded		
1/2"	EA.	0.286
3/4"	"	0.320
1"	"	0.320
1-1/4"	"	0.400
1-1/2"	"	0.444
Balancing valve, with meter connections, circuit setter		
1/2"	EA.	0.320
3/4"	"	0.364
1"	"	0.400
1-1/4"	"	0.444
1-1/2"	"	0.533
2"	"	0.667
2-1/2"	"	0.800
3"	"	1.000
4"	"	1.333
Balancing valve, straight type		
1/2"	EA.	0.320
3/4"	"	0.320
Angle type		
1/2"	EA.	0.320
3/4"	"	0.320
Square head cock, 125 lb, bronze body		
1/2"	EA.	0.267
3/4"	"	0.320
1"	"	0.364
1-1/4"	"	0.400
Radiator temp control valve, with control and sensor		
1/2" valve	EA.	0.500
1" valve	"	0.500

15430.60   VALVES

PLUMBING	UNIT	MAN/HOURS
Pressure relief valve, 1/2", bronze		
Low pressure	EA.	0.320
High pressure	"	0.320
Press and temperature relief valve		
Bronze, 3/4"	EA.	0.320
Cast iron, 3/4"		
High pressure	EA.	0.320
Temperature relief	"	0.320
Pressure & temp relief valve	"	0.320
Pressure reducing valve, bronze, threaded, 250 lb		
1/2"	EA.	0.500
3/4"	"	0.500
1"	"	0.500
1-1/4"	"	0.571
1-1/2"	"	0.667
Pressure regulating valve, bronze, class 300		
1"	EA.	0.500
1-1/2"	"	0.615
2"	"	0.800
3"	"	1.143
4"	"	1.600
5"	"	2.000
6"	"	2.667
Solar water temperature regulating valve		
3/4"	EA.	0.667
1"	"	0.800
1-1/4"	"	0.889
1-1/2"	"	1.000
2"	"	1.143
2-1/2"	"	2.000
Tempering valve, threaded		
3/4"	EA.	0.267
1"	"	0.320
1-1/4"	"	0.400
1-1/2"	"	0.400
2"	"	0.500
2-1/2"	"	0.667
3"	"	0.800
4"	"	1.143
Thermostatic mixing valve, threaded		
1/2"	EA.	0.286
3/4"	"	0.320
1"	"	0.348
1-1/2"	"	0.400
2"	"	0.500
Sweat connection		
1/2"	EA.	0.286
3/4"	"	0.320
Mixing valve, sweat connection		
1/2"	EA.	0.286
3/4"	"	0.320
Liquid level gauge, aluminum body		
3/4"	EA.	0.320
125 psi, pvc body		

PLUMBING	UNIT	MAN/HOURS
15430.60 VALVES		
3/4"	EA.	0.320
150 psi, crs body		
3/4"	EA.	0.320
1"	"	0.320
175 psi, bronze body, 1/2"	"	0.286
15430.65 VACUUM BREAKERS		
Vacuum breaker, atmospheric, threaded connection		
3/4"	EA.	0.320
1"	"	0.320
Anti-siphon, brass		
3/4"	EA.	0.320
1"	"	0.320
1-1/4"	"	0.400
1-1/2"	"	0.444
2"	"	0.500
Air eliminators, purger, cast iron, threaded		
1"	EA.	0.320
1-1/4"	"	0.400
1-1/2"	"	0.400
2"	"	0.500
2-1/2"	"	1.000
3"	"	1.333
Airtrol fitting, 3/4"	"	0.320
Air eliminator, air vents, 1/4"	"	0.320
Air vent for hot water	"	0.286
15430.68 STRAINERS		
Strainer, Y pattern, 125 psi, cast iron body, threaded		
3/4"	EA.	0.286
1"	"	0.320
1-1/4"	"	0.400
1-1/2"	"	0.400
2"	"	0.500
250 psi, brass body, threaded		
3/4"	EA.	0.320
1"	"	0.320
1-1/4"	"	0.400
1-1/2"	"	0.400
2"	"	0.500
Cast iron body, threaded		
3/4"	EA.	0.320
1"	"	0.320
1-1/4"	"	0.400
1-1/2"	"	0.400
2"	"	0.500
15430.70 DRAINS, ROOF & FLOOR		
Floor drain, cast iron, with cast iron top		
2"	EA.	0.667
3"	"	0.667
4"	"	0.667

PLUMBING	UNIT	MAN/HOURS
15430.70 DRAINS, ROOF & FLOOR		
6"	EA.	0.800
Roof drain, cast iron		
2"	EA.	0.667
3"	"	0.667
4"	"	0.667
5"	"	0.800
6"	"	0.800
15430.80 TRAPS		
Bucket trap, threaded		
3/4"	EA.	0.500
1"	"	0.533
1-1/4"	"	0.615
1-1/2"	"	0.727
Inverted bucket steam trap, threaded		
3/4"	EA.	0.500
1"	"	0.500
1-1/4"	"	0.444
1-1/2"	"	0.667
With stainless interior		
1/2"	EA.	0.500
3/4"	"	0.500
1"	"	0.500
1-1/4"	"	0.571
Brass interior		
3/4"	EA.	0.500
1"	"	0.533
1-1/4"	"	0.571
Cast steel body, threaded, high temperature		
3/4"	EA.	0.500
1"	"	0.571
1-1/4	"	0.615
1-1/2"	"	0.667
2"	"	0.800
Float trap, 15 psi		
3/4"	EA.	0.500
1"	"	0.533
1-1/4"	"	0.571
1-1/2"	"	0.667
2"	"	0.800
30 psi		
3/4"	EA.	0.500
1"	"	0.533
1-1/4"	"	0.667
1-1/2"	"	0.800
75 psi		
3/4"	EA.	0.500
1"	"	0.533
1-1/4"	"	0.571
1-1/2"	"	0.667
125 psi		
3/4"	EA.	0.500
1"	"	0.533

PLUMBING

15430.80 — TRAPS

	UNIT	MAN/HOURS
1-1/4	EA.	0.571
1-1/2	"	0.667
Float and thermostatic trap, 15 psi		
3/4"	EA.	0.500
1"	"	0.533
1-1/4"	"	0.571
1-1/2"	"	0.667
2"	"	0.800
30 psi		
3/4"	EA.	0.500
1"	"	0.533
1-1/4"	"	0.571
1-1/2"	"	0.667
75 psi		
3/4"	EA.	0.500
1"	"	0.533
1-1/4"	"	0.571
1-1/2"	"	0.667
Steam trap, cast iron body, threaded, 125 psi		
3/4"	EA.	0.500
1"	"	0.533
1-1/4"	"	0.571
1-1/2"	"	0.667
Thermostatic trap, low pressure, angle type, 25 psi		
1/2"	EA.	0.500
3/4"	"	0.500
1"	"	0.533
50 psi		
1/2"	EA.	0.500
3/4"	"	0.500
1"	"	0.533
Cast iron body, threaded, 125 psi		
3/4"	EA.	0.500
1"	"	0.571
1-1/4"	"	0.615
1-1/2"	"	0.667
Low pressure, 25 psi, swivel type, 1/2"	"	0.500
Straightway type, 3/4"	"	0.500
Vertical type, 1/2"	"	0.500
Medium pressure, 50 psi, angle type, 1/2"	"	0.500
High pressure, 125 psi, angle type		
1/2"	EA.	0.500
3/4"	"	0.500
1"	"	0.571
Straightway type		
1/2"	EA.	0.500
Thermo disc trap		
3/4"	EA.	0.500
1"	"	0.533
Drip pan ell, cast iron		
2-1/2"	EA.	1.000
3"	"	1.143
4"	"	1.333
5"	"	1.600

PLUMBING

15430.80 — TRAPS

	UNIT	MAN/HOURS
6"	EA.	2.000
8"	"	2.667
Steel		
2-1/2"	EA.	1.000
3"	"	1.143
4"	"	1.333
5"	"	1.600
6"	"	2.000
8"	"	2.667

PLUMBING FIXTURES

15440.10 — BATHS

	UNIT	MAN/HOURS
Bath tub, 5' long		
Minimum	EA.	2.667
Average	"	4.000
Maximum	"	8.000
6' long		
Minimum	EA.	2.667
Average	"	4.000
Maximum	"	8.000
Square tub, whirlpool, 4'x4'		
Minimum	EA.	4.000
Average	"	8.000
Maximum	"	10.000
5'x5'		
Minimum	EA.	4.000
Average	"	8.000
Maximum	"	10.000
6'x6'		
Minimum	EA.	4.000
Average	"	8.000
Maximum	"	10.000
For trim and rough-in		
Minimum	EA.	2.667
Average	"	4.000
Maximum	"	8.000

15440.12 — DISPOSALS & ACCESSORIES

	UNIT	MAN/HOURS
Continuous feed		
Minimum	EA.	1.600
Average	"	2.000
Maximum	"	2.667
Batch feed, 1/2 hp		
Minimum	EA.	1.600
Average	"	2.000

PLUMBING FIXTURES	UNIT	MAN/ HOURS
15440.12 DISPOSALS & ACCESSORIES		
Maximum	EA.	2.667
Hot water dispenser		
Minimum	EA.	1.600
Average	"	2.000
Maximum	"	2.667
Epoxy finish faucet	"	1.600
Lock stop assembly	"	1.000
Mounting gasket	"	0.667
Tailpipe gasket	"	0.667
Stopper assembly	"	0.800
Switch assembly, on/off	"	1.333
Tailpipe gasket washer	"	0.400
Stop gasket	"	0.444
Tailpipe flange	"	0.400
Tailpipe	"	0.500
15440.15 FAUCETS		
Kitchen		
Minimum	EA.	1.333
Average	"	1.600
Maximum	"	2.000
Bath		
Minimum	EA.	1.333
Average	"	1.600
Maximum	"	2.000
Lavatory, domestic		
Minimum	EA.	1.333
Average	"	1.600
Maximum	"	2.000
Hospital, patient rooms		
Minimum	EA.	2.000
Average	"	2.667
Maximum	"	4.000
Operating room		
Minimum	EA.	2.000
Average	"	2.667
Maximum	"	4.000
Washroom		
Minimum	EA.	1.333
Average	"	1.600
Maximum	"	2.000
Handicapped		
Minimum	EA.	1.600
Average	"	2.000
Maximum	"	2.667
Shower		
Minimum	EA.	1.333
Average	"	1.600
Maximum	"	2.000
For trim and rough-in		
Minimum	EA.	1.600
Average	"	2.000
Maximum	"	4.000

PLUMBING FIXTURES	UNIT	MAN/ HOURS
15440.18 HYDRANTS		
Wall hydrant		
8" thick	EA.	1.333
12" thick	"	1.600
18" thick	"	1.778
24" thick	"	2.000
Ground hydrant		
2' deep	EA.	1.000
4' deep	"	1.143
6' deep	"	1.333
8' deep	"	2.000
15440.20 LAVATORIES		
Lavatory, counter top, porcelain enamel on cast iron		
Minimum	EA.	1.600
Average	"	2.000
Maximum	"	2.667
Wall hung, china		
Minimum	EA.	1.600
Average	"	2.000
Maximum	"	2.667
Handicapped		
Minimum	EA.	2.000
Average	"	2.667
Maximum	"	4.000
For trim and rough-in		
Minimum	EA.	2.000
Average	"	2.667
Maximum	"	4.000
15440.30 SHOWERS		
Shower, fiberglass, 36"x34"x84"		
Minimum	EA.	5.714
Average	"	8.000
Maximum	"	8.000
Steel, 1 piece, 36"x36"		
Minimum	EA.	5.714
Average	"	8.000
Maximum	"	8.000
Receptor, molded stone, 36"x36"		
Minimum	EA.	2.667
Average	"	4.000
Maximum	"	6.667
For trim and rough-in		
Minimum	EA.	3.636
Average	"	4.444
Maximum	"	8.000
15440.40 SINKS		
Service sink, 24"x29"		
Minimum	EA.	2.000
Average	"	2.667
Maximum	"	4.000
Kitchen sink, single, stainless steel, single bowl		
Minimum	EA.	1.600

PLUMBING FIXTURES		UNIT	MAN/HOURS
15440.40	**SINKS**		
Average		EA.	2.000
Maximum		"	2.667
Double bowl			
Minimum		EA.	2.000
Average		"	2.667
Maximum		"	4.000
Porcelain enamel, cast iron, single bowl			
Minimum		EA.	1.600
Average		"	2.000
Maximum		"	2.667
Double bowl			
Minimum		EA.	2.000
Average		"	2.667
Maximum		"	4.000
Mop sink, 24"x36"x10"			
Minimum		EA.	1.600
Average		"	2.000
Maximum		"	2.667
Washing machine box			
Minimum		EA.	2.000
Average		"	2.667
Maximum		"	4.000
For trim and rough-in			
Minimum		EA.	2.667
Average		"	4.000
Maximum		"	5.333
15440.50	**URINALS**		
Urinal, flush valve, floor mounted			
Minimum		EA.	2.000
Average		"	2.667
Maximum		"	4.000
Wall mounted			
Minimum		EA.	2.000
Average		"	2.667
Maximum		"	4.000
For trim and rough-in			
Minimum		EA.	2.000
Average		"	4.000
Maximum		"	5.333
15440.60	**WATER CLOSETS**		
Water closet flush tank, floor mounted			
Minimum		EA.	2.000
Average		"	2.667
Maximum		"	4.000
Handicapped			
Minimum		EA.	2.667
Average		"	4.000
Maximum		"	8.000
Bowl, with flush valve, floor mounted			
Minimum		EA.	2.000
Average		"	2.667
Maximum		"	4.000

PLUMBING FIXTURES		UNIT	MAN/HOURS
15440.60	**WATER CLOSETS**		
Wall mounted			
Minimum		EA.	2.000
Average		"	2.667
Maximum		"	4.000
For trim and rough-in			
Minimum		EA.	2.000
Average		"	2.667
Maximum		"	4.000
15440.70	**WATER HEATERS**		
Water heater, electric			
6 gal		EA.	1.333
10 gal		"	1.333
15 gal		"	1.333
20 gal		"	1.600
30 gal		"	1.600
40 gal		"	1.600
52 gal		"	2.000
66 gal		"	2.000
80 gal		"	2.000
100 gal		"	2.667
120 gal		"	2.667
Oil fired			
20 gal		EA.	4.000
50 gal		"	5.714
15440.90	**MISCELLANEOUS FIXTURES**		
Electric water cooler			
Floor mounted		EA.	2.667
Wall mounted		"	2.667
Wash fountain			
Wall mounted		EA.	4.000
Circular, floor supported		"	8.000
Deluge shower and eye wash		"	4.000
15440.95	**FIXTURE CARRIERS**		
Water fountain, wall carrier			
Minimum		EA.	0.800
Average		"	1.000
Maximum		"	1.333
Lavatory, wall carrier			
Minimum		EA.	0.800
Average		"	1.000
Maximum		"	1.333
Sink, industrial, wall carrier			
Minimum		EA.	0.800
Average		"	1.000
Maximum		"	1.333
Toilets, water closets, wall carrier			
Minimum		EA.	0.800
Average		"	1.000
Maximum		"	1.333

PLUMBING FIXTURES	UNIT	MAN/HOURS
15440.95 FIXTURE CARRIERS		
Floor support		
Minimum	EA.	0.667
Average	"	0.800
Maximum	"	1.000
Urinals, wall carrier		
Minimum	EA.	0.800
Average	"	1.000
Maximum	"	1.333
Floor support		
Minimum	EA.	0.667
Average	"	0.800
Maximum	"	1.000
15450.30 PUMPS		
In-line pump, bronze, centrifugal		
5 gpm, 20' head	EA.	0.500
20 gpm, 40' head	"	0.500
50 gpm		
50' head	EA.	1.000
100' head	"	1.000
70 gpm, 100' head	"	1.333
100 gpm, 80' head	"	1.333
250 gpm, 150' head	"	2.000
Cast iron, centrifugal		
50 gpm, 200' head	EA.	1.000
100 gpm		
100' head	EA.	1.333
200' head	"	1.333
200 gpm		
100' head	EA.	2.000
200' head	"	2.000
Centrifugal, close coupled, c.i., single stage		
50 gpm, 100' head	EA.	1.000
100 gpm, 100' head	"	1.333
Base mounted		
50 gpm, 100' head	EA.	1.000
100 gpm, 50' head	"	1.333
200 gpm, 100' head	"	2.000
300 gpm, 175' head	"	2.000
Suction diffuser, flanged, strainer		
3" inlet, 2-1/2" outlet	EA.	1.000
3" outlet	"	1.000
4" inlet		
3" outlet	EA.	1.333
4" outlet	"	1.333
6" inlet		
4" outlet	EA.	1.600
5" outlet	"	1.600
6" Outlet	"	1.600
8" inlet		
6" outlet	EA.	2.000
8" outlet	"	2.000
10" inlet		
8" outlet	EA.	2.667

PLUMBING FIXTURES	UNIT	MAN/HOURS
15450.30 PUMPS		
Vertical turbine		
Single stage, C.I., 3550 rpm, 200 gpm, 50'head	EA.	2.667
Multi stage, 3550 rpm		
50 gpm, 100' head	EA.	2.000
100 gpm		
100' head	EA.	2.000
200 gpm		
50' head	EA.	2.667
100' head	"	2.667
Bronze		
Single stage, 3550 rpm, 100 gpm, 50' head	EA.	2.000
Multi stage, 3550 rpm, 50 gpm, 100' head	"	2.000
100 gpm		
100' head	EA.	2.000
200 gpm		
50' head	EA.	2.667
100' head	"	2.667
Sump pump, bronze, 1750 rpm, 25 gpm		
20' head	EA.	10.000
150' head	"	13.333
50 gpm		
100' head	EA.	10.000
100 gpm		
50' head	EA.	10.000
Condensate pump, simplex		
1000 sf EDR, 2 gpm	EA.	6.667
2000 sf EDR, 3 gpm	"	6.667
4000 sf EDR, 6 gpm	"	7.273
6000 sf EDR, 9 gpm	"	7.273
Duplex, bronze		
8000 sf EDR, 12 gpm	EA.	7.273
10,000 sf EDR, 15 gpm	"	10.000
15,000 sf EDR, 23 gpm	"	11.429
20,000 sf EDR, 30 gpm	"	16.000
25,000 sf EDR, 38 gpm	"	16.000
30,000 sf EDR, 45 gpm	"	17.778
40,000 sf EDR, 60 gpm	"	10.000
50,000 sf EDR, 75 gpm	"	11.429
75,000 sf EDR, 112 gpm	"	13.333
100,000 sf EDR, 150 gpm	"	20.000
15450.40 STORAGE TANKS		
Hot water storage tank, cement lined		
10 gallon	EA.	2.667
70 gallon	"	4.000
200 gallon	"	5.714
900 gallon	"	10.000
1100 gallon	"	10.000
2000 gallon	"	10.000
15480.10 SPECIAL SYSTEMS		
Air compressor, air cooled, two stage		
5.0 cfm, 175 psi	EA.	16.000

PLUMBING FIXTURES	UNIT	MAN/ HOURS
15480.10 SPECIAL SYSTEMS		
10 cfm, 175 psi	EA.	17.778
20 cfm, 175 psi	"	19.048
50 cfm, 125 psi	"	21.053
80 cfm, 125 psi	"	22.857
Single stage, 125 psi		
1.0 cfm	EA.	11.429
1.5 cfm	"	11.429
2.0 cfm	"	11.429
Automotive compressor, hose reel, air and water, 50' hose	"	6.667
Lube equipment, 3 reel, with pumps	"	32.000
Tire changer		
Truck	EA.	11.429
Passenger car	"	6.154
Air hose reel, includes, 50' hose	"	6.154
Hose reel, 5 reel, motor oil, gear oil, lube, air & water	"	32.000
Water hose reel, 50' hose	"	6.154
Pump, air operated, for motor or gear oil, fits 55 gal drum	"	0.800
For chassis lube	"	0.800
Fuel dispensing pump, lighted dial, one product		
One hose	EA.	6.667
Two hose	"	6.667
Two products, two hose	"	6.667

HEATING & VENTILATING	UNIT	MAN/ HOURS
15555.10 BOILERS		
Cast iron, gas fired, hot water		
115 mbh	EA.	20.000
175 mbh	"	21.818
235 mbh	"	24.000
940 mbh	"	48.000
1600 mbh	"	60.000
3000 mbh	"	80.000
6000 mbh	"	120
Steam		
115 mbh	EA.	20.000
175 mbh	"	21.818
235 mbh	"	24.000
940 mbh	"	48.000
1600 mbh	"	60.000
3000 mbh	"	80.000
6000 mbh	"	120
Electric, hot water		
115 mbh	EA.	12.000
175 mbh	"	12.000
235 mbh	"	12.000
940 mbh	"	24.000

HEATING & VENTILATING	UNIT	MAN/ HOURS
15555.10 BOILERS		
1600 mbh	EA.	48.000
3000 mbh	"	60.000
6000 mbh	"	80.000
Steam		
115 mbh	EA.	12.000
175 mbh	"	12.000
235 mbh	"	12.000
940 mbh	"	24.000
1600 mbh	"	48.000
3000 mbh	"	60.000
6000 mbh	"	80.000
Oil fired, hot water		
115 mbh	EA.	16.000
175 mbh	"	18.462
235 mbh	"	21.818
940 mbh	"	40.000
1600 mbh	"	48.000
3000 mbh	"	60.000
6000 mbh	"	120
Steam		
115 mbh	EA.	16.000
175 mbh	"	18.462
235 mbh	"	21.818
940 mbh	"	40.000
1600 mbh	"	48.000
3000 mbh	"	60.000
6000 mbh	"	120
15610.10 FURNACES		
Electric, hot air		
40 mbh	EA.	4.000
60 mbh	"	4.211
80 mbh	"	4.444
100 mbh	"	4.706
125 mbh	"	4.848
160 mbh	"	5.000
200 mbh	"	5.161
400 mbh	"	5.333
Gas fired hot air		
40 mbh	EA.	4.000
60 mbh	"	4.211
80 mbh	"	4.444
100 mbh	"	4.706
125 mbh	"	4.848
160 mbh	"	5.000
200 mbh	"	5.161
400 mbh	"	5.333
Oil fired hot air		
40 mbh	EA.	4.000
60 mbh	"	4.211
80 mbh	"	4.444
100 mbh	"	4.706
125 mbh	"	4.848

HEATING & VENTILATING

15610.10   FURNACES

	UNIT	MAN/HOURS
160 mbh	EA.	5.000
200 mbh	"	5.161
400 mbh	"	5.333

REFRIGERATION

15670.10   CONDENSING UNITS

	UNIT	MAN/HOURS
Air cooled condenser, single circuit		
3 ton	EA.	1.333
5 ton	"	1.333
7.5 ton	"	3.810
20 ton	"	4.000
25 ton	"	4.000
30 ton	"	4.000
40 ton	"	5.714
50 ton	"	5.714
60 ton	"	5.000
With low ambient dampers		
3 ton	EA.	2.000
5 ton	"	2.000
7.5 ton	"	4.000
20 ton	"	5.333
25 ton	"	5.333
30 ton	"	5.333
40 ton	"	6.667
50 ton	"	7.273
60 ton	"	7.273
Dual circuit		
10 ton	EA.	4.000
15 ton	"	5.714
20 ton	"	5.714
25 ton	"	5.714
30 ton	"	5.714
40 ton	"	6.667
50 ton	"	6.667
60 ton	"	6.667
80 ton	"	8.889
100 ton	"	8.889
120 ton	"	8.889
With low ambient dampers		
15 ton	EA.	5.714
20 ton	"	5.714
25 ton	"	5.714
30 ton	"	5.714
40 ton	"	6.667
50 ton	"	6.667

REFRIGERATION

15670.10   CONDENSING UNITS

	UNIT	MAN/HOURS
60 ton	EA.	6.667
80 ton	"	8.889
100 ton	"	8.889
120 ton	"	8.889

15680.10   CHILLERS

	UNIT	MAN/HOURS
Chiller, reciprocal		
Air cooled, remote condenser, starter		
20 ton	EA.	8.000
25 ton	"	8.000
30 ton	"	8.000
40 ton	"	12.000
50 ton	"	13.333
60 ton	"	14.118
80 ton	"	21.818
100 ton	"	24.000
120 ton	"	26.667
150 ton	"	30.000
180 ton	"	34.286
200 ton	"	40.000
Water cooled, with starter		
20 ton	EA.	8.000
25 ton	"	8.000
30 ton	"	12.000
40 ton	"	12.000
50 ton	"	13.333
60 ton	"	14.118
80 ton	"	21.818
100 ton	"	24.000
120 ton	"	26.667
150 ton	"	30.000
180 ton	"	34.286
200 ton	"	40.000
Packaged, air cooled, with starter		
20 ton	EA.	6.000
25 ton	"	6.000
30 ton	"	6.000
40 ton	"	6.000
50 ton	"	8.000
60 ton	"	8.000
80 ton	"	12.000
100 ton	"	12.000
120 ton	"	12.000
Heat recovery, air cooled, with starter		
40 ton	EA.	12.000
50 ton	"	12.000
60 ton	"	16.000
75 ton	"	24.000
100 ton	"	24.000
Water cooled, with starter		
40 ton	EA.	12.000
50 ton	"	12.000
60 ton	"	16.000
75 ton	"	24.000

REFRIGERATION

	UNIT	MAN/HOURS
15680.10 CHILLERS		
100 ton	EA.	26.667
Centrifugal, single bundle condenser, with starter		
80 ton	EA.	34.286
130 ton	"	40.000
160 ton	"	43.636
180 ton	"	48.000
230 ton	"	53.333
280 ton	"	60.000
360 ton	"	60.000
460 ton	"	80.000
560 ton	"	85.714
670 ton	"	96.000
15710.10 COOLING TOWERS		
Cooling tower, propeller type		
100 ton	EA.	8.000
200 ton	"	12.000
300 ton	"	20.000
400 ton	"	24.000
600 ton	"	34.286
800 ton	"	48.000
1000 ton	"	60.000
Centrifugal		
100 ton	EA.	8.000
200 ton	"	12.000
300 ton	"	20.000
400 ton	"	24.000
600 ton	"	34.286
800 ton	"	48.000
1000 ton	"	60.000

HEAT TRANSFER

	UNIT	MAN/HOURS
15780.10 COMPUTER ROOM A/C		
Air cooled, alarm, high efficiency filter, elec. heat		
3 ton	EA.	6.154
5 ton	"	6.667
7.5 ton	"	8.000
10 ton	"	10.000
15 ton	"	11.429
Steam heat		
3 ton	EA.	6.154
5 ton	"	6.667
7.5 ton	"	8.000
10 ton	"	10.000
15 ton	"	11.429
Hot water heat		

HEAT TRANSFER

	UNIT	MAN/HOURS
15780.10 COMPUTER ROOM A/C		
3 ton	EA.	6.154
5 ton	"	6.667
7.5 ton	"	8.000
10 ton	"	10.000
15 ton	"	11.429
Air cooled condenser, low ambient damper		
3 ton	EA.	1.600
5 ton	"	2.000
7.5 ton	"	4.000
10 ton	"	5.714
15 ton	"	4.706
Water cooled, high efficiency filter, alarm, elec. heat		
3 ton	EA.	5.714
5 ton	"	6.667
7.5 ton	"	10.000
10 ton	"	11.429
15 ton	"	13.333
Steam heat		
3 ton	EA.	5.714
5 ton	"	6.667
7.5 ton	"	10.000
10 ton	"	11.429
15 ton	"	13.333
Hot water heat		
3 ton	EA.	5.714
5 ton	"	6.667
7.5 ton	"	10.000
10 ton	"	11.429
15 ton	"	13.333
Chilled water, alarm, high eff. filter, elec. heat		
7.5 ton	EA.	7.273
10 ton	"	8.889
15 ton	"	10.000
Steam heat		
7.5 ton	EA.	7.273
10 ton	"	8.889
15 ton	"	10.000
Hot water heat		
7.5 ton	EA.	7.273
10 ton	"	8.889
15 ton	"	10.000
15780.20 ROOFTOP UNITS		
Packaged, single zone rooftop unit, with roof curb		
2 ton	EA.	8.000
3 ton	"	8.000
4 ton	"	10.000
5 ton	"	13.333
7.5 ton	"	16.000
15830.10 RADIATION UNITS		
Baseboard radiation unit		
1.7 mbh/lf	L.F.	0.320

HEAT TRANSFER

15830.10 RADIATION UNITS

	UNIT	MAN/HOURS
2.1 mbh/lf	L.F.	0.400
Enclosure only		
Two tier	L.F.	0.133
Three tier	"	0.133
Copper element only, 3/4" dia.		
Two tier	L.F.	0.200
Three tier	"	0.267
Fin-tube, 16 ga, sloping cover, 1-1/4" steel		
One tier	L.F.	0.267
Two tier	"	0.320
2" steel		
Two tier	L.F.	0.320
Three tier	"	0.400
1-1/4" copper		
Two tier	L.F.	0.267
18 ga flat cover, 1-1/4" steel		
One tier	L.F.	0.267
Two tier	"	0.320
Three tier	"	0.400
2" steel		
One tier	L.F.	0.267
Two tier	"	0.320
Three tier	"	0.400
1-1/4" copper		
One tier	L.F.	0.267
Two tier	"	0.320
Three tier	"	0.400

15830.20 FAN COIL UNITS

	UNIT	MAN/HOURS
Fan coil unit, 2 pipe, complete		
200 cfm ceiling hung	EA.	2.667
Floor mounted	"	2.000
300 cfm, ceiling hung	"	3.200
Floor mounted	"	2.667
400 cfm, ceiling hung	"	3.810
Floor mounted	"	2.667
500 cfm, ceiling hung	"	4.000
Floor mounted	"	3.077
600 cfm, ceiling hung	"	4.420
Floor mounted	"	3.636
800 cfm, ceiling hung	"	5.000
Floor mounted	"	3.810
1000 cfm, ceiling hung	"	5.714
Floor mounted	"	4.211
1200 cfm ceiling hung	"	6.667
Floor mounted	"	5.000

15830.70 UNIT HEATERS

	UNIT	MAN/HOURS
Steam unit heater, horizontal		
12,500 btuh, 200 cfm	EA.	1.333
17,000 btuh, 300 cfm	"	1.333
40,000 btuh, 500 cfm	"	1.333

HEAT TRANSFER

15830.70 UNIT HEATERS

	UNIT	MAN/HOURS
60,000 btuh, 700 cfm	EA.	1.333
70,000 btuh, 1000 cfm	"	2.000
Vertical		
12,500 btuh, 200 cfm	EA.	1.333
17,000 btuh, 300 cfm	"	1.333
40,000 btuh, 500 cfm	"	1.333
60,000 btuh, 700 cfm	"	1.333
70,000 btuh, 1000 cfm	"	1.333
Gas unit heater, horizontal		
27,400 btuh	EA.	3.200
38,000 btuh	"	3.200
56,000 btuh	"	3.200
82,200 btuh	"	3.200
103,900 btuh	"	5.000
125,700 btuh	"	5.000
133,200 btuh	"	5.000
149,000 btuh	"	5.000
172,000 btuh	"	5.000
190,000 btuh	"	5.000
225,000 btuh	"	5.000
Hot water unit heater, horizontal		
12,500 btuh, 200 cfm	EA.	1.333
17,000 btuh, 300 cfm	"	1.333
25,000 btuh, 500 cfm	"	1.333
30,000 btuh, 700 cfm	"	1.333
50,000 btuh, 1000 cfm	"	2.000
60,000 btuh, 1300 cfm	"	2.000
Vertical		
12,500 btuh, 200 cfm	EA.	1.333
17,000 btuh, 300 cfm	"	1.333
25,000 btuh, 500 cfm	"	1.333
30,000 btuh, 700 cfm	"	1.333
50,000 btuh, 1000 cfm	"	1.333
60,000 btuh, 1300 cfm	"	1.333
Cabinet unit heaters, ceiling, exposed, hot water		
200 cfm	EA.	2.667
300 cfm	"	3.200
400 cfm	"	3.810
600 cfm	"	4.211
800 cfm	"	5.000
1000 cfm	"	5.714
1200 cfm	"	6.667
2000 cfm	"	8.889

AIR HANDLING

15855.10 AIR HANDLING UNITS

	UNIT	MAN/HOURS
Air handling unit, medium pressure, single zone		
1500 cfm	EA.	5.000
3000 cfm	"	8.889
4000 cfm	"	10.000
5000 cfm	"	10.667
6000 cfm	"	11.429
7000 cfm	"	12.308
8500 cfm	"	13.333
10,500 cfm	"	16.000
12,500 cfm	"	17.778
15,500 cfm	"	22.857
17,500 cfm	"	26.667
20,500 cfm	"	32.000
25,000 cfm	"	40.000
31,500 cfm	"	53.333
Rooftop air handling units		
4950 cfm	EA.	8.889
7370 cfm	"	11.429
9790 cfm	"	13.333
14,300 cfm	"	11.429
21,725 cfm	"	11.429
33,000 cfm	"	13.333

15870.20 EXHAUST FANS

	UNIT	MAN/HOURS
Belt drive roof exhaust fans		
640 cfm, 2618 fpm	EA.	1.000
940 cfm, 2604 fpm	"	1.000
1050 cfm, 3325 fpm	"	1.000
1170 cfm, 2373 fpm	"	1.000
2440 cfm, 4501 fpm	"	1.000
2760 cfm, 4950 fpm	"	1.000
3890 cfm, 6769 fpm	"	1.000
2380 cfm, 3382 fpm	"	1.000
2880 cfm, 3859 fpm	"	1.000
3200 cfm, 4173 fpm	"	1.333
3660 cfm, 3437 fpm	"	1.333
4070 cfm, 3694 fpm	"	1.333
5030 cfm, 3251 fpm	"	1.333
5830 cfm, 6932 fpm	"	1.600
6380 cfm, 3817 fpm	"	1.600
8460 cfm, 6721 fpm	"	1.600
10,970 cfm, 5906 fpm	"	2.000
12,470 cfm, 6620 fpm	"	2.667
7000 cfm, 3449 fpm	"	2.000
13,000 cfm, 5456 fpm	"	2.000
11,250 cfm, 4854 fpm	"	2.000
18,490 cfm, 7405 fpm	"	3.636
11,300 cfm, 3232 fpm	"	3.478
18,330 cfm, 4488 fpm	"	3.478
21,720 cfm, 5131 fpm	"	3.478
31,110 cfm, 6965 fpm	"	4.000
Direct drive fans		
60 to 390 cfm	EA.	1.000
145 to 590 cfm	"	1.000

AIR HANDLING

15870.20 EXHAUST FANS

	UNIT	MAN/HOURS
295 to 860 cfm	EA.	1.000
235 to 1300 cfm	"	1.000
415 to 1630 cfm	"	1.000
590 to 2045 cfm	"	1.000
805 cfm, 3235 fpm	"	1.000
1455 cfm, 4360 fpm	"	1.000
1385 cfm, 3655 fpm	"	1.000
2260 cfm, 4930 fpm	"	1.000
1720 cfm, 3870 fpm	"	1.000
2700 cfm, 5220 fpm	"	1.000
Terminal blenders and cooling		
400 cfm	EA.	1.600
800 cfm	"	1.600
1200 cfm	"	2.000
2000 cfm	"	2.000

AIR DISTRIBUTION

15890.10 METAL DUCTWORK

	UNIT	MAN/HOURS
Rectangular duct		
Galvanized steel		
Minimum	Lb.	0.073
Average	"	0.089
Maximum	"	0.133
Aluminum		
Minimum	Lb.	0.160
Average	"	0.200
Maximum	"	0.267
Fittings		
Minimum	EA.	0.267
Average	"	0.400
Maximum	"	0.800

15890.30 FLEXIBLE DUCTWORK

	UNIT	MAN/HOURS
Flexible duct, 1.25" fiberglass		
5" dia.	L.F.	0.040
6" dia.	"	0.044
7" dia.	"	0.047
8" dia.	"	0.050
10" dia.	"	0.057
12" dia.	"	0.062
14" dia.	"	0.067
16" dia.	"	0.073
Flexible duct connector, 3" wide fabric	"	0.133

AIR DISTRIBUTION

15895.10 — ROOF CURBS

	UNIT	MAN/HOURS
8" high, insulated, with liner and raised can		
15" x 15"	EA.	0.400
17" x 17"	"	0.400
19" x 19"	"	0.400
21" x 21"	"	0.400
25" x 25"	"	0.500
28" x 28"	"	0.533
32" x 32"	"	0.571
36" x 36"	"	0.571
40" x 40"	"	0.571
44" x 44"	"	0.615
48" x 48"	"	0.615
52" x 52"	"	0.667
56" x 56"	"	0.667
60" x 60"	"	0.800
64" x 64"	"	0.800
68" x 68"	"	0.889
72" x 72"	"	1.000

15910.10 — DAMPERS

	UNIT	MAN/HOURS
Horizontal parallel aluminum backdraft damper		
12" x 12"	EA.	0.200
16" x 16"	"	0.229
20" x 20"	"	0.286
24" x 24"	"	0.400
28" x 28"	"	0.444
32" x 32"	"	0.500
36" x 36"	"	0.571
40" x 40"	"	0.667
44" x 44"	"	0.727
48" x 48"	"	0.800
"Up", parallel dampers		
12" x 12"	EA.	0.200
16" x 16"	"	0.229
20" x 20"	"	0.286
24" x 24"	"	0.400
28" x 28"	"	0.444
32" x 32"	"	0.500
36" x 36"	"	0.571
40" x 40"	"	0.667
44" x 44"	"	0.727
48" x 48"	"	0.800
"Down", parallel dampers		
12" x 12"	EA.	0.200
16" x 16"	"	0.229
20" x 20"	"	0.286
24" x 24"	"	0.400
28" x 28"	"	0.444
32" x 32"	"	0.500
36" x 36"	"	0.571
40" x 40"	"	0.667
44" x 44"	"	0.727
48" x 48"	"	0.800
Fire damper, 1.5 hr rating		

AIR DISTRIBUTION

15910.10 — DAMPERS

	UNIT	MAN/HOURS
12" x 12"	EA.	0.400
16" x 16"	"	0.400
20" x 20"	"	0.400
24" x 24"	"	0.400
28" x 28"	"	0.571
32" x 32"	"	0.667
36" x 36"	"	0.800
40" x 40"	"	0.889
44" x 44"	"	1.000
48" x 48"	"	1.143

15940.10 — DIFFUSERS

	UNIT	MAN/HOURS
Ceiling diffusers, round, baked enamel finish		
6" dia.	EA.	0.267
8" dia.	"	0.333
10" dia.	"	0.333
12" dia.	"	0.333
14" dia.	"	0.364
16" dia.	"	0.364
18" dia.	"	0.400
20" dia.	"	0.400
Rectangular		
6x6"	EA.	0.267
9x9"	"	0.400
12x12"	"	0.400
15x15"	"	0.400
18x18"	"	0.400
21x21"	"	0.500
24x24"	"	0.500
Lay in, flush mounted, perforated face, with grid		
6x6/24x24	EA.	0.320
8x8/24x24	"	0.320
9x9/24x24	"	0.320
10x10/24x24	"	0.320
12x12/24x24	"	0.320
15x15/24x24	"	0.320
18x6/24x24	"	0.320
18x18/24x24	"	0.320
Two-way slot diffuser with balancing damper, 4'	"	0.800

15940.20 — RELIEF VENTILATORS

	UNIT	MAN/HOURS
Intake ventilator, aluminum, with screen, no curbs		
12" x 12"	EA.	0.667
16" x 16"	"	0.800
20" x 20"	"	0.800
30" x 30"	"	1.143
36" x 36"	"	1.333
42" x 42"	"	1.333
48" x 48"	"	1.600

15940.40 — REGISTERS AND GRILLES

	UNIT	MAN/HOURS
Lay in flush mounted, perforated face, return		
6x6/24x24	EA.	0.320

AIR DISTRIBUTION

15940.40   REGISTERS AND GRILLES

	UNIT	MAN/HOURS
8x8/24x24	EA.	0.320
9x9/24x24	"	0.320
10x10/24x24	"	0.320
12x12/24x24	"	0.320
Rectangular, ceiling return, single deflection		
10x10	EA.	0.400
12x12	"	0.400
14x14	"	0.400
16x8	"	0.400
16x16	"	0.400
18x8	"	0.400
20x20	"	0.400
24x12	"	0.400
24x18	"	0.400
36x24	"	0.444
36x30	"	0.444
Wall, return air register		
12x12	EA.	0.200
16x16	"	0.200
18x18	"	0.200
20x20	"	0.200
24x24	"	0.200
Ceiling, return air grille		
6x6	EA.	0.267
8x8	"	0.320
10x10	"	0.320
Ceiling, exhaust grille, aluminum egg crate		
6x6	EA.	0.267
8x8	"	0.320
10x10	"	0.320
12x12	"	0.400
14x14	"	0.400
16x16	"	0.400
18x18	"	0.400

15940.80   PENTHOUSE LOUVERS

	UNIT	MAN/HOURS
Penthouse louvers		
12" high, extruded aluminum, 4" louver		
6' perimeter	EA.	2.000
8' perimeter	"	2.000
10' perimeter	"	2.000
12' perimeter	"	2.000
14' perimeter	"	2.667
16' perimeter	"	3.200
18' perimeter	"	4.444
20' perimeter	"	5.333
16" high x 4' perimeter	"	2.000
6' perimeter	"	2.000
8' perimeter	"	2.000
10' perimeter	"	2.000
12' perimeter	"	2.000
14' perimeter	"	2.667
16' perimeter	"	3.200
18' perimeter	"	4.444

AIR DISTRIBUTION

15940.80   PENTHOUSE LOUVERS

	UNIT	MAN/HOURS
20' perimeter	EA.	5.333
22' perimeter	"	6.667
24' perimeter	"	8.889
20" high x 4' perimeter	"	2.000
6' perimeter	"	2.000
8' perimeter	"	2.000
10' perimeter	"	2.000
12' perimeter	"	2.000
14' perimeter	"	2.667
16' perimeter	"	3.200
18' perimeter	"	4.444
20' perimeter	"	5.333
22' perimeter	"	6.667
24' perimeter	"	8.889
24" high x 4' perimeter	"	2.000
6' perimeter	"	2.000
8' perimeter	"	2.000
10' perimeter	"	2.000
12' perimeter	"	2.000
16' perimeter	"	3.200
18' perimeter	"	4.444
20' perimeter	"	5.333
22' perimeter	"	6.667
24' perimeter	"	8.889

CONTROLS

15950.10   HVAC CONTROLS

	UNIT	MAN/HOURS
Pressure gauge, direct reading gage cock and siphon	EA.	0.500
Control valve, 1", modulating		
2-way	EA.	0.667
3-way	"	1.000
Self contained control valve with sensing element, 3/4"	"	0.500
Control dampers, round		
6" dia.	EA.	0.320
8" dia	"	0.320
10" dia	"	0.320
12" dia	"	0.320
12" dia	"	0.400
18" dia	"	0.400
20" dia	"	0.400
Rectangular, parallel blade standard leakage		
12" x 12"	EA.	0.400
16" x 16"	"	0.400
20" x 20"	"	0.400
28" x 28"	"	0.500
32" x 32"	"	0.500

CONTROLS	UNIT	MAN/ HOURS
15950.10 **HVAC CONTROLS**		
36" x 36"	EA.	0.667
40" x 40"	"	0.800
44" x 44"	"	1.000
48" x 48"	"	1.143
48" x 52"	"	1.333
48" x 56"	"	1.333
48" x 60"	"	1.333
48" x 64"	"	1.333
48" x 68"	"	1.333
48" x 72"	"	1.333
Low leakage		
12" x 12"	EA.	0.400
16" x 16"	"	0.400
20" x 20"	"	0.400
24" x 24"	"	0.400
28" x 28"	"	0.500
32" x 32"	"	0.571
36" x 36"	"	0.667
40" x 40"	"	0.800
44" x 44"	"	1.000
48" x 48"	"	1.143
48" x 56"	"	1.333
48" x 60"	"	1.333
48" x 64"	"	1.333
48" x 68"	"	1.333
48" x 72"	"	1.333
Rectangular, opposed horizontal blade		
12" x 12"	EA.	0.400
16" x 16"	"	0.400
20" x 20"	"	0.400
24" x 24"	"	0.400
28" x 28"	"	0.500
32" x 32"	"	0.533
36" x 36"	"	0.667
40" x 40"	"	0.800
44" x 44"	"	1.000
48" x 48"	"	1.143
48" x 52"	"	1.143
48" x 56"	"	1.333
48" x 60"	"	1.333
48" x 64"	"	1.333
48" x 68"	"	1.333
48" x 72"	"	1.333

BASIC MATERIALS	UNIT	MAN/HOURS
16050.30 BUS DUCT		
Bus duct, 100a, plug-in		
10', 600v	EA.	2.759
With ground	"	4.211
10', 277/480v	"	2.759
With ground	"	4.211
Cable tap box	"	2.500
End closure	"	0.400
Edgewise hanger	"	0.727
Flatwise hanger	"	0.727
Outside elbow	"	0.800
Inside elbow	"	0.800
Outside tee	"	1.100
Inside tee	"	1.100
Outlet cover	"	0.400
Wall flange	"	0.400
Circuit breakers, with enclosure		
1 pole		
15a-60a	EA.	1.000
70a-100a	"	1.250
2 pole		
15a-60a	EA.	1.100
70a-100a	"	1.301
3 pole		
15a-60a	EA.	1.159
70a-100a	"	1.509
Bus duct, copper feeder duct, 277/480v, 4 wire		
800a	L.F.	0.400
1000a	"	0.500
1200a	"	0.533
1350a	"	0.615
1600a	"	0.727
2000a	"	0.800
2500a	"	0.851
3000a	"	0.952
Weatherproof		
800a	L.F.	0.444
1000a	"	0.533
1350a	"	0.667
1600a	"	0.727
2000a	"	0.833
2500a	"	0.899
3000a	"	0.976
4000a	"	1.509
5000a	"	1.818
Plug-in feeder duct, 277/480v, 4 wire		
400a	L.F.	0.400
600a	"	0.444
800a	"	0.500
1000a	"	0.500
1200a	"	0.533
1350a	"	0.615
1600a	"	0.727
2000a	"	0.800
2500a	"	0.851

BASIC MATERIALS	UNIT	MAN/HOURS
16050.30 BUS DUCT		
3000a	L.F.	0.952
Copper flanged ends, 277/480v, 4 wire		
225a	EA.	2.500
400a	"	2.759
600a	"	2.963
800a	"	3.077
1000a	"	3.200
1200a	"	3.265
1350a	"	3.333
1600a	"	3.478
2000a	"	3.478
2500a	"	3.636
3000a	"	3.810
4000a	"	4.444
5000a	"	4.706
Bus duct, copper elbows, 277/480v-4w		
225a-1000a	EA.	2.105
1200a-3000a	"	2.500
4000a-5000a	"	2.963
Tees, 277/480v-4w		
225a-1000a	EA.	2.222
1200a-3000a	"	2.581
4000a-5000a	"	2.963
Crosses, 277/480v-4w		
225a-1000a	EA.	2.222
1200a-3000a	"	2.581
4000a-5000a	"	2.963
Copper end closures, 277/480v-4w		
225a-1000a	EA.	0.899
1200a-3000a	"	1.194
4000a-5000a	"	1.667
Tap boxes, 277/480v-4w		
225a	EA.	3.478
400a	"	4.444
600a	"	7.273
800a	"	8.000
1000a	"	10.000
1200a	"	11.004
1350a	"	13.008
1600a	"	14.011
2000a	"	16.985
2500a	"	22.989
3000a	"	27.972
4000a	"	37.915
5000a	"	44.944
Circuit breaker, adapter cubicle		
225a	EA.	1.509
200a	"	1.600
600a	"	1.702
800a	"	1.818
1000a	"	1.905
1200a	"	2.000
1600a	"	2.105
2000a	"	2.222

BASIC MATERIALS	UNIT	MAN/ HOURS
16050.30 BUS DUCT		
Transformer taps, 1 phase 277/480v		
600a	EA.	7.273
800a	"	8.000
1000a	"	10.000
1200a	"	11.004
1350a	"	13.008
1600a	"	14.011
2000a	"	16.985
2500a	"	22.989
3000a	"	27.972
4000a	"	37.915
5000a	"	45.977
3 phase, 480v, 3 wire		
600a	EA.	7.273
800a	"	8.000
1000a	"	10.000
1200a	"	11.004
1350a	"	13.008
1600a	"	14.011
2000a	"	16.985
2500a	"	22.989
3000a	"	27.972
4000a	"	37.915
5000a	"	45.977
3 phase, 4 wire, 277/480v		
600a	EA.	7.273
800a	"	8.000
1000a	"	10.000
1200a	"	11.004
1350a	"	13.008
1600a	"	14.011
2000a	"	16.985
2500a	"	22.989
3000a	"	27.972
4000a	"	40.816
5000a	"	45.977
Transformer connection, 4 wire, 277/480v		
600a	EA.	2.759
800a	"	2.857
1000a	"	2.963
1200a	"	3.077
1350a	"	3.200
1600a	"	3.333
2000a	"	3.478
2500a	"	3.636
3000a	"	3.810
4000a	"	4.444
5000a	"	4.706
Unfused reducers, 3 wire, 480v, 3 phase		
400a	EA.	2.500
600a	"	3.810
800a	"	4.706
1000a	"	5.000
1200a	"	5.333

BASIC MATERIALS	UNIT	MAN/ HOURS
16050.30 BUS DUCT		
1350a	EA.	5.714
1600a	"	6.154
2000a	"	6.400
2500a	"	6.667
3000a	"	7.273
4000a	"	8.753
5000a	"	10.796
Circuit breaker reducers, 4 wire, 277/480v		
400a	EA.	2.222
600a	"	3.478
800a	"	4.211
1000a	"	4.706
1350a	"	5.333
1600a	"	5.714
2000a	"	6.154
2500a	"	6.667
3000a	"	6.957
4000a	"	7.273
5000a	"	10.000
Expansion fittings, 4 wire, 277/480v		
225a	EA.	2.500
400a	"	3.810
600a	"	4.706
800a	"	5.000
1000a	"	5.333
1200a	"	5.714
1350a	"	5.926
1600a	"	6.154
2000a	"	6.667
2500a	"	7.273
3000a	"	8.753
4000a	"	10.796
5000a	"	11.994
Wall flanges		
225a-2500a	EA.	4.000
3000a-5000a	"	6.154
Weather seals	"	1.000
Roof flanges	"	4.000
Fire barriers	"	1.509
Spring hangers	"	1.739
Sway brace collars	"	1.250
Hook sticks		
8'	EA.	
14'	"	
Fusible switches, 240v, 3 phase		
30a	EA.	1.000
60a	"	1.250
100a	"	1.509
200a	"	2.105
400a	"	4.000
600a	"	6.154
208v, 4 wire		
30a	EA.	1.194
60a	"	1.356

BASIC MATERIALS

16050.30 — BUS DUCT

	UNIT	MAN/HOURS
100a	EA.	1.818
200a	"	2.759
400a	"	5.000
600a	"	8.000
600v		
30a	EA.	1.000
60a	"	1.250
100a	"	1.509
200a	"	2.105
400a	"	4.000
600a	"	6.154
800a	"	6.667
1000a	"	8.000
1200a	"	11.004
1600a	"	11.994
480v, 4 wire		
30a	EA.	1.194
60a	"	1.356
100a	"	1.818
200a	"	2.759
400a	"	5.000
600a	"	8.000
800a	"	8.247
1000a	"	11.004
1200a	"	11.994
1600a	"	14.011
Fusible combination starters, 600v, 3 phase		
Size 0	EA.	1.290
Size 1	"	1.600
Size 2	"	1.818
Size 3	"	2.581
Circuit breaker combination starters, 600v, 3 phase		
Size 0	EA.	1.290
Size 1	"	1.600
Size 2	"	1.818
Size 3	"	2.581
Fusible combination contactors, 600v, 3 phase		
30a	EA.	1.290
60a	"	1.600
100a	"	1.818
200a	"	2.581
Circuit breaker, combination contactors, 600v, 3 phase		
30a	EA.	1.290
60a	"	1.600
100a	"	1.818
200a	"	2.581
Fusible contactor electrically held, 480v, 4 wire		
30a	EA.	1.290
60a	"	1.600
100a	"	1.818
200a	"	2.759
Mechanically held		
30a	EA.	1.290
60a	"	1.600

BASIC MATERIALS

16050.30 — BUS DUCT

	UNIT	MAN/HOURS
100a	EA.	1.860
200a	"	2.759
Circuit breakers, 240v, 3 phase		
15a-60a	EA.	1.159
70a-100a	"	1.600
600v, 3 phase		
15a-60a	EA.	1.159
125a-225a	"	2.286
250a-400a	"	4.211
500a-600a	"	5.333
700a-800a	"	8.000
900a-1000a	"	10.000
1200a-1600a	"	11.004
120/208v, 4 wire		
15a-60a	EA.	1.290
70a-100a	"	1.860
277/480v, 4 wire		
15a-60a	EA.	1.290
70a-100a	"	1.905
125a-225a	"	2.759
250a-400a	"	5.000
500a-600a	"	8.000
700a-800a	"	8.000
900a-1000a	"	11.004
1200a-1600a	"	14.011
600v, 3 phase, 65,000 aic.		
60a	EA.	1.159
70a-100a	"	1.600
125a-225a	"	2.286
250a-400a	"	4.211
500a-600a	"	6.154
700a-800a	"	8.000
900a-1000a	"	10.000
277/480v, 4 wire, 65,000 aic.		
15a-60a	EA.	1.290
70a-100a	"	1.905
125a-225a	"	2.759
250a-400a	"	5.000
500a-600a	"	6.154
700a-800a	"	8.000
900a-1000a	"	10.349
600v, 3 phase, current limiting		
15a-60a	EA.	1.159
70a-100a	"	1.600
125a-225a	"	2.286
250a-400a	"	4.211
500a-600a	"	6.154
700a-800a	"	8.000
900a-1000a	"	10.000
277/480v, 4 wire, current limiting		
15a-60a	EA.	1.290
70a-100a	"	40.000
125a-225a	"	2.759
250a-400a	"	5.000

BASIC MATERIALS		UNIT	MAN/HOURS
16050.30	**BUS DUCT**		
500a-600a		EA.	6.154
700a-800a		"	8.000
900a-1000a		"	11.004
Capacitors, 3 phase, 240v			
5 kvar		EA.	5.000
7.5 kvar		"	6.154
10 kvar		"	8.097
15 kvar		"	9.195
480v			
2.5 kvar		EA.	2.759
5 kvar		"	4.706
7.5 kvar		"	5.714
10 kvar		"	8.097
15 kvar		"	8.999
20 kvar		"	11.494
25 kvar		"	13.008
30 kvar		"	14.210
Transformers, 3 phase, 480v			
1.0 kva		EA.	1.739
1.5 kva		"	2.000
2 kva		"	2.500
3 kva		"	2.759
5 kva		"	4.000
7.5 kva		"	5.000
10 kva		"	5.333
16110.12	**CABLE TRAY**		
Cable tray, 6"		L.F.	0.059
Ventilated cover		"	0.030
Solid cover		"	0.030
Flat 90		EA.	0.500
Outside 90		"	0.500
Inside 90		"	0.500
Flat 45		"	0.500
Outside 45		"	0.500
Inside 45		"	0.500
Adjustable elbow		"	0.500
Support riser		"	0.500
Adjustable riser		"	0.500
Tee		"	1.739
Cross		"	1.818
Blind end		"	0.296
Expansion joint		"	0.500
Box connector		"	2.500
Standard dropout		"	0.500
2"		"	0.615
3"		"	0.800
4"		"	1.000
Cable tray, 9"		L.F.	0.070
Ventilated cover		"	0.040
Solid cover		"	0.040
Flat 90		EA.	0.533
Outside 90		"	0.533

BASIC MATERIALS		UNIT	MAN/HOURS
16110.12	**CABLE TRAY**		
Inside 90		EA.	0.533
Flat 45		"	0.533
Outside 45		"	0.533
Inside 45		"	0.533
Adjustable elbow		"	0.533
Support riser		"	0.533
Adjustable riser		"	0.533
Tee		"	1.818
Cross		"	1.818
Blind end		"	0.320
Expansion joint		"	0.533
Box connector		"	2.581
Standard dropout		"	0.533
2"		"	0.667
3"		"	0.667
4"		"	1.096
Cable tray, 12"		L.F.	0.080
Ventilated cover		"	0.050
Solid cover		"	0.050
Flat 90		EA.	0.533
Outside 90		"	0.533
Inside 90		"	0.533
Flat 45		"	0.533
Outside 45		"	0.533
Inside 45		"	0.533
Adjustable elbow		"	0.533
Support riser		"	0.533
Adjustable riser		"	0.533
Tee		"	2.000
Cross		"	2.000
Blind end		"	0.348
Expansion joint		"	0.615
Box connector		"	2.759
Standard dropout		"	0.615
2"		"	0.727
3"		"	0.899
4"		"	1.159
Cable tray, 18"		L.F.	0.100
Ventilated cover		"	0.059
Solid cover		"	0.059
Flat 90		EA.	0.727
Outside 90		"	0.727
Inside 90		"	0.727
Flat 45		"	0.727
Outside 45		"	0.727
Inside 45		"	0.727
Adjustable elbow		"	0.727
Support riser		"	0.727
Adjustable riser		"	0.727
Tee		"	2.105
Cross		"	2.105
Blind end		"	0.400
Expansion joint		"	0.727
Box connector		"	2.963

Left Column

BASIC MATERIALS	UNIT	MAN/HOURS
16110.12 CABLE TRAY		
Standard dropout	EA.	0.667
2"	"	0.727
3"	"	0.952
4"	"	1.194
Cable tray, 24"	L.F.	0.123
Ventilated cover	"	0.070
Solid cover	"	0.070
Flat 90	EA.	0.727
Outside 90	"	0.727
Inside 90	"	0.727
Flat 45	"	0.727
Outside 45	"	0.727
Inside 45	"	0.727
Adjustable elbow	"	0.727
Support riser	"	0.727
Adjustable riser	"	0.727
Tee	"	2.222
Cross	"	2.222
Blind end	"	0.444
Expansion joint	"	0.727
Box connector	"	3.478
Standard dropout	"	0.667
2"	"	0.800
3"	"	0.976
4"	"	1.250
Cable tray, 36"	L.F.	0.145
Ventilated cover	"	0.080
Solid cover	"	0.080
Flat 90	EA.	0.851
Outside 90	"	0.851
Inside 90	"	0.851
Flat 45	"	0.851
Outside 45	"	0.851
Inside 45	"	0.851
Adjustable elbow	"	0.851
Support riser	"	0.851
Adjustable riser	"	0.851
Tee	"	2.286
Cross	"	2.963
Blind end	"	0.471
Expansion joint	"	0.727
Box connector	"	3.810
Standard dropout	"	0.727
2"	"	0.800
3"	"	1.000
4"	"	1.290
Reducers		
9" - 6"	EA.	0.500
12" - 9"	"	0.500
18" - 12"	"	0.615
24" - 18"	"	0.727
36" - 18"	"	0.800
36" - 24"	"	0.899
Conduit dropouts		

Right Column

BASIC MATERIALS	UNIT	MAN/HOURS
16110.12 CABLE TRAY		
3/4"	EA.	0.348
1"	"	0.348
1-1/4"	"	0.400
1-1/2"	"	0.500
2"	"	0.533
2-1/2"	"	0.727
3"	"	0.800
Wall brackets		
6"	EA.	0.145
9"	"	0.145
12"	"	0.200
18"	"	0.250
24"	"	0.296
36"	"	0.400
16110.15 FIBERGLASS CABLE TRAY		
Fiberglass cable tray, 6"	L.F.	0.040
Tray cover	"	0.030
Horizontal		
90	EA.	0.276
45	"	0.276
30	"	0.276
Inside		
90	EA.	0.276
45	"	0.276
30	"	0.276
Horizontal tee	"	0.727
Horizontal cross	"	1.096
Splice plate	"	0.145
Floor flange	"	0.348
Panel flange	"	1.250
End plate	"	0.145
Nylon rivet	"	0.050
Barrier strip	"	0.050
Hold down clamp	"	0.050
Drop out	"	0.533
Cover stand off	"	0.050
Wall bracket	"	0.200
Sealer	"	
Outside		
90	EA.	0.276
45	"	0.276
30	"	0.276
Fiberglass cable tray, 9"	L.F.	0.050
Tray cover	"	0.040
Horizontal		
90	EA.	0.276
45	"	0.276
30	"	0.276
Inside		
90	EA.	0.276
45	"	0.276
30	"	0.276
Horizontal tee	"	0.727

BASIC MATERIALS	UNIT	MAN/HOURS
16110.15 FIBERGLASS CABLE TRAY		
Horizontal cross	EA.	1.096
Splice plate	"	0.145
Floor flange	"	0.348
Panel flange	"	1.250
End plate	"	0.145
Nylon rivet	"	0.050
Barrier strap	"	0.050
Hold down clamp	"	0.050
Drop out	"	0.533
Cover stand off	"	0.050
Wall bracket	"	0.200
Sealer	"	
Outside		
90	EA.	0.276
45	"	0.276
30	"	0.276
Fiberglass cable tray, 12"	L.F.	0.059
Tray cover	"	0.050
Horizontal		
90	EA.	0.348
45	"	0.348
30	"	0.348
Inside		
90	EA.	0.348
45	"	0.348
Inside 30	"	0.348
Horizontal tee	"	0.952
Horizontal cross	"	1.290
Splice plate	"	0.160
Floor flange	"	0.364
Panel flange	"	1.600
End plate	"	0.160
Nylon rivet	"	0.050
Barrier strip	"	0.050
Hold down clamp	"	0.050
Drop out	"	0.533
Cover stand off	"	0.050
Wall bracket	"	0.200
Sealer	"	
Outside		
90	EA.	0.348
45	"	0.348
30	"	0.348
Fiberglass cable tray, 18"	L.F.	0.070
Tray cover	"	0.059
Horizontal		
90	EA.	0.400
45	"	0.400
30	"	0.400
Inside		
90	EA.	0.400
45	"	0.400
30	"	0.400
Horizontal tee	"	1.538

BASIC MATERIALS	UNIT	MAN/HOURS
16110.15 FIBERGLASS CABLE TRAY		
Horizontal cross	EA.	1.356
Splice plate	"	0.160
Floor flange	"	0.381
Panel flange	"	1.702
End plate	"	0.170
Nylon rivet	"	0.050
Barrier strip	"	0.050
Hold down clamp	"	0.050
Drop out	"	0.533
Cover stand off	"	0.050
Wall bracket	"	0.200
Sealer	"	
Outside		
90	EA.	0.400
45	"	0.400
30	"	0.400
Fiberglass cable tray, 24"	L.F.	0.080
Tray cover	"	0.070
Horizontal		
90	EA.	0.500
45	"	0.500
30	"	0.500
Inside		
90	EA.	0.500
45	"	0.500
30	"	0.500
Horizontal tee	"	1.000
Horizontal cross	"	1.509
Splice plate	"	0.160
Floor flange	"	0.400
Panel flange	"	1.818
End plate	"	0.170
Nylon Rivet	"	0.050
Barrier strip	"	0.050
Hold down clamp	"	0.050
Drop out	"	0.533
Cover stand off	"	0.050
Wall bracket	"	0.200
Sealer	"	
Outside		
90	EA.	0.500
45	"	0.500
30	"	0.500
Fiberglass cable tray, 30"	L.F.	0.089
Tray cover	"	0.080
Horizontal		
90	EA.	0.615
45	"	0.615
30	"	0.615
Inside		
90	EA.	0.615
45	"	0.615
30	"	0.615
Horizontal tee	"	1.096

BASIC MATERIALS

16110.15 FIBERGLASS CABLE TRAY

	UNIT	MAN/HOURS
Horizontal cross	EA.	1.600
Splice plate	"	0.160
Floor flange	"	0.400
Panel flange	"	1.905
End plate	"	0.200
Nylon rivet	"	0.050
Barrier strip	"	0.050
Hold down clamp	"	0.050
Dropout	"	0.615
Cover stand off	"	0.050
Wall bracket	"	0.200
Sealer	"	
Outside		
90	EA.	0.615
45	"	0.615
30	"	0.615
Fiberglass cable tray, 36"	L.F.	0.100
Tray cover	"	0.089
Horizontal		
90	EA.	0.667
45	"	0.667
30	"	0.667
Inside		
90	EA.	0.667
45	"	0.667
30	"	0.667
Horizontal tee	"	1.194
horizontal cross	"	1.702
Splice plate	"	0.160
Floor flange	"	0.444
Panel flange	"	2.105
End plate	"	0.200
Nylon rivet	"	0.050
Barrier strip	"	0.050
Hold down clamp	"	0.050
Drop out	"	0.667
5Cover stand off	"	0.050
Wall bracket	"	0.200
Sealer	"	
Outside		
90	EA.	0.667
45	"	0.667
30	"	0.667
Reducers		
12" - 6"	EA.	0.296
12" - 9"	"	0.296
18" - 6"	"	0.348
18" - 9"	"	0.348
18" - 12"	"	0.400
24" - 6"	"	0.400
24" - 9"	"	0.400
24" - 12"	"	0.400
24" - 18"	"	0.444
30" - 9"	"	0.400

BASIC MATERIALS

16110.15 FIBERGLASS CABLE TRAY

	UNIT	MAN/HOURS
30" - 12"	EA.	0.400
30" - 18"	"	0.444
30" - 24"	"	0.500
36" - 9"	"	0.500
36" - 12"	"	0.500
36" - 18"	"	0.533
36" - 24"	"	0.615
36" - 30"	"	0.727

16110.20 CONDUIT SPECIALTIES

	UNIT	MAN/HOURS
Rod beam clamp, 1/2"	EA.	0.050
Hanger rod		
3/8"	L.F.	0.040
1/2"	"	0.050
All thread rod		
1/4"	L.F.	0.030
3/8"	"	0.040
1/2"	"	0.050
5/8"	"	0.080
Hanger channel, 1-1/2"		
No holes	EA.	0.030
Holes	"	0.030
Channel strap		
1/2"	EA.	0.050
3/4"	"	0.050
1"	"	0.050
1-1/4"	"	0.080
1-1/2"	"	0.080
2"	"	0.080
2-1/2"	"	0.123
3"	"	0.123
3-1/2"	"	0.123
4"	"	0.145
5"	"	0.145
6"	"	0.145
Conduit penetrations, roof and wall, 8" thick		
1/2"	EA.	0.615
3/4"	"	0.615
1"	"	0.800
1-1/4"	"	0.800
1-1/2"	"	0.800
2"	"	1.600
2-1/2"	"	1.600
3"	"	1.600
3-1/2"	"	2.000
4"	"	2.000
Plastic duct bank conduit spacer, 3" separation		
2"	EA.	0.050
3"	"	0.050
4"	"	0.050
5"	"	0.050
6"	"	0.050
Intermediate, 3" separation		
2"	EA.	0.050

Left Column

BASIC MATERIALS	UNIT	MAN/HOURS
16110.20 CONDUIT SPECIALTIES		
3"	EA.	0.050
4"	"	0.050
5"	"	0.050
6"	"	0.050
Base with 1-1/2" separation		
2"	EA.	0.050
3"	"	0.160
4"	"	0.160
5"	"	0.160
6"	"	0.160
Intermediate, 1-1/2" separation		
2"	EA.	0.160
3"	"	0.160
3-1/2"	"	0.160
4"	"	0.160
5"	"	0.160
6"	"	0.160
OD beam clamp, 1/4"	"	0.200
Threaded rod couplings		
1/4"	EA.	0.050
3/8"	"	0.050
1/2"	"	0.050
5/8"	"	0.050
3/4"	"	0.050
Hex nuts		
1/4"	EA.	0.050
3/8"	"	0.050
1/2"	"	0.050
5/8"	"	0.050
3/4"	"	0.050
Square nuts		
1/4"	EA.	0.050
3/8"	"	0.050
3/8"	"	0.050
5/8"	"	0.050
3/4"	"	0.050
Channel closure strip	L.F.	0.133
Channel end cap	EA.	0.133
Li-channel trapeze hangers		
12" long	EA.	0.145
18" long	"	0.145
24" long	"	0.145
30" long	"	0.250
36" long	"	0.250
42" long	"	0.296
Channel spring nuts		
1/4"	EA.	0.059
3/8"	"	0.080
1/2"	"	0.100
Fireproofing, for conduit penetrations		
1/2"	EA.	0.500
3/4"	"	0.500
1"	"	0.500
1-1/4"	"	0.727

Right Column

BASIC MATERIALS	UNIT	MAN/HOURS
16110.20 CONDUIT SPECIALTIES		
1-1/2"	EA.	0.727
2"	"	0.727
2-1/2"	"	0.899
3"	"	0.899
3-1/2"	"	1.250
4"	"	1.509
16110.21 ALUMINUM CONDUIT		
Aluminum conduit		
1/2"	L.F.	0.030
3/4"	"	0.040
1"	"	0.050
1-1/4"	"	0.059
1-1/2"	"	0.080
2"	"	0.089
2-1/2"	"	0.100
3"	"	0.107
3-1/2"	"	0.123
4"	"	0.145
5"	"	0.182
6"	"	0.200
90 deg. elbow		
1/2"	EA.	0.190
3/4"	"	0.250
1"	"	0.308
1-1/4"	"	0.381
1-1/2"	"	0.400
2"	"	0.444
2-1/2"	"	0.571
3"	"	0.667
3-1/2"	"	0.800
4"	"	0.889
5"	"	1.143
6"	"	2.222
Coupling		
1/2"	EA.	0.050
3/4"	"	0.059
1"	"	0.080
1-1/4"	"	0.089
1-1/2"	"	0.100
2"	"	0.107
2-1/2"	"	0.123
3"	"	0.123
3-1/2"	"	0.145
4"	"	0.160
5"	"	0.160
6"	"	0.190
16110.22 EMT CONDUIT		
EMT conduit		
1/2"	L.F.	0.030
3/4"	"	0.040
1"	"	0.050

BASIC MATERIALS	UNIT	MAN/ HOURS
16110.22 EMT CONDUIT		
1-1/4"	L.F.	0.059
1-1/2"	"	0.080
2"	"	0.089
2-1/2"	"	0.100
3"	"	0.123
3-1/2"	"	0.145
4"	"	0.182
90 deg. elbow		
1/2"	EA.	0.089
3/4"	"	0.100
1"	"	0.107
1-1/4"	"	0.123
1-1/2"	"	0.145
2"	"	0.190
2-1/2"	"	0.211
3"	"	0.242
3-1/2"	"	0.258
4"	"	0.286
Connector, steel compression		
1/2"	EA.	0.089
3/4"	"	0.089
1"	"	0.089
1-1/4"	"	0.107
1-1/2"	"	0.145
2"	"	0.190
2-1/2"	"	0.250
3"	"	0.286
3-1/2"	"	0.308
4"	"	0.333
Coupling, steel, compression		
1/2"	EA.	0.059
3/4"	"	0.059
1"	"	0.059
1-1/4"	"	0.089
1-1/2"	"	0.107
2"	"	0.145
2-1/2"	"	0.222
3"	"	0.250
3-1/2"	"	0.286
4"	"	0.308
1 hole strap, steel		
1/2"	EA.	0.040
3/4"	"	0.040
1"	"	0.040
1-1/4"	"	0.050
1-1/2"	"	0.050
2"	"	0.050
2-1/2"	"	0.059
3"	"	0.059
3-1/2"	"	0.059
4"	"	0.059
Connector, steel set screw		
1/2"	EA.	0.070
3/4"	"	0.070

BASIC MATERIALS	UNIT	MAN/ HOURS
16110.22 EMT CONDUIT		
1"	EA.	0.070
1-1/4"	"	0.107
1-1/2"	"	0.145
2"	"	0.182
2-1/2"	"	0.242
3"	"	0.267
3-1/2"	"	0.296
4"	"	0.348
Insulated throat		
1/2"	EA.	0.070
3/4"	"	0.070
1"	"	0.070
1-1/4"	"	0.107
1-1/2"	"	0.145
2"	"	0.182
2-1/2"	"	0.242
3"	"	0.267
3-1/2"	"	0.296
4"	"	0.348
Connector, die cast set screw		
1/2"	EA.	0.059
3/4"	"	0.059
1"	"	0.059
1-1/4"	"	0.089
1-1/2"	"	0.107
2"	"	0.145
2-1/2"	"	0.200
3"	"	0.222
3-1/2"	"	0.250
4"	"	0.286
Insulated throat		
1/2"	EA.	0.059
3/4"	"	0.059
1"	"	0.059
1-1/4"	"	0.089
1-1/2"	"	0.107
2"	"	0.145
2-1/2"	"	0.200
3"	"	0.222
3-1/2"	"	0.250
4"	"	0.286
Coupling, steel set screw		
1/2"	EA.	0.040
3/4"	"	0.040
1"	"	0.040
1-1/4"	"	0.050
1-1/2"	"	0.080
2"	"	0.107
2-1/2"	"	0.160
3"	"	0.190
3-1/2"	"	0.222
4"	"	0.250
Diecast set screw		
1/2"	EA.	0.040

BASIC MATERIALS	UNIT	MAN/HOURS
16110.22 EMT CONDUIT		
3/4"	EA.	0.040
1"	"	0.040
1-1/4"	"	0.050
1-1/2"	"	0.080
2"	"	0.107
2-1/2"	"	0.160
3"	"	0.186
3-1/2"	"	0.222
4"	"	0.250
1 hole malleable straps		
1/2"	EA.	0.040
3/4"	"	0.040
1"	"	0.040
1-1/4"	"	0.050
1-1/2"	"	0.050
2"	"	0.050
2-1/2"	"	0.059
3"	"	0.059
3-1/2"	"	0.059
4"	"	0.059
EMT to rigid compression coupling		
1/2"	EA.	0.100
3/4"	"	0.100
1"	"	0.150
Set screw couplings		
1/2"	EA.	0.100
3/4"	"	0.100
1"	"	0.145
Set screw offset connectors		
1/2"	EA.	0.100
3/4"	"	0.100
1"	"	0.145
Compression offset connectors		
1/2"	EA.	0.100
3/4"	"	0.100
1"	"	0.145
Type "LB" set screw condulets		
1/2"	EA.	0.229
3/4"	"	0.296
1"	"	0.381
1-1/4"	"	0.444
1-1/2"	"	0.533
2"	"	0.615
2-1/2"	"	0.727
3"	"	1.000
3-1/2"	"	1.333
4"	"	1.600
Type "T" set screw condulets		
1/2"	EA.	0.296
3/4"	"	0.400
1"	"	0.444
1-1/4"	"	0.533
1-1/2"	"	0.615
2"	"	0.667

BASIC MATERIALS	UNIT	MAN/HOURS
16110.22 EMT CONDUIT		
Type "C" set screw condulets		
1/2"	EA.	0.250
3/4"	"	0.296
1"	"	0.381
1-1/4"	"	0.444
1-1/2"	"	0.533
2"	"	0.381
Type "LL" setscrew condulets		
1/2"	EA.	0.250
3/4"	"	0.296
1"	"	0.381
1-1/4"	"	0.444
1-1/2"	"	0.533
2"	"	0.615
Type "LR" set screw condulets		
1/2"	EA.	0.250
3/4"	"	0.296
1"	"	0.381
1-1/4"	"	0.444
1-1/2"	"	0.533
2"	"	0.615
Type "LB" compression condulets		
1/2"	EA.	0.296
3/4"	"	0.500
1"	"	0.500
Type "T" compression condulets		
1/2"	EA.	0.400
3/4"	"	0.444
1"	"	0.615
Condulet covers		
1/2"	EA.	0.123
3/4"	"	0.123
1"	"	0.123
1-1/4"	"	0.123
1-1/2"	"	0.145
2"	"	0.145
2-1/2"	"	0.145
3"	"	0.182
3-1/2"	"	0.182
4"	"	0.182
Clamp type entrance caps		
1/2"	EA.	0.250
3/4"	"	0.296
1"	"	0.400
1-1/4"	"	0.533
1-1/2"	"	0.615
2"	"	0.899
2-1/2"	"	1.000
3"	"	1.509
3-1/2"	"	1.739
4"	"	2.222
Slip fitter type entrance caps		
1/2"	EA.	0.250
3/4"	"	0.296

BASIC MATERIALS

16110.22   EMT CONDUIT

	UNIT	MAN/HOURS
1"	EA.	0.400
1-1/4"	"	0.533
1-1/2"	"	0.615
2"	"	0.899
2-1/2"	"	1.000
3"	"	1.509
3-1/2"	"	1.739
4"	"	2.222

16110.23   FLEXIBLE CONDUIT

	UNIT	MAN/HOURS
Flexible conduit, steel		
3/8"	L.F.	0.030
1/2	"	0.030
3/4"	"	0.040
1"	"	0.040
1-1/4"	"	0.050
1-1/2"	"	0.059
2"	"	0.080
2-1/2"	"	0.089
3"	"	0.107
Flexible conduit, liquid tight		
3/8"	L.F.	0.030
1/2"	"	0.030
3/4"	"	0.040
1"	"	0.040
1-1/4"	"	0.050
1-1/2"	"	0.059
2"	EA.	0.080
2-1/2"	"	0.089
3"	"	0.107
4"	"	0.145
Connector, straight		
3/8"	EA.	0.080
1/2"	"	0.080
3/4"	"	0.089
1"	"	0.100
1-1/4"	"	0.107
1-1/2"	"	0.123
2"	"	0.145
2-1/2"	"	0.182
3"	"	0.190
Straight insulated throat connectors		
3/8"	EA.	0.123
1/2"	"	0.123
3/4"	"	0.145
1"	"	0.145
1-1/4"	"	0.182
1-1/2"	"	0.211
2"	"	0.229
2-1/2"	"	0.267
3"	"	0.333
4"	"	0.421
90 deg connectors		
3/8"	EA.	0.148

BASIC MATERIALS

16110.23   FLEXIBLE CONDUIT

	UNIT	MAN/HOURS
1/2"	EA.	0.148
3/4"	"	0.170
1"	"	0.182
1-1/4"	"	0.229
1-1/2"	"	0.250
2"	"	0.267
2-1/2"	"	0.333
3"	"	0.381
4"	"	0.444
90 degree insulated throat connectors		
3/8"	EA.	0.145
1/2"	"	0.145
3/4"	"	0.170
1"	"	0.178
1-1/4"	"	0.229
1-1/2"	"	0.250
2"	"	0.267
2-1/2"	"	0.333
3"	"	0.381
4"	"	0.444
Flexible aluminum conduit		
3/8"	L.F.	0.030
1/2"	"	0.030
3/4"	"	0.040
1"	"	0.040
1-1/4"	"	0.050
1-1/2"	"	0.059
2"	"	0.080
2-1/2"	"	0.089
3"	"	0.107
3-1/2"	"	0.123
4"	"	0.145
Connector, straight		
3/8"	EA.	0.100
1/2"	"	0.100
3/4"	"	0.107
1"	"	0.123
1-1/4"	"	0.145
1-1/2"	"	0.182
2"	"	0.190
2-1/2"	"	0.222
3"	"	0.276
4"	"	0.348
Straight insulated throat connectors		
3/8"	EA.	0.089
1/2"	"	0.089
3/4"	"	0.089
1"	"	0.100
1-1/4"	"	0.107
1-1/2"	"	0.123
2"	"	0.145
2-1/2"	"	0.182
3"	"	0.190
3-1/2"	"	0.222

BASIC MATERIALS	UNIT	MAN/HOURS
16110.23 **FLEXIBLE CONDUIT**		
4"	EA.	0.276
90 deg connectors		
3/8"	EA.	0.145
1/2"	"	0.145
3/4"	"	0.145
1"	"	0.170
1-1/4"	"	0.182
1-1/2"	"	0.200
2"	"	0.211
2-1/2"	"	0.229
3"	"	0.267
90 deg insulated throat connectors		
3/8"	EA.	0.145
1/2"	"	0.145
3/4"	"	0.145
1"	"	0.170
1-1/4"	"	0.182
1-1/2"	"	0.200
2"	"	0.211
2-1/2"	"	0.229
3"	"	0.267
3-1/2"	"	0.333
4"	"	0.421
16110.24 **GALVANIZED CONDUIT**		
Galvanized rigid steel conduit		
1/2"	L.F.	0.040
3/4"	"	0.050
1"	"	0.059
1-1/4"	"	0.080
1-1/2"	"	0.089
2"	"	0.100
2-1/2"	"	0.145
3"	"	0.182
3-1/2"	"	0.190
4"	"	0.211
5"	"	0.286
6"	"	0.381
90 degree ell		
1/2"	EA.	0.250
3/4"	"	0.308
1"	"	0.381
1-1/4"	"	0.444
1-1/2"	"	0.500
2"	"	0.533
2-1/2"	"	0.667
3"	"	0.889
3-1/2"	"	1.000
4"	"	1.333
5"	"	2.222
6"	"	3.333
Couplings, with set screws		
1/2"	EA.	0.050

BASIC MATERIALS	UNIT	MAN/HOURS
16110.24 **GALVANIZED CONDUIT**		
3/4"	EA.	0.059
1"	"	0.080
1-1/4"	"	0.100
1-1/2"	"	0.123
2"	"	0.145
2-1/2"	"	0.190
3"	"	0.250
3-1/2"	"	0.286
4"	"	0.308
5"	"	0.444
6"	"	0.500
Split couplings		
1/2"	EA.	0.190
3/4"	"	0.250
1"	"	0.276
1-1/4"	"	0.308
1-1/2"	"	0.381
2"	"	0.571
2-1/2"	"	0.571
3"	"	0.727
3-1/2"	"	1.000
4"	"	1.333
5"	"	1.633
6"	"	2.051
Erickson couplings		
1/2"	EA.	0.444
3/4"	"	0.500
1"	"	0.615
1-1/4"	"	0.889
1-1/2"	"	1.000
2"	"	1.333
2-1/2"	"	1.860
3"	"	2.105
3-1/2"	"	2.500
4"	"	2.667
5"	"	2.963
6"	"	3.200
Seal fittings		
1/2"	EA.	0.667
3/4"	"	0.800
1"	"	1.000
1-1/4"	"	1.143
1-1/2"	"	1.333
2"	"	1.600
2-1/2"	"	1.905
3"	"	2.105
3-1/2"	"	2.500
4"	"	2.963
5"	"	4.444
6"	"	5.000
Entrance fitting, (weather head), threaded		
1/2"	EA.	0.444
3/4"	"	0.500
1"	"	0.571

BASIC MATERIALS	UNIT	MAN/HOURS
16110.24 GALVANIZED CONDUIT		
1-1/4"	EA.	0.727
1-1/2"	"	0.800
2"	"	0.889
2-1/2"	"	1.000
3"	"	1.333
3-1/2"	"	1.739
4"	"	2.500
5"	"	3.478
6"	"	4.444
Locknuts		
1/2"	EA.	0.050
3/4"	"	0.050
1"	"	0.050
1-1/4"	"	0.050
1-1/2"	"	0.059
2"	"	0.059
2-1/2"	"	0.080
3"	"	0.080
3-1/2"	"	0.080
4"	"	0.089
5"	"	0.089
6"	"	0.089
Plastic conduit bushings		
1/2"	EA.	0.123
3/4"	"	0.145
1"	"	0.190
1-1/4"	"	0.222
1-1/2"	"	0.250
2"	"	0.308
2-1/2"	"	0.500
3"	"	0.667
3-1/2"	"	0.800
4"	"	0.889
5"	"	1.143
6"	"	1.600
Conduit bushings, steel		
1/2"	EA.	0.123
3/4"	"	0.145
1"	"	0.190
1-1/4"	"	0.222
1-1/2"	"	0.250
2"	"	0.308
2-1/2"	"	0.500
3"	"	0.667
3-1/2"	"	0.800
4"	"	0.889
5"	"	1.143
6"	"	1.600
Pipe cap		
1/2"	EA.	0.050
3/4"	"	0.050
1"	"	0.050
1-1/4"	"	0.080
1-1/2"	"	0.080

BASIC MATERIALS	UNIT	MAN/HOURS
16110.24 GALVANIZED CONDUIT		
2"	EA.	0.080
2-1/2"	"	0.089
3"	"	0.089
3-1/2"	"	0.089
4"	"	0.107
5"	"	0.145
6"	"	0.200
GRS elbows, 36" radius		
2"	EA.	0.667
2-1/2"	"	0.808
3"	"	1.053
3-1/2"	"	1.250
4"	"	1.509
5"	"	2.500
6"	"	3.810
42" radius		
2"	EA.	0.808
2-1/2"	"	1.000
3"	"	1.250
3-1/2"	"	1.509
4"	"	1.739
5"	"	2.857
6"	"	4.000
48" radius		
2"	EA.	0.930
2-1/2"	"	1.127
3"	"	1.429
3-1/2"	"	1.739
4"	"	2.162
5"	"	3.077
6"	"	4.444
Threaded couplings		
1/2"	EA.	0.050
3/4"	"	0.059
1"	"	0.080
1-1/4"	"	0.089
1-1/2"	"	0.100
2"	"	0.107
2-1/2"	"	0.123
3"	"	0.145
3-1/2"	"	0.145
4"	"	0.160
5"	"	0.182
6"	"	0.190
Threadless couplings		
1/2"	EA.	0.100
3/4"	"	0.123
1"	"	0.145
1-1/4"	"	0.190
1-1/2"	"	0.250
2"	"	0.308
2-1/2"	"	0.500
3"	"	0.615
3-1/2"	"	0.808

BASIC MATERIALS	UNIT	MAN/HOURS
16110.24 GALVANIZED CONDUIT		
4"	EA.	1.000
5"	"	1.250
6"	"	5.333
Threadless connectors		
1/2"	EA.	0.100
3/4"	"	0.123
1"	"	0.145
1-1/4"	"	0.190
1-1/2"	"	0.250
2"	"	0.308
2-1/2"	"	0.500
3"	"	0.615
3-1/2"	"	0.808
4"	"	1.000
5"	"	1.250
6"	"	1.509
Setscrew connectors		
1/2"	EA.	0.080
3/4"	"	0.089
1"	"	0.100
1-1/4"	"	0.123
1-1/2"	"	0.145
2"	"	0.190
2-1/2"	"	0.250
3"	"	0.308
3-1/2"	"	0.381
4"	"	0.500
5"	"	0.615
6"	"	0.808
Clamp type entrance caps		
1/2"	EA.	0.308
3/4"	"	0.381
1"	"	0.444
1-1/4"	"	0.500
1-1/2"	"	0.615
2"	"	0.727
3-1/2"	"	0.941
3"	"	1.127
3-1/2"	"	1.379
4"	"	2.424
"LB" condulets		
1/2"	EA.	0.308
3/4"	"	0.381
1"	"	0.444
1-1/4"	"	0.500
1-1/2"	"	0.615
2"	"	0.727
2-1/2"	"	1.000
3"	"	1.379
3-1/2"	"	1.739
4"	"	2.105
"T" condulets		
1/2"	EA.	0.381
3/4"	"	0.444

BASIC MATERIALS	UNIT	MAN/HOURS
16110.24 GALVANIZED CONDUIT		
1"	EA.	0.500
1-1/4"	"	0.571
1-1/2"	"	0.615
2"	"	0.727
2-1/2"	"	1.127
3"	"	1.509
3-1/2"	"	1.860
4"	"	2.222
"X" condulets		
1/2"	EA.	0.444
3/4"	"	0.500
1"	"	0.571
1-1/4"	"	0.615
1-1/2"	"	0.667
2"	"	0.879
Blank steel condulet covers		
1/2"	EA.	0.100
3/4"	"	0.100
1"	"	0.100
1-1/4"	"	0.123
1-1/2"	"	0.123
2"	"	0.123
2-1/2"	"	0.145
3"	"	0.145
3-1/2"	"	0.145
4"	"	0.200
Solid condulet gaskets		
1/2"	EA.	0.050
3/4"	"	0.050
1"	"	0.050
1-1/4"	"	0.080
1-1/2"	"	0.080
2"	"	0.080
2-1/2"	"	0.100
3"	"	0.100
3-1/2"	"	0.100
4"	"	0.145
One-hole malleable straps		
1/2"	EA.	0.040
3/4"	"	0.040
1"	"	0.040
1-1/4"	"	0.050
1-1/2"	"	0.050
2"	"	0.050
2-1/2"	"	0.059
3"	"	0.059
3-1/2"	"	0.059
4"	"	0.080
5"	"	0.080
6"	"	0.080
One-hole steel straps		
1/2"	EA.	0.040
3/4"	"	0.040
1"	"	0.040

BASIC MATERIALS	UNIT	MAN/HOURS
16110.24 GALVANIZED CONDUIT		
1-1/4"	EA.	0.050
1-1/2"	"	0.050
2"	"	0.050
2-1/2"	"	0.059
3"	"	0.059
3-1/2"	"	0.059
4"	"	0.080
Bushed chase nipples		
1/2"	EA.	0.059
3/4"	"	0.070
1"	"	0.089
1-1/4"	"	0.100
1-1/2"	"	0.123
2"	"	0.145
2-1/2"	"	0.145
3"	"	0.182
3-1/2"	"	0.250
4"	"	0.296
Offset nipples		
1/2"	EA.	0.059
3/4"	"	0.070
1"	"	0.089
1-1/4"	"	0.107
1-1/2"	"	0.123
2"	"	0.145
3"	"	0.182
Short elbows		
1/2"	EA.	0.145
3/4"	"	0.200
1"	"	0.250
1-1/4"	"	0.296
1-1/2"	"	0.348
2"	"	0.400
Pulling elbows, female to female		
1/2"	EA.	0.250
3/4"	"	0.296
1"	"	0.400
1-1/4"	"	0.533
1-1/2"	"	0.727
2"	"	0.851
Grounding locknuts		
1/2"	EA.	0.080
3/4"	"	0.080
1"	"	0.080
1-1/4"	"	0.089
1-1/2"	"	0.089
2"	"	0.089
2-1/2"	"	0.100
3"	"	0.100
3-1/2"	"	0.100
4"	"	0.145
Insulated grounding metal bushings		
1/2"	EA.	0.190
3/4"	"	0.222

BASIC MATERIALS	UNIT	MAN/HOURS
16110.24 GALVANIZED CONDUIT		
1"	EA.	0.250
1-1/4"	"	0.308
1-1/2"	"	0.381
2"	"	0.444
2-1/2"	"	0.667
3"	"	0.808
3-1/2"	"	0.941
4"	"	1.053
5"	"	1.569
6"	"	1.739
Nipples		
1/2" x		
4"	EA.	0.145
6"	"	0.145
8"	"	0.145
10"	"	0.145
12"	"	0.145
3/4" x		
4"	EA.	0.145
6"	"	0.145
8"	"	0.145
10"	"	0.145
12"	"	0.145
1" x		
4"	EA.	0.145
6"	"	0.145
8"	"	0.145
10"	"	0.145
12"	"	0.145
1-1/4" x		
4"	EA.	0.250
6"	"	0.250
8"	"	0.250
10"	"	0.250
12"	"	0.250
1-1/2" x		
4"	EA.	0.250
6"	"	0.250
8"	"	0.250
10"	"	0.250
12"	"	0.250
2" x		
4"	EA.	0.250
6"	"	0.250
8"	"	0.250
10"	"	0.250
12"	"	0.250
2-1/2" x		
6"	EA.	0.300
8"	"	0.300
10"	"	0.300
12"	"	0.300
3" x		
6"	EA.	0.300

BASIC MATERIALS	UNIT	MAN/HOURS
16110.24 GALVANIZED CONDUIT		
8"	EA.	0.300
10"	"	0.300
12"	"	0.300
3-1/2" x		
6"	EA.	0.300
8"	"	0.300
10"	"	0.300
12"	"	0.300
4" x		
8"	EA.	0.400
10"	"	0.400
12"	"	0.400
5" x		
8"	EA.	0.400
10"	"	0.400
12"	"	0.400
6" x		
8"	EA.	0.400
10"	"	0.400
12"	"	0.400
16110.25 PLASTIC CONDUIT		
PVC conduit, schedule 40		
1/2"	L.F.	0.030
3/4"	"	0.030
1"	"	0.040
1-1/4"	"	0.040
1-1/2"	"	0.050
2"	"	0.050
2-1/2"	"	0.059
3"	"	0.059
3-1/2"	"	0.080
4"	"	0.080
5"	"	0.089
6"	"	0.100
Couplings		
1/2"	EA.	0.050
3/4"	"	0.050
1"	"	0.050
1-1/4"	"	0.059
1-1/2"	"	0.059
2"	"	0.059
2-1/2"	"	0.059
3"	"	0.080
3-1/2"	"	0.080
4"	"	0.100
5"	"	0.100
6"	"	0.100
90 degree elbows		
1/2"	EA.	0.100
3/4"	"	0.123
1"	"	0.123
1-1/4"	"	0.145
1-1/2"	"	0.190

BASIC MATERIALS	UNIT	MAN/HOURS
16110.25 PLASTIC CONDUIT		
2"	EA.	0.222
2-1/2"	"	0.250
3"	"	0.308
3-1/2"	"	0.381
4"	"	0.500
5"	"	0.615
6"	"	0.727
Terminal adapters		
1/2"	EA.	0.100
3/4"	"	0.100
1"	"	0.100
1-1/4"	"	0.160
1-1/2"	"	0.160
2"	"	0.160
2-1/2"	"	0.222
3"	"	0.222
3-1/2"	"	0.222
4"	"	0.381
5"	"	0.381
6"	"	0.381
End bells		
1"	EA.	0.100
1-1/4"	"	0.160
1-1/2"	"	0.160
2"	"	0.160
2-1/2"	"	0.222
3"	"	0.222
3-1/2"	"	0.222
4"	"	0.381
5"	"	0.381
6"	"	0.381
LB conduit body		
1/2"	EA.	0.190
3/4"	"	0.190
1	"	0.190
1-1/4"	"	0.308
1-1/2"	"	0.308
2"	"	0.308
2-1/2"	"	0.444
3"	"	0.533
3-1/2"	"	0.615
4"	"	0.727
Direct burial, conduit		
2"	L.F.	0.050
3"	"	0.059
4"	"	0.080
5"	"	0.089
6"	"	0.100
Encased burial conduit		
2"	L.F.	0.050
3"	"	0.059
4"	"	0.080
5"	"	0.089
6"	"	0.100

16110.25 PLASTIC CONDUIT

BASIC MATERIALS	UNIT	MAN/HOURS
"EB" and "DB" duct, 90 degree elbows		
1-1/2"	EA.	0.145
2"	"	0.229
3"	"	0.381
3-1/2"	"	0.444
4"	"	0.533
5"	"	0.667
6"	"	0.899
45 degree elbows		
1-1/2"	EA.	0.229
2"	"	0.229
3"	"	0.381
3-1/2"	"	0.444
4"	"	0.533
5"	"	0.667
6"	"	0.899
Couplings		
1-1/2"	EA.	0.059
2"	"	0.059
3"	"	0.080
3-1/2"	"	0.080
4"	"	0.100
5"	"	0.100
6"	"	0.160
Bell ends		
1-1/2"	EA.	0.160
2"	"	0.160
3"	"	0.222
3-1/2"	"	0.222
4"	"	0.381
5"	"	0.381
6"	"	0.381
Female adapters, 1-1/2"	"	0.200
5 degree couplings		
1-1/2"	EA.	0.070
2"	"	0.070
3"	"	0.100
4"	"	0.145
5"	"	0.145
6"	"	0.145
45 degree elbows		
1/2"	EA.	0.123
3/4"	"	0.145
1"	"	0.145
1-1/4"	"	0.182
1-1/2"	"	0.229
2"	"	0.267
2-1/2"	"	0.296
3"	"	0.381
3-1/2"	"	0.444
4"	"	0.615
5"	"	0.727
6"	"	0.899
Female adapters		

16110.25 PLASTIC CONDUIT

BASIC MATERIALS	UNIT	MAN/HOURS
1/2"	EA.	0.123
3/4"	"	0.123
1"	"	0.123
1-1/4"	"	0.200
1-1/2"	"	0.200
2"	"	0.200
2-1/2"	"	0.267
3"	"	0.267
3-1/2"	"	0.267
4"	"	0.444
5"	"	0.444
6"	"	0.444
Expansion couplings		
1/2"	EA.	0.123
3/4"	"	0.123
1"	"	0.145
1-1/4"	"	0.200
1-1/2"	"	0.200
2"	"	0.200
2-1/2"	"	0.296
3"	"	0.296
3-1/2"	"	0.296
4"	"	0.444
5"	"	0.444
6"	"	0.444
Plugs		
2"	EA.	0.200
3"	"	0.296
3-1/2"	"	0.296
4"	"	0.444
5"	"	0.444
6"	"	0.500
Type "T" condulets		
1/2"	EA.	0.296
3/4"	"	0.296
1"	"	0.296
1-1/4"	"	0.500
1-1/2"	"	0.500
2"	"	0.500
EB & DB female adapters		
2"	EA.	0.250
3"	"	0.381
3-1/2"	"	0.615
4"	"	0.727
5"	"	1.000
6"	"	1.600

16110.27 PLASTIC COATED CONDUIT

BASIC MATERIALS	UNIT	MAN/HOURS
Rigid steel conduit, plastic coated		
1/2"	L.F.	0.050
3/4"	"	0.059
1"	"	0.080
1-1/4"	"	0.100
1-1/2"	"	0.123

BASIC MATERIALS	UNIT	MAN/HOURS
16110.27 PLASTIC COATED CONDUIT		
2"	L.F.	0.145
2-1/2"	"	0.190
3"	"	0.222
3-1/2"	"	0.250
4"	"	0.308
5"	"	0.381
90 degree elbows		
1/2"	EA.	0.308
3/4"	"	0.381
1"	"	0.444
1-1/4"	"	0.500
1-1/2"	"	0.615
2"	"	0.800
2-1/2"	"	1.143
3"	"	1.333
3-1/2"	"	1.633
4"	"	2.000
5"	"	2.500
Couplings		
1/2"	EA.	0.059
3/4"	"	0.080
1"	"	0.089
1-1/4"	"	0.107
1-1/2"	"	0.123
2"	"	0.145
2-1/2"	"	0.182
3"	"	0.190
3-1/2"	"	0.200
4"	"	0.222
5"	"	0.250
1 hole conduit straps		
3/4"	EA.	0.050
1"	"	0.050
1-1/4"	"	0.059
1-1/2"	"	0.059
2"	"	0.059
3"	"	0.080
3-1/2"	"	0.080
4"	"	0.100
"L.B." condulets with covers		
1/2"	EA.	0.500
3/4"	"	0.500
1"	"	0.615
1-1/4"	"	0.727
1-1/2"	"	0.879
2"	"	1.000
2-1/2"	"	1.379
3"	"	1.739
3-1/2"	"	2.162
4"	"	2.500
"T" condulets with covers		
1/2"	EA.	0.571
3/4"	"	0.615
1"	"	0.667

BASIC MATERIALS	UNIT	MAN/HOURS
16110.27 PLASTIC COATED CONDUIT		
1-1/4"	EA.	0.808
1-1/2"	"	0.941
2"	"	1.053
2-1/2"	"	1.509
3-1/2"	"	2.222
4"	"	2.667
5"	"	3.333
16110.28 STEEL CONDUIT		
Intermediate metal conduit (IMC)		
1/2"	L.F.	0.030
3/4"	"	0.040
1"	"	0.050
1-1/4"	"	0.059
1-1/2"	"	0.080
2"	"	0.089
2-1/2"	"	0.119
3"	"	0.145
3-1/2"	"	0.182
4"	"	0.190
90 degree ell		
1/2"	EA.	0.250
3/4"	"	0.308
1"	"	0.381
1-1/4"	"	0.444
1-1/2"	"	0.500
2"	"	0.571
2-1/2"	"	0.667
3"	"	0.889
3-1/2"	"	1.143
4"	"	1.333
Couplings		
1/2"	EA.	0.050
3/4"	"	0.059
1"	"	0.080
1-1/4"	"	0.089
1-1/2"	"	0.100
2"	"	0.107
2-1/2"	"	0.123
3"	"	0.145
3-1/2"	"	0.145
4"	"	0.160
16110.32 FLEXIBLE WIRING SYSTEMS		
Single circuit cables		
5'	EA.	0.059
10'	"	0.100
15'	"	0.145
20'	"	0.200
25'	"	0.267
30'	"	0.296
40'	"	0.400

BASIC MATERIALS	UNIT	MAN/HOURS
16110.32 FLEXIBLE WIRING SYSTEMS		
Two circuit cables		
5'	EA.	0.059
10'	"	0.100
15'	"	0.145
20'	"	0.200
25'	"	0.267
30'	"	0.296
40'	"	0.400
Two wire switch and receptacle cables		
5'	EA.	0.059
10'	"	0.100
15'	"	0.145
20'	"	0.200
25'	"	0.267
30'	"	0.296
40'	"	0.348
Three wire switch		
5'	EA.	0.059
10'	"	0.100
15'	"	0.145
20'	"	0.200
25'	"	0.267
30'	"	0.296
40'	"	0.348
Distribution boxes		
2 circuit	EA.	0.533
3 circuit	"	0.667
4 circuit	"	0.800
6 circuit	"	1.096
12 circuit	"	2.222
18 circuit	"	2.963
Tap boxes		
1 single pole switch	EA.	0.400
2 single pole switches	"	0.533
1 3 way switch	"	0.444
1 4 way switch	"	0.533
1 receptacle	"	0.400
2 receptacles	"	0.533
4 receptacles	"	0.800
1 clock	"	0.400
2 clocks	"	0.533
4 clocks	"	0.800
Dust cap	"	0.100
Cable coupler	"	0.200
Reversing connector	"	0.296
16110.35 WIREMOLD		
Wiremold raceway with fittings, surface mounted		
#200	L.F.	0.030
#500	"	0.030
#700	"	0.040
#800	"	0.040
Fittings, #200, 90 degree flat elbow	EA.	0.050

BASIC MATERIALS	UNIT	MAN/HOURS
16110.35 WIREMOLD		
Internal elbow	EA.	0.050
Extension adapter	"	0.059
#200, #500, #700		
Single pole switch and box	EA.	0.400
Duplex receptacle with box	"	0.308
#500, #700		
90 deg. flat elbow	EA.	0.080
Internal elbow	"	0.080
Junction box	"	0.133
Fixture box	"	0.133
Shallow switch and receptacle box	"	0.080
#800		
90 deg. flat elbow	EA.	0.080
Internal elbow	"	0.080
Junction box	"	0.123
#1500 series		
Raceway	L.F.	0.050
Wire clip	EA.	0.040
2 hole clip	"	0.059
Flat elbow	"	0.200
Internal elbow	"	0.200
External elbow	"	0.200
Adapter fitting	"	0.050
Duplex receptacle box	"	0.145
#2000 series		
Raceway	L.F.	0.050
Wire clip	EA.	0.040
Coupling	"	0.050
Supporting clip	"	0.040
Entrance end	"	0.145
Flat elbow	"	0.145
Tee	"	0.200
Single pole switch	"	0.200
#2100 series		
Raceway	L.F.	0.050
Cover	"	0.010
Wireclip	EA.	0.040
Coupling	"	0.050
Entrance end	"	0.200
Flat elbow	"	0.200
Tee	"	0.200
Device box	"	0.145
#2220 series		
Raceway	L.F.	0.100
Cover	"	0.020
Wire cup	EA.	0.040
Coupling	"	0.050
Entrance end	"	0.200
Flat elbow	"	0.200
Wall box connection	"	0.200
External elbow	"	0.145
#2600 series		
Raceway	L.F.	0.100
Wireclip	EA.	0.040

BASIC MATERIALS

16110.35 WIREMOLD

	UNIT	MAN/HOURS
Bushing	EA.	0.040
Flat elbow	"	0.200
Junction box	"	0.200
#3000 series		
Raceway	L.F.	0.100
Wire clip	EA.	0.040
Coupling	"	0.040
Supporting clip	"	0.050
Device bracket	"	0.059
Entrance end	"	0.200
Flat elbow	"	0.250
Tee	"	0.348
Device box	"	0.200
Circuit breaker housing	"	0.250
Panel connection	"	0.500
Conduit connector	"	0.250
#4000 series		
Raceway	L.F.	0.100
Cover	"	0.050
Divider	"	0.050
Wire clip	EA.	0.050
Coupling	"	0.059
Device bracket	"	0.059
Entrance end	"	0.308
Flat elbow	"	0.250
Tee	"	0.308
Internal elbow	"	0.308
External elbow	"	0.308
Panel connector	"	0.727
#6000 series		
Raceway	L.F.	0.148
Cover	"	0.050
Wire cup	EA.	0.050
Coupling	"	0.100
Device bracket	"	0.100
Entrance end	"	0.444
Flat elbow	"	0.400
Tee	"	0.500
Internal elbow	"	0.400
External elbow	"	0.444
Panel connector	"	0.899
Telepower poles		
# 21TP2	EA.	0.727
# 21TP4	"	1.000
# 30TP4	"	1.000
Fittings		
Telephone	EA.	0.200
T-bar clamp	"	0.200
C-hanger clamp	"	0.200
Tap-off fitting	"	0.200
Take-off connector	"	0.727

BASIC MATERIALS

16110.40 UNDERFLOOR DUCT

	UNIT	MAN/HOURS
Underfloor blank duct, insert duct		
7/8"	L.F.	0.050
1-3/8"	"	0.050
1-7/8"	"	0.050
Box opening plugs	EA.	0.145
Duct end plugs	"	0.145
Sleeve couplings	"	0.348
Expansion couplings	"	0.348
Vertical elbow	"	0.348
Offset elbow	"	0.348
Horizontal elbow	"	0.348
Adjustable elbow	"	0.348
Cabinet connector	"	1.194
Y-take off	"	0.615
Underfloor duct leveling legs	"	0.145
Conduit adapters		
1/2"	EA.	0.250
3/4"	"	0.250
1"	"	0.250
1-1/4"	"	0.296
2"	"	0.348
Reducer bushings		
1-1/4" x 3/4"	EA.	0.200
1-1/4" x 1"	"	0.200
2" x 1-1/2"	"	0.250
Support couplers		
1 standard	EA.	0.250
2 standard	"	0.296
3 standard	"	0.348
Supports		
1 duct	EA.	0.145
2 duct	"	0.170
3 duct	"	0.190
4 duct	"	0.250
5 duct	"	0.296
Single level junction box		
1 standard	EA.	0.800
2 standard	"	1.509
3 standard	"	2.963
4 standard	"	4.000
Two level junction boxes		
1 standard	EA.	1.000
2 standard	"	2.000
Sealing compound	"	
Insert adapters	"	0.145
Ellipsoids	"	0.348
Insert closing cap	"	0.145
Marker Screws	"	0.145
Access boxes	"	1.000
Closing caps	"	0.145
Afterset markers	"	0.145
Cell markers	"	0.145
Tie down straps	"	0.145
Plastic grommets	"	0.400

BASIC MATERIALS	UNIT	MAN/HOURS
16110.40 **UNDERFLOOR DUCT**		
Metal grommets	EA.	0.400
Receptacle		
Duplex, 20a	EA.	0.500
Single		
30a	EA.	0.500
50a	"	0.615
Double duplex	"	0.615
Single, 20a	"	0.444
Double single	"	0.500
Twist lock	"	0.444
1 conduit opening	"	0.444
2 conduit openings	"	0.500
1 bushed opening	"	0.444
2 bushed openings	"	0.500
Amphenol connector		
1"	EA.	0.500
2"	"	0.533
5"	"	0.615
Standpipes		
Aluminum	EA.	0.250
Brass	"	0.250
Abandonment plates		
Aluminum	EA.	0.250
Brass	"	0.250
Split bell caps		
Aluminum	EA.	0.250
Brass	"	0.250
Flush floor receptacles		
Aluminum	EA.	0.727
Brass	"	0.727
Flush floor telephone		
Aluminum	EA.	0.500
Brass	"	0.500
Super underfloor duct blank duct		
1/2"	L.F.	0.059
7/8"	"	0.059
1-3/8"	"	0.059
1-7/8"	"	0.059
Box opening plugs	EA.	0.145
End plugs	"	0.145
Conduit adapters	"	0.727
Sleeve coupling	"	0.444
Expansion coupling	"	0.444
Reducing coupling	"	0.444
Vertical elbow	"	0.444
Offset elbow	"	0.444
Horizontal elbow	"	0.444
Adjustable elbow	"	0.444
Cabinet connector	"	1.509
Super underfloor duct Y-take off	"	0.727
Leveling legs	"	0.145
Support couplers		
1 super	EA.	0.250
2 super	"	0.296

BASIC MATERIALS	UNIT	MAN/HOURS
16110.40 **UNDERFLOOR DUCT**		
1 super, 1 standard	EA.	0.296
2 super, 2 standard	"	0.348
Single level junction boxes		
1 super	EA.	1.509
2 super	"	2.000
4 super	"	3.478
Double level junction boxes		
1 super	EA.	1.194
2 super	"	1.509
16110.50 **WALL DUCT**		
Lay-in wall duct, 10"	L.F.	0.059
Horizontal elbow	EA.	0.727
Edgewise elbow	"	0.727
Tee	"	0.899
Cross	"	1.096
Cabinet connector	"	1.600
Reverse elbow	"	0.533
Sweep elbow	"	0.727
Partition	"	0.059
Straight tunnel	"	0.276
Elbow tunnel	"	0.348
Tee kit	"	0.400
Ceiling dropout	"	1.000
Coupling device	"	0.145
End cap	"	0.276
Lay-in wall duct, 18"	L.F.	0.080
Horizontal elbow	EA.	0.800
Edgeware elbow	"	1.000
Tee	"	1.096
Cross	"	1.250
Reverse elbow	"	0.727
Sweep elbow	"	0.800
Partition	"	0.100
Straight tunnel	"	0.400
Elbow tunnel	"	0.500
Tee kit	"	0.533
Ceiling dropout	"	1.096
Coupling device	"	0.200
Reducer coupling	"	0.400
Cabinet connector	"	2.000
End cap	"	0.400
16110.60 **TRENCH DUCT**		
Trench duct, with cover		
9"	L.F.	0.170
12"	"	0.200
18"	"	0.267
24"	"	0.348
30"	"	0.400
36"	"	0.571
Tees		

16110.60 TRENCH DUCT

BASIC MATERIALS	UNIT	MAN/HOURS
9"	EA.	1.739
12"	"	2.000
18"	"	2.222
24"	"	2.500
30"	"	2.963
36"	"	3.376
Vertical elbows		
9"	EA.	0.800
12"	"	1.096
18"	"	1.356
24"	"	1.667
30"	"	2.000
36"	"	2.500
Cabinet connectors		
9"	EA.	2.000
12"	"	2.105
18"	"	2.424
24"	"	2.500
30"	"	2.759
36"	"	2.963
End closers		
9"	EA.	0.615
12"	"	0.667
18"	"	0.800
24"	"	1.096
30"	"	1.290
36"	"	1.455
Horizontal elbows		
9"	EA.	1.509
12"	"	1.739
18"	"	2.105
24"	"	2.500
30"	"	2.857
36"	"	3.200
Crosses		
9"	EA.	2.000
12"	"	2.222
18"	"	2.500
24"	"	2.759
30"	"	3.200
36"	"	3.478

16110.80 WIREWAYS

BASIC MATERIALS	UNIT	MAN/HOURS
Wireway, hinge cover type		
2-1/2" x 2-1/2"		
1' section	EA.	0.154
2'	"	0.190
3'	"	0.250
5'	"	0.381
10'	"	0.667
4" x 4"		
1'	EA.	0.250
2'	"	0.250

16110.80 WIREWAYS

BASIC MATERIALS	UNIT	MAN/HOURS
3'	EA.	0.308
4'	"	0.308
10'	"	0.800
6" x 6"		
1'	EA.	0.381
2'	"	0.381
3'	"	0.444
4'	"	0.444
5'	"	0.571
10'	"	0.889
8" x 8"		
1'	EA.	0.444
2'	"	0.444
3'	"	0.500
4'	"	0.500
5'	"	0.615
12" x 12"		
1'	EA.	0.615
2'	"	0.615
3'	"	0.727
4'	"	0.727
5'	"	0.889
Fittings		
2-1/2" x 2-1/2"		
Drop hanger	EA.	0.123
Bracket hanger	"	0.123
Panel adapter	"	0.500
End plate	"	0.123
U-connector	"	0.123
Tee	"	0.200
Cross	"	0.250
90 degree elbow	"	0.200
Sweep elbow	"	0.200
45 degree elbow	"	0.200
Lay-in adapter	"	0.145
4" x 4"		
Drop hanger	EA.	0.145
Bracket hanger	"	0.145
Panel adapter	"	0.615
End plate	"	0.145
U-connector	"	0.145
Tee	"	0.250
Cross	"	0.348
90 degree elbow	"	0.250
Sweep elbow	"	0.250
45 degree elbow	"	0.250
Lay-in adapter	"	0.200
6" x 6"		
Drop hanger	EA.	0.145
Bracket hanger	"	0.145
Reducing bushing	"	0.145
Panel adapter	"	0.727
End plate	"	0.145
U-connector	"	0.145

BASIC MATERIALS

16110.80 WIREWAYS

BASIC MATERIALS	UNIT	MAN/HOURS
Tee	EA.	0.250
Cross	"	0.348
90 degree elbow	"	0.250
Sweep elbow	"	0.250
45 degree elbow	"	0.250
Lay-in adapter	"	0.250
8" x 8"		
Drop hanger	EA.	0.145
Bracket hanger	"	0.145
Reducing bushing	"	0.145
Panel adapter	"	0.899
End plate	"	0.145
U-connector	"	0.145
Tee	"	0.400
Cross	"	0.444
90 degree elbow	"	0.400
Sweep elbow	"	0.400
45 degree elbow	"	0.400
Lay-in adapter	"	0.250
10" x 10"		
Drop hanger	EA.	0.145
Bracket hanger	"	0.145
Reducing bushing	"	0.145
Panel adapter	"	1.000
End plate	"	0.145
U-connector	"	0.145
Tee	"	0.444
Cross	"	0.500
90 degree elbow	"	0.444
Sweep elbow	"	0.444
45 degree elbow	"	0.444
Lay-in adapter	"	0.296
12" x 12"		
Drop hanger	EA.	0.145
Bracket hanger	"	0.145
Reducing bushing	"	0.145
Panel adapter	"	1.250
End plate	"	0.145
U-connector	"	0.615
Tee	"	0.533
Cross	"	0.615
90 degree elbow	"	0.615
Sweep elbow	"	0.615
45 degree elbow	"	0.615
Lay-in adapter	"	0.533
Raintight wireway, 4" x 4"		
1' section	EA.	0.400
5'	"	0.400
10'	"	1.000
Fittings		
90 degree elbow	EA.	0.400
Tee	"	0.444
Cross	"	0.500
Panel adapter	"	0.727

BASIC MATERIALS

16110.80 WIREWAYS

BASIC MATERIALS	UNIT	MAN/HOURS
End plate	EA.	0.145
Gusset bracket	"	0.145
6" x 6"		
1' section	EA.	0.500
5'	"	0.727
10'	"	1.509
Fittings		
90 degree elbow	EA.	0.500
Tee	"	0.500
Cross	"	0.727
Panel adapter	"	1.000
End plate	"	0.145
Gusset bracket	"	0.145

16120.41 ALUMINUM CONDUCTORS

	UNIT	MAN/HOURS
Type XHHW, stranded aluminum, 600v		
#8	L.F.	0.005
#6	"	0.006
#4	"	0.008
#2	"	0.009
1/0	"	0.011
2/0	"	0.012
3/0	"	0.014
4/0	"	0.015
300 MCM	"	0.020
350 MCM	"	0.023
400 MCM	"	0.028
500 MCM	"	0.033
600 MCM	"	0.040
700 MCM	"	0.047
750 MCM	"	0.052
THW, stranded		
#8	L.F.	0.005
#6	"	0.006
#4	"	0.008
#3	"	0.009
#1	"	0.010
1/0	"	0.011
2/0	"	0.012
3/0	"	0.012
4/0	"	0.015
250 MCM	"	0.018
300 MCM	"	0.020
350 MCM	"	0.023
400 MCM	"	0.028
500 MCM	"	0.033
600 MCM	"	0.040
700 MCM	"	0.047
750 MCM	"	0.052
XLP, stranded		
#6	L.F.	0.005
#4	"	0.008
#2	"	0.009

BASIC MATERIALS	UNIT	MAN/HOURS
16120.41 ALUMINUM CONDUCTORS		
#1	L.F.	0.010
1/0	"	0.011
2/0	"	0.012
3/0	"	0.014
4/0	"	0.015
250 MCM	"	0.016
300 MCM	"	0.020
350 MCM	"	0.023
400 MCM	"	0.028
500 MCM	"	0.033
600 MCM	"	0.040
700 MCM	"	0.047
750 MCM	"	0.052
1000 MCM	"	0.057
Bare stranded aluminum wire		
#4	L.F.	0.008
#2	"	0.009
1/0	"	0.011
2/0	"	0.012
3/0	"	0.014
4/0	"	0.015
Triplex XLP cable		
#4	L.F.	0.015
#2	"	0.020
1/0	"	0.030
4/0	"	0.048
Aluminum quadruplex XLP cable		
#4	L.F.	0.018
#2	"	0.023
1/0	"	0.032
2/0	"	0.042
4/0	"	0.064
Triplexed URD-XLP cable		
#6	L.F.	0.011
#4	"	0.014
#2	"	0.018
1/0	"	0.028
2/0	"	0.033
3/0	"	0.040
4/0	"	0.047
250 MCM	"	0.055
350 MCM	"	0.057
Type S.E.U. cable		
#8/3	L.F.	0.025
#6/3	"	0.028
#4/3	"	0.035
#2/3	"	0.038
#1/3	"	0.040
1/0-3	"	0.042
2/0-3	"	0.044
3/0-3	"	0.052
4/0-3	"	0.057
Type S.E.R. cable with ground		
#8/3	L.F.	0.028

BASIC MATERIALS	UNIT	MAN/HOURS
16120.41 ALUMINUM CONDUCTORS		
#6/3	L.F.	0.035
#4/3	"	0.038
#2/3	"	0.040
#1/3	"	0.044
1/0-3	"	0.050
2/0-3	"	0.055
3/0-3	"	0.059
4/0-3	"	0.067
#6/4	"	0.038
#4/4	"	0.044
#2/4	"	0.044
#1/4	"	0.050
1/0-4	"	0.052
2/0-4	"	0.057
3/0-4	"	0.064
4/0-4	"	0.076
16120.43 COPPER CONDUCTORS		
Copper conductors, type THW, solid		
#14	L.F.	0.004
#12	"	0.005
#10	"	0.006
Stranded		
#14	L.F.	0.004
#12	"	0.005
#10	"	0.006
#8	"	0.008
#6	"	0.009
#4	"	0.010
#3	"	0.010
#2	"	0.012
#1	"	0.014
1/0	"	0.016
2/0	"	0.020
3/0	"	0.025
4/0	"	0.028
250 MCM	"	0.030
300 MCM	"	0.033
350 MCM	"	0.040
400 MCM	"	0.044
500 MCM	"	0.052
600 MCM	"	0.059
750 MCM	"	0.067
1000 MCM	"	0.076
THHN-THWN, solid		
#14	L.F.	0.004
#12	"	0.005
#10	"	0.006
Stranded		
#14	L.F.	0.004
#12	"	0.005
#10	"	0.006
#8	"	0.008
#6	"	0.009

BASIC MATERIALS	UNIT	MAN/HOURS
16120.43 COPPER CONDUCTORS		
#4	L.F.	0.010
#2	"	0.012
#1	"	0.014
1/0	"	0.016
2/0	"	0.020
3/0	"	0.025
4/0	"	0.028
250 MCM	"	0.030
350 MCM	"	0.040
XHHW		
#14	L.F.	0.004
#10	"	0.006
#8	"	0.008
#6	"	0.009
#4	"	0.009
#2	"	0.011
#1	"	0.014
1/0	"	0.016
2/0	"	0.019
3/0	"	0.025
XLP, 600v		
#12	L.F.	0.005
#10	"	0.006
#8	"	0.008
#6	"	0.009
#4	"	0.010
#3	"	0.011
#2	"	0.012
#1	"	0.014
1/0	"	0.016
2/0	"	0.020
3/0	"	0.026
4/0	"	0.028
250 MCM	"	0.030
300 MCM	"	0.033
350 MCM	"	0.039
400 MCM	"	0.044
500 MCM	"	0.052
600 MCM	"	0.059
750 MCM	"	0.067
1000 MCM	"	0.076
Bare solid wire		
#14	L.F.	0.004
#12	"	0.005
#10	"	0.006
#8	"	0.008
#6	"	0.009
#4	"	0.010
#2	"	0.012
Bare stranded wire		
#8	L.F.	0.008
#6	"	0.010
#4	"	0.010
#2	"	0.011

BASIC MATERIALS	UNIT	MAN/HOURS
16120.43 COPPER CONDUCTORS		
#1	L.F.	0.014
1/0	"	0.018
2/0	"	0.020
3/0	"	0.025
4/0	"	0.028
250 MCM	"	0.030
300 MCM	"	0.033
350 MCM	"	0.040
400 MCM	"	0.044
500 MCM	"	0.052
Type "BX" solid armored cable		
#14/2	L.F.	0.025
#14/3	"	0.028
#14/4	"	0.031
#12/2	"	0.028
#12/3	"	0.031
#12/4	"	0.035
#10/2	"	0.031
#10/3	"	0.035
#10/4	"	0.040
#8/2	"	0.035
#8/3	"	0.040
Steel type, metal clad cable, solid, with ground		
#14/2	L.F.	0.018
#14/3	"	0.020
#14/4	"	0.023
#12/2	"	0.020
#12/3	"	0.025
#12/4	"	0.030
#10/2	"	0.023
#10/3	"	0.028
#10/4	"	0.033
Metal clad cable, stranded, with ground		
#8/2	L.F.	0.028
#8/3	"	0.035
#8/4	"	0.042
#6/2	"	0.030
#6/3	"	0.038
#6/4	"	0.044
#4/2	"	0.040
#4/3	"	0.044
#4/4	"	0.055
#3/3	"	0.050
#3/4	"	0.059
#2/3	"	0.057
#2/4	"	0.067
#1/3	"	0.076
#4	"	0.035
16120.45 FLAT CONDUCTOR CABLE		
Flat conductor cable, with shield, 3 conductor		
#12 awg	L.F.	0.059
#10 awg	"	0.059
4 conductor		

16120.45 FLAT CONDUCTOR CABLE

BASIC MATERIALS	UNIT	MAN/HOURS
#12 awg	L.F.	0.080
#10 awg	"	0.080
Transition boxes		
#12 awg	L.F.	0.089
#10 awg	"	0.089
Flat conductor cable communication, with shield		
10 conductor	L.F.	0.059
16 conductor	"	0.070
24 conductor	"	0.100
Power and communication heads, duplex receptacle	EA.	0.800
Double duplex receptacle	"	0.952
Telephone	"	0.800
Receptacle and telephone	"	0.952
Blank cover	"	0.145
Transition boxes		
Surface	EA.	0.727
Flush	"	1.000
Flat conductor cable fittings		
End caps	EA.	0.145
Insulators	"	0.296
Splice connectors	"	0.444
Tap connectors	"	0.444
Cable connectors	"	0.444
Terminal blocks	"	0.615
Tape	"	

16120.47 SHEATHED CABLE

BASIC MATERIALS	UNIT	MAN/HOURS
Non-metallic sheathed cable		
Type NM cable with ground		
#14/2	L.F.	0.015
#12/2	"	0.016
#10/2	"	0.018
#8/2	"	0.020
#6/2	"	0.025
#14/3	"	0.026
#12/3	"	0.027
#10/3	"	0.027
#8/3	"	0.028
#6/3	"	0.028
#4/3	"	0.032
#2/3	"	0.035
Type U.F. cable with ground		
#14/2	L.F.	0.016
#12/2	"	0.019
#10/2	"	0.020
#8/2	"	0.023
#6/2	"	0.027
#14/3	"	0.020
#12/3	"	0.022
#10/3	"	0.025
#8/3	"	0.028
#6/3	"	0.032
Type S.F.U. cable, 3 conductor		

16120.47 SHEATHED CABLE

BASIC MATERIALS	UNIT	MAN/HOURS
#8	L.F.	0.028
#6	"	0.031
#3	"	0.040
#2	"	0.044
#1	"	0.050
#1/0	"	0.055
#2/0	"	0.064
#3/0	"	0.070
#4/0	"	0.076
Type SER cable, 4 conductor		
#6	L.F.	0.036
#4	"	0.039
#3	"	0.044
#2	"	0.048
#1	"	0.055
#1/0	"	0.064
#2/0	"	0.067
#3/0	"	0.076
#4/0	"	0.084
Flexible cord, type STO cord		
#18/2	L.F.	0.004
#18/3	"	0.005
#18/4	"	0.006
#16/2	"	0.004
#16/3	"	0.004
#16/4	"	0.005
#14/2	"	0.005
#14/3	"	0.006
#14/4	"	0.007
#12/2	"	0.006
#12/3	"	0.007
#12/4	"	0.008
#10/2	"	0.007
#10/3	"	0.008
#10/4	"	0.009
#8/2	"	0.008
#8/3	"	0.009
#8/4	"	0.010

16130.10 FLOOR BOXES

BASIC MATERIALS	UNIT	MAN/HOURS
Adjustable floor boxes, steel	EA.	0.533
Cast bronze round	"	0.727
1 gang	"	0.800
2 gang	"	0.952
3 gang	"	1.000
Aluminum round	"	0.727
1 gang	"	0.800
2 gang	"	0.952
3 gang	"	1.000
Steel plate single recept	"	0.145
Duplex receptacle	"	0.182
Twist lock receptacle	"	0.182
Plug, 3/4"	"	0.145
1" plug	"	0.145

BASIC MATERIALS	UNIT	MAN/HOURS
16130.10 FLOOR BOXES		
Carpet flange	EA.	0.145
Adjustable bronze plates for round cast boxes		
1/2" plug	EA.	0.145
3/4" plug	"	0.145
1" plug	"	0.145
1-1/4" plug	"	0.182
2" plug	"	0.200
Combination plug	"	0.200
Duplex receptacle plug	"	0.200
Adjustable aluminum plates for round cast boxes		
1/2" plug	EA.	0.145
3/4" plug	"	0.145
1" plug	"	0.145
1-1/4" plug	"	0.182
2" plug	"	0.200
Combination plug	"	0.200
Duplex receptacle plug	"	0.200
Adjustable bronze plates for gang type boxes		
1/2" plug	EA.	0.145
3/4" plug	"	0.145
1" plug	"	0.145
1-1/4" plug	"	0.182
2" plug	"	0.200
Carpet plate		
1 gang	EA.	0.145
2 gang	"	0.145
3 gang	"	0.200
Adjustable aluminum plates for gang type boxes		
1/2" plug	EA.	0.145
3/4" plug	"	0.145
1" plug	"	0.145
1-1/4" plug	"	0.182
2" plug	"	0.200
Duplex recept	"	0.200
Carpet plate		
1 gang	EA.	0.145
2 gang	"	0.145
3 gang	"	0.200
4 gang carpet plate	"	0.571
Telephone	"	0.500
Floor box nozzles, horizontal		
Duplex recept	EA.	0.533
Single recept	"	0.533
Double duplex recept	"	0.727
Vertical with duplex recept	"	0.615
Double duplex recept	"	0.727
Floor box bell nozzles split bell	"	0.250
One piece bell	"	0.250
Floor box standpipe		
1/2" x 3"	EA.	0.145
1/2" x 1"	"	0.145
Poke thru floor outlets		
2" floor	EA.	1.000
3" floor	"	1.194

BASIC MATERIALS	UNIT	MAN/HOURS
16130.10 FLOOR BOXES		
4" floor	EA.	1.290
7" floor	"	1.509
9" floor	"	1.600
11" floor	"	1.818
13" floor	"	2.000
16130.40 BOXES		
Round cast box, type SEH		
1/2"	EA.	0.348
3/4"	"	0.421
SEHC		
1/2"	EA.	0.348
3/4"	"	0.421
SEHL		
1/2"	EA.	0.348
3/4"	"	0.444
SEHT		
1/2"	EA.	0.421
3/4"	"	0.500
SEHX		
1/2"	EA.	0.500
3/4"	"	0.615
Blank cover	"	0.145
1/2", hub cover	"	0.145
Cover with gasket	"	0.178
Rectangle, type FS boxes		
1/2"	EA.	0.348
3/4"	"	0.400
1"	"	0.500
FSA		
1/2"	EA.	0.348
3/4"	"	0.400
FSC		
1/2"	EA.	0.348
3/4"	"	0.421
1"	"	0.500
FSL		
1/2"	EA.	0.348
3/4"	"	0.400
FSR		
1/2"	EA.	0.348
3/4"	"	0.400
FSS		
1/2"	EA.	0.348
3/4"	"	0.400
FSLA		
1/2"	EA.	0.348
3/4"	"	0.400
FSCA		
1/2"	EA.	0.348
3/4"	"	0.400
FSCC		
1/2"	EA.	0.400

BASIC MATERIALS	UNIT	MAN/HOURS
16130.40 BOXES		
3/4"	EA.	0.500
FSCT		
1/2"	EA.	0.400
3/4"	"	0.500
1"	"	0.571
FST		
1/2"	EA.	0.500
3/4"	"	0.571
FSX		
1/2"	EA.	0.615
3/4"	"	0.727
FSCD boxes		
1/2"	EA.	0.615
3/4"	"	0.727
Rectangle, type FS, 2 gang boxes		
1/2"	EA.	0.348
3/4"	"	0.400
1"	"	0.500
FSC, 2 gang boxes		
1/2"	EA.	0.348
3/4"	"	0.400
1"	"	0.500
FSS, 2 gang boxes		
3/4"	EA.	0.400
FS, tandem boxes		
1/2"	EA.	0.400
3/4"	"	0.444
FSC, tandem boxes		
1/2"	EA.	0.400
3/4"	"	0.444
FS, three gang boxes		
3/4"	EA.	0.444
1"	"	0.500
FSS, three gang boxes, 3/4"	"	0.500
Weatherproof cast aluminum boxes, 1 gang, 3 outlets		
1/2"	EA.	0.400
3/4"	"	0.500
2 gang, 3 outlets		
1/2"	EA.	0.500
3/4"	"	0.533
1 gang, 4 outlets		
1/2"	EA.	0.615
3/4"	"	0.727
2 gang, 4 outlets		
1/2"	EA.	0.615
3/4"	"	0.727
1 gang, 5 outlets		
1/2"	EA.	0.727
3/4"	"	0.800
2 gang, 5 outlets		
1/2"	EA.	0.727
3/4"	"	0.800
2 gang, 6 outlets		
1/2"	EA.	0.851

BASIC MATERIALS	UNIT	MAN/HOURS
16130.40 BOXES		
3/4"	EA.	0.899
2 gang, 7 outlets		
1/2"	EA.	1.000
3/4"	"	1.096
Weatherproof and type FS box covers, blank, 1 gang	"	0.145
Tumbler switch, 1 gang	"	0.145
1 gang, single recept	"	0.145
Duplex recept	"	0.145
Despard	"	0.145
Red pilot light	"	0.145
SW and		
Single recept	EA.	0.200
Duplex recept	"	0.200
2 gang		
Blank	EA.	0.182
Tumbler switch	"	0.182
Single recept	"	0.182
Duplex recept	"	0.182
3 gang		
Blank	EA.	0.200
Tumbler switch	"	0.200
4 gang		
Tumbler switch	EA.	0.250
Explosion proof boxes type E		
1/2"	EA.	0.348
3/4"	"	0.400
1"	"	0.500
1-1/4"	"	0.571
1-1/2"	"	0.615
Type L.B.		
1/2"	EA.	0.400
3/4"	"	0.500
1"	"	0.571
1-1/4"	"	0.667
1-1/2"	"	0.727
2"	"	0.800
Type C		
1/2"	EA.	0.400
3/4"	"	0.500
1"	"	0.571
1-1/4"	"	0.667
1-1/2"	"	0.727
2"	"	0.800
Type CA		
1/2"	EA.	0.571
3/4"	"	0.727
Type L		
1/2"	EA.	0.400
3/4"	"	0.500
1"	"	0.571
1-1/4"	"	0.667
1-1/2"	"	0.727
2"	"	0.800
Type N		

BASIC MATERIALS	UNIT	MAN/HOURS
16130.40 BOXES		
1/2"	EA.	0.400
3/4"	"	0.500
1"	"	0.615
1-1/4"	"	0.667
Type T		
1/2"	EA.	0.533
3/4"	"	0.727
1"	"	0.851
1-1/4"	"	1.000
1-1/2"	"	1.159
2"	"	1.290
Type TA		
1/2"	EA.	0.727
3/4"	"	0.800
Type X		
1/2"	EA.	0.727
3/4"	"	0.851
1"	"	1.000
1-1/4"	"	1.159
1-1/2"	"	1.290
2"	"	1.455
With union hubs		
1/2"	EA.	0.727
3/4"	"	0.800
Box covers		
Surface	EA.	0.200
Sealing	"	0.200
Dome	"	0.200
1/2" nipple	"	0.200
3/4" nipple	"	0.200
16130.45 EXPLOSION PROOF FITTINGS		
Flexible couplings with female unions		
1/2" x 18"	EA.	0.200
3/4" x 18"	"	0.276
1" x 18"	"	0.348
1-1/4" x 18"	"	0.421
1-1/2" x 18"	"	0.500
2" x 18"	"	0.571
1/2" x 24"	"	0.250
3/4" x 24"	"	0.296
1" x 24"	"	0.400
1-1/4" x 24"	"	0.444
1-1/2" x 24"	"	0.571
2" x 24"	"	0.615
Female seal-offs		
1/2"	EA.	0.571
3/4"	"	0.667
1"	"	0.727
1-1/4"	"	0.851
1-1/2"	"	1.000
2"	"	1.159
2-1/2"	"	1.739
3"	"	2.162

BASIC MATERIALS	UNIT	MAN/HOURS
16130.45 EXPLOSION PROOF FITTINGS		
4"	EA.	2.667
Conduit plugs		
1/2"	EA.	0.145
3/4"	"	0.145
1"	"	0.145
1-1/4"	"	0.250
1-1/2"	"	0.250
2"	"	0.296
2-1/2"	"	0.296
3"	"	0.348
4"	"	0.348
Sealing cement		
1 pound	EA.	
5 pound	"	
Fibre		
1 ounce	EA.	
8 ounce	"	
Male unions		
1/2"	EA.	0.200
3/4"	"	0.242
1"	"	0.276
1-1/4"	"	0.296
1-1/2"	"	0.348
2"	"	0.421
2-1/2"	"	0.500
3"	"	0.727
4"	"	0.899
Female unions		
1/2"	EA.	0.200
3/4"	"	0.242
1"	"	0.276
1-1/4"	"	0.296
1-1/2"	"	0.348
2"	"	0.421
2-1/2"	"	0.500
3"	"	0.727
4"	"	0.899
Male elbows		
1/2"	EA.	0.250
3/4"	"	0.296
1"	"	0.348
1-1/4"	"	0.444
Female elbows		
1/2"	EA.	0.250
3/4"	"	0.296
1"	"	0.348
1-1/4"	"	0.444
Pulling elbows		
1/2"	EA.	0.348
3/4"	"	0.444
1"	"	0.500
1-1/4"	"	0.615
1-1/2"	"	0.727
2"	"	1.905

BASIC MATERIALS	UNIT	MAN/HOURS
16130.45 EXPLOSION PROOF FITTINGS		
2-1/2"	EA.	2.500
3"	"	2.963
3-1/2"	"	3.478
4"	"	4.211
Male expansion couplings		
1/2"	EA.	0.250
3/4"	"	0.296
1"	"	0.444
Female expansion couplings		
1/2"	EA.	0.250
3/4"	"	0.296
1"	"	0.444
16130.60 PULL AND JUNCTION BOXES		
4"		
Octagon box	EA.	0.114
Box extension	"	0.059
Plaster ring	"	0.059
Cover blank	"	0.059
Square box	"	0.114
Box extension	"	0.059
Plaster ring	"	0.059
Cover blank	"	0.059
4-11/16"		
Square box	EA.	0.114
Box extension	"	0.059
Plaster ring	"	0.059
Cover blank	"	0.059
Switch and device boxes		
2 gang	EA.	0.114
3 gang	"	0.114
4 gang	"	0.160
Device covers		
2 gang	EA.	0.059
3 gang	"	0.059
4 gang	"	0.059
Handy box	"	0.114
Extension	"	0.059
Switch cover	"	0.059
Switch box with knockout	"	0.145
Weatherproof cover, spring type	"	0.080
Cover plate, dryer receptacle 1 gang plastic	"	0.100
For 4" receptacle, 2 gang	"	0.100
Duplex receptacle cover plate, plastic	"	0.059
4", vertical bracket box, 1-1/2" with		
RMX clamps	EA.	0.145
BX clamps	"	0.145
4", octagon device cover		
1 switch	EA.	0.059
1 duplex recept	"	0.059
4", octagon swivel hanger box, 1/2" hub	"	0.059
3/4" hub	"	0.059
4" octagon adjustable bar hangers		
18-1/2"	EA.	0.050

BASIC MATERIALS	UNIT	MAN/HOURS
16130.60 PULL AND JUNCTION BOXES		
26-1/2"	EA.	0.050
With clip		
18-1/2"	EA.	0.050
26-1/2"	"	0.050
4", square face bracket boxes, 1-1/2"		
RMX	EA.	0.145
BX	"	0.145
4" square to round plaster rings	"	0.059
2 gang device plaster rings	"	0.059
Surface covers		
1 gang switch	EA.	0.059
2 gang switch	"	0.059
1 single recept	"	0.059
1 20a twist lock recept	"	0.059
1 30a twist lock recept	"	0.059
1 duplex recept	"	0.059
2 duplex recept	"	0.059
Switch and duplex recept	"	0.059
4-11/16" square to round plaster rings	"	0.059
2 gang device plaster rings	"	0.059
Surface covers		
1 gang switch	EA.	0.059
2 gang switch	"	0.059
1 single recept	"	0.059
1 20a twist lock recept	"	0.059
1 30a twist lock recept	"	0.059
1 duplex recept	"	0.059
2 duplex recept	"	0.059
Switch and duplex recept	"	0.059
4" plastic round boxes, ground straps		
Box only	EA.	0.145
Box w/clamps	"	0.200
Box w/16" bar	"	0.229
Box w/24" bar	"	0.250
4" plastic round box covers		
Blank cover	EA.	0.059
Plaster ring	"	0.059
4" plastic square boxes		
Box only	EA.	0.145
Box w/clamps	"	0.200
Box w/hanger	"	0.250
Box w/nails and clamp	"	0.250
4" plastic square box covers		
Blank cover	EA.	0.059
1 gang ring	"	0.059
2 gang ring	"	0.059
Round ring	"	0.059
16130.65 PULL BOXES AND CABINETS		
Galvanized pull boxes, screw cover		
4x4x4	EA.	0.190
4x6x4	"	0.190
6x6x4	"	0.190

16130.65 PULL BOXES AND CABINETS

BASIC MATERIALS	UNIT	MAN/HOURS
6x8x4	EA.	0.190
8x8x4	"	0.250
8x10x4	"	0.242
8x12x4	"	0.250
Screw cover		
10x10x4	EA.	0.308
12x12x6	"	0.444
12x15x6	"	0.444
12x18x6	"	0.500
15x18x6	"	0.571
18x24x6	"	0.615
18x30x6	"	0.727
24x36x6	"	0.727
Cast iron junction box, unflanged		
6x6x4		
3/4" tap	EA.	0.500
1" tap	"	0.500
Two 1/2" taps	"	0.500
3/4" taps	"	0.500
6" adapter plate	"	0.348
6" exterior collar	"	0.348
Screw cover cabinet		
12x12x4	EA.	0.615
12x16x4	"	0.615
12x16x6	"	0.615
12x18x4	"	0.667
12x18x6	"	0.667
18x18x4	"	1.000
18x18x6	"	1.000
18x24x6	"	1.143
24x24x6	"	1.333
24x36x6	"	1.667
36x48x6	"	2.500
NEMA 3R, rain tight screw cover enclosures		
6x6x4	EA.	0.211
8x6x4	"	0.296
8x8x4	"	0.296
10x8x4	"	0.400
10x10x4	"	0.400
12x8x4	"	0.444
12x12x4	"	0.444
15x12x4	"	0.533
8x8x6	"	0.400
10x8x6	"	0.444
10x10x6	"	0.444
12x8x6	"	0.533
12x10x6	"	0.548
12x12x6	"	0.548
18x12x6	"	0.702

16130.80 RECEPTACLES

BASIC MATERIALS	UNIT	MAN/HOURS
Contractor grade duplex receptacles, 15 a 120v		
Duplex	EA.	0.200

16130.80 RECEPTACLES

BASIC MATERIALS	UNIT	MAN/HOURS
125 volt, 20a, duplex, grounding type, standard grade	EA.	0.200
Ground fault interrupter type	"	0.296
250 volt, 20a, 2 pole, single receptacle, ground type	"	0.200
120/208v, 4 pole, single receptacle, twist lock		
20a	EA.	0.348
50a	"	0.348
125/250v, 3 pole, flush receptacle		
30a	EA.	0.296
50a	"	0.296
60a	"	0.348
277 v, 20a, 2 pole, grounding type, twist lock	"	0.200
Dryer receptacle, 250v, 30a/50a, 3 wire	"	0.296
Clock receptacle, 2 pole, grounding type	"	0.200
125v, 20a single recept. grounding type		
Standard grade	EA.	0.200
Specification	"	0.200
Hospital	"	0.200
Isolated ground orange	"	0.250
Duplex		
Specification grade	EA.	0.200
Hospital	"	0.200
Isolated ground orange	"	0.250
250v, 20a, duplex, 2 pole, grounding, spec. grade	"	0.200
Combination recepts, 20a, 125v and 250v, duplex	"	0.200
GFI hospital grade recepts, 20a, 125v, duplex	"	0.296
125/250v, 3 pole, 3 wire surface recepts		
30a	EA.	0.296
50a	"	0.296
60a	"	0.348
Cord set, 3 wire, 6' cord		
30a	EA.	0.296
50a	"	0.296
125/250v, 3 pole, 3 wire cap		
30a	EA.	0.400
50a	"	0.400
60a	"	0.444

16198.10 ELECTRIC MANHOLES

BASIC MATERIALS	UNIT	MAN/HOURS
Precast, handhole, 4' deep		
2'x2'	EA.	3.478
3'x3'	"	5.556
4'x4'	"	10.256
Power manhole, complete, precast, 8' deep		
4'x4'	EA.	14.035
6'x6'	"	20.000
8'x8'	"	21.053
6' deep, 9' x 12'	"	25.000
Cast in place, power manhole, 8' deep		
4'x4'	EA.	14.035
6'x6'	"	20.000
8'x8'	"	21.053

BASIC MATERIALS

16199.10 UTILITY POLES & FITTINGS

	UNIT	MAN/HOURS
Wood pole, creosoted		
25'	EA.	2.353
30'	"	2.963
35'	"	3.478
40'	"	3.791
45'	"	6.957
50'	"	7.207
55'	"	7.547
Treated, wood preservative, 6"x6"		
8'	EA.	0.500
10'	"	0.800
12'	"	0.889
14'	"	1.333
16'	"	1.600
18'	"	2.000
20'	"	2.000
Aluminum, brushed, no base		
8'	EA.	2.000
10'	"	2.667
15'	"	2.759
20'	"	3.200
25'	"	3.810
30'	"	4.396
35'	"	5.000
40'	"	6.250
Steel, no base		
10'	EA.	2.500
15'	"	2.963
20'	"	3.810
25'	"	4.520
30'	"	5.096
35'	"	6.250
Concrete, no base		
13'	EA.	5.517
16'	"	7.273
18'	"	8.791
25'	"	10.000
30'	"	12.121
35'	"	14.035
40'	"	16.000
45'	"	17.021
50'	"	18.182
55'	"	19.048
60'	"	20.000
Pole line hardware		
Wood crossarm		
4'	EA.	1.333
8'	"	1.667
10'	"	2.051
Angle steel brace		
1 piece	EA.	0.250
2 piece	"	0.348
Eye nut, 5/8"	"	0.050
Bolt (14-16"), 5/8"	"	0.200

BASIC MATERIALS

16199.10 UTILITY POLES & FITTINGS

	UNIT	MAN/HOURS
Transformer, ground connection	EA.	0.250
Stirrup	"	0.308
Secondary lead support	"	0.400
Spool insulator	"	0.200
Guy grip, preformed		
7/16"	EA.	0.145
1/2"	"	0.145
Hook	"	0.250
Strain insulator	"	0.364
Wire		
5/16"	L.F.	0.005
7/16"	"	0.006
1/2"	"	0.008
Soft drawn ground, copper, #8	"	0.008
Ground clamp	EA.	0.308
Perforated strapping for conduit, 1-1/2"	L.F.	0.145
Hot line clamp	EA.	0.800
Lightning arrester		
3kv	EA.	1.000
10kv	"	1.600
30kv	"	2.000
36kv	"	2.500
Fittings		
Plastic molding	L.F.	0.145
Molding staples	EA.	0.050
Ground wires staples	"	0.030
Copper butt plate	"	0.296
Anchor bond clamp	"	0.145
Guy wire		
1/4"	L.F.	0.030
3/8"	"	0.050
Guy grip		
1/4"	EA.	0.050
3/8"	"	0.050

POWER GENERATION

16210.10 GENERATORS

	UNIT	MAN/HOURS
Diesel generator, with auto transfer switch		
30kw	EA.	30.769
50kw	"	30.769
75kw	"	42.105
100kw	"	47.059
125kw	"	50.000
150kw	"	57.143
175kw	"	66.667
200kw	"	80.000

POWER GENERATION		UNIT	MAN/HOURS
16210.10	**GENERATORS**		
250kw		EA.	88.889
300kw		"	100
350kw		"	114
400kw		"	133
450kw		"	145
500kw		"	160
600kw		"	200
750kw		"	200
16230.10	**CAPACITORS**		
Three phase capacitors			
240v			
1.5 kvar		EA.	2.500
2.5 kvar		"	3.200
3.0 kvar		"	4.000
4 kvar		"	5.000
5 kvar		"	5.333
6 kvar		"	5.714
7.5 kvar		"	6.154
10 kvar		"	8.000
15 kvar		"	9.501
20 kvar		"	11.994
25 kvar		"	13.008
40 kvar		"	18.018
50 kvar		"	20.997
60 kvar		"	21.505
75 kvar		"	25.000
100 kvar		"	29.963
480v			
1.5 kvar		EA.	2.500
2.5 kvar		"	3.200
3 kvar		"	4.000
4 kvar		"	5.000
5 kvar		"	5.333
6 kvar		"	5.714
7.5 kvar		"	6.154
10 kvar		"	8.000
12.5 kvar		"	9.501
15 kvar		"	11.994
18 kvar		"	12.500
20 kvar		"	13.008
22.5 kvar		"	13.491
25 kvar		"	14.842
30 kvar		"	14.842
35 kvar		"	16.000
40 kvar		"	18.018
45 kvar		"	20.000
50 kvar		"	20.997
60 kvar		"	21.978
70 kvar		"	24.024
75 kvar		"	25.000
80 kvar		"	27.027
90 kvar		"	28.986
100 kvar		"	29.963

POWER GENERATION		UNIT	MAN/HOURS
16230.10	**CAPACITORS**		
125 kvar		EA.	33.058
150 kvar		"	37.037
16320.10	**TRANSFORMERS**		
Floor mounted, single phase, int. dry, 480v-120/240v			
3 kva		EA.	1.818
5 kva		"	3.077
7.5 kva		"	3.478
10 kva		"	3.810
15 kva		"	4.301
25 kva		"	7.547
37.5 kva		"	9.412
50 kva		"	10.256
75 kva		"	10.667
100 kva		"	11.594
Three phase, 480v-120/280v			
15 kva		EA.	6.015
30 kva		"	9.412
45 kva		"	10.811
75 kva		"	10.959
112.5 kva		"	12.698
150 kva		"	13.559
225 kva		"	15.385
Single phase, dry type, 2400v			
167 kva		EA.	22.472
250 kva		"	29.963
333 kva		"	37.559
5000v			
167 kva		EA.	22.472
250 kva		"	29.963
333 kva		"	37.559
8660v			
167 kva		EA.	27.491
250 kva		"	34.934
333 kva		"	67.797
1500v			
167 kva		EA.	27.491
250 kva		"	34.934
333 kva		"	42.553
Three phase, dry type transformer, 2400v			
225 kva		EA.	25.000
300 kva		"	27.491
500 kva		"	42.553
750 kva		"	52.632
5000v			
225.0 kva		EA.	25.000
300 kva		"	27.491
500 kva		"	42.553
750 kva		"	52.632
8660v			
225.0 kva		EA.	29.963
300 kva		"	32.520
500 kva		"	47.619
750 kva		"	57.554

POWER GENERATION	UNIT	MAN/HOURS
16320.10 TRANSFORMERS		
1500v		
225 kva	EA.	29.963
300 kva	"	32.520
500 kva	"	47.619
750 kva	"	57.554
Buck boost transformers		
.25 kva	EA.	1.000
.50 kva	"	1.250
.75 kva	"	1.509
1.00 kva	"	1.739
1.50 kva	"	2.000
2.00 kva	"	2.500
3.00 kva	"	2.963
16350.10 CIRCUIT BREAKERS		
Molded case, 240v, 15-60a, bolt-on		
1 pole	EA.	0.250
2 pole	"	0.348
70-100a, 2 pole	"	0.533
15-60a, 3 pole	"	0.400
70-100a, 3 pole	"	0.615
480v, 2 pole		
15-60a	EA.	0.296
70-100a	"	0.400
3 pole		
15-60a	EA.	0.400
70-100a	"	0.444
70-225a	"	0.615
Draw out air circuit breakers		
600a	EA.	16.000
800a	"	18.182
1600a	"	24.242
2000a	"	27.586
3000a	"	32.000
4000a	"	38.095
Load center circuit breakers, 240v		
1 pole, 10-60a	EA.	0.250
2 pole		
10-60a	EA.	0.400
70-100a	"	0.667
110-150a	"	0.727
3 pole		
10-60a	EA.	0.500
70-100a	"	0.727
Load center, G.F.I. breakers, 240v		
1 pole, 15-30a	EA.	0.296
2 pole, 15-30a	"	0.400
Key operated breakers, 240v, 1 pole, 10-30a	"	0.296
Tandem breakers, 240v		
1 pole, 15-30a	EA.	0.400
2 pole, 15-30a	"	0.533
Bolt-on, G.F.I. breakers, 240v, 1 pole, 15-30a	"	0.348
Enclosed breaker, 120v, 1 pole, 15-50a, NEMA 1	"	0.800

POWER GENERATION	UNIT	MAN/HOURS
16350.10 CIRCUIT BREAKERS		
240v, 2 pole		
15-60a, NEMA 1	EA.	1.250
70-100a, NEMA 1	"	1.739
3 pole		
15-60a, NEMA 1	EA.	1.509
70-100a, NEMA 1	"	2.222
Enclosed circuit breakers		
120v, 1 pole, NEMA 3R, 15-50a	EA.	0.899
240v, 2 pole, NEMA 3R		
15-60a	EA.	1.250
70-100a	"	1.739
3 pole, NEMA 3R		
15-60a	EA.	1.509
70-100a	"	2.222
480v, NEMA 1		
1 pole, 15-50a	EA.	0.800
2 pole, 15-60a	"	1.250
70-100a	"	1.509
3 pole, NEMA 1		
15-60a	EA.	1.509
70-100a	"	2.222
480v, 1 pole, 15-50a, NEMA 3R	"	1.000
2 pole		
2 pole, 15-60a, NEMA 3R	EA.	1.250
70-100a, NEMA 3R	"	1.739
3 pole		
15-60a, NEMA 3R	EA.	1.509
70-100a, NEMA 3R	"	2.222
70-100a, NEMA 3R	"	1.250
70-100a, NEMA 1	"	1.739
3 pole		
15-60a, NEMA 1	EA.	1.739
70-100a, NEMA 1	"	2.222
Enclosed breakers, 600v, 2 phase, NEMA 3R		
15-60a	EA.	1.250
70-100a	"	1.739
3 phase, NEMA 3R		
15-60a	EA.	1.509
70-100a	"	2.222
600v, 3 phase, NEMA 1		
125a	EA.	2.222
150a	"	2.963
175a	"	2.963
200a	"	2.963
225a	"	2.963
250a	"	6.154
300a	"	6.154
350a	"	6.154
400a	"	6.154
500a	"	9.744
600a	"	9.744
700a	"	10.753
800a	"	10.753
900a	"	15.009

POWER GENERATION		UNIT	MAN/HOURS
16350.10	**CIRCUIT BREAKERS**		
100a		EA.	15.009
1200a		"	18.519
1400a		"	18.519
1600a		"	24.024
1800a		"	29.963
2000a		"	29.963
600v, 3 phase, NEMA 3R			
125-225a		EA.	2.222
250-400a		"	5.714
500-600a		"	9.744
700-800a		"	11.004
900-1000a		"	15.009
1000-1200a		"	19.002
1400-1600a		"	24.024
1800-2000a		"	29.963
16360.10	**SAFETY SWITCHES**		
Fused, 3 phase, 30 amp, 600v, heavy duty			
NEMA 1		EA.	1.143
NEMA 3r		"	1.143
NEMA 4		"	1.600
NEMA 12		"	1.739
60a			
NEMA 1		EA.	1.143
NEMA 3r		"	1.143
NEMA 4		"	1.600
NEMA 12		"	1.739
100a			
NEMA 1		EA.	1.739
NEMA 3r		"	1.739
NEMA 4		"	2.000
NEMA 12		"	2.500
200a			
NEMA 1		EA.	2.500
NEMA 3r		"	2.500
NEMA 4		"	2.759
NEMA 12		"	3.478
400a			
NEMA 1		EA.	5.517
NEMA 3r		"	5.517
NEMA 4		"	5.755
NEMA 12		"	7.018
600a			
NEMA 1		EA.	8.000
NEMA 3r		"	8.000
NEMA 4		"	8.989
NEMA 12		"	12.308
Non-fused, 240-600v, heavy duty, 3 phase, 30 amp			
NEMA 1		EA.	1.143
NEMA 3r		"	1.143
NEMA 4		"	1.739
NEMA 12		"	1.739
60a			

POWER GENERATION		UNIT	MAN/HOURS
16360.10	**SAFETY SWITCHES**		
NEMA1		EA.	1.143
NEMA 3r		"	1.143
NEMA 4		"	1.739
NEMA 12		"	1.739
100a			
NEMA 1		EA.	1.739
NEMA 3r		"	1.739
NEMA 4		"	2.500
NEMA 12		"	2.500
200a, NEMA 1		"	2.500
600a, NEMA 12		"	12.308
Bolt-on hubs			
3/4" - 1-1/2"		EA.	0.250
2"		"	0.296
2-1/2"		"	0.296
3"		"	0.348
3-1/2"		"	0.400
4"		"	0.400
Watertight hubs			
1/2"		EA.	0.250
3/4"		"	0.296
1"		"	0.400
1-1/4"		"	0.444
1-1/2"		"	0.471
2"		"	0.500
2-1/2"		"	0.533
3"		"	0.615
3-1/2"		"	0.800
4"		"	0.851
Non-fused, 600v, 3 pole, NEMA 7			
600a		EA.	2.222
100a		"	3.200
225a		"	4.000
NEMA 9			
60a		EA.	2.500
100a		"	3.333
225a		"	4.211
Fusible bolted pressure switches, 600v/3 pole, NEMA 1			
800a		EA.	16.000
1200a		"	21.978
1600a		"	25.000
2000a		"	29.963
2500a		"	34.934
3000a		"	44.944
4000a		"	51.948
Non-fusible			
800a		EA.	14.493
1200a		"	20.000
1600a		"	22.989
2000a		"	27.972
2500a		"	34.934
3000a		"	44.944
4000a		"	51.948
Fusible load interrupter switches, 4.16 kv, NEMA 1			

POWER GENERATION	UNIT	MAN/HOURS
16360.10 SAFETY SWITCHES		
200a	EA.	29.963
600a	"	70.175
Fusible load interrupter switch, 13.8 kv		
NEMA 1, 600a	EA.	100
NEMA 3R, 600a	"	100
4.16 kv, NEMA 3R		
200a	EA.	29.963
600a	"	70.175
Non-fused load interrupter switch, 4.16 kv, NEMA 1		
200a	EA.	29.963
600a	"	70.175
13.8 kv, NEMA 1, 600a	"	100
4.16 kv, NEMA 3R		
200a	EA.	29.963
600a	"	70.175
13.8 kv, NEMA 3R, 600a	"	100
Interrupter switch accessories, strip heater	"	
Cable lugs	"	
Key interlock	"	
Auxiliary switch	"	
Lightning arrester		
5 kva	EA.	
15 kv	"	
16365.10 FUSES		
Fuse, one-time, 250v		
30a	EA.	0.050
60a	"	0.050
100a	"	0.050
200a	"	0.050
400a	"	0.050
600a	"	0.050
600v		
30a	EA.	0.050
60a	"	0.050
100a	"	0.050
200a	"	0.050
400a	"	0.050
Fusetron, 600v		
200a	EA.	0.050
400a	"	0.050
Fuse, amp-trap, K1, 250v		
30a	EA.	0.050
60a	"	0.050
100a	"	0.050
200a	"	0.050
400a	"	0.050
600a	"	0.050
600v		
30a	EA.	0.050
60a	"	0.050
100a	"	0.050
200a	"	0.050

POWER GENERATION	UNIT	MAN/HOURS
16365.10 FUSES		
400a	EA.	0.050
K5, 250v		
30a	EA.	0.050
60a	"	0.050
100a	"	0.050
200a	"	0.050
400a	"	0.050
600a	"	0.050
600v		
30a	EA.	0.050
60a	"	0.050
100a	"	0.050
200a	"	0.050
400a	"	0.050
600a	"	0.050
J, 600v		
30a	EA.	0.050
60a	"	0.050
100a	"	0.050
200a	"	0.050
400a	"	0.050
L, 600v		
1200a	EA.	0.400
1600a	"	0.400
2000a	"	0.400
2500a	"	0.400
3000a	"	0.400
4000a	"	0.400
5000a	"	0.400
Fuse cl-ay 250v		
600a	EA.	0.296
1200a	"	0.296
1600a	"	0.296
2000a	"	0.296
600v		
1200a	EA.	0.296
1600a	"	0.296
2000a	"	0.296
Reducers, 600v		
60a-30a	EA.	0.145
100a-30a	"	0.145
100a-60a	"	0.145
200a-60a	"	0.250
200a-100a	"	0.250
400a-100a	"	0.348
400a-200a	"	0.348
600a-100a	"	0.400
600a-200a	"	0.400
600a-400a	"	0.400
16395.10 GROUNDING		
Ground rods, copper clad, 1/2" x		
6'	EA.	0.667

POWER GENERATION		UNIT	MAN/HOURS
16395.10	**GROUNDING**		
8'		EA.	0.727
10'		"	1.000
5/8" x			
5'		EA.	0.615
6'		"	0.727
8'		"	1.000
10'		"	1.250
3/4" x			
8'		EA.	0.727
10'		"	0.800
Ground rod clamp			
5/8"		EA.	0.123
3/4"		"	0.123
Coupling, on threaded rods, 3/4"		"	0.050
Ground receptacles		"	0.250
Bus bar, copper, 2" x 1/4"		L.F.	0.145
Copper braid, 1" x 1/8", for door ground		EA.	0.100
Brazed connection for			
#6 wire		EA.	0.500
#2 wire		"	0.800
#2/0 wire		"	1.000
#4/0 wire		"	1.143
Ground rod couplings			
1/2"		EA.	0.100
5/8"		"	0.100
Ground rod, driving stud			
1/2"		EA.	0.100
5/8"		"	0.100
3/4"		"	0.100
Ground rod clamps, #8-2 to			
1" pipe		EA.	0.200
2" pipe		"	0.250
3" pipe		"	0.296
5" pipe		"	0.348
6" pipe		"	0.444
#4-4/0 to			
1" pipe		EA.	0.200
2" pipe		"	0.250
3" pipe		"	0.296
3" pipe		"	0.348
8 pipe		"	0.444
8 pipe		"	0.667
10 pipe		"	0.952
12 pipe		"	1.290

SERVICE AND DISTRIBUTION	UNIT	MAN/HOURS
16425.10 **SWITCHBOARDS**		
Switchboard, 90" high, no main disconnect, 208/120v		
400a	EA.	7.921
600a	"	8.000
1000a	"	8.000
1200a	"	10.000
1600a	"	11.940
2000a	"	14.035
2500a	"	16.000
240/480v		
600a	EA.	8.163
800a	"	8.163
1600a	"	11.940
2000a	"	14.035
2500a	"	16.000
3000a	"	27.586
4000a	"	29.630
Main breaker sections, 600v		
1200a, GFI	EA.	16.985
1600a, GFI	"	19.512
2000a, GFI	"	20.000
2500a, GFI	"	25.000
3000a, GFI	"	29.963
4000a, GFI	"	34.934
Switchboard meter sections, 600v		
400a	EA.	8.000
600a	"	10.000
800a	"	11.004
1000a	"	13.491
2000a	"	16.000
2500a	"	20.000
3000a	"	25.000
4000a	"	29.963
Insulated case, draw out compartment, 208/120v		
800a	EA.	2.500
1600a	"	2.963
2000a	"	3.478
2500a	"	3.478
3000a	"	4.000
4000a	"	4.790
Accessories for power trip breakers		
Shunt trip	EA.	0.500
Key interlock	"	2.222
Lifting and transport truck	"	4.494
Lifting device	"	1.333
Bus duct connection, 3 phase, 4 wire		
225a	EA.	2.963
400a	"	2.963
600a	"	3.333
800a	"	4.000
2500a	"	6.015
3000a	"	7.477
4000a	"	8.791
Provision for mounting current transformers		
800a & below primary	EA.	2.963

SERVICE AND DISTRIBUTION	UNIT	MAN/HOURS
16425.10 SWITCHBOARDS		
1000 to 1500a primary	EA.	2.963
2000 to 6000a primary	"	2.963
Provision for mounting potential transformers		
2000a max	EA.	3.810
Switchboard instruments		
Voltmeter	EA.	1.000
Ammeter, incoming line	"	1.000
Wattmeter	"	1.000
Varmeter	"	1.000
Power factor meter	"	1.000
Frequency meter	"	1.000
Recording voltmeter	"	2.000
Wattmeter	"	2.000
Power factor meter	"	2.000
Frequency meter	"	2.000
Instrument phase select switch	"	0.500
Enclosure, 90" high, 3 phase, 4 wire		
1000a	EA.	6.838
1200a	"	7.018
1600a	"	8.602
2000a	"	13.333
5500a	"	15.686
3000a	"	18.182
4000a	"	23.529
Circuit breakers, 600v, 100a, frame		
15-30a, 1 pole	EA.	0.296
15-60a, 2 pole	"	0.348
70-100a, 2 pole	"	0.400
15-60a, 3 pole	"	0.444
70-100a, 3 pole	"	0.500
Bolt on breakers, 600v, 225a frame, 110-225a		
2 pole	EA.	0.615
3 pole	"	1.096
400a frame, 250-400a, 2 pole	"	1.250
800a frame		
450-600a, 2 pole	EA.	1.905
700-800a, 2 pole	"	2.500
450-600a, 3 pole	"	4.211
700-800a, 3 pole	"	4.444
Bolt on branch breakers, 600v		
1000-2000a, 2 pole	EA.	5.333
2500a, 2 pole	"	10.753
1000-2000a, 3 pole	"	8.000
2500a, 3 pole	"	11.004
3000a, 3 pole	"	20.000
Metal clad substation switch board, selector switch		
600a, 5kv	EA.	42.105
600a, 15kv	"	47.904
Fused switch, 600a		
5kv	EA.	34.934
15kv	"	34.934
1200a		
5kv	EA.	40.000
15kv	"	40.000

SERVICE AND DISTRIBUTION	UNIT	MAN/HOURS
16425.10 SWITCHBOARDS		
Oil cutout switch		
5 kv	EA.	15.009
15 kv	"	18.018
Liquid air terminal section	"	8.000
Dry air terminal section	"	8.502
Auxiliary compartment	"	29.963
16430.20 METERING		
Outdoor wp meter sockets, 1 gang, 240v, 1 phase		
Includes sealing ring, 100a	EA.	1.509
150a	"	1.778
200a	"	2.000
Die cast hubs, 1-1/4"	"	0.320
1-1/2"	"	0.320
2"	"	0.320
Indoor meter center, main switch single phase, 240v		
400a	EA.	8.000
600a	"	11.004
800a	"	11.696
Main breaker		
400a	EA.	8.000
600a	"	11.004
800a	"	11.696
1000a	"	16.000
Terminal box	"	16.495
1600a	"	18.018
Terminal box		
800a	EA.	10.000
1600a	"	18.018
Main switch, three phase, 208v		
400a	EA.	8.502
600a	"	11.994
800a	"	13.491
Main breaker		
400a	EA.	8.502
600a	"	11.994
800a	"	13.491
1000a	"	16.985
1200a	"	18.018
1600a	"	20.997
Terminal box		
800a	EA.	13.008
1600a	"	20.997
Indoor meter center		
2 meters	EA.	5.000
3 meters	"	6.154
4 meters	"	7.273
5 meters	"	8.000
6 meters	"	8.999
Plug on breakers, single phase, 208v		
60a	EA.	0.250
70a	"	0.250
80a	"	0.250
90a	"	0.250

SERVICE AND DISTRIBUTION

16430.20 METERING

	UNIT	MAN/HOURS
100a	EA.	0.348
Indoor meter center, single phase, 125a breakers		
3 meters	EA.	6.154
4 meters	"	7.273
5 meters	"	8.000
6 meters	"	8.502
7 meters	"	10.000
8 meters	"	11.004
10 meters	"	11.994
150a breakers		
3 meters	EA.	6.154
4 meters	"	7.273
6 meters	"	8.000
7 meters	"	10.000
8 meters	"	11.004
200a breakers		
3 meters	EA.	6.154
4 meters	"	7.273
6 meters	"	8.000
7 meters	"	10.000
8 meters	"	11.004
Indoor meter center, three phase, 125a breakers		
3 meters	EA.	6.154
4 meters	"	7.273
5 meters	"	8.000
6 meters	"	8.999
7 meters	"	10.000
8 meters	"	11.004
10 meters	"	11.994
150a breakers		
3 meters	EA.	6.154
4 meters	"	7.273
6 meters	"	8.502
7 meters	"	11.004
8 meters	"	11.994
200a breakers		
3 meters	EA.	6.667
4 meters	"	7.273
6 meters	"	8.999
7 meters	"	11.004
8 meters	"	11.994
NEMA 3R, meter center, main switch, 1 phase, 240v		
400a	EA.	8.000
600a	"	10.000
800a	"	11.004
Main breaker		
400a	EA.	8.000
600a	"	10.000
800a	"	12.308
1000a	"	15.009
1200a	"	16.000
Terminal box		
225a	EA.	7.273
800a	"	11.494

SERVICE AND DISTRIBUTION

16430.20 METERING

	UNIT	MAN/HOURS
1600a	EA.	18.018
NEMA 3R, three phase, 280v		
400a	EA.	8.502
600a	"	11.994
800a	"	13.008
Main breaker		
400a	EA.	8.502
600a	"	11.994
800a	"	13.008
1000a	"	16.985
1200a	"	18.018
Terminal box		
225a	EA.	8.000
800a	"	13.008
1600a	"	20.997
NEMA 3R meter center, single phase, 208v, 100a		
2 meters	EA.	5.000
3 meters	"	6.154
4 meters	"	7.273
5 meters	"	8.000
6 meters	"	8.999
4 meters	"	7.273
6 meters	"	8.000
7 meters	"	10.000
8 meters	"	11.004
125a, 3 meters	"	6.154
4 meters	"	7.273
6 meters	"	8.239
7 meters	"	10.000
8 meters	"	11.004
150a, 3 meters	"	6.154
4 meters	"	7.273
6 meters	"	8.502
7 meters	"	10.000
8 meters	"	11.004
NEMA 3R center, 3 phase, 208v, 125a breakers		
3 meters	EA.	6.154
4 meters	"	7.273
6 meters	"	8.502
7 meters	"	10.000
8 meters	"	11.004
150a		
3 meters	EA.	6.667
4 meters	"	7.273
6 meters	"	8.999
7 meters	"	10.499
8 meters	"	11.494
200a		
3 meters	EA.	6.667
4 meters	"	7.273
6 meters	"	8.999
7 meters	"	11.004
8 meters	"	11.494
NEMA 3R, center plug-on breakers, 208v, 1 phase		

SERVICE AND DISTRIBUTION		UNIT	MAN/ HOURS
16430.20	**METERING**		
60a		EA.	0.250
70a		"	0.250
90a		"	0.250
100a		"	0.348
125a		"	0.400
16460.10	**TRANSFORMERS**		
Pad mounted, single phase, dry type, 480v-120/240v			
15 kva		EA.	8.000
25 kva		"	8.989
37.5 kva		"	10.000
50 kva		"	10.959
3 phase			
225 kva		EA.	25.000
300 kva		"	30.769
500 kva		"	38.095
750 kva		"	47.059
1000 kva		"	50.000
1500 kva		"	57.143
Substation transformers, outdoor, 5 kv - 208v			
112.5 kva		EA.	21.978
150 kva		"	24.024
225 kva		"	27.972
300 kva		"	29.963
500 kva		"	44.944
750 kva		"	55.172
1000 kva		"	65.041
15 kv, 208v			
112 kva		EA.	27.972
150 kva		"	29.963
225 kva		"	34.934
300 kva		"	40.000
500 kva		"	50.000
750 kva		"	60.150
1000 kva		"	70.175
5kv, 480v			
112kva		EA.	21.978
150 kva		"	24.024
225 kva		"	27.972
300 kva		"	29.963
500 kva		"	44.944
750 kva		"	55.172
1000 kva		"	65.041
1500 kva		"	74.766
2000 kva		"	89.888
2500 kva		"	110
15 kv, 480v			
112.5 kva		EA.	27.972
150 kva		"	29.963
225 kva		"	34.934
300 kva		"	40.000
500 kva		"	50.000
750 kva		"	60.150

SERVICE AND DISTRIBUTION		UNIT	MAN/ HOURS
16460.10	**TRANSFORMERS**		
1000 kva		EA.	70.175
1500 kva		"	80.000
2000 kva		"	89.888
2500 kva		"	119
Pad mounted 3 phase, 15 kv outdoor			
50 kva		EA.	10.256
75 kva		"	11.765
112 kva		"	12.903
150 kva		"	14.545
225 kva		"	15.385
300 kva		"	17.021
500 kva		"	27.586
750 kva		"	36.364
1000 kva		"	44.444
1500 kva		"	53.333
Dry type, for power gear, 5 kv indoor			
75 kva		EA.	16.000
112.5 kva		"	18.605
150 kva		"	21.053
225 kva		"	23.529
300 kva		"	25.000
500 kva		"	27.586
750 kva		"	36.364
16470.10	**PANELBOARDS**		
Indoor load center, 1 phase 240v main lug only			
30a - 2 spaces		EA.	2.000
100a - 8 spaces		"	2.424
150a - 16 spaces		"	2.963
200a - 24 spaces		"	3.478
200a - 42 spaces		"	4.000
Main circuit breaker			
100a - 8 spaces		EA.	2.424
100a - 16 spaces		"	2.759
150a - 16 spaces		"	2.963
150a - 24 spaces		"	3.200
200a - 24 spaces		"	3.478
200a - 42 spaces		"	3.636
3 phase, 480/277v, main lugs only, 120a, 30 circuits		"	3.478
277/480v, 4 wire, flush surface			
225a, 30 circuits		EA.	4.000
400a, 30 circuits		"	5.000
600a, 42 circuits		"	6.015
208/120v, main circuit breaker, 3 phase, 4 wire			
100a			
12 circuits		EA.	5.096
20 circuits		"	6.299
30 circuits		"	7.018
225a			
30 circuits		EA.	7.767
42 circuits		"	9.524
400a			
30 circuits		EA.	14.815

SERVICE AND DISTRIBUTION	UNIT	MAN/ HOURS
16470.10 PANELBOARDS		
42 circuits	EA.	16.000
600a, 42 circuits	"	18.182
120/208v, flush, 3 ph., 4 wire, main only		
100a		
12 circuits	EA.	5.096
20 circuits	"	6.299
30 circuits	"	7.018
225a		
30 circuits	EA.	7.767
42 circuits	"	9.524
400a		
30 circuits	EA.	14.815
42 circuits	"	16.000
600a, 42 circuits	"	18.182
Panelboard accessories		
Grounding bus	EA.	0.348
Handle lock device	"	0.145
Factory assembled panel		
1 pole space	EA.	0.348
2 pole space	"	0.145
3 pole space	"	0.133
Panelboards 1 phase, 240/120v main circuit breaker		
Single phase, 3 wire, 120/240v flush		
100a, 20 circuits	EA.	3.478
225a, 30 circuits	"	4.000
240/120v, main lugs only		
100a		
8 circuits	EA.	2.963
12 circuits	"	2.963
20 circuits	"	2.963
225a		
24 circuits	EA.	3.478
30 circuits	"	3.810
42 circuits	"	3.810
Distribution panelboards, 3 ph, main breaker		
225a	EA.	16.000
400a	"	18.018
600a	"	21.978
800a	"	24.024
1000a	"	27.972
1200a	"	29.963
Single phase		
225a	EA.	14.011
400a	"	16.000
600a	"	20.000
800a	"	24.024
1000a	"	27.972
1200a	"	29.963
Fusible distribution panelboards, 3 phase, 600v		
100a	EA.	14.011
200a	"	16.000
400a	"	20.000
600a	"	24.024
800a	"	27.972

SERVICE AND DISTRIBUTION	UNIT	MAN/ HOURS
16470.10 PANELBOARDS		
Single phase		
100a	EA.	11.994
200a	"	14.011
400a	"	18.018
600a	"	21.978
800a	"	25.974
Hospital panels, operating room		
3kv - 208v	EA.	6.154
3kv - 277v	"	6.154
5kv - 208v	"	6.154
5kv - 277v	"	6.154
Coronary care		
3kv - 208v	EA.	7.273
3kv - 277v	"	7.273
5kv - 208v	"	7.273
5kv - 277v	"	7.273
Intensive care		
3kv - 208v	EA.	8.000
3kv - 277v	"	8.000
5kv - 208v	"	8.000
5kv - 277v	"	8.000
15kv - 208v	"	11.994
15kv - 277v	"	11.994
25kv - 208v	"	16.000
25kv - 277v	"	16.000
Explosion proof, 240v, m.l.b. 20a, single phase		
6 breakers	EA.	11.004
8 breakers	"	11.747
10 breakers	"	12.500
12 breakers	"	13.245
14 breakers	"	14.011
16 breakers	"	14.011
18 breakers	"	15.504
20 breakers	"	16.260
22 breakers	"	16.985
24 breakers	"	17.738
16480.10 MOTOR CONTROLS		
Motor generator set, 3 phase, 480/277v, w/controls		
10kw	EA.	27.586
15kw	"	30.769
20kw	"	32.000
25kw	"	34.783
30kw	"	36.364
40kw	"	38.095
50kw	"	40.000
60kw	"	44.444
75kw	"	50.000
100kw	"	61.538
125kw	"	66.667
150kw	"	66.667
200kw	"	72.727
250kw	"	72.727
300kw	"	80.000

SERVICE AND DISTRIBUTION	UNIT	MAN/ HOURS
16480.10 MOTOR CONTROLS		
2 pole, 230 volt starter, w/NEMA-1		
1 hp, 9 amp, size 00	EA.	1.000
2 hp, 18amp, size 0	"	1.000
3 hp, 27amp, size 1	"	1.000
5 hp, 45amp, size 1p	"	1.000
7-1/2 hp, 45a, size 2	"	1.000
15 hp, 90a, size 3	"	1.000
2 pole, w/NEMA-4 enclosure		
2 hp, 18a, size 1	EA.	1.600
5 hp, 45amp, size 1p	"	1.600
7-1/2 hp, 45a, size 2	"	1.600
3 pole, 2 hp, 9a, 200-575v starter		
W/NEMA-1, size 00	EA.	1.333
W/NEMA-4 enclosure, size 00	"	1.739
5hp, 18a		
W/NEMA-1 enclosure, size 0	EA.	1.333
W/NEMA-4 enclosure, size 0	"	1.739
7.5-10hp, 27a		
7.5-10hp 27a, w/NEMA-1 enclosure, size 1	EA.	1.333
W/NEMA-4 enclosure size 1	"	1.739
10-25hp, 45a		
W/NEMA-1 enclosure, size 2	EA.	1.333
W/NEMA-4 enclosure, size 2	"	1.739
25-50hp, 90a		
W/NEMA-1 enclosure, size 3	EA.	1.739
W/NEMA-4 enclosure, size 3	"	2.500
40-100hp, 135a		
W/NEMA-1 enclosure, size 4	EA.	2.500
W/NEMA-4 enclosure, size 4	"	3.478
75-200hp, 270a		
W/NEMA-1 enclosure, size 5	EA.	5.517
W/NEMA-4 enclosure, size 5	"	7.018
Magnetic starter accessories		
On-off-auto selector switch kit	EA.	0.320
With pilot light	"	0.348
Control center main lug only, 208v, 3 phase		
600a	EA.	11.994
1200a	"	16.000
Main circuit breakers, 208v, 3 phase		
400a	EA.	10.000
600a	"	14.011
800a	"	16.000
1000a	"	18.018
1200a	"	20.000
Non-reversing starters		
Size 1	EA.	0.727
Size 2	"	1.250
Size 3	"	1.509
Size 4	"	1.739
Reversing starters		
Size 1	EA.	0.727
Size 2	"	1.096
Fusible switch, non-revolving starters		
Size 1	EA.	0.727

SERVICE AND DISTRIBUTION	UNIT	MAN/ HOURS
16480.10 MOTOR CONTROLS		
Size 2	EA.	1.250
Size 3	"	1.509
Size 4	"	1.739
Reversing starters		
Size 1	EA.	0.727
Size 2	"	1.096
Two speed, non-reversing starter		
Size 1	EA.	0.727
Size 2	"	1.096
Magnetic starter, 600v, 2 pole, NEMA 3R		
Size 0, 2 hp	EA.	1.000
Size 1, 5hp	"	1.096
NEMA 3R		
Size 2, 7.5 hp	EA.	1.143
Size 3, 15 hp	"	1.194
NEMA 7		
Size 0, 2 hp	EA.	1.739
Size 1, 5 hp	"	2.000
Size 2, 7.5 hp	"	2.222
NEMA 12		
Size 0, 2 hp	EA.	1.509
Size 1, 5 hp	"	1.739
Size 2, 7.5 hp	"	2.000
Size 3, 15 hp	"	2.222
3 pole, NEMA 1		
Size 6	EA.	10.000
Size 7	"	11.994
Size 8	"	16.000
NEMA 4		
Size 6	EA.	14.011
Size 7	"	16.000
Size 8	"	20.000
NEMA 3R		
Size 0	EA.	1.250
Size 1	"	1.356
Size 2	"	1.818
Size 3	"	1.905
Size 4	"	2.759
NEMA 7		
Size 0	EA.	2.000
Size 1	"	2.162
Size 2	"	2.222
Size 3	"	2.759
Size 4	"	4.444
Size 5	"	9.744
Size 6	"	16.000
NEMA 12		
Size 00	EA.	1.739
Size 0	"	1.860
Size 1	"	1.905
Size 2	"	2.000
Size 3	"	2.500
Size 4	"	3.478
Size 5	"	7.273

SERVICE AND DISTRIBUTION

16480.10 — MOTOR CONTROLS

Description	UNIT	MAN/HOURS
Size 6	EA.	12.500
Size 7	"	14.011
Size 8	"	20.000
Reversing magnetic starters, 600v, 3 pole, NEMA 1		
Size 00	EA.	1.250
Size 0	"	1.290
Size 1	"	1.356
Size 2	"	1.509
Size 3	"	1.739
Size 4	"	2.000
Size 5	"	5.333
Size 6	"	9.501
Size 7	"	11.004
Size 8	"	18.018
NEMA 4		
Size 0	EA.	1.739
Size 4	"	1.818
Size 2	"	1.905
Size 3	"	2.000
Size 4	"	2.500
Size 5	"	7.273
Size 6	"	12.500
Size 7	"	15.009
Size 8	"	20.000
NEMA 7		
Size 0	EA.	2.000
Size 1	"	2.222
Size 2	"	2.500
Size 3	"	2.963
NEMA 12		
Size 0	EA.	1.739
Size 1	"	1.905
Size 2	"	2.000
Size 3	"	2.222
Size 4	"	2.500
Size 5	"	7.273
Size 6	"	14.011
Size 7	"	16.000
Size 8	"	20.000
Electrically held lighting contactors, NEMA 1, 20a		
2 pole	EA.	1.000
3 pole	"	1.250
4 pole	"	1.509
6 pole	"	2.000
8 pole	"	2.500
10 pole	"	2.963
12 pole	"	3.478
30a		
2 pole	EA.	1.000
3 pole	"	1.250
4 pole	"	1.509
5 pole	"	1.739
60a		
2 pole	EA.	1.000

SERVICE AND DISTRIBUTION

16480.10 — MOTOR CONTROLS

Description	UNIT	MAN/HOURS
3 pole	EA.	1.250
4 pole	"	1.509
5 pole	"	1.739
100a		
2 pole	EA.	1.250
3 pole	"	1.739
4 pole	"	2.222
5 pole	"	2.759
200a		
2 pole	EA.	2.759
3 pole	"	2.963
4 pole	"	3.200
300a		
2 pole	EA.	4.211
3 pole	"	5.333
400a		
2 pole	EA.	4.211
3 pole	"	5.333
600a		
2 pole	EA.	6.667
3 pole	"	9.249
800a		
2 pole	EA.	8.000
3 pole	"	11.004
Mechanically held lighting contactors, NEMA 1, 20a		
2 pole	EA.	1.000
3 pole	"	1.250
4 pole	"	1.509
6 pole	"	2.000
8 pole	"	2.500
10 pole	"	2.963
30a		
2 pole	EA.	1.000
3 pole	"	1.250
4 pole	"	1.509
5 pole	"	1.739
60a		
2 pole	EA.	1.000
3 pole	"	1.250
4 pole	"	1.509
5 pole	"	1.739
100a		
2 pole	EA.	1.250
3 pole	"	1.739
4 pole	"	2.000
5 pole	"	2.500
200a		
2 pole	EA.	1.739
3 pole	"	2.500
4 pole	"	3.200
300a		
2 pole	EA.	4.211
3 pole	"	5.333
400a		

SERVICE AND DISTRIBUTION	UNIT	MAN/HOURS
16480.10 MOTOR CONTROLS		
2 pole	EA.	4.211
3 pole	"	5.333
600a		
2 pole	EA.	6.667
3 pole	"	8.889
800a		
2 pole	EA.	8.000
3 pole	"	11.429
AC relays, control type open, 15a, 600v		
2 pole	EA.	1.000
3 pole	"	1.250
4 pole	"	1.509
6 pole	"	2.000
8 pole	"	2.500
10 pole	"	2.963
12 pole	"	3.478
16490.10 SWITCHES		
Oil switches, medium voltage, bus components		
Switches, 277/120v, toggle device only	EA.	1.600
With oil 35kv, g&w gram 44, 4 way switch	"	8.000
Weatherproof enclosure		
3 way switch	EA.	10.000
4 way switch	"	10.959
Fused interrupter load, 35kv		
20A		
1 pole	EA.	16.000
2 pole	"	17.021
3 way	"	17.021
4 way	"	18.182
30a, 1 pole	"	16.000
3 way	"	17.021
4 way	"	18.182
Weatherproof switch, including box & cover, 20a		
1 pole	EA.	16.000
2 pole	"	17.021
3 way	"	18.182
4 way	"	18.182
3 way, oil switch, 15kv enclosure	"	11.940
Pedestal for 35kv double breaker switch	"	5.000
Bus terminal connector, 2	"	2.500
2 to 3	"	2.500
Support connector, 3	"	1.600
Tee connector, 2 to 3	"	2.000
Flexible bus stud connector	"	1.739
End cap 3	"	1.333
Weldment connection, 3	"	1.000
Plate switch, 1 gang	"	0.050
Start stop stations, manual motor starters	"	0.727
Lockout switch	"	0.250
Forward-reverse switch	"	0.727
On-off switch	"	0.727
Open-close switch	"	0.727
Forward-reverse-stop switch	"	1.000

SERVICE AND DISTRIBUTION	UNIT	MAN/HOURS
16490.10 SWITCHES		
Standard 3 button switch any standard legend	EA.	1.000
Standard 3 button with lockout	"	1.000
Manual motor starters, tog, 115/230v		
Size 1 gp	EA.	1.000
Size 2	"	1.000
Button		
Size 0	EA.	1.000
Size 1	"	1.000
Size 2	"	1.000
3-phase		
Size 0	EA.	1.333
Size 1	"	1.333
Time & float switches	"	1.600
Astronomical time switch, 40a, 240v	"	1.000
Timer switch 0-5 minute, with box	"	0.500
Single pole/single throw time, 277v, NEMA-1	"	0.727
Single toggle switch, 20a, 120v, with pilot	"	0.250
3-way toggle	"	0.296
Photo electric switches		
1000 watt		
105-135v	EA.	0.727
208-277v	"	0.727
3000 watt, 105-130v		
Double throw	EA.	1.000
Single throw	"	1.000
Double pole/single throw, 210-250v	"	1.333
Dimmer switch and switch plate		
600 w	EA.	0.308
1000 w	"	0.348
Dimmer switch incandescent		
1500w	EA.	0.702
2000w	"	0.748
Fluorescent		
12 lamps	EA.	0.500
20 lamps	"	0.552
30 lamps	"	0.602
40 lamps	"	0.702
Time clocks with skip, 40a, 120v		
SPST	EA.	0.748
SPDT	"	0.748
DPST	"	0.748
DPDT	"	1.000
SPST	"	1.000
Astronomic time clocks with skip, 40a, 120v		
DPST	EA.	0.748
SPST	"	1.000
SPDT	"	0.748
Raintight time clocks, 40a, 120v		
SPDT	EA.	1.000
DPST	"	1.000
Contractor grade wall switch 15a, 120v		
Single pole	EA.	0.160
Three way	"	0.200
Four way	"	0.267

SERVICE AND DISTRIBUTION	UNIT	MAN/HOURS
16490.10 SWITCHES		
Specification grade toggle switches, 20a, 120-277v		
Single pole	EA.	0.200
Double pole	"	0.296
3 way	"	0.250
4 way	"	0.296
30a, 120-277v		
Single pole	EA.	0.200
Double pole	"	0.296
3 way	"	0.250
Specification grade key switches, 20a, 120-277v		
Single pole	EA.	0.200
Double pole	"	0.296
3 way	"	0.250
4 way	"	0.296
Red pilot light handle switches, 20a, 120-277v		
Single pole	EA.	0.200
Double pole	"	0.296
3 way	"	0.250
30a, 120-277v		
Single pole	EA.	0.200
Double pole	"	0.296
3 way	"	0.250
Momentary contact switches, 20a		
SPDT, ivory	EA.	0.250
SPDT, locking	"	0.296
Maintained contact switches		
SPDT ivory	EA.	0.250
DPDT ivory	"	0.250
SPDT locking	"	0.296
DPDT locking	"	0.348
Mercury switch, 3 way	"	0.250
Door switches, open on or off	"	0.500
Combination switch and pilot light, single pole	"	0.296
3 way	"	0.348
Combination switch and receptacle, single pole	"	0.296
3 way	"	0.296
Combination two switches, single pole/single pole	"	0.250
3 way	"	0.400
Switch plates, plastic ivory		
1 gang	EA.	0.080
2 gang	"	0.100
3 gang	"	0.119
4 gang	"	0.145
5 gang	"	0.160
6 gang	"	0.182
Stainless steel		
1 gang	EA.	0.080
2 gang	"	0.100
3 gang	"	0.123
4 gang	"	0.145
5 gang	"	0.160
6 gang	"	0.182
Brass		
1 gang	EA.	0.080

SERVICE AND DISTRIBUTION	UNIT	MAN/HOURS
16490.10 SWITCHES		
2 gang	EA.	0.100
3 gang	"	0.123
4 gang	"	0.145
5 gang	"	0.160
6 gang	"	0.182
16490.20 TRANSFER SWITCHES		
Automatic transfer switch 600v, 3 pole		
30a	EA.	3.478
60a	"	3.478
100a	"	4.762
150a	"	6.015
225a	"	8.000
260a	"	8.000
400a	"	10.000
600a	"	15.094
800a	"	18.182
1000a	"	21.053
1200a	"	22.857
1600a	"	25.000
2000a	"	29.630
2600a	"	42.105
Automatic transfer switches, 600v, 3 phase, 3000a	"	50.000
16490.80 SAFETY SWITCHES		
Safety switch, 600v, 3 pole, heavy duty, NEMA-1		
30a	EA.	1.000
60a	"	1.143
100a	"	1.600
200a	"	2.500
400a	"	5.517
600a	"	8.000
800a	"	10.526
1200a	"	14.286

LIGHTING	UNIT	MAN/HOURS
16510.05 INTERIOR LIGHTING		
Recessed fluorescent fixtures, 2'x2'		
2 lamp	EA.	0.727
4 lamp	"	0.727
2 lamp w/flange	"	1.000
4 lamp w/flange	"	1.000
1'x4'		
2 lamp	EA.	0.667

LIGHTING	UNIT	MAN/HOURS
16510.05 INTERIOR LIGHTING		
3 lamp	EA.	0.667
2 lamp w/flange	"	0.727
3 lamp w/flange	"	0.727
2'x4'		
2 lamp	EA.	0.727
3 lamp	"	0.727
4 lamp	"	0.727
2 lamp w/flange	"	1.000
3 lamp w/flange	"	1.000
4 lamp w/flange	"	1.000
4'x4'		
4 lamp	EA.	1.000
6 lamp	"	1.000
8 lamp	"	1.000
4 lamp w/flange	"	1.509
6 lamp w/flange	"	1.509
8 lamp, w/flange	"	1.509
Surface mounted incandescent fixtures		
40w	EA.	0.667
75w	"	0.667
100w	"	0.667
150w	"	0.667
Pendant		
40w	EA.	0.800
75w	"	0.800
100w	"	0.800
150w	"	0.800
Contractor grade recessed down lights		
100 watt housing only	EA.	1.000
150 watt housing only	"	1.000
100 watt trim	"	0.500
150 watt trim	"	0.500
Recessed incandescent fixtures		
40w	EA.	1.509
75w	"	1.509
100w	"	1.509
150w	"	1.509
Exit lights, 120v		
Recessed	EA.	1.250
Back mount	"	0.727
Universal mount	"	0.727
Emergency battery units, 6v-120v, 50 unit	"	1.509
With 1 head	"	1.509
With 2 heads	"	1.509
Mounting bucket	"	0.727
Light track single circuit		
2'	EA.	0.500
4'	"	0.500
8'	"	1.000
12'	"	1.509
Fittings and accessories		
Dead end	EA.	0.145
Starter kit	"	0.250
Conduit feed	"	0.145

LIGHTING	UNIT	MAN/HOURS
16510.05 INTERIOR LIGHTING		
Straight connector	EA.	0.145
Center feed	"	0.145
L-connector	"	0.145
T-connector	"	0.145
X-connector	"	0.200
Cord and plug	"	0.100
Rigid corner	"	0.145
Flex connector	"	0.145
2 way connector	"	0.200
Spacer clip	"	0.050
Grid box	"	0.145
T-bar clip	"	0.050
Utility hook	"	0.145
Fixtures, square		
R-20	EA.	0.145
R-30	"	0.145
40w flood	"	0.145
40w spot	"	0.145
100w flood	"	0.145
100w spot	"	0.145
Mini spot	"	0.145
Mini flood	"	0.145
Quartz, 500w	"	0.145
R-20 sphere	"	0.145
R-30 sphere	"	0.145
R-20 cylinder	"	0.145
R-30 cylinder	"	0.145
R-40 cylinder	"	0.145
R-30 wall wash	"	0.145
R-40 wall wash	"	0.145
Explosion proof, incan., surface mounted		
100w - 200w	EA.	1.739
300w	"	1.739
500w	"	1.739
With guard		
100w-200w	EA.	2.222
300w	"	2.222
500w	"	2.222
Reflectors for incan. light fixtures, dome	"	0.250
Angle	"	0.250
Highbay	"	0.296
Explosion proof fluor. fixtures, 800 ms.		
1 lamp	EA.	2.222
2 lamp	"	2.667
3 lamp	"	2.963
4 lamp	"	3.200
Explosion proof hp sodium fixtures		
50w-70w	EA.	2.222
100w	"	2.222
150w	"	2.500
200w	"	2.500
250w	"	2.500
310w	"	2.500
400w	"	2.500

LIGHTING	UNIT	MAN/HOURS
16510.05 INTERIOR LIGHTING		
With guard		
50w-70w	EA.	2.500
100w	"	2.500
150w	"	2.759
200w	"	2.759
250w	"	2.759
310w	"	2.759
400w	"	2.759
Explosion proof metal halide fixtures		
175w	EA.	2.500
250w	"	2.500
400w	"	2.500
With guard, 175w	"	2.759
250w	"	2.759
400w	"	2.759
Energy saving rapid start fluor. lamps		
F30cw	EA.	0.100
F40 cw	"	0.100
F40 cwx	"	0.100
F30 ww	"	0.100
F40 ww	"	0.100
F40 wwx	"	0.100
Slimline		
F48 cw	EA.	0.145
F96 cwx	"	0.145
F48 ww	"	0.145
F96 ww	"	0.145
F96 wwx	"	0.145
High output	"	0.145
F96 cwx	"	0.145
F96 cw	"	0.145
Power groove, F48 cw	"	0.145
Circle		
Fc6 cw	EA.	0.100
Fc8 cw	"	0.100
Fc12 cw	"	0.100
Fc16 cw	"	0.100
Fc6 ww	"	0.100
Fc8 ww	"	0.100
Fc12 ww	"	0.100
Fc16 ww	"	0.100
Incandescent lamps		
200w	EA.	0.100
300w	"	0.100
500w	"	0.100
750w	"	0.145
1000w	"	0.200
1500w	"	0.200
Energy saving reflector floodlight lamps		
25w	EA.	0.100
30w	"	0.100
50w	"	0.100
75w	"	0.100
120w	"	0.100

LIGHTING	UNIT	MAN/HOURS
16510.05 INTERIOR LIGHTING		
150w	EA.	0.100
200w	"	0.100
300w	"	0.100
500w	"	0.145
750w	"	0.145
Reflector spotlight		
75w	EA.	0.100
100w	"	0.100
125w	"	0.100
150w	"	0.100
250w	"	0.100
300w	"	0.100
400w	"	0.145
500w	"	0.145
1000w	"	0.200
Medium par flood lamps		
75w	EA.	0.100
100w	"	0.100
150w	"	0.100
200w	"	0.100
300w	"	0.100
500w	"	0.145
Medium par spot lamps		
75w	EA.	0.100
120w	"	0.100
150w	"	0.100
Tubular quartz lamps		
100w	EA.	0.145
150w	"	0.145
200w	"	0.145
400w	"	0.145
500w	"	0.200
750w	"	0.200
1000w	"	0.250
1250w	"	0.250
1500w	"	0.250
Ballast replacements rapid start fluor		
1f-40-120v	EA.	0.727
1f-40-277v	"	0.727
1f-96-120v	"	0.727
1f-96-277v	"	0.727
2f-40-120v	"	0.727
2f-40-277v	"	0.727
2f-96-120v	"	0.727
2f-96-277v	"	0.727
Circline, 1fc6-1fc16	"	0.727
Very high output, 1500ma		
1f48-120v	EA.	0.727
1f48-277v	"	0.727
1f96-120v	"	0.727
1f96-277v	"	0.727
2f48-120v	"	0.727
2f48-277v	"	0.727
2f96-120v	"	0.727

LIGHTING	UNIT	MAN/HOURS
16510.05 INTERIOR LIGHTING		
2f96-277v	EA.	0.727
Mercury, multi tap		
475w	EA.	1.000
100w	"	1.000
175w	"	1.000
250w	"	1.000
400w	"	1.000
1000W	"	1.000
Metal halide, multi tap		
175w	EA.	1.000
250w	"	1.000
400w	"	1.000
1000w	"	1.000
1500w	"	1.000
High pressure sodium		
70w	EA.	1.000
100w	"	1.000
150w	"	1.000
250w	"	1.000
400w	"	1.000
1000w	"	1.000
Surface mounted fluorescent, wrap around lens		
1 lamp	EA.	0.800
2 lamps	"	0.889
4 lamps	"	1.000
Wall mounted fluorescent		
2-20w lamps	EA.	0.500
2-30w lamps	"	0.500
2-40w lamps	"	0.667
Indirect, with wood shielding, 2049w lamps		
4'	EA.	1.000
8'	"	1.600
Industrial fluorescent, 2 lamp		
4'	EA.	0.727
8'	"	1.333
Strip fluorescent		
4'		
1 lamp	EA.	0.667
2 lamps	"	0.667
8'		
1 lamp	EA.	0.727
2 lamps	"	0.889
Wire guard for strip fixture, 4' long	"	0.348
Strip fluorescent, 8' long, two 4' lamps	"	1.333
With four 4' lamps	"	1.600
Wet location fluorescent, plastic housing		
4' long		
1 lamp	EA.	1.000
2 lamps	"	1.333
8' long		
2 lamps	EA.	1.600
4 lamps	"	1.739
Parabolic troffer, 2'x2'		
With 2 "U" lamps	EA.	1.000

LIGHTING	UNIT	MAN/HOURS
16510.05 INTERIOR LIGHTING		
With 3 "U" lamps	EA.	1.143
2'x4'		
With 2 40w lamps	EA.	1.143
With 3 40w lamps	"	1.333
With 4 40w lamps	"	1.333
1'x4'		
With 1 T-12 lamp, 9 cell	EA.	0.727
With 2 T-12 lamps	"	0.889
With 1 T-12 lamp, 20 cell	"	0.727
With 2 T-12 lamps	"	0.889
Steel sided surface fluorescent, 2'x4'		
3 lamps	EA.	1.333
4 lamps	"	1.333
Outdoor sign fluor., 1 lamp, remote ballast		
4' long	EA.	6.015
6' long	"	8.000
Recess mounted, commercial, 2'x2', 13" high		
100w	EA.	4.000
250w	"	4.494
High pressure sodium, hi-bay open		
400w	EA.	1.739
1000w	"	2.424
Enclosed		
400w	EA.	2.424
1000w	"	2.963
Metal halide hi-bay, open		
400w	EA.	1.739
1000w	"	2.424
Enclosed		
400w	EA.	2.424
1000w	"	2.963
High pressure sodium, low bay, surface mounted		
100w	EA.	1.000
150w	"	1.143
250w	"	1.333
400w	"	1.600
Metal halide, low bay, pendant mounted		
175w	EA.	1.333
250w	"	1.600
400w	"	2.222
Indirect luminare, square, metal halide, freestanding		
175w	EA.	1.000
250w	"	1.000
400w	"	1.000
High pressure sodium		
150w	EA.	1.000
250w	"	1.000
400w	"	1.000
Round, metal halide		
175w	EA.	1.000
250w	"	1.000
400w	"	1.000
High pressure sodium		
150w	EA.	1.000

LIGHTING	UNIT	MAN/HOURS
16510.05 INTERIOR LIGHTING		
250w	EA.	1.000
400w	"	1.000
Wall mounted, metal halide		
175w	EA.	2.500
250w	"	2.500
400w	"	3.200
High pressure sodium		
150w	EA.	2.500
250w	"	2.500
400w	"	3.200
Wall pack lithonia, high pressure sodium		
35w	EA.	0.889
55w	"	1.000
150w	"	1.600
250w	"	1.739
Low pressure sodium		
35w	EA.	1.739
55w	"	2.000
Wall pack hubbell, high pressure sodium		
35w	EA.	0.889
150w	"	1.600
250w	"	1.739
Compact fluorescent		
2-7w	EA.	1.000
2-13w	"	1.333
1-18w	"	1.333
Handball & racquet ball court, 2'x2', metal halide		
250w	EA.	2.500
400w	"	2.759
High pressure sodium		
250w	EA.	2.500
400w	"	2.759
Bollard light, 42" w/found., high pressure sodium		
70w	EA.	2.581
100w	"	2.581
150w	"	2.581
Light fixture lamps		
Lamp		
20w med. bipin base, cool white, 24"	EA.	0.145
30w cool white, rapid start, 36"	"	0.145
40w cool white "U", 3"	"	0.145
40w cool white, rapid start, 48"	"	0.145
70w high pressure sodium, mogul base	"	0.200
75w slimline, 96"	"	0.200
100w		
Incandescent, 100a, inside frost	EA.	0.100
Mercury vapor, clear, mogul base	"	0.200
High pressure sodium, mogul base	"	0.200
150w		
Par 38 flood or spot, incandescent	EA.	0.100
High pressure sodium, 1/2 mogul base	"	0.200
175w		
Mercury vapor, clear, mogul base	EA.	0.200
Mercury halide, clear, mogul base	"	0.200

LIGHTING	UNIT	MAN/HOURS
16510.05 INTERIOR LIGHTING		
250w		
High pressure sodium, mogul base	EA.	0.200
Mercury vapor, clear, mogul base	"	0.200
Metal halide, clear, mogul base	"	0.200
High pressure sodium, mogul base	"	0.200
400w		
Mercury vapor, clear, mogul base	EA.	0.200
Metal halide, clear, mogul base	"	0.200
High pressure sodium, mogul base	"	0.200
1000w		
Mercury vapor, clear, mogul base	EA.	0.250
High pressure sodium, mogul base	"	0.250
16510.30 EXTERIOR LIGHTING		
Exterior light fixtures		
Rectangle, high pressure sodium		
70w	EA.	2.500
100w	"	2.581
150w	"	2.581
250w	"	2.759
400w	"	3.478
Flood, rectangular, high pressure sodium		
70w	EA.	2.500
100w	"	2.581
150w	"	2.581
400w	"	3.478
1000w	"	4.494
Round		
400w	EA.	3.478
1000w	"	4.494
Round, metal halide		
400w	EA.	3.478
1000w	"	4.494
Light fixture arms, cobra head, 6', high press. sodium		
100w	EA.	2.000
150w	"	2.500
250w	"	2.500
400w	"	2.963
Flood, metal halide		
400w	EA.	3.478
1000w	"	4.494
1500w	"	6.015
Mercury vapor		
250w	EA.	2.759
400w	"	3.478
Incandescent		
300w	EA.	1.739
500w	"	2.000
1000w	"	3.200
16510.90 POWER LINE FILTERS		
Heavy duty power line filter, 240v		
100a	EA.	10.000

LIGHTING

	UNIT	MAN/HOURS
16510.90 POWER LINE FILTERS		
300a	EA.	16.000
600a	"	24.242
16600.20 CENTRAL INVERTER SYSTEMS		
Central inverter systems		
500va	EA.	2.963
1000va	"	4.000
1500va	"	5.333
2400va	"	6.667
3000va	"	8.502
4500va	"	10.000
6000va	"	11.004
7500va	"	14.011
10,000va	"	16.000
16,600va	"	22.989
25,000va	"	34.934
16610.30 UNINTERRUPTIBLE POWER		
Uninterruptible power systems, (U.P.S.)	EA.	8.000
5 kva	"	11.004
7.5 kva	"	16.000
10 kva	"	21.978
15 kva	"	22.857
20 kva	"	24.024
25 kva	"	25.000
30 kva	"	25.974
35 kva	"	27.027
40 kva	"	27.972
45 kva	"	28.986
50 kva	"	29.963
62.5 kva	"	32.000
75 kva	"	34.934
100 kva	"	36.036
150 kva	"	50.000
200 kva	"	55.172
300 kva	"	74.766
400 kva	"	89.888
500 kva	"	110
16670.10 LIGHTNING PROTECTION		
Lightning protection		
Copper point, nickel plated, 12'		
1/2" dia.	EA.	1.000
5/8" dia.	"	1.000

COMMUNICATIONS

	UNIT	MAN/HOURS
16720.10 FIRE ALARM SYSTEMS		
Master fire alarm box, pedestal mounted	"	16.000
Master fire alarm box	"	6.015
Box light	"	0.500
Ground assembly for box	"	0.667
Bracket for pole type box	"	0.727
Pull station		
Waterproof	EA.	0.500
Manual	"	0.400
Horn, waterproof	"	1.000
Interior alarm	"	0.727
Coded transmitter, automatic	"	2.000
Control panel, 8 zone	"	8.000
Battery charger and cabinet	"	2.000
Batteries, nickel cadmium or lead calcium	"	5.000
CO2 pressure switch connection	"	0.727
Annunciator panels		
Fire detection annunciator, remote type, 8-zone	EA.	1.818
12-zone	"	2.000
16-zone	"	2.500
Fire alarm systems		
Bell	EA.	0.615
Weatherproof bell	"	0.667
Horn	"	0.727
Siren	"	2.000
Chime	"	0.615
Audio/visual	"	0.727
Strobe light	"	0.727
Smoke detector	"	0.667
Heat detection	"	0.500
Thermal detector	"	0.500
Ionization detector	"	0.533
Duct detector	"	2.759
Test switch	"	0.500
Remote indicator	"	0.571
Door holder	"	0.727
Telephone jack	"	0.296
Fireman phone	"	1.000
Speaker	"	0.800
Remote fire alarm annunciator panel		
24 zone	EA.	6.667
48 zone	"	13.008
Control panel		
12 zone	EA.	2.963
16 zone	"	4.444
24 zone	"	6.667
48 zone	"	16.000
Power supply	"	1.509
Status command	"	5.000
Printer	"	1.509
Transponder	"	0.899
Transformer	"	0.667
Transceiver	"	0.727
Relays	"	0.500
Flow switch	"	2.000

COMMUNICATIONS

16720.10 FIRE ALARM SYSTEMS

	UNIT	MAN/HOURS
Tamper switch	EA.	2.963
End of line resistor	"	0.348
Printed ckt. card	"	0.500
Central processing unit	"	6.154
UPS backup to c.p.u.	"	8.999
Smoke detector, fixed temp. & rate of rise comb.	"	1.600

16720.50 SECURITY SYSTEMS

	UNIT	MAN/HOURS
Sensors		
Balanced magnetic door switch, surface mounted	EA.	0.500
With remote test	"	1.000
Flush mounted	"	1.860
Mounted bracket	"	0.348
Mounted bracket spacer	"	0.348
Photoelectric sensor, for fence		
6 beam	EA.	2.759
9 beam	"	4.255
Photoelectric sensor, 12 volt dc		
500' range	EA.	1.600
800' range	"	2.000
Capacitance wire grid kit		
Surface	EA.	1.000
Duct	"	1.600
Tube grid kit	"	0.500
Vibration sensor, 30 max per zone	"	0.500
Audio sensor, 30 max per zone	"	0.500
Inertia sensor		
Outdoor	EA.	0.727
Indoor	"	0.500
Ultrasonic transmitter, 20 max per zone		
Omni-directional	EA.	1.600
Directional	"	1.333
Transceiver		
Omni-directional	EA.	1.000
Directional	"	1.000
Passive infra-red sensor, 20 max per zone	"	1.600
Access/secure control unit, balanced magnetic switch	"	1.600
Photoelectric sensor	"	1.600
Photoelectric fence sensor	"	1.600
Capacitance sensor	"	1.739
Audio and vibration sensor	"	1.600
Inertia sensor	"	1.600
Ultrasonic sensor	"	1.739
Infra-red sensor	"	2.000
Monitor panel, with access/secure tone, standard	"	1.739
High security	"	2.000
Emergency power indicator	"	0.500
Monitor rack with 115v power supply		
1 zone	EA.	1.000
10 zone	"	2.500
Monitor cabinet, wall mounted		
1 zone	EA.	1.000
5 zone	"	1.600
10 zone	"	1.739

COMMUNICATIONS

16720.50 SECURITY SYSTEMS

	UNIT	MAN/HOURS
20 zone	EA.	2.000
Floor mounted, 50 zone	"	4.000
Security system accessories		
Tamper assembly for monitor cabinet	EA.	0.444
Monitor panel blank	"	0.348
Audible alarm	"	0.500
Audible alarm control	"	0.348
Termination screw, terminal cabinet		
25 pair	EA.	1.600
50 pair	"	2.500
150 pair	"	5.000
Universal termination, cabinets & panel		
Remote test	EA.	1.739
No remote test	"	0.727
High security line supervision termination	"	1.000
Door cord for capacitance sensor, 12"	"	0.500
Insulation block kit for capacitance sensor	"	0.348
Termination block for capacitance sensor	"	0.348
Guard alert display	"	0.615
Uninterrupted power supply		
Plug-in 40kva transformer		
12 volt	EA.	0.348
18 volt	"	0.348
24 volt	"	0.348
Test relay	"	0.348
Coaxial cable, 50 ohm	L.F.	0.006
Door openers	EA.	0.500
Push buttons		
Standard	EA.	0.348
Weatherproof	"	0.444
Bells	"	0.727
Horns		
Standard	EA.	1.000
Weatherproof	"	1.250
Chimes	"	0.667
Flasher	"	0.615
Motion detectors	"	1.509
Intercom units	"	0.727
Remote annunciator	"	5.000

16730.20 CLOCK SYSTEMS

	UNIT	MAN/HOURS
Clock systems		
Single face	EA.	0.800
Double face	"	0.800
Skeleton	"	2.759
Master	"	5.000
Signal generator	"	4.000
Elapsed time indicator	"	0.800
Controller	"	0.533
Clock and speaker	"	1.096
Bell		
Standard	EA.	0.533
Weatherproof	"	0.800

COMMUNICATIONS

16730.20   CLOCK SYSTEMS

	UNIT	MAN/HOURS
Horn		
Standard	EA.	0.727
Weatherproof	"	0.952
Chime	"	0.533
Buzzer	"	0.533
Flasher	"	0.615
Control Board	"	3.478
Program unit	"	5.000
Block back box	"	0.500
Double clock back box	"	0.667
Wire guard	"	0.200

16740.10   TELEPHONE SYSTEMS

	UNIT	MAN/HOURS
Communication cable		
25 pair	L.F.	0.026
100 pair	"	0.029
150 pair	"	0.033
200 pair	"	0.040
300 pair	"	0.042
400 pair	"	0.044
Cable tap in manhole or junction box		
25 pair cable	EA.	3.810
50 pair cable	"	7.547
75 pair cable	"	11.268
100 pair cable	"	15.094
150 pair cable	"	22.222
200 pair cable	"	29.630
300 pair cable	"	44.444
400 pair cable	"	61.538
Cable terminations, manhole or junction box		
25 pair cable	EA.	3.756
50 pair cable	"	7.477
100 pair cable	"	15.094
150 pair cable	"	22.222
200 pair cable	"	29.630
300 pair cable	"	44.444
400 pair cable	"	61.538
Telephones, standard		
1 button	EA.	2.963
2 button	"	3.478
6 button	"	5.333
12 button	"	7.619
18 button	"	8.889
Hazardous area		
Desk	EA.	7.273
Wall	"	5.000
Accessories		
Standard ground	EA.	1.600
Push button	"	1.600
Buzzer	"	1.600
Interface device	"	0.800
Long cord	"	0.800
Interior jack	"	0.400

COMMUNICATIONS

16740.10   TELEPHONE SYSTEMS

	UNIT	MAN/HOURS
Exterior jack	EA.	0.615
Hazardous area		
Selector switch	EA.	3.200
Bell	"	3.200
Horn	"	4.211
Horn relay	"	3.077

16740.30   CALL SYSTEMS

	UNIT	MAN/HOURS
Call systems, single bed station	EA.	0.533
Double bed station	"	0.727
Call-in cord	"	0.200
Pull cord	"	0.200
Pillow speaker	"	0.276
Dome light	"	0.533
Zone light	"	0.533
Stake station	"	0.615
Duty station	"	0.500
Utility station	"	0.615
Nurses station	"	0.533
Surgical station	"	0.727
Master station	"	2.500
Control station	"	8.000
Annunciator	"	2.000
Power supply	"	1.538
Speakers	"	0.800
Foot switch	"	0.296
Code blue systems		
Bed station	EA.	0.727
Dome light	"	0.667
Zone light	"	0.727
Pull cord	"	0.250
Nurses station	"	0.533
Annunciator	"	2.000
Power supply	"	1.455
Nurse station indicator, alarm annunciators, flush		
4 circuit	EA.	4.000
6 circuit	"	8.000
12 circuit	"	12.012
Desktop		
4 circuit	EA.	3.478
6 circuit	"	3.478
12 circuit	"	5.000

16750.20   SIGNALING SYSTEMS

	UNIT	MAN/HOURS
Signaling systems		
4" bell	EA.	0.602
6" bell	"	0.650
10" bell	"	0.748
Buzzer		
Size 0	EA.	0.444
Size 1	"	0.444
Size 2	"	0.500
Size 3	"	0.533
Horn	"	0.615

16750.20 SIGNALING SYSTEMS

COMMUNICATIONS	UNIT	MAN/HOURS
Chime	EA.	0.533
Push button		
Standard	EA.	0.400
Weatherproof	"	0.500
Door opener		
Mortise	EA.	0.500
Rim	"	0.400
Transformer	"	0.444
Contractor grade doorbell chime kit		
Chime	EA.	1.000
Doorbutton	"	0.320
Transformer	"	0.500

16770.30 SOUND SYSTEMS

COMMUNICATIONS	UNIT	MAN/HOURS
Power amplifiers	EA.	3.478
Pre-amplifiers	"	2.759
Tuner	"	1.455
Horn		
Equilizer	EA.	1.600
Mixer	"	2.222
Tape recorder	"	1.860
Microphone	"	1.000
Cassette Player	"	2.162
Record player	"	1.905
Equipment rack	"	1.290
Speaker		
Ceiling	EA.	1.356
Wall	"	4.000
Paging	"	0.800
Column	"	0.533
Single	"	0.615
Double	"	4.444
Volume control	"	0.533
Plug-in	"	0.800
Desk	"	0.400
Outlet	"	0.400
Stand	"	0.296
Console	"	8.000
Power supply	"	1.290

16780.10 ANTENNAS AND TOWERS

COMMUNICATIONS	UNIT	MAN/HOURS
Guy cable, alumaweld		
1x3, 7/32"	L.F.	0.050
1x3, 1/4"	"	0.050
1x3, 25/64"	"	0.059
1x19, 1/2"	"	0.070
1x7, 35/64"	"	0.080
1x19, 13/16"	"	0.100
Preformed alumaweld end grip		
1/4" cable	EA.	0.100
3/8" cable	"	0.100
1/2" cable	"	0.145
9/16" cable	"	0.200

16780.10 ANTENNAS AND TOWERS

COMMUNICATIONS	UNIT	MAN/HOURS
5/8" cable	EA.	0.250
Fiberglass guy rod, white epoxy coated		
1/4" dia.	L.F.	0.145
3/8" dia	"	0.145
1/2" dia	"	0.200
5/8" dia	"	0.250
Preformed glass grip end grip, guy rod		
1/4" dia.	EA.	0.145
3/8" dia.	"	0.200
1/2" dia.	"	0.250
5/8" dia.	"	0.250
Spelter socket end grip, 1/4" dia. guy rod		
Standard strength	EA.	0.500
High performance	"	0.500
3/8" dia. guy rod		
Standard strength	EA.	0.348
High performance	"	0.500
Timber pole, Douglas Fir		
80-85 ft	EA.	19.512
90-95 ft	"	22.222
Southern yellow pine		
35-45 ft	EA.	10.959
50-55 ft	"	14.035

16780.50 TELEVISION SYSTEMS

COMMUNICATIONS	UNIT	MAN/HOURS
TV outlet, self terminating, w/cover plate	EA.	0.308
Thru splitter	"	1.600
End of line	"	1.333
In line splitter multitap		
4 way	EA.	1.818
2 way	"	1.702
Equipment cabinet	"	1.600
Antenna		
Broad band uhf	EA.	3.478
Lightning arrester	"	0.727
TV cable	L.F.	0.005
Coaxial cable rg	"	0.005
Cable drill, with replacement tip	EA.	0.500
Cable blocks for in-line taps	"	0.727
In-line taps ptu-series 36 tv system	"	1.143
Control receptacles	"	0.449
Coupler	"	2.424
Head end equipment	"	6.667
TV camera	"	1.667
TV power bracket	"	0.800
TV monitor	"	1.455
Video recorder	"	2.105
Console	"	8.502
Selector switch	"	1.379
TV controller	"	1.404

RESISTANCE HEATING	UNIT	MAN/HOURS
16850.10 ELECTRIC HEATING		
Baseboard heater		
2', 375w	EA.	1.000
3', 500w	"	1.000
4', 750w		1.143
5', 935w	"	1.333
6', 1125w	"	1.600
7', 1310w	"	1.818
8', 1500w	"	2.000
9', 1680w	"	2.222
10', 1875w	"	2.286
Unit heater wall mounted		
1500w	EA.	1.600
2500w	"	1.818
4000w	"	2.286
Thermostat		
Integral	EA.	0.500
Line voltage	"	0.500
Electric heater connection	"	0.250
Fittings		
Inside corner	EA.	0.400
Outside corner	"	0.400
Receptacle section	"	0.400
Blank section	"	0.400
Unit heater, wall mounted		
750w	EA.	1.600
1500w	"	1.667
2000w	"	1.739
3000w	"	2.000
Infrared heaters		
600w	EA.	1.000
2000w	"	1.194
3000w	"	2.000
4000w	"	2.500
Controller	"	0.667
Wall bracket	"	0.727
Radiant ceiling heater panels		
500w	EA.	1.000
750w	"	1.000
Unit heaters, suspended, single phase		
3.0 kw	EA.	2.759
5.0 kw	"	2.759
7.5 kw	"	3.200
10.0 kw	"	3.810
Three phase		
5 kw	EA.	2.759
7.5 kw	"	3.200
10 kw	"	3.810
15 kw	"	4.211
20 kw	"	5.333
25 kw	"	6.400
30 kw	"	8.000
35 kw	"	8.000
Unit heater thermostat	"	0.533
Mounting bracket	"	0.727

RESISTANCE HEATING	UNIT	MAN/HOURS
16850.10 ELECTRIC HEATING		
Relay	EA.	0.615
Duct heaters, three phase		
10 kw	EA.	3.810
15 kw	"	3.810
17.5 kw	"	4.000
20 kw	"	6.154

CONTROLS	UNIT	MAN/HOURS
16910.40 CONTROL CABLE		
Control cable, 600v, #14 THWN, PVC jacket		
2 wire	L.F.	0.008
4 wire	"	0.010
6 wire	"	0.131
8 wire	"	0.145
10 wire	"	0.160
12 wire	"	0.182
14 wire	"	0.211
16 wire	"	0.222
18 wire	"	0.242
20 wire	"	0.250
22 wire	"	0.286
Audio cables, shielded, #24 gauge		
3 conductor	L.F.	0.004
4 conductor	"	0.006
5 conductor	"	0.007
6 conductor	"	0.009
7 conductor	"	0.011
8 conductor	"	0.012
9 conductor	"	0.014
10 conductor	"	0.015
15 conductor	"	0.018
20 conductor	"	0.023
25 conductor	"	0.027
30 conductor	"	0.030
40 conductor	"	0.036
50 conductor	"	0.042
#22 gauge		
3 conductor	L.F.	0.004
4 conductor	"	0.006
#20 gauge		
3 conductor	L.F.	0.004
10 conductor	"	0.015
15 conductor	"	0.018
#18 gauge		
3 conductor	L.F.	0.004
4 conductor	"	0.006

CONTROLS	UNIT	MAN/ HOURS
16910.40 CONTROL CABLE		
Microphone cables, #24 gauge		
2 conductor	L.F.	0.004
3 conductor	"	0.005
#20 gauge		
1 conductor	L.F.	0.004
2 conductor	"	0.004
2 conductor	"	0.004
3 conductor	"	0.006
4 conductor	"	0.007
5 conductor	"	0.009
7 conductor	"	0.011
8 conductor	"	0.012
Computer cables shielded, #24 gauge		
1 pair	L.F.	0.004
2 pair	"	0.004
3 pair	"	0.006
4 pair	"	0.007
5 pair	"	0.009
6 pair	"	0.011
7 pair	"	0.012
8 pair	"	0.014
50 pair	"	0.039
Coaxial cables		
RG 6/u	L.F.	0.006
RG 6a/u	"	0.006
RG 8/u	"	0.006
RG 8a/u	"	0.006
RG 9/u	"	0.006
RG 11/u	"	0.006
RG 58/u	"	0.006
RG 59/u	"	0.006
RG 62/u	"	0.006
RG 174/u	"	0.006
RG 213/u	"	0.006
MATV and CCTV camera cables		
1 conductor	L.F.	0.004
2 conductor	"	0.005
4 conductor	"	0.006
7 conductor	"	0.009
12 conductor	"	0.015
13 conductor	"	0.016
14 conductor	"	0.018
28 conductor	"	0.027
Fire alarm cables, #22 gauge		
6 conductor	L.F.	0.010
9 conductor	"	0.015
12 conductor	"	0.016
#18 gauge		
2 conductor	L.F.	0.005
4 conductor	"	0.007
#16 gauge		
2 conductor	L.F.	0.007
4 conductor	"	0.008
#14 gauge		

CONTROLS	UNIT	MAN/ HOURS
16910.40 CONTROL CABLE		
2 conductor	L.F.	0.008
#12 gauge		
2 conductor	L.F.	0.010
Plastic jacketed thermostat cable		
2 conductor	L.F.	0.004
3 conductor	"	0.005
4 conductor	"	0.006
5 conductor	"	0.008
6 conductor	"	0.009
7 conductor	"	0.012
8 conductor	"	0.013

BNi Building News

Supporting Construction Reference Data

This section contains information, text, charts and tables on various aspects of construction. The intent is to provide the user with a better understanding of unfamiliar areas in order to be able to estimate better. This information includes actual takeoff data for some areas and also selected explanations of common construction materials, methods and common practices.

TYPICAL BUILDING COST BROKEN DOWN
BY CSI FORMAT
(Commercial Construction)

Division		New Construction	Remodeling Construction
1.	General Requirements	6 to 8%	Up to 30%
2.	Sitework	4 to 6%	
3.	Concrete	15 to 20%	
4.	Masonry	8 to 12%	
5.	Metals	5 to 7%	
6.	Wood And Plastics	1 to 5%	
7.	Thermal And Moisture Protection	4 to 6%	
8.	Doors And Windows	5 to 7%	Up to 30%
9.	Finishes	8 to 12%	
10.	Specialties		
11.	Architectural Equipment		
12.	Furnishings	6 to 10%	
13.	Special Construction		
14.	Conveying Systems		
15.	Mechanical	15 to 25%	Up to 40%
16.	Electrical	8 to 12%	
	Total Cost	100%	

CONVERSION FACTORS

Change	To	Multiply By
Atmospheres	Pounds per square inch	14.696
Atmospheres	Inches of mercury	29.92
Atmospheres	Feet of water	34
Barrels, oil	Gallons, of oil	42
Barrels, cement	Pounds of cement	376
Bags or sacks, cement	Pounds of cement	94
Btu/min.	Foot-pounds/sec.	12.96
Btu/min.	Horse-power	0.02356
Btu/min.	Kilowatts	0.01757
Btu/min.	Watts	17.57
Centimeters	Inches	0.3937
Centimeters of mercury	Atmospheres	0.01316
Centimeters of mercury	Feet of water	0.4461
Cubic inches	Cubic feet	0.00058
Cubic feet	Cubic inches	1728
cubic feet	Cubic yards	0.03703
Cubic yards	Cubic feet	27
Cubic inches	Gallons	0.00433
Cubic feet	Gallons	7.48
Feet	Inches	12
Feet	Yards	0.3333
Yards	Feet	3
Feet of water	Atmospheres	0.02950
Feet of water	Inches of mercury	0.8826
Gallons	Cubic Inches	231
Gallons	Cubic feet	0.1337
Gallons	Pounds of water	8.33
Gallons	Quarts	4
Gallons per min.	Cubic feet sec.	0.002228
Gallons per min.	Cubic feet hour	8.0208
Gallons water per min.	Tons water/24 hours	6.0086
Horse-power	Foot-lbs./sec.	550
Inches	Centimeters	2.540
Inches	Feet	0.0833
Inches	Millimeters	25.4
Inches of water	Pounds per Sq. inch	0.0361
Inches of water	Inches of mercury	0.0735
Inches of water	Ounces per square inch	0.578
Inches of water	Ounces per square foot	5.2
Inches of mercury	Inches of water	13.6
Inches of mercury	Feet of water	1.1333
Inches of mercury	Pounds per square inch	0.4914
Kilometers	Miles	0.6214
Meters	Inches	39.37
Miles	Feet	5280
Millimeters	Centimeters	0.1
Millimeters	Inches	0.03937
Ounces (fluid)	Cubic inches	1.805
Ounces	Pounds	0.0625
Pounds	Ounces	16
Pounds per square inch	Inches of water	27.72
Pounds per square inch	Feet of water	2.310
Pounds per square inch	Inches of mercury	2.04
Pounds per square inch	Atmospheres	0.0681
Quarts	Cubic Inches	67.20
Square Inches	Square feet	0.00694
Square Feet	Square inches	144
Square Feet	Square yards	0.11111
Square yards	Square feet	9
Square miles	Acres	640
Short tons	Pounds	2000
Short tons	Long tons	0.89285
Tons of water/24 hours	Gallons per minute	0.16643
Yards	Feet	3
Yards	Centimeters	91.44
Yards	Inches	36

CONVERSION CALCULATIONS

Commercial Measure

16 drams	= 1 ounce
16 ounces	= 1 pound
2,000 pounds	= 1 ton

Long Measure

12 inches	= 1 foot
3 feet	= 1 yard
16½ feet	= 1 rod
40 rods	= 1 furlong
8 furlongs (5,280 ft) =	= 1 mile
3 miles	= 1 league

Square Measure

144 square inches	= 1 square foot
9 square feet	= 1 square yard
30¼ square yards	= 1 square rod
160 square rods	= 1 acre
4840 square yards	= 1 acre
640 acres	= 1 square mile
36 square miles	= 1 township

Surveyors Measure

7.92 inches	= 1 link
25 links	= 1 rod
4 rods (66 ft.)	= 1 chain
10 chains	= 1 furlong
8 furlongs	= 1 mile
1 square mile	= 1 section

Cubic Measure

1728 cubic inches	= 1 cubic foot
27 cubic feet	= 1 cubic yard
128 cubic feet	= 1 cord (wood/stone)
231 cubic inches	= 1 U.S. gallon
7.48 U.S. Gallons	= 1 cubic foot
2150.4 cubic inches	= 1 U.S. bushel

Liquid Measure

4 fluid ounces	= 1 gill
4 gills	= 1 pint
2 pints	= 1 quart
4 quarts	= 1 gallon
9 gallons	= 1 firkin
31½ gallons	= 1 barrel
2 barrels	= 1 hogshead

Dry Measure

2 pints	= 1 quart
8 quarts	= 1 peck
4 pecks	= 1 bushel
2150.42 cubic inches	= 1 bushel

SQUARE

$$A = a^2$$

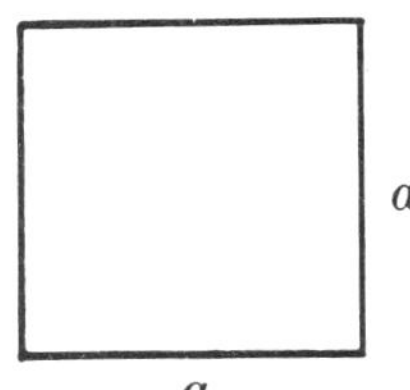

RECTANGLE

$$A = bh$$

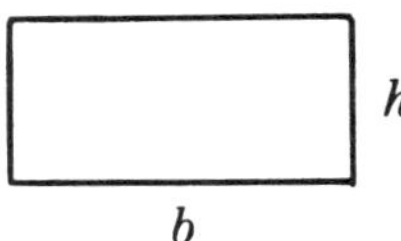

TRIANGLE

$$A = \frac{1}{2}bh$$

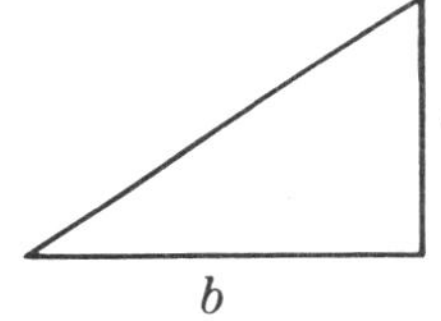
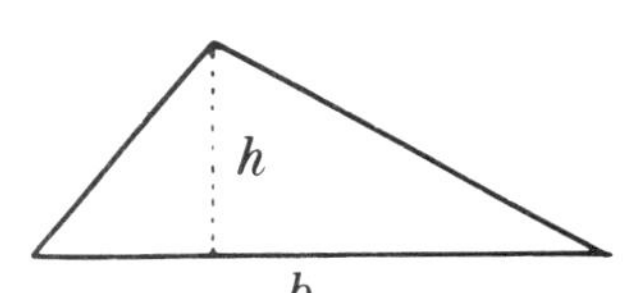

PARALLELOGRAM

$$A = bh = ab\ Sin\phi$$

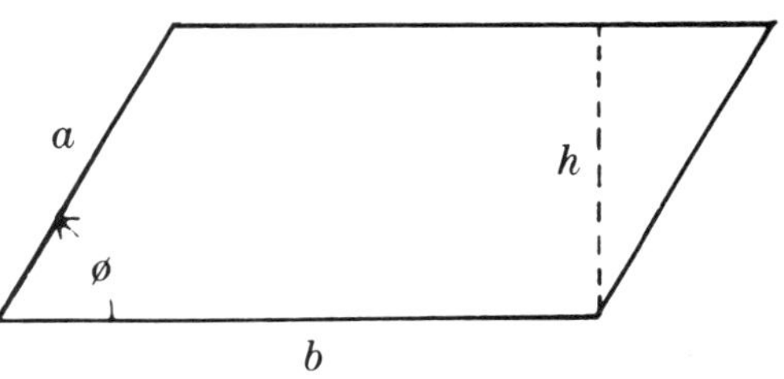

TRAPEZOID

$$A = \left(\frac{a+b}{2}\right)h$$

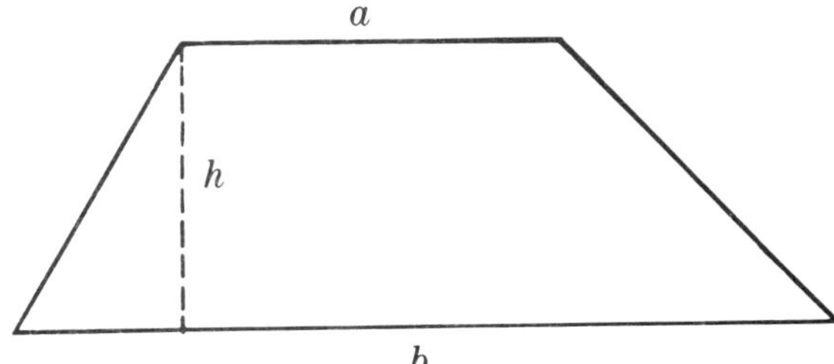

CIRCLE

$$A = \pi r^2 = \frac{\pi d^2}{4}$$

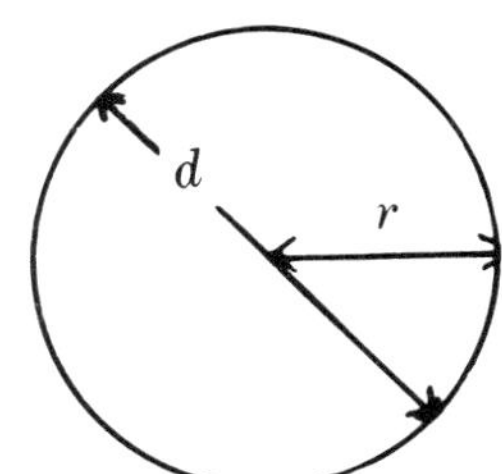

$$Circumference = C = 2\pi r = \pi d$$

ELLIPSE
$$A = 0.7854\, ab$$

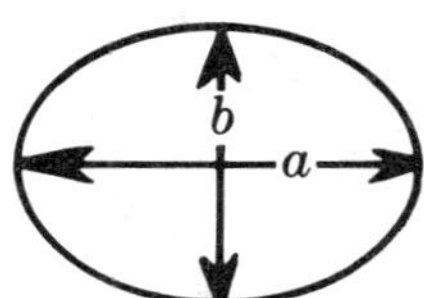

PARABOLA
$$A = \frac{2}{3} bh$$

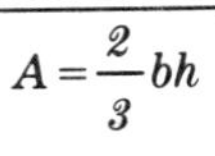
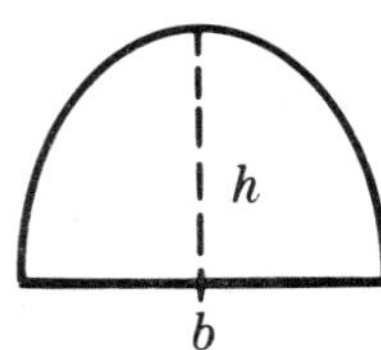

VOLUMES

CUBE
$$V = a^3$$

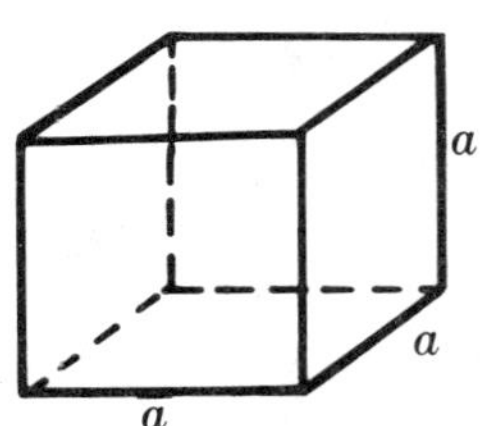

CYLINDER
$$V = \pi r^2 h$$

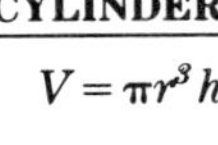

PYRAMID
$$V = \frac{1}{3}(Base)\, h$$

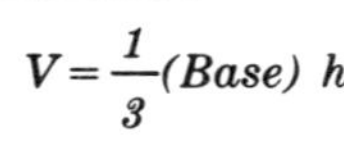
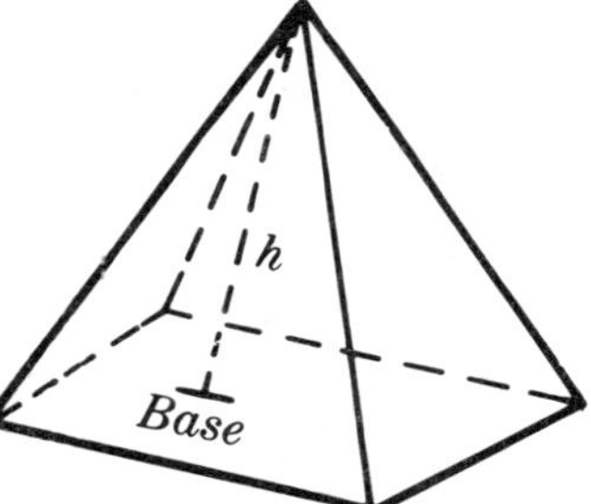

CONE
$$V = \frac{1}{3} \pi r^2 h$$

SPHERE
$$V = \frac{4}{3} \pi r^3 = \frac{1}{6} \pi d^3$$

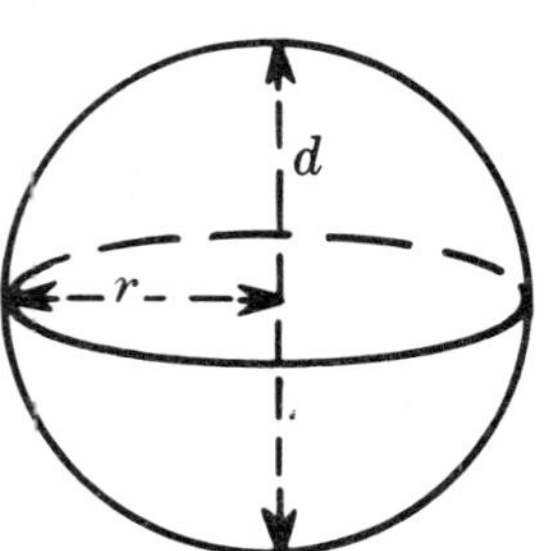

WEDGE
$$V = \frac{1}{2} abc$$

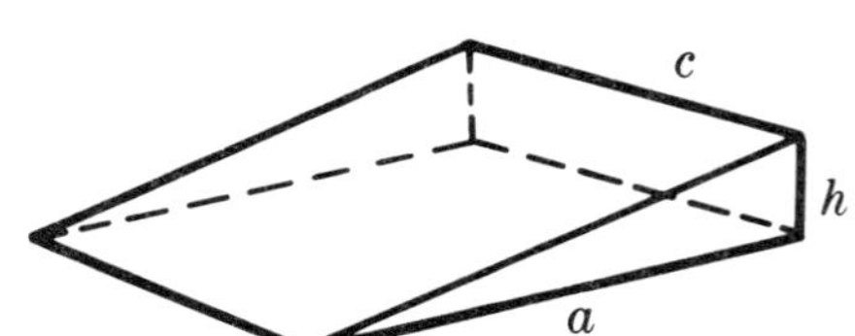

CONVERSION FACTORS
ENGLISH TO SI (SYSTEM INTERNATIONAL)

To Convert from	To	Multiply by
LENGTH		
Inches	Millimetres	25.4[a]
Feet	Metres	0.3048[a]
Yards	Metres	0.9144[a]
Miles (statute)	Kilometres	1.609
AREA		
Square inches	Square millimetres	645.2
Square feet	Square metres	0.0929
Square yards	Square metres	0.8361
VOLUME		
Cubic inches	Cubic millimetres	16.387
Cubic feet	Cubic metres	0.02832
Cubic yards	Cubic metres	0.7646
Gallons (U.S. liquid)[b]	Cubic metres[c]	0.003785
Gallons (Canadian liquid)[b]	Cubic metres[c]	0.004546
Ounces (U.S. liquid)[b]	Millilitres[c, d]	29.57
Quarts (U.S. liquid)[b]	Litres[c, d]	0.9464
Gallons (U.S. liquid)[b]	Litres[c]	3.785
FORCE		
Kilograms force	Newtons	9.807
Pounds force	Newtons	4.448
Pounds force	Kilograms force[d]	0.4536
Kips	Newtons	4448
Kips	Kilograms force[d]	453.6
PRESSURE, STRESS, STRENGTH (FORCE PER UNIT AREA)		
Kilograms force per sq. centimetre	Megapascals	0.09807
Pounds force per square inch (psi)	Megapascals	6895
Kips per square inch	Megapascals	6.895
Pounds force per square inch (psi)	Kilograms force per square centimetre[d]	0.07031
Pounds force per square foot	Pascals	47.88
Pounds force per square foot	Kilograms force per square metre[d]	4.882
BENDING MOMENT OR TORQUE		
Inch-pounds force	Metre-kilog. force[d]	0.01152
Inch-pounds force	Newton-metres	0.1130
Foot-pounds force	Metre-kilog. force[d]	0.1383
Foot-pounds force	Newton-metres	1.356
Metre-kilograms force	Newton-metres	9.807
MASS		
Ounce (avoirdupois)	Grams	28.35
Pounds (avoirdupois)	Kilograms	0.4536
Tons (metric)	Kilograms	1000[a]
Tons, short (2000 pounds)	Kilograms	907.2
Tons, short (2000 pounds)	Megagrams[e]	0.9072
MASS PER UNIT VOLUME		
Pounds mass per cubic foot	Kilog. per cubic metre	16.02
Pounds mass per cubic yard	Kilog. per cubic metre	0.5933
Pds. mass per gallon (U.S. liquid)[b]	Kilog. per cubic metre	119.8
Pds. mass p/gal. (Canadian liquid)[b]	Kilog. per cubic metre	99.78
TEMPERATURE		
Degrees Fahrenheit	Degrees Celsius	$tK = (1F - 32)/1.8$
Degrees Fahrenheit	Degrees Kelvin	$tK = (1F + 459.67)/1.8$
Degree Celsius	Degree Kelvin	$tK = 1C + 273.15$

[a] The factor given is exact.
[b] One U.S. gallon equals 0.8327 Canadian gallon.
[c] 1 litre = 1000 millilitres = 10,000 cubic centimetres = 1 cubic decimetre = 0.001 cubic metre.
[d] Metric but not SI unit.
[e] Called "tonne" in England. Called "metric ton" in other metric systems.

TRENCH BRACING

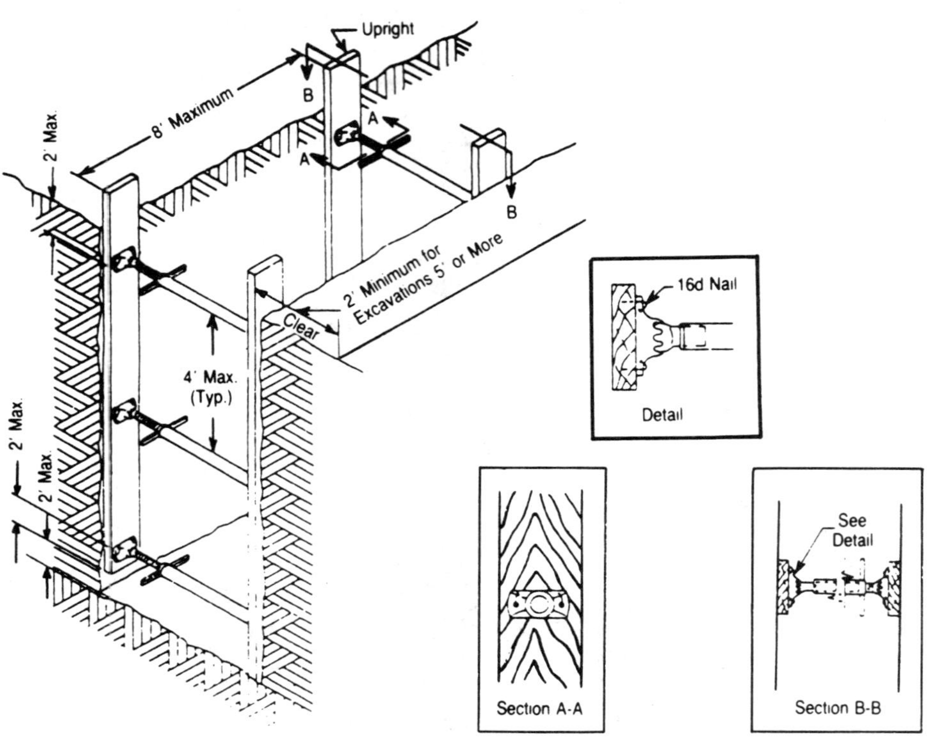

CLOSED VERTICAL SHEETING

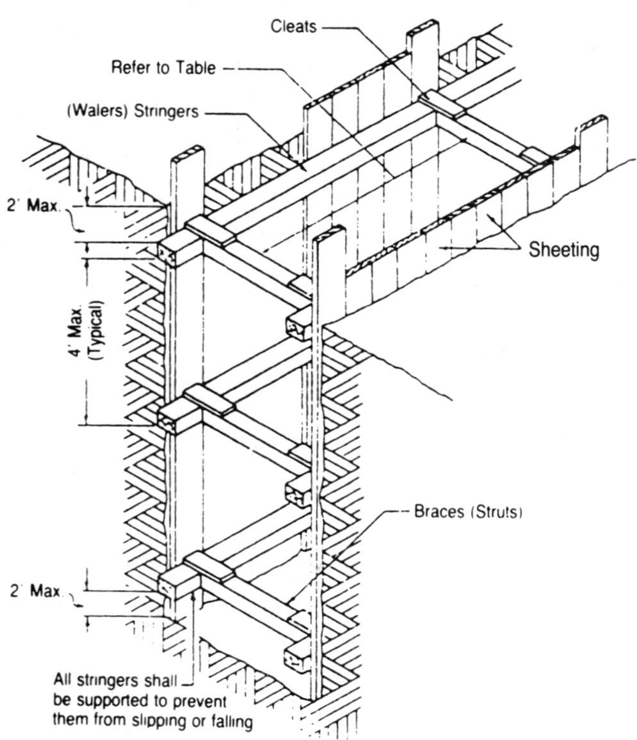

STANDARD NOMENCLATURE
FOR STREET CONSTRUCTION

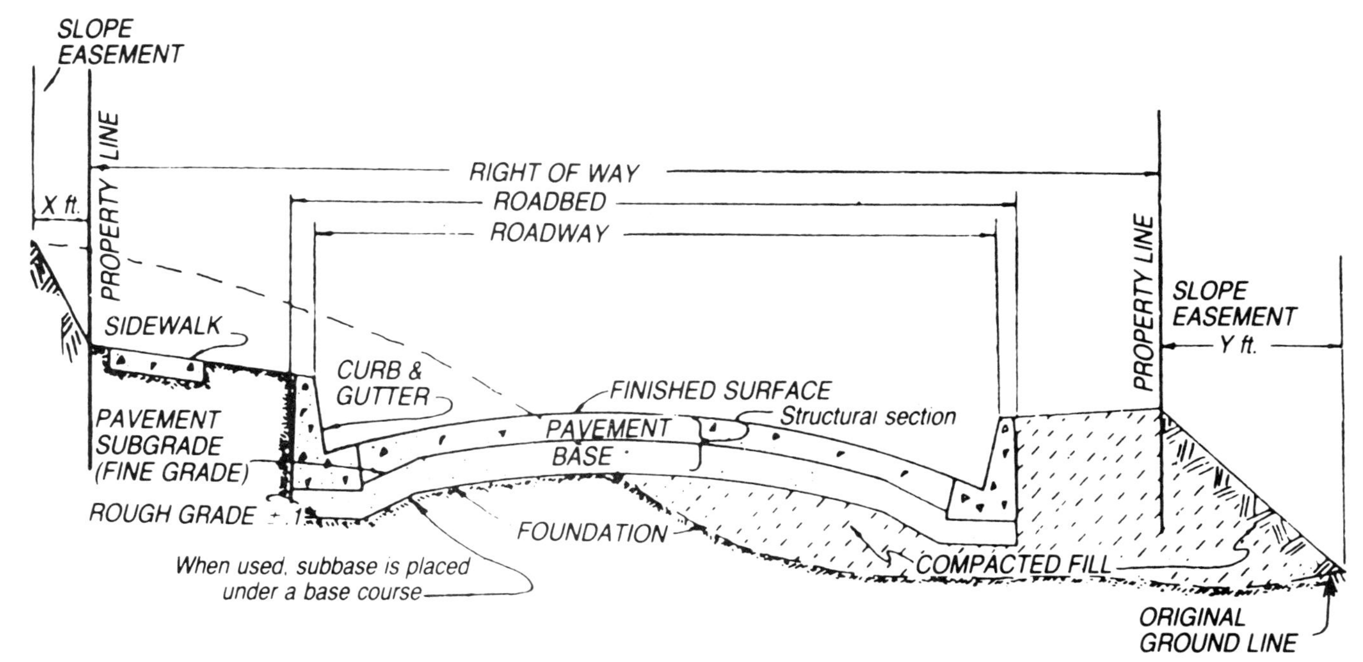

PIPE

Clay Pipe. Clay pipe is manufactured by blending various clays together, milling, mixing, extruding and firing in a kiln to obtain vitrification. The physical properties of the pipe can be changed by varying the proportions of the several clays used. The pipe is supplied in two basic styles: spigot and socket; and plain end.

Spigot and Socket Pipe has a spigot on one end and a socket on the other, and is commonly referred to as "bell and spigot" pipe. The plans generally specify the type of joint to be used from the several types of jointing methods available. This type of pipe is manufactured with matching polyurethane gaskets molded on the spigot and socket which form a tight seal when the pipe is jointed.

Plain End Pipe is without a socket on either end and is joined with special couplings. The coupling consists of a circular rubber sleeve, two stainless steel compression bands with tightening devices and a corrosion resistant shear ring. Sometimes this joint is supplied with a cardboard form, open at the top, which is filled with portland cement mortar to resist shear and prevent future corrosion of the bands.

Concrete Pipe. Unreinforced and reinforced concrete pipe is manufactured by casting in stationary or revolving metal molds. At the present time, the design practice is to specify reinforced concrete pipe for all purposes.

Unreinforced Concrete Pipe is cast in vertical steel molds, usually in pipe sizes of 21 inches or less, and is of the spigot and socket type. No steel reinforcement is used and the pipe is usually intended for use in irrigation systems and under light loading conditions.

Reinforced Concrete Pipe (RCP) is made in a number of different manufacturing processes and for a wide variety of pressure and non-pressure classes. It is available in standard sizes or it can be made to order to any diameter desired. Some of the larger diameters include diameters of 12 and 14 feet. A large variety of joint details are used with RCP. Tongue and groove joints are used for storm drain pipelines.

Reinforced concrete pipe for wastewater pipeline projects is supplied with gasketed joints and a polyvinyl chloride (PVC) plastic liner cast into the pipe.

(a) **Cast Pipe** is cast vertically in steel forms with the reinforcing cage securely held in place. The reinforcement is generally elliptical in shape to provide the maximum structural strength to resist the loads imposed on the pipe by the backfill and other stresses. Consolidation of the concrete is obtained by the use of external form vibrators.

(b) **Centrifugally Spun Pipe** is manufactured by introducing concrete into a spinning horizontal steel cylinder into which the reinforcement cage has been previously installed and which is equipped with end dams to provide the proper pipe wall thickness. The speed of rotation of the mold is increased and the centrifugal force produces a smooth, dense concrete pipe.

(c) **Pressure Pipe** may be cast or centrifugally spun pipe but it usually has a circular steel reinforcement cage (or cages) designed not only to resist the trench loading, but also the internal pressures exerted on the pipe from the fluid under pressure in the line.

Concrete Cylinder Pipe. This class of pipe is generally used for high pressure water lines and sewer force mains and is available in sizes ranging from 10 inches to 60 inches and larger in special cases.

A sheet steel cylinder is wrapped with the designed steel reinforcement and a concrete lining is centrifugally spun in the interior of the steel cylinder. An exterior coating of concrete is applied generally by the gunite process, while the cylinder is slowly rotated. These coatings vary in thickness from ½ to ¾ of an inch. The joints are commonly of the steel ring and rubber gasket type, but are generally designed for the special purpose for which the pipe line is intended.

Definition of Terms. In general, the terms used to designate types of reinforced concrete pipe refer to the process used in manufacture.

Cast RCP (Cast Reinforced Concrete Pipe). A concrete pipe having one or more cylindrical or elliptical cages of reinforcement steel embedded in it, the concrete for which is cast with the forms in a vertical position.

CSRCP (Centrifugally Spun Reinforced Concrete Pipe). A concrete pipe having one or more cylindrical or elliptical cages of reinforcement steel embedded in it, and cast in a horizontal position while the forms are spinning rapidly. This type of pipe may be designated as Spun RCP or as CCP (Centrifugal Concrete Pipe).

RCP (Reinforced Concrete Pipe). A reinforced concrete pipe manufactured by either the casting or spinning method.

Steel Reinforcement. Steel for reinforcing concrete pipe is generally furnished in large coils which will permit the use of machines to fabricate the "cages." The continuous steel rod is wound spirally at a prescribed pitch on a drum of the proper diameter. Where the rod crosses a longitudinal spacer rod, it is electrically welded to it so that the complete cage is relatively rigid.

Reinforcement cages for pipe designed for external loading are generally elliptical in shape to take full advantage of the steel in tension. Pipe to be used with relatively small external loads or pipe designed for pressure lines will have circular cages.

Reinforcement cages must be rigidly fixed in the forms so that the placement of concrete or the effects of centrifugal spinning will not result in distortion or displacement of the steel. The orientation of an elliptical cage must be marked on the forms to assure that the minor axis can be located after the concrete is placed.

CAPACITIES FOR SEPTIC TANKS SERVING
AN INDIVIDUAL DWELLING

No. of bedrooms	Capacity of tank (gals.)
2 or less	750
3	900
4	1,000

STANDARD SIZES OF STEEL REINFORCEMENT BARS

STANDARD REINFORCEMENT BARS				
		Nominal Dimensions		
Bar Designation Number*	Nominal Weight, lb. per ft.	Diameter, in.	Cross Sectional Area, sq. in.	Perimeter, in.
3	0.376	0.375	0.11	1.178
4	0.668	0.500	0.20	1.571
5	1.043	0.625	0.31	1.963
6	1.502	0.750	0.44	2.356
7	2.044	0.875	0.60	2.749
8	2.670	1.000	0.79	3.142
9	3.400	1.128	1.00	3.544
10	4.303	1.270	1.27	3.990
11	5.313	1.410	1.56	4.430
14	7.65	1.693	2.25	5.32
18	13.60	2.257	4.00	7.09

*The bar numbers are based on the number of ⅛ inches included in the nominal diameter of the bar.

Type of Steel and ASTM Specification No.	Size Nos. Inclusive	Grade	Tensile Strength Min., psi	Yield (a) Min., psi
Billet Steel A 615	3-11	40	70,000	40,000
	3-11 14, 18	60	90,000	60,000
	11, 14, 18	75	100,000	75,000

WELDED WIRE FABRIC – COMMON STOCK STYLES OF WELDED WIRE FABRIC

Style Designation	Steel Area sq. in. per ft.		Weight Approx. lbs. per 100 sq. ft.
	Longit.	Transv.	
Rolls			
6x6—W1.4xW1.4	.03	.03	21
6x6—W2xW2	.04	.04	29
6x6—W2.9xW2.9	.06	.06	42
6x6—W4xW4	.08	.08	58
4x4—W1.4xW1.4	.04	.04	31
4x4—W2xW2	.06	.06	43
4x4—W2.9xW2.9	.09	.09	62
4x4—W4xW4	.12	.12	86
Sheets			
6x6—W2.9xW2.9	.06	.06	42
6x6—W4xW4	.08	.08	58
6x6—W5.5xW5.5	.11	.11	80
4x4—W4xW4	.12	.12	86

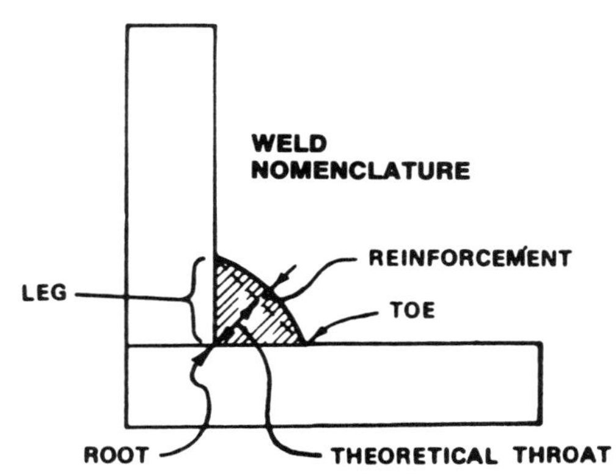

WELDED JOINTS

SQUARE BUTT

SINGLE VEE BUTT

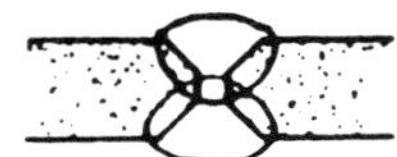

DOUBLE VEE BUTT

SINGLE U BUTT

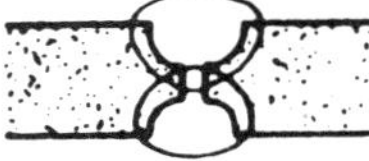

DOUBLE U BUTT

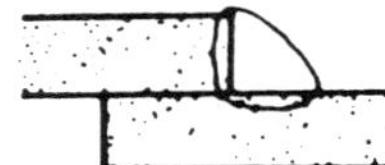

SINGLE FILLET LAP

DOUBLE FILLET LAP

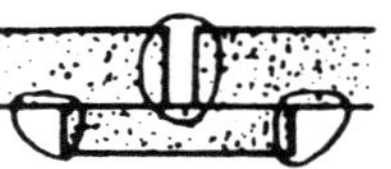

STRAP JOINT

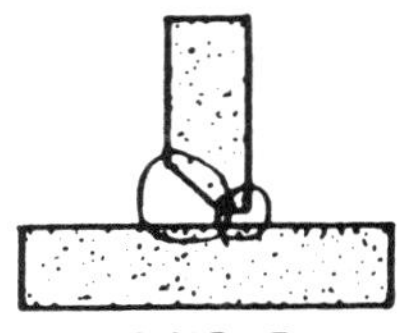

SINGLE
BEVEL TEE

DOUBLE
BEVEL TEE

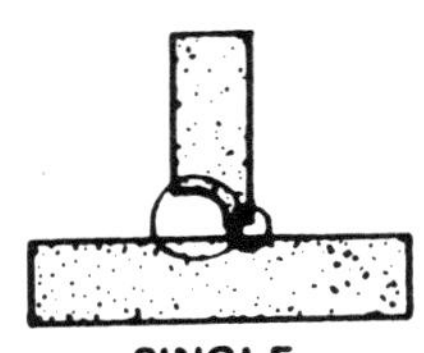

SINGLE
J TEE

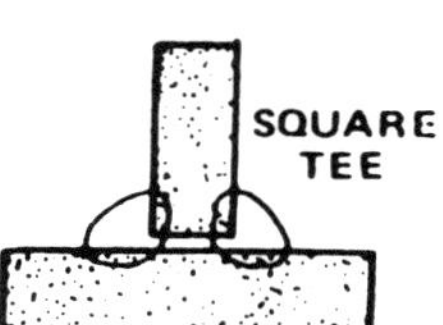

DOUBLE J TEE

CLOSED CORNER
(FLUSH) JOINT

HALF OPEN
CORNER JOINT

WELDING POSITIONS

FLAT (F)

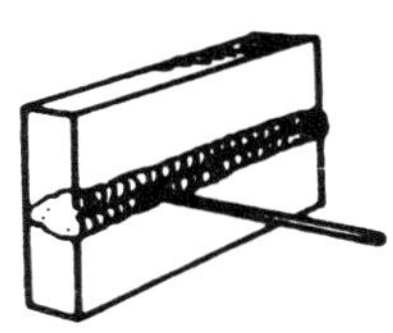

HORIZONTAL (H)

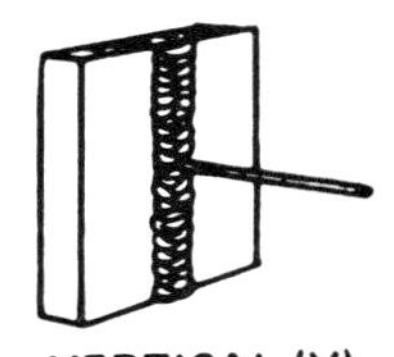

VERTICAL (V)

OVERHEAD (OH)

BOLTS IN COMMON USAGE

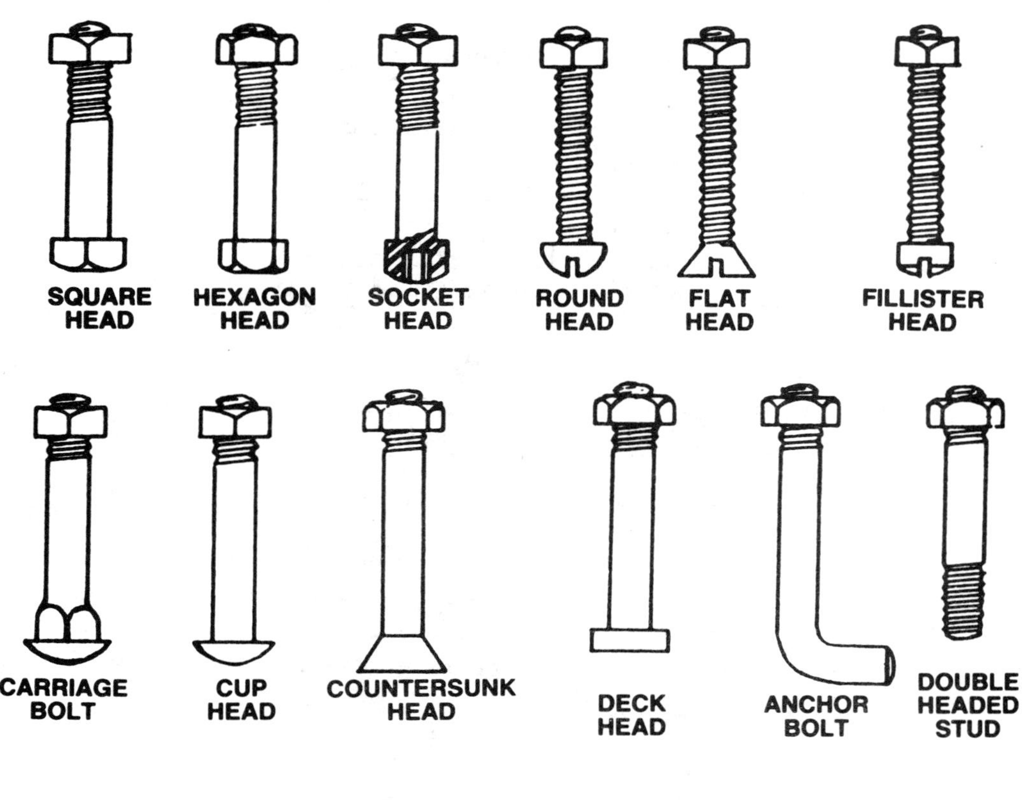

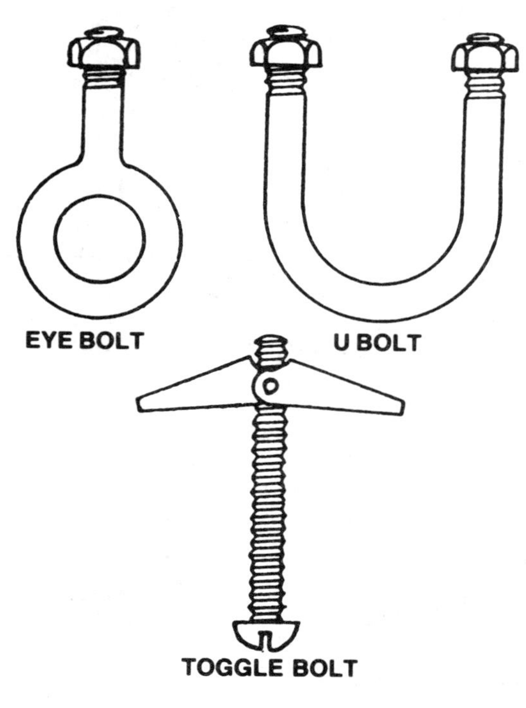

DOWNSPOUT/VERTICAL LEADER CALCULATIONS

Roof Type	Slope	S.F. Roof/ Sq. In. Leader
Gravel	Less than ¼″ per foot	300
Gravel	Greater than ¼″ per foot	250
Metal or Shingle	Any	200

Alternate calculations:

$$\text{Diameter of downspout/leader} = 1.128 \sqrt{\frac{\text{Area of drainage}}{\text{SF Roof/Sq. Inch}}}$$

TYPICAL MINIMUM SIZE OF VERTICAL CONDUCTORS AND LEADERS

Size of leader or conductor (Inches)	Maximum projected roof area (Square feet)
2	544
2½	987
3	1,610
4	3,460
5	6,280
6	10,200
8	22,000

TYPICAL MINIMUM SIZE OF ROOF GUTTERS

Diameter gutter (Inches)	Maximum projected roof area for gutters of various slopes			
	1/16 in. per Ft. slope (Sq. ft.)	1/8 in. per Ft. slope (Sq. ft.)	1/4 in. per Ft. slope (Sq. ft.)	1/2 in. per Ft. slope (Sq. ft.)
3	170	240	340	480
4	360	510	720	1020
5	625	880	1250	1770
6	960	1360	1920	2770
7	1380	1950	2760	3900
8	1990	2800	3980	5600
10	3600	5100	7200	10000

PIPE WEIGHTS

CAST IRON PIPE

SERVICE WEIGHT

Size, Inches	Weight of Pipe	Weight of Water	Total Weight-Lbs.
2″	3.8	1.45	5.3
3″	5.6	3.2	8.8
4″	7.5	5.5	13.0
5″	9.8	8.7	18.5
6″	12.4	12.5	24.9
8″	18.5	21.7	40.2

EXTRA HEAVY

Size, Inches	Weight of Pipe	Weight of Water	Total Weight-Lbs.
2″	4.3	1.45	5.8
3″	8.3	3.2	11.5
4″	10.8	5.5	16.3
5″	13.3	8.7	22.0
6″	16.0	12.5	28.5
8″	26.5	21.7	48.2

STEEL PIPE

Pipe Size	W/40	H_2O/Lbs.	Total Lbs./L.F.
2″	3.65	1.45	5.1
2½″	5.79	2.07	7.86
3″	7.57	3.2	10.77
3½″	9.11	4.28	13.39
4″	10.8	5.51	16.31
5″	14.6	8.66	23.26
6″	18.0	12.5	30.5
8″	28.6	21.66	50.26
10″	40.5	34.15	74.65
12″			

SPRINKLER AREA CALCULATIONS

Typical maximum floor area allowed per system riser:

Light Hazard	**Ordinary Hazard**	**Extra Hazard**
52,000 S.F.	40,000-52,000 S.F.	25,000 S.F.

Typical maximum floor area coverage allowed per sprinkler head:

Light Hazard	**Ordinary Hazard**	**Extra Hazard**
130-200 S.F.	100-130 S.F.	90 S.F.

Typical maximum spacing between lines and sprinkler heads:

Light Hazard	**Ordinary Hazard**	**Extra Hazard**
12-15 feet	12-15 feet	12 feet

Note: This data is for estimating purposes only. Check all applicable codes and regulations for specific requirements.

SPRINKLER HEAD CALCULATIONS
Typical maximum quantity of sprinkler heads allowed by pipe size.

Light Hazard:

For sprinklers below ceiling:

Steel		Copper	
1 in. pipe	2 sprinklers	1 in. tube	2 sprinklers
1¼ in. pipe	3 sprinklers	1¼ in. tube	3 sprinklers
1½ in. pipe	5 sprinklers	1½ in. tube	5 sprinklers
2 in. pipe	10 sprinklers	2 in. tube	12 sprinklers
2½ in. pipe	30 sprinklers	2½ in. tube	40 sprinklers
3 in. pipe	60 sprinklers	3 in. tube	65 sprinklers
3½ in. pipe	100 sprinklers	3½ in. tube	115 sprinklers

For sprinklers above and below ceiling:

Steel		Copper	
1 in.	2 sprinklers	1 in.	2 sprinklers
1¼ in.	4 sprinklers	1¼ in.	4 sprinklers
1½ in.	7 sprinklers	1½ in.	7 sprinklers
2 in.	15 sprinklers	2 in.	18 sprinklers
2½ in.	50 sprinklers	2½ in.	65 sprinklers

Ordinary Hazard:

For sprinklers above ceiling:

Steel		Copper	
1 in. pipe	2 sprinklers	1 in. tube	2 sprinklers
1¼ in. pipe	3 sprinklers	1¼ in. tube	3 sprinklers
1½ in. pipe	5 sprinklers	1½ in. tube	5 sprinklers
2 in. pipe	10 sprinklers	2 in. tube	12 sprinklers
2½ in. pipe	20 sprinklers	2½ in. tube	25 sprinklers
3 in. pipe	40 sprinklers	3 in. tube	45 sprinklers
3½ in. pipe	65 sprinklers	3½ in. tube	75 sprinklers
4 in. pipe	100 sprinklers	4 in. tube	115 sprinklers
5 in. pipe	160 sprinklers	5 in. tube	180 sprinklers
6 in. pipe	275 sprinklers	6 in. tube	300 sprinklers

For sprinklers above and below ceiling:

Steel		Copper	
1 in.	2 sprinklers	1 in.	2 sprinklers
1¼ in.	4 sprinklers	1¼ in.	4 sprinklers
1½ in.	7 sprinklers	1½ in.	7 sprinklers
2 in.	15 sprinklers	2 in.	18 sprinklers
2½ in.	30 sprinklers	2½ in.	40 sprinklers
3 in.	60 sprinklers	3 in.	65 sprinklers

Extra Hazard:

For sprinklers below ceiling:

Steel		Copper	
1 in. pipe	1 sprinkler	1 in. tube	1 sprinkler
1¼ in. pipe	2 sprinklers	1¼ in. tube	2 sprinklers
1½ in. pipe	5 sprinklers	1½ in. tube	5 sprinklers
2 in. pipe	8 sprinklers	2 in. tube	8 sprinklers
2½ in. pipe	15 sprinklers	2½ in. tube	20 sprinklers
3 in. pipe	27 sprinklers	3 in. tube	30 sprinklers
3½ in. pipe	40 sprinklers	3½ in. tube	45 sprinklers
4 in. pipe	55 sprinklers	4 in. tube	65 sprinklers
5 in. pipe	90 sprinklers	5 in. tube	100 sprinklers
6 in. pipe	150 sprinklers	6 in. tube	170 sprinklers

Note: This data is for estimating purposes only. Check all applicable codes and regulations for specific requirements.

SPRINKLER HAZARD OCCUPANCIES

Typical Light Hazard Occupancies:

Churches
Clubs
Eaves and overhangs, if
  combustible construction with
  no combustible beneath
Educational
Hospitals
Institutional
Libraries, except large stack
  rooms
Museums
Nursing or Convalescent Homes
Office, including Data Processing
Residential
Restaurant seating areas
Theaters seating areas
Theaters and Auditoriums
  excluding stages and
  prosceniums
Unused attics

Typical Ordinary Hazard Occupancies (Group 1):

Automobile parking garages
Bakeries
Beverage manufacturing
Canneries
Dairy products manufacturing
  and processing
Electronic plants
Glass and glass products
  manufacturing
Laundries
Restaurant service areas

Typical Ordinary Hazard Occupancies (Group 2):

Cereal mills
Chemical plants — Ordinary
Cold Storage warehouses
Confectionery products
Distilleries
Leather goods mfg.
Libraries-large stack room areas
Mercantiles
Machine shops
Metal working
Printing and publishing
Textile mfg.
Tobacco Products mfg.
Wood product assembly

Typical Ordinary Hazard Occupancies (Group 3):

Feed mills
Paper and pulp mills
Paper process plants
Piers and wharves
Repair garages
Tire manufacturing
Warehouses (having moderate
  to higher combustibility of
  content, such as paper,
  household furniture, paint,
  general storage, whiskey, etc.)[1]
Wood machining

Typical Extra Hazard Occupancies (Group 1):

Combustible Hydraulic Fluid
  use areas
  Die Casting
  Metal Extruding
Plywood and particle board
  manufacting
Printing (using inks with below
  100°F [37.8°C] flash points)
Rubber reclaiming, compounding,
  drying, milling, vulcanizing
Saw Mills
Textile picking, opening,
  blending, garnetting, carding,
  combining of cotton, synthetics,
  wool shoddy or burlap
Upholstering with plastic foams

Typical Extra Hazard Occupancies (Group 2):

Asphalt saturating
Flammable liquids spraying
Flow coating
Mobile Home or Modular Building
  assemblies (where finished en-
closure is present and has
  combustible interiors)
Open Oil quenching
Solvent cleaning
Varnish and paint dipping

TYPICAL MINIMUM SIZE OF HORIZONTAL BUILDING STORM DRAINS AND BUILDING STORM SEWERS

Diameter of drain	Maximum projected area in square feet for various slopes		
	⅛ inch per feet slope	¼ inch per feet slope	½ inch per feet slope
3	822	1160	1644
4	1880	2650	3760
5	3340	4720	6680
6	5350	7550	10700
8	11500	16300	23000
10	20700	29200	41400
12	33300	47000	66600
15	59500	84000	119000

PLUMBING BUDGETS

Budget estimates for plumbing can be determined as a percentage of total building costs depending on building type. For example:

Apartments	9 to 12 percent
Assembly	4 to 7 percent
Banks	3 to 6 percent
Dormitories	7 to 10 percent
Factories	4 to 8 percent
Hospitals	8 to 12 percent
Motels	9 to 12 percent
Office Buildings	4 to 7 percent
Retail (small)	4 to 7 percent
Retail (large)	3 to 6 percent
Schools	3 to 6 percent
Warehouses	3 to 7 percent

MECHANICAL / PLUMBING VENTS   15410

TYPICAL SIZE AND LENGTH OF PLUMBING VENTS

Diameter of soil or waste stack (in.)	Total fixture units connected to stack (dfu)	DIAMETER OF VENT PIPE										
		$1\frac{1}{4}$	1	2	$2\frac{1}{2}$	3	4	5	6	8	10	12
$1\frac{1}{4}$	2	30										
$1\frac{1}{2}$	8	50	150									
$1\frac{1}{2}$	10	30	100									
2	12	30	75	200								
2	20	26	50	150								
$2\frac{1}{2}$	42		30	100	300							
3	10		42	150	360	1040						
3	21		32	110	270	810						
3	53		27	94	230	680						
3	102		25	86	210	620						
4	43			35	85	250	980					
4	140			27	65	200	750					
4	320			23	55	170	640					
4	540			21	50	150	580					
5	190				28	82	320	990				
5	490				21	63	250	760				
5	940				18	53	210	670				
5	1400				16	49	190	590				
6	500					33	130	400	1000			
6	1100					26	100	310	780			
6	2000					22	84	260	660			
6	2900					20	77	240	600			
8	1800						31	95	240	940		
8	3400						24	73	190	720		
8	5600						20	62	160	610		
8	7600						18	56	140	560		

TYPICAL SIZES OF FIXTURE WATER SUPPLY PIPES

Fixture	Nominal pipe size (inches)
Bath tubs	½
Combination sink and tray	½
Drinking fountain	⅜
Dishwasher (domestic)	½
Kitchen sink, residential	½
Kitchen sink, commercial	¾
Lavatory	⅜
Laundry tray, 1, 2 or 3 compartments	½
Shower (single head)	½
Sinks (service, slop)	½
Sinks flushing rim	¾
Urinal (flash tank)	½
Urinal (direct flush valve)	¾
Water closet (tank type)	⅜
Water closet (flush valve type)	1
Hose bibs	½
Wall hydrant	½

TYPICAL VENTILATION AIR REQUIREMENTS
FOR SPECIAL USES

Occupancy Classification	Required ventilation air in cfm per human occupant
Special areas	
Lockers	2* (or 30 per locker)
Wardrobes	2*
Public bathrooms	40**
Private bathrooms	25**
Swimming pools	15 (per occupant)
Exitways and corridors	1½*

*Per square foot floor area.
**Per water closet or urinal.

TYPICAL VENTILATION AIR REQUIREMENTS
FOR RETAIL USES

Occupancy Classification	Required ventilation air in cfm per human occupant
Mercantile	
Sales floors and showrooms (basement & grade floors)	7
Sales and showrooms (upper floors)	7
Storage areas	5
Dressing rooms	7
Malls	7
Shipping areas	15
Elevators	7
Supermarkets	
Meat processing rooms	5
Drugs stores	
Pharmacists' work rooms	20
Specialty shops	
Pet shops	1.0*
Florists	5
Greenhouses	5

*cfm per sq. ft. floor area.

TYPICAL VENTILATION AIR REQUIREMENTS FOR RESIDENTIAL USES

Occupancy Classification	Required ventilation air in cfm per human occupant
Residential	
General living areas	5
Bedrooms	5
Kitchens	20
Basements, utility rooms	5
Mobile homes	5
Hotels, motels	
Bedrooms (single, double)	7
Living rooms (suites)	10
Corridors	5
Lobbies	7
Conference rooms (small)	20
Assembly rooms (large)	15

TYPICAL VENTILATION AIR REQUIREMENTS FOR STORAGE USES

Occupancy Classification	Required ventilation air in cfm per human occupant
Storage	
Garages, service stations, parking garages (enclosed)	1.5*
Auto repair shops	1.5**
Warehouses	
General	7

*cfm/s.f. floor area.
**Must have positive engine exhaust system.

TYPICAL VENTILATION AIR REQUIREMENTS
FOR FACTORY AND INDUSTRIAL USES

Occupancy Classification	Required ventilation air in cfm per human occupant
Factory and industrial	
Metalworking & finishing	35
Automotive engine test	Require
Paint spray booths	Special
Picking, etching and plating lines	
Degreasing booths	Exhaust
Sandblasting booths	Systems
Chemicals and pharmaceuticals	
Dusty operations	30
Rooms containing potential gas emitters	20
Drying oven rooms	15
Fermentation rooms	15
Pillmaking booths	10
Packaging areas	10
Utility rooms	7
Computer rooms	7
Textiles-clothes manufacturer	15
Electronics and aerospace circuit board and soldering rooms	20
Wood products, papermaking	20
Brewing, distilling, wineries, bottling	20*
Food processing	20
Tobacco processing	20
Power plants	
Control rooms	10
Boiler rooms	35
Generator rooms	20
Sewage treatment plants	
Control rooms	10
Compressor/blower motor rooms	20
Glass and ceramic manufacturer	20
Agricultural	20

TYPICAL VENTILATION AIR REQUIREMENTS
FOR BUSINESS USES

Occupancy Classification	Required ventilation air in cfm per human occupant
Business	
Banks	
(see offices)	
Vaults	5
Barber, beauty and health services	
Beauty shops (hair dressers)	25
Reducing salons	25
Sauna baths, steam rooms	5
barber shops	7
Photo studios	
Camera rooms, stages	5
Dark rooms	
Shoe repair shops	
Workrooms/trade areas	10
Offices	
General office space and showrooms	15
Conference rooms	25
Drafting/art rooms	7
Doctor's consultation rooms	10
Waiting rooms	10
Lithographing rooms	7
Diazo printing rooms	7
Computer rooms	5
Keypunch rooms	7
Communication	
TV/radio broadcasting booths, studios	30
Motion picture and TV stages	30
Pressrooms	15
Composing rooms	7
Engraving rooms	7
Telephone switchboard rooms (manual)	7
Telephone switchgear rooms (automatic)	7
Teletypewriter/facsimile rooms	5
Research institutes	
Laboratories:	
Light duty; non-chemical	15
Chemical	15
Heavy-duty	15
Radioisotope, chemical & biologically toxic	15
Machine shops	15
Dark rooms, spectroscopy rooms	10
Animal rooms	40
Veterinary hospitals	
Kennels, stalls	25
Operating rooms	25
Reception rooms	10

TYPICAL VENTILATION AIR REQUIREMENTS
FOR INSTITUTIONAL USES

Occupancy Classification	Required ventilation air in cfm per human occupant
Institutional	
Prisons	
Cell blocks	7
Eating halls	15
Guard stations	7

AIR CONDITIONING
RECOMMENDED SHEET METAL GAUGES AND CONSTRUCTION FOR RECTANGULAR DUCT

LOW PRESSURE — LOW VELOCITY = 2″ W.G. MAX

Plate No.	Dimension of Longest Side of Duct	Steel Metal Gauges: Steel	Steel Metal Gauges: Aluminum	Plain "S" Slip (B) / Pocket Lock (K) / Drive Slip (A)	Hemmed "S" Slip (C) / Bar Slip (E) / Standing Seam (1)	Reinforced Bar Slip (G)	Angle Slip (H) / Alternate Bar Slip (F) / Angle RFD Pocket (L)	Companion Angles (M) / Angle Reinforced Standing Seam (J)	Reinforcing Between Joints
6	Thru 12″	26	24 (.020)	A-B-K	—	—			
6	13″ thru 18″	24	22 (.025)	A-B-K	—	—			
7 / 7A	19″ thru 30″	24	22 (.025)	K @ 5′ cc A	C-E-@ 5′ cc C-E-@ 10′ cc	—			1″x1″x⅛″ @ 5′ cc
8	31″ thru 42″	22	20 (.032)	K @ 5′ cc	E-G-K @ 5′ cc E-G-K @ 10′ cc	—			1″x1″x⅛″ @ 5′ cc
9	43″ thru 54″	22	20 (.032)	K @ 4′ cc K @ 8′ cc	E-@ 4′ cc E-@ 8′ cc	G -@ 4′ cc G -@ 8′ cc			1½″x1½″x⅛″ @ 4′ cc
9	55″ thru 60″	20	18 (.040)	K @ 4′ cc K @ 8′ cc	E-@ 4′ cc E-@ 8′ cc	G -@ 4′ cc G -@ 8′ cc			1½″x1½″x⅛″ @ 4′ cc
10	61″ thru 84″	20	18 (.040)	—	—	G -@ 4′ cc G -@ 5′ cc	H-@ 4′ cc F-@ 4′ cc L-@ 4′ cc H-@ 5′ cc F-@ 5′ cc L-@ 5′ cc	J-@ 2′ cc	1½″x1½″x⅛″ @ 2′ cc 1½″x1½″x⅛″ @ 2′-6″ cc
11	85″ thru 96″	18	16 (.051)	—	—	—	H-@ 4′ cc L-@ 4′ cc H-@ 5′ cc L-@ 5′ cc	M-@ 4′ cc M-@ 5′ cc J-@ 2′ cc	1½″x1½″x3/16″ @ 2′ cc 1½″x1½″x3/16″ @ 2′-6″ cc 1½″x1½″x3/16″ @ 2′ cc
12	Over 96″	18	16 (.051)	—	—	—	H-@ 4′ cc L-@ 4′ cc H-@ 5′ cc L-@ 5′ cc	M-@ 4′ cc M-@ 5′ cc J-@ 2′ cc	2″x2″x¼″ @ 2′ cc 2″x2″x¼″ @ 2′-6″ cc 2″x2″x¼″ @ 2′ cc

H (height dimension)—up to 42″ = 1″
H (height dimension)—43″ to 96″ = 1½″
H (height dimension)—over 96″ = 2″

AIR CONDITIONING
TYPICAL DUCT CONNECTIONS
CROSS JOINTS FOR SHEET METAL DUCTWORK
(NOT TO SCALE)

H=HEIGHT REFERRED TO IN DIMENSIONS

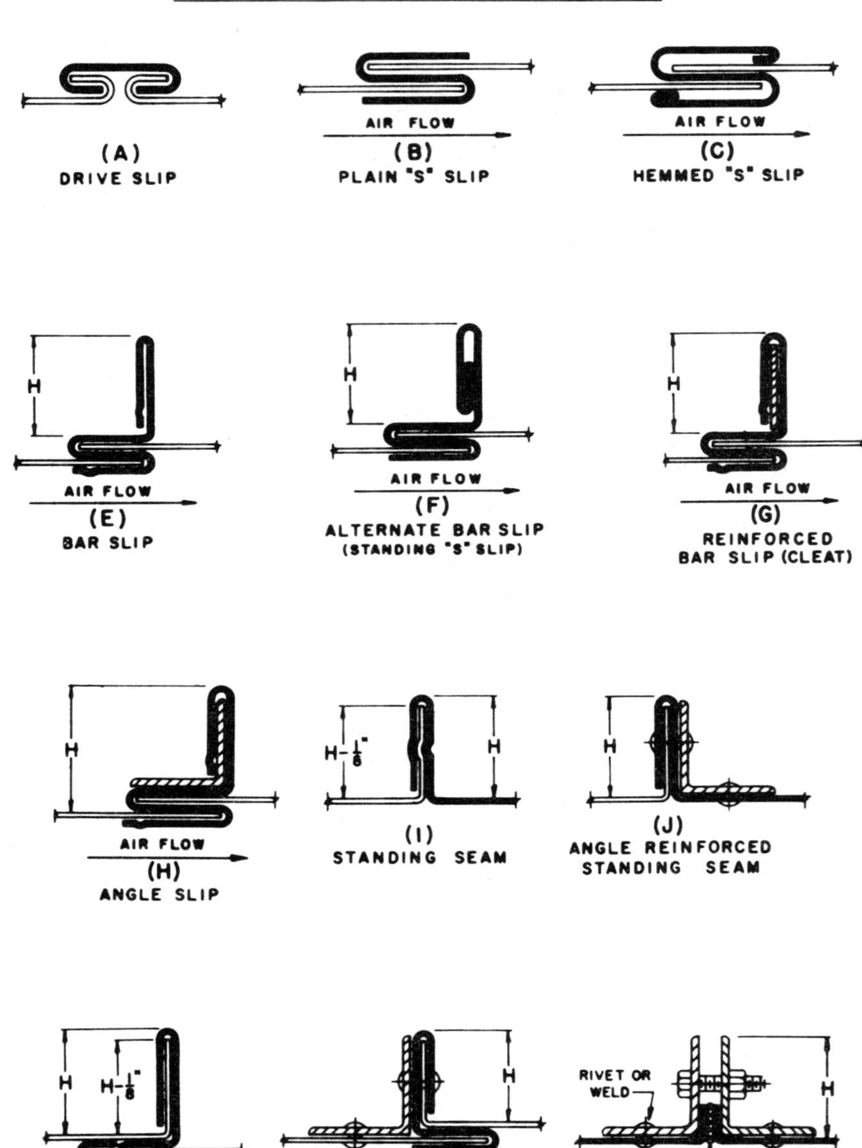

AIR CONDITIONING (Cont.)
LONGITUDINAL SEAMS
FOR SHEET METAL DUCTWORK

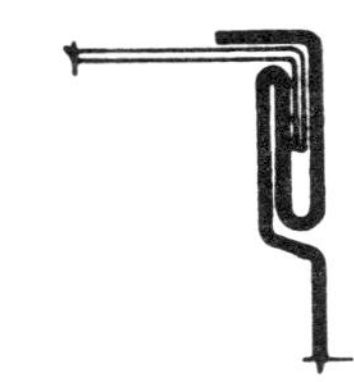

Fig. "N"
PITTSBURGH LOCK

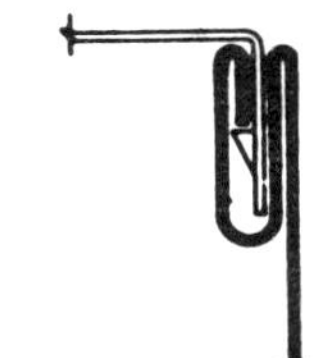

Fig. "Z"
BUTTON PUNCH SNAP LOCK

Fig. "O"
ACME LOCK-GROOVED SEAM

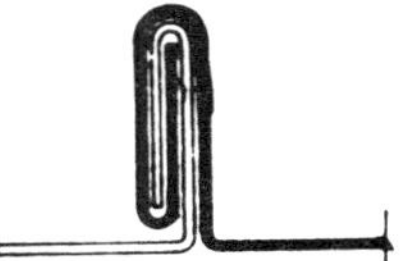

Fig. "T"
DOUBLE SEAM

Approximately 2″ Spacing
Between "Buttons"

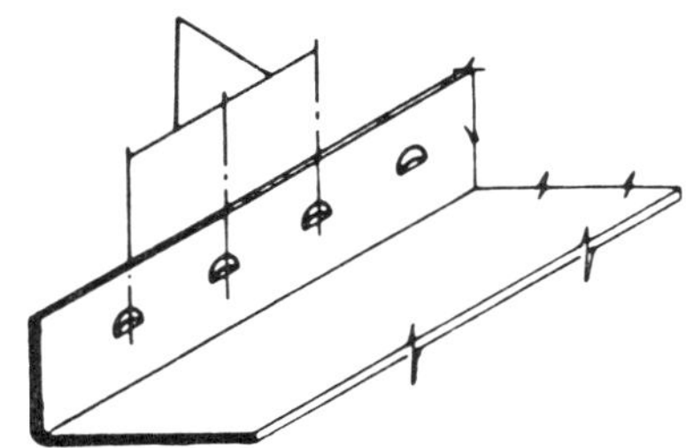

DETAIL NO. 1
MALE PIECE-SNAP LOCK

TYPICAL DUCT CONSTRUCTION SHEET METAL GAGES
IN ONE- AND TWO-FAMILY DWELLINGS

Metal Gauges (duct not enclosed in partitions)		
ROUND DUCTS		
Diameter, inches	Minimum thickness galvanized sheet gage	Minimum thickness aluminum B&S gage
Less than 12	30	26
12-14	28	26
15-18	26	24
Over 18	24	22
RECTANGULAR DUCTS		
Width, inches	Minimum thickness galvanized sheet gage	Minimum thickness aluminum B&S gage
Less than 14	28	24
14-24	26	22
25-30	24	22
Over 30	22	20
Metal Gauges (duct enclosed in partitions)		
Width, inches	Minimum thickness galvanized sheet gage	Minimum thickness aluminum B&S gage
14 or less	30	26
Over 14	28	24

TYPICAL DUCT CONSTRUCTION SHEET METAL GAGES
(All uses except 1- and 2-family dwellings)

RECTANGULAR DUCTS		
Maximum side inches	Steel min. Galv. Sheet Gage	Aluminum Min. B&S Gage
Through 12	26 (0.022 in.)	24 (0.020 in.)
13 through 30	24 (0.028 in.)	22 (0.025 in.)
31 through 54	22 (0.034 in.)	20 (0.032 in.)
55 through 84	20 (0.040 in.)	18 (0.040 in.)
Over 84	18 (0.052 in.)	16 (0.051 in.)

ROUND DUCTS			
Diameter inches	Spiral seam duct — Steel min. Galv. Sht. Gage	Longitudinal seam duct — Steel min. Galv. Sht. Gage	Fittings — Steel min. Galv. Sht. Gage
Through 12	28.(0.019 in.)	26 (0.022 in.)	26 (0.022 in.)
13 through 18	26 (0.022 in.)	24 (0.028 in.)	24 (0.028 in.)
19 through 28	24 (0.028 in.)	22 (0.034 in.)	22 (0.034 in.)
29 through 36	22 (0.034 in.)	20 (0.040 in.)	20 (0.040 in.)
37 through 52	20 (0.040 in.)	18 (0.052 in.)	18 (0.052 in.)

APPLICATIONS FOR CONDUCTORS USED FOR GENERAL WIRING

	AMBIENT TEMPERATURE								FEATURES
	60°C 140°F	75°C 167°F	85°C 185°F	90°C 194°F	110°C 230°F	200°C 392°F	Dry	Dry or Wet	
R	X						X		Code Rubber
RH		X					X		Heat Resistant
RHH				X			X		More Heat Resistant
RW	X							X	Moisture Resistant
RH-RW	X							X	Moisture and Heat Resistant
		X					X		Moisture and Heat Resistant
RHW		X						X	Moisture and Heat Resistant
RU	X						X		Latex Rubber
RUH		X					X		Heat Resistant
RUW	X							X	Moisture Resistant
T	X						X		Thermoplastic
TW	X							X	Moisture Resistant
THHN				X			X		Heat Resistant
THW		X						X	Moisture and Heat Resistant
THWN		X						X	Moisture and Heat Resistant
MI			X					X	Mineral Insulated Metal Sheathed
V			X				X		Varnished Cambric
AVA					X		X		With Asbestos
AVB				X			X		With Asbestos
AVL					X			X	With Asbestos

This table does not include special condition conductors, thickness of conductor insulation, or reference to all outer protective coverings.

GENERAL CLASSIFICATION OF INSULATIONS:

A .... Asbestos
H .... Heat Resistant
MI .... Mineral Insulation
R .... Rubber
RU ....Latex Rubber
V ....Varnished Cambric
T ....Thermoplastic
W ....(Water) Moisture Resistant

WIRE AND SHEET METAL GAGES
(In Decimals of an Inch)

Name of Gage	American Wire Gage (A.W.G.) (Corresponds to Brown & Sharpe Gage)	Birmingham Iron Wire Gage (B.W.G.)	United States Standard Gage (U.S.S.G.)	
Principal Use	Electrical Wire & Non-Ferrous Sheet Metal	Iron or Steel Wire	Ferrous Sheet Metal	
Gage No.				Gage No.
00 00000				00 00000
0 00000	.5800			0 00000
00000	.5165	.500		00000
0000	.4600	.454		0000
000	.4096	.425		000
00	.3648	.380		00
0	.3249	.340		0
1	.2893	.300		1
2	.2576	.284		2
3	.2294	.259	23.91	3
4	.2043	.238	.2242	4
5	.1819	.220	.2092	5
6	.1620	.203	.1943	6
7	.1443	.180	.1793	7
8	.1285	.165	.1644	8
9	.1144	.148	.1495	9
10	.1019	.134	.1345	10
11	.0907	.120	.1196	11
12	.0808	.109	.1046	12
13	.0720	.095	.0897	13
14	.0641	.083	.0747	14
15	.0571	.072	.0673	15
16	.0508	.065	.0598	16
17	.0453	.058	.0538	17
18	.0403	.049	.0478	18
19	.0359	.042	.0418	19
20	.0320	.035	.0359	20
21	.0285	.032	.0329	21
22	.0253	.028	.0299	22
23	.0226	.025	.0269	23
24	.0201	.022	.0239	24
25	.0179	.020	.0209	25
26	.0159	.018	.0179	26
27	.0142	.016	.0164	27
28	.0126	.014	.0149	28
29	.0113	.013	.0135	29
30	.0100	.012	.0120	30
31	.0089	.010	.0105	31
32	.0080	.009	.0097	32
33	.0071	.008	.0090	33
34	.0063	.007	.0082	34
35	.0056	.005	.0075	35
36	.0050	.004	.0067	36
37	.0045		.0064	37
38	.0040		.0060	38
39	.0035			39
40	.0031			40

Geographic Cost Modifiers

The costs as presented in this book attempt to represent national averages. Costs, however, vary among regions, states and even between adjacent localities.

In order to more closely approximate the probable costs for specific locations throughout the U.S., this table of Geographic Cost Modifiers is provided. These adjustment factors are used to modify costs obtained from this book to help account for regional variations of construction costs and to provide a more accurate estimate for specific areas. The factors are formulated by comparing costs in a specific area to the costs as presented in the Costbook pages. An example of how to use these factors is shown below. Whenever local current costs are known, whether material prices or labor rates, they should be used when more accuracy is required.

$$\begin{matrix} \text{Cost obtained} \\ \text{from Costbook} \\ \text{pages} \end{matrix} \times \begin{matrix} \text{Location Cost} \\ \text{Adjustment} \\ \text{Factor} \end{matrix} = \text{Adjusted Cost}$$

For example, a project estimated to cost $125,000 using the Costbook pages can be adjusted to more closely approximate the cost in Los Angeles: $125,000 $\times$ 1.07 = $133,750.

Geographic Cost Modifiers

ALABAMA
BIRMINGHAM	0.79
HUNTSVILLE	0.77
MOBILE	0.81
MONTGOMERY	0.75
TUSCALOOSA	0.75

ALASKA
ANCHORAGE	1.30
JUNEAU	1.33
FAIRBANKS	1.37
NOME	1.43

ARIZONA
FLAGSTAFF	0.91
PHOENIX	0.89
PRESCOTT	0.91
TUCSON	0.89
YUMA	0.87

ARKANSAS
FAYETTEVILLE	0.73
FORT SMITH	0.74
LITTLE ROCK	0.78
PINE BLUFF	0.75

CALIFORNIA
ANAHEIM	1.02
BAKERSFIELD	0.97
LOS ANGELES	1.07
REDDING	0.92
RIVERSIDE	0.97
SACRAMENTO	1.00
SAN DIEGO	1.02
SAN JOSE	1.07
SAN FRANCISCO	1.12
SANTA BARBARA	1.07

COLORADO
BOULDER	0.92
COLORADO SPRINGS	0.92
DENVER	0.95
GRAND JUNCTION	0.90
PUEBLO	0.90

CONNECTICUT
BRIDGEPORT	1.03
HARTFORD	1.01
NEW LONDON	0.99
STAMFORD	1.06
WATERBURY	0.96

DELAWARE
DOVER	0.89
WILMINGTON	0.91

FLORIDA
JACKSONVILLE	0.80
MIAMI	0.86
ORLANDO	0.80
TAMPA	0.82
WEST PALM BEACH	0.84

GEORGIA
ATLANTA	0.83
AUGUSTA	0.75
COLUMBUS	0.75
MACON	0.77
SAVANNAH	0.79

HAWAII
HILO	1.31
HONOLULU	1.25
MAUI	1.28

IDAHO
BOISE	0.92
LEWISTON	0.90
POCATELLO	0.88
TWIN FALLS	0.88

ILLINOIS
CHICAGO	1.03
MOLINE	0.89
PEORIA	0.92
ROCKFORD	0.92
SPRINGFIELD	0.89

INDIANA
FORT WAYNE	0.88
EVANSVILLE	0.88
GARY	0.99
INDIANAPOLIS	0.95
TERRE HAUTE	0.86

IOWA
COUNCIL BLUFFS	0.85
DAVENPORT	0.87
DES MOINES	0.89
SIOUX CITY	0.85
WATERLOO	0.87

KANSAS
DODGE CITY	0.77
SALINA	0.79
TOPEKA	0.83
WICHITA	0.81

KENTUCKY
BOWLING GREEN	0.83
LEXINGTON	0.85
LOUISVILLE	0.89
PADUCAH	0.87

LOUISIANA
BATON ROUGE	0.84
LAKE CHARLES	0.82
MONROE	0.78
NEW ORLEANS	0.86
SHREVEPORT	0.78

MAINE
AUGUSTA	0.85
BANGOR	0.83
LEWISTON	0.87
PORTLAND	0.89

MARYLAND
ANNAPOLIS	0.92
BALTIMORE	0.90
HAGERSTOWN	0.88
ROCKVILLE	0.95

MASSACHUSETTS
BOSTON	1.01
LOWELL	0.94
FALL RIVER	0.91
SPRINGFIELD	0.94
WORCESTER	0.96

MICHIGAN
DETROIT	0.95
GRAND RAPIDS	0.88
LANSING	0.88
MARQUETTE	0.86
SAGINAW	0.90

MINNESOTA
DULUTH	0.93
MINNEAPOLIS	0.98
ROCHESTER	0.93
SAINT PAUL	0.98

MISSISSIPPI
COLUMBUS	0.77
GULFPORT	0.79
JACKSON	0.81
VICKSBURG	0.79

MISSOURI
JOPLIN	0.84
KANSAS CITY	0.86
SAINT LOUIS	0.90
SAINT JOSEPH	0.84
SPRINGFIELD	0.84

MONTANA
BILLINGS	0.87
BUTTE	0.85
GREAT FALLS	0.87
HELENA	0.85
MISSOULA	0.85

Geographic Cost Modifiers

NEBRASKA	
GRAND ISLAND	0.81
LINCOLN	0.85
NORTH PLATTE	0.81
OMAHA	0.87

NEVADA	
CARSON CITY	0.95
LAS VEGAS	0.97
SPARKS	0.97
RENO	1.00

NEW HAMPSHIRE	
CONCORD	0.86
MANCHESTER	0.88
NASHUA	0.90

NEW JERSEY	
ATLANTIC CITY	0.97
CAMDEN	0.95
NEWARK	1.02
PATERSON	0.99
TRENTON	0.97

NEW MEXICO	
ALBUQUERQUE	0.86
CARLSBAD	0.84
LAS CRUCES	0.82
SANTA FE	0.92

NEW YORK	
ALBANY	0.87
BINGHAMTON	0.83
BUFFALO	0.90
LONG ISLAND	1.01
NEW YORK CITY	1.15
ROCHESTER	0.90
SYRACUSE	0.87
WATERTOWN	0.83
WHITE PLAINS	1.01

NORTH CAROLINA	
ASHEVILLE	0.75
CHARLOTTE	0.79
GREENSBORO	0.75
RALEIGH	0.77
WILMINGTON	0.74

NORTH DAKOTA	
BISMARK	0.86
FARGO	0.90
GRAND FORKS	0.88
MINOT	0.84

OHIO	
CINCINNATI	0.83
CLEVELAND	0.91
COLUMBUS	0.89
TOLEDO	0.89
YOUNGSTOWN	0.85

OKLAHOMA	
BARTLESVILLE	0.77
ENID	0.77
LAWTON	0.77
OKLAHOMA CITY	0.83
TULSA	0.81

OREGON	
EUGENE	0.90
MEDFORD	0.90
PORTLAND	0.95
SALEM	0.92

PENNSYLVANIA	
ALLENTOWN	0.91
HARRISBURG	0.85
PHILADELPHIA	0.98
PITTSBURGH	0.89
SCRANTON	0.87

RHODE ISLAND	
PAWTUCKET	0.95
PROVIDENCE	0.98
NEWPORT	0.98
WESTERLY	0.95
WOONSOCKET	0.95

SOUTH CAROLINA	
CHARLESTON	0.79
COLUMBIA	0.81
FLORENCE	0.79
GREENVILLE	0.81

SOUTH DAKOTA	
ABERDEEN	0.79
PIERRE	0.79
RAPID CITY	0.81
SIOUX FALLS	0.83
WATERTOWN	0.81

TENNESSEE	
CHATTANOOGA	0.77
JOHNSON CITY	0.75
KNOXVILLE	0.79
MEMPHIS	0.81
NASHVILLE	0.83

TEXAS	
AUSTIN	0.80
DALLAS	0.82
HOUSTON	0.86
LUBBOCK	0.78
SAN ANTONIO	0.80

UTAH	
LOGAN	0.83
OGDEN	0.85
PROVO	0.83
SALT LAKE	0.87

VERMONT	
BRATTLEBORO	0.85
BURLINGTON	0.94
RUTLAND	0.87

VIRGINIA	
ALEXANDRIA	0.94
LYNCHBURG	0.81
NORFOLK	0.83
RICHMOND	0.85
ROANOKE	0.81

WASHINGTON	
BELLINGHAM	0.99
SEATTLE	1.06
SPOKANE	0.99
TACOMA	1.03
YAKIMA	0.96

WASHINGTON, D.C.	
DISTRICT	0.95

WEST VIRGINIA	
CHARLESTON	0.87
HUNTINGTON	0.89
MORGANTOWN	0.83
PARKERSBURG	0.83

WISCONSIN	
EAU CLAIRE	0.90
GREEN BAY	0.92
MADISON	0.90
MILWAUKEE	0.97
WAUSAU	0.90

WYOMING	
CASPER	0.89
CHEYENNE	0.91
ROCK SPRINGS	0.87
SHERIDAN	0.85

Square Foot Tables

The following Square Foot Tables list hundreds of actual projects for dozens of building types each with associated building size, total square foot building cost and percentage of project costs for total mechanical and electrical components. This data provides an overview of construction costs by building type. These costs are for actual projects. The variations within similar building types may be due, among other factors, to size, location, quality and specified components, materials and processes. Depending upon all such factors, specific building costs can vary significantly and may not necessarily fall within the range of costs as presented.

The data has been updated to reflect current construction costs and is an expanded summary of the projects and costs presented in the **Building News Square Foot 1995 Costbook.** The source of this data is **Design Cost & Data** magazine which has over 26 years of history in tracking, collecting and publishing construction cost data.

SQUARE FOOT TABLES

COMMERCIAL

AUTO DEALERSHIP

Project Size Gross S.F.	Project Cost $/S.F.	% Cost Mechanical	% Cost Electrical
7,700	76.80	16.7	9.4
16,100	42.80	10.2	15.9
20,000	44.70	12.9	23.4
26,300	45.70	12.5	22.0
43,600	35.95	19.4	13.2
53,600	62.00	12.5	11.5

BUSINESS CENTER

Project Size Gross S.F.	Project Cost $/S.F.	% Cost Mechanical	% Cost Electrical
3,900	46.60	12.0	9.2
9,900	43.80	9.1	7.6
54,400	29.80	3.4	12.2
135,000	32.80	8.2	1.5

CINEMA

Project Size Gross S.F.	Project Cost $/S.F.	% Cost Mechanical	% Cost Electrical
18,000	105.50	10.9	6.7
22,500 A	65.25	6.6	4.2

MALL/PLAZA

Project Size Gross S.F.	Project Cost $/S.F.	% Cost Mechanical	% Cost Electrical
9,700	32.90	15.0	13.3
10,500	53.30	8.0	13.5
16,300	48.60	9.4	10.6
26,900	48.00	15.0	7.0
36,000	40.80	10.0	11.0
36,300	46.37	12.4	8.2
44,720	48.20	18.3	12.4
59,100	46.50	9.8	9.5
60,000 R	48.30	10.0	9.5
64,100	48.50	22.4	18.4
66,000	56.00	13.5	11.5
67,400	40.00	21.0	15.0
73,500	58.70	14.9	6.9

For more information see **Building News Square Foot 1995 Costbook.**

MALL/PLAZA (Cont.)

Project Size Gross S.F.	Project Cost $/S.F.	% Cost Mechanical	% Cost Electrical
142,000	37.00	7.1	8.0
220,000	98.30	11.0	6.4
223,700	28.30	9.0	9.3
321,200	33.50	7.3	7.4
379,900	43.70	11.2	6.2
405,100	45.50	13.9	6.0
482,000	77.00	10.5	9.4
630,000	51.30	12.2	12.4

RESTAURANT

Project Size Gross S.F.	Project Cost $/S.F.	% Cost Mechanical	% Cost Electrical
4,300 R	85.40	6.9	8.7
4,400 R	125.60	14.6	8.5
5,800	100.90	28.0	10.6
6,800 A	118.90	7.0	11.1
7,360 R	127.05	16.0	6.5
9,600	126.70	24.7	13.1
10,000 R	131.60	21.0	10.0
10,100	113.50	28.6	18.4
10,600	205.30	20.4	6.4
22,900 R	136.40	15.8	16.9

RETAIL STORE

Project Size Gross S.F.	Project Cost $/S.F.	% Cost Mechanical	% Cost Electrical
1,000	125.60	12.8	6.7
3,000 R	112.40	14.3	10.5
12,300	154.30	14.0	10.0
30,000	78.00	15.6	26.2
61,300	39.80	13.3	13.0
115,000	55.90	14.6	11.3
154,700	79.60	11.2	12.4
314,700 R	66.60	13.8	9.4

A = Addition   R = Remodel

SQUARE FOOT TABLES

RESIDENTIAL

APARTMENTS

Project Size Gross S.F.	Project Cost $/S.F.	% Cost Mechanical	% Cost Electrical
3,700	47.10	14.9	4.4
13,900	63.80	7.9	7.2
19,200 R	101.70	45.3	7.5
19,700	59.70	7.4	10.4
23,700	65.30	10.6	4.3
26,500	61.70	25.2	12.8
35,100	50.05	16.4	5.6
54,000	87.90	23.3	13.1
62,700	63.80	17.0	9.0
67,300	58.10	13.8	8.4
70,600	32.30	18.1	7.8
75,300	72.20	13.2	8.1
75,600	70.30	14.5	8.9
72,200	61.30	18.4	10.9
77,600	79.70	26.9	14.3
88,100	73.80	15.3	9.3
89,500	68.50	10.7	11.1
94,100	40.95	8.5	6.8
96,000	50.90	17.0	13.3
102,000	89.40	17.8	12.5
103,200	40.40	12.1	8.9
103,600	65.40	19.1	9.0
105,200	83.70	14.9	8.9
106,200	60.30	12.8	9.3
110,900	57.05	15.6	8.4
111,800	81.20	17.3	7.4
115,900	61.50	12.5	8.6
117,200	37.30	12.9	8.0
119,000	52.70	15.6	8.4

APARTMENTS (Cont.)

Project Size Gross S.F.	Project Cost $/S.F.	% Cost Mechanical	% Cost Electrical
119,400	34.20	18.6	8.7
144,300	76.30	15.0	8.6
176,300	65.30	19.1	9.0
192,300	36.60	10.1	6.0
210,900	75.50	19.1	9.5
220,200	78.00	14.8	7.5
253,900	101.40	20.6	7.7
369,500	73.40	15.8	9.0

CONDOS/TOWNHOUSES

Project Size Gross S.F.	Project Cost $/S.F.	% Cost Mechanical	% Cost Electrical
8,600	66.70	8.5	4.4
16,700	52.80	13.2	6.5
18,000	113.90	14.6	9.7
18,400	58.10	12.9	5.8
74,800	51.80	13.8	5.2
111,700	66.80	9.2	7.1
150,300	73.20	15.9	7.9
278,800	115.70	14.1	7.9
1,109,900	53.20	9.8	4.7

SINGLE-FAMILY HOMES

Project Size Gross S.F.	Project Cost $/S.F.	% Cost Mechanical	% Cost Electrical
600 R	44.60	16.3	2.0
900	59.90	33.0	3.0
2,100	69.40	8.8	4.0
2,200	149.90	23.8	3.4
2,500	72.70	6.1	7.1
2,900	70.10	7.7	2.8
3,000	49.00	9.5	6.8
3,100	109.90	15.5	4.6
3,600	69.00	8.5	3.0
3,700	96.00	9.8	3.4

A = Addition   R = Remodel

SQUARE FOOT TABLES

RESIDENTIAL (Cont.)

SINGLE-FAMILY HOMES (Cont.)

Project Size Gross S.F.	Project Cost $/S.F.	% Cost Mechanical	% Cost Electrical
4,200	84.00	15.5	5.7
4,600	134.30	9.0	4.7
5,200	174.50	8.3	6.0
5,700	101.00	7.4	3.7
5,700	68.80	8.7	12.6
21,300*	48.80	26.0	4.0
22,700*	44.60	27.0	4.6
45,000*	68.90	7.8	2.5
51,458*	49.40	11.0	5.0

*Townhouses.

EDUCATIONAL

ADMINISTRATION (OFFICES)

Project Size Gross S.F.	Project Cost $/S.F.	% Cost Mechanical	% Cost Electrical
53,700	120.10	15.4	9.0

ATHLETIC FACILITY

38,100	116.90	16.8	9.8
44,100	107.50	16.8	8.4
100,000	73.10	19.2	5.3
160,000	132.90	13.7	7.9
247,500	114.60	11.2	9.0
271,000	106.30	13.2	7.6
283,100	128.70	14.6	6.0

AUDITORIUM/PERFORMING ARTS

9,900	184.50	17.5	29.2
17,800	173.10	11.4	16.1
29,200	169.10	16.2	10.5
62,700	125.80	13.0	12.9

CLASSROOM

Project Size Gross S.F.	Project Cost $/S.F.	% Cost Mechanical	% Cost Electrical
35,400	171.30	10.2	6.8
70,000	77.80	24.3	13.1
78,900	147.40	13.3	14.3
80,100	118.90	16.9	10.5
100,000	120.10	21.1	9.8
166,000	80.50	16.1	11.2
298,400	76.10	11.3	11.9

COMPLETE COLLEGE FACILITIES

95,300	121.95	19.4	11.7
450,000	146.90	18.6	10.8

ELEMENTARY SCHOOL

18,000	88.00	19.1	13.8
30,800	81.60	16.0	10.0
31,600	69.30	12.1	11.3
35,700	100.90	17.3	8.7
40,000	86.90	18.5	13.4
40,500	73.50	22.1	11.3
57,000	66.50	22.1	7.5
69,700	102.90	20.6	9.3
91,400	83.20	18.2	7.6

HIGH SCHOOL

116,400	91.40	18.1	12.9
133,000	79.00	17.8	10.7
184,000	135.60	23.0	10.0
217,200 R	73.90	26.4	10.6
254,000 R	58.50	17.2	14.6
431,700	88.10	13.1	9.6

A = Addition   R = Remodel

SQUARE FOOT TABLES

EDUCATIONAL (Cont.)

JUNIOR HIGH SCHOOL

Project Size Gross S.F.	Project Cost $/S.F.	% Cost Mechanical	% Cost Electrical
26,000	114.10	9.5	9.3
28,100	70.70	11.9	9.5
52,800	91.30	18.3	8.6
91,600	113.20	21.2	11.0
123,700	92.50	29.1	9.1

LABORATORY/RESEARCH

Project Size Gross S.F.	Project Cost $/S.F.	% Cost Mechanical	% Cost Electrical
9,200	197.20	25.4	4.5
80,300	157.90	20.2	15.5

LIBRARY

Project Size Gross S.F.	Project Cost $/S.F.	% Cost Mechanical	% Cost Electrical
6,900	106.20	17.9	10.3
8,200	97.40	18.8	10.6
12,000	126.80	19.4	15.9
15,000	110.85	15.7	18.7
16,300	91.90	11.3	7.4
28,600	83.40	15.7	9.0
30,100	113.10	13.0	11.0
37,700	88.90	14.6	8.0
43,500	75.30	16.6	6.7
47,900	132.12	29.7	9.0
51,400	134.00	13.1	12.4
63,400 A	100.83	13.5	8.3
64,000	93.90	10.9	11.4
74,000	107.50	17.2	8.0
75,600	143.80	12.5	16.7
176,000	79.20	11.2	9.8

SPECIAL NEEDS FUNCTION

Project Size Gross S.F.	Project Cost $/S.F.	% Cost Mechanical	% Cost Electrical
15,200	89.70	16.6	8.4
27,900	108.10	19.4	11.2

STUDENT CENTER/MULTIPURPOSE

Project Size Gross S.F.	Project Cost $/S.F.	% Cost Mechanical	% Cost Electrical
90,000	136.50	16.2	10.0
49,600	100.90	18.2	9.2
187,700	151.90	15.3	8.8
194,800	80.50	17.2	8.5

HOTEL/MOTEL

CONVENTION/CONFERENCE CENTER

Project Size Gross S.F.	Project Cost $/S.F.	% Cost Mechanical	% Cost Electrical
8,600 A	118.70	20.7	16.3
71,900	131.70	12.5	13.8
433,800	68.80	22.1	8.3

HOTEL

Project Size Gross S.F.	Project Cost $/S.F.	% Cost Mechanical	% Cost Electrical
19,900 A	67.70	16.8	5.8
25,875 R	61.70	11.8	10.5
48,400 A	122.00	23.6	8.2
64,300 R	163.50	20.3	10.1
104,200 A	76.20	15.0	8.8
108,040	65.10	13.0	7.0
110,100	84.10	18.4	10.5
132,000	158.10	16.4	5.4
135,900 A	96.80	15.2	7.7
144,100 A	110.80	19.3	11.5
231,000	122.10	15.2	8.4
449,800 A	68.15	13.1	7.0

HOTEL/INN

Project Size Gross S.F.	Project Cost $/S.F.	% Cost Mechanical	% Cost Electrical
57,400	78.40	13.0	10.0
73,000	50.30	24.4	18.1
75,900	67.60	16.5	7.6
162,000	81.30	17.5	8.0
197,000	78.30	15.7	7.7
277,900	59.00	18.8	9.4

A = Addition   R = Remodel

SQUARE FOOT TABLES

INDUSTRIAL

MANUFACTURING

Project Size Gross S.F.	Project Cost $/S.F.	% Cost Mechanical	% Cost Electrical
14,300	69.10	13.0	7.0
18,500	102.90	20.5	13.5
26,600	31.10	6.1	12.1
31,400	81.20	18.9	17.7
33,400	47.00	23.1	15.8
37,300	23.40	3.0	24.0
43,400	45.15	14.7	11.7
45,400	76.05	15.9	15.0
79,800	80.00	16.0	14.5
81,100	84.30	8.2	8.4
137,400	52.80	41.6	13.6
179,600	60.90	26.3	13.6
186,000	56.90	19.5	11.7

RESEARCH AND DEVELOPMENT

Project Size Gross S.F.	Project Cost $/S.F.	% Cost Mechanical	% Cost Electrical
89,140	107.75	20.8	9.0
100,400	142.70	36.1	25.1
114,200	138.50	21.4	9.8
125,000	95.20	20.8	8.4
140,000	132.90	20.1	11.2

WAREHOUSE W/OFFICE

Project Size Gross S.F.	Project Cost $/S.F.	% Cost Mechanical	% Cost Electrical
14,000	28.60	6.5	9.2
19,000	17.90	2.3	1.8
19,700	36.90	9.5	7.0
31,200	29.00	7.0	7.0
40,500	36.43	6.6	10.5
62,000	48.20	10.8	10.0
96,200	22.30	2.2	6.3
105,000	21.10	5.3	11.1
149,800	22.50	14.5	8.6

WAREHOUSE W/OFFICE (Cont.)

Project Size Gross S.F.	Project Cost $/S.F.	% Cost Mechanical	% Cost Electrical
168,600	29.15	11.4	6.3
209,600	26.60	4.9	10.3
402.400	36.75	14.9	8.3

MEDICAL

EDUCATION CENTER

Project Size Gross S.F.	Project Cost $/S.F.	% Cost Mechanical	% Cost Electrical
35,400	182.20	10.2	6.8

HOSPITALS

Project Size Gross S.F.	Project Cost $/S.F.	% Cost Mechanical	% Cost Electrical
9,300 R	183.30	29.5	12.0
15,900 A	133.50	27.4	15.3
16,600 R	45.65	14.7	9.6
22,000	307.80	31.6	17.5
39,100	152.20	23.3	7.9
63,800	121.50	23.4	7.8
98,000 A	175.30	20.5	17.0
100,200 A	280.60	22.2	9.6
103,900 A	155.20	31.1	11.7
109,300	153.70	20.7	15.7
148,700	120.90	25.2	11.8
154,700	196.30	19.3	10.5
165,484	170.60	21.5	14.8
165,700 A	165.90	26.2	17.7
179,400 A	101.30	28.9	20.3
182,800	150.86	19.1	14.0
265,000	179.90	31.2	14.1
281,100	172.36	24.1	14.6
435,000	178.70	21.5	13.7
694,300	99.00	28.9	9.7
772,300	180.80	31.5	13.4

A = Addition   R = Remodel

SQUARE FOOT TABLES

MEDICAL (Cont.)

MEDICAL OFFICES/CENTERS

Project Size Gross S.F.	Project Cost $/S.F.	% Cost Mechanical	% Cost Electrical
3,000	76.40	13.7	11.5
5,500	79.55	8.5	18.7
10,000	100.60	13.9	9.4
10,600	123.35	13.2	8.9
16,300	101.90	21.8	13.2
18,300	55.00	7.1	13.4
20,600	79.05	14.3	6.4
24,900	140.50	21.0	15.1
27,000	140.35	19.7	10.1
28,400	96.20	13.6	9.3
30,500	107.40	17.2	12.7
32,000	86.45	18.1	10.5
44,300	86.15	24.4	16.2
50,200	48.60	9.8	4.9
51,200	124.30	21.0	12.0
64,600	59.05	13.2	6.4
66,000	55.10	9.8	8.8
80,000	35.20	8.2	8.3
137,175	79.40	12.8	6.6

NURSING HOMES

Project Size Gross S.F.	Project Cost $/S.F.	% Cost Mechanical	% Cost Electrical
11,600 A	232.20	53.2	7.9
16,800	150.75	33.3	7.8
31,900 A	114.20	22.0	11.0
64,100	94.45	20.6	11.1
290,000	134.35	16.1	13.6

RESEARCH

Project Size Gross S.F.	Project Cost $/S.F.	% Cost Mechanical	% Cost Electrical
34,600	127.95	18.1	3.0

PUBLIC FACILITIES

ANIMAL CENTER

Project Size Gross S.F.	Project Cost $/S.F.	% Cost Mechanical	% Cost Electrical
20,000	163.30	20.7	4.6
39,100	122.65	22.9	6.8
44,300	116.20	8.1	5.8

AUTO DEALERSHIP

7,700	77.20	16.7	9.4

BROADCASTING

20,000	225.00	20.0	13.0
29,500	173.15	16.6	15.0
45,000 R	125.85	15.0	13.0

CIVIC CENTER

6,000	124.40	9.2	2.8
23,900	155.45	12.3	10.9
34,400	70.80	3.5	17.8
69,800 A	187.10	17.7	8.8
206,500	100.50	15.0	11.0

CORRECTION FACILITIES

44,600	145.85	20.9	13.1
66,000	90.25	15.0	23.0
257,800 A	157.90	20.9	10.7
360,000	104.60	32.7	13.2

FIRE STATION

6,900	132.15	12.4	9.7
7,600	100.70	16.0	9.5
8,430	137.35	12.8	9.6
9,600	121.70	13.5	11.8

A = Addition   R = Remodel

SQUARE FOOT TABLES

PUBLIC FACILITIES (Cont.)

GOVERNMENT BUILDINGS

Project Size Gross S.F.	Project Cost $/S.F.	% Cost Mechanical	% Cost Electrical
12,300	94.35	13.5	10.7
23,500	157.35	11.1	15.6
27,300	121.95	19.5	9.3
31,600	137.30	27.1	11.3
46,600	129.25	17.4	13.8
72,100	140.50	24.4	10.7
78,200	116.45	19.7	15.2
332,900	113.55	16.2	14.2
364,100	147.95	14.6	12.9
771,000	163.20	17.6	11.3

MUSEUM

Project Size Gross S.F.	Project Cost $/S.F.	% Cost Mechanical	% Cost Electrical
27,600	115.00	17.8	14.1
30,100	131.25	18.6	7.9
43,264	122.20	11.1	8.3
63,000	132.35	8.8	18.1

PARKING GARAGE

Project Size Gross S.F.	Project Cost $/S.F.	% Cost Mechanical	% Cost Electrical
66,000	33.20	2.8	3.1
169,000	20.70	10.3	3.7
562,700	18.50	2.4	6.2

TRANSPORTATION

Project Size Gross S.F.	Project Cost $/S.F.	% Cost Mechanical	% Cost Electrical
7,300	211.10	15.1	3.2
14,300	172.40	13.3	16.6
23,000	118.50	9.6	13.7
35,500	99.70	1.0	19.0
49,100	146.50	35.5	11.6

TRANSPORTATION (Cont.)

Project Size Gross S.F.	Project Cost $/S.F.	% Cost Mechanical	% Cost Electrical
288,100	84.70	8.3	13.3
2,160,000	139.65	23.3	11.5

OFFICES

BANKS

Project Size Gross S.F.	Project Cost $/S.F.	% Cost Mechanical	% Cost Electrical
2,900	198.35	5.3	4.3
3,100	88.05	5.9	6.7
3,300	124.75	10.3	16.4
3,600	96.10	10.3	12.2
4,000	109.75	8.0	9.0
4,100	108.35	21.4	13.0
4,200	116.00	8.6	14.21
4,400	134.00	12.2	12.7
4,500	84.70	12.4	13.5
4,900	133.95	11.7	11.2
5,900	99.10	9.3	13.8
6,000	112.00	11.6	7.3
6,100	149.60	11.0	8.0
7,000	191.95	6.0	9.0
7,300	123.40	11.9	11.3
7,700	136.05	8.0	7.5
7,800	151.70	11.0	11.7
8,000	76.85	10.0	14.0
9,200	121.95	9.9	12.1
9,400	88.35	11.7	11.8
10,200	184.35	12.6	12.6
12,600	65.80	7.0	18.0

A = Addition   R = Remodel

SQUARE FOOT TABLES

OFFICES (Cont.)

BANKS (Cont.)

Project Size Gross S.F.	Project Cost $/S.F.	% Cost Mechanical	% Cost Electrical
13,300	115.65	9.5	8.3
13,800	103.60	10.0	9.7
15,000	89.00	15.4	12.4
15,200	70.90	9.2	12.9
15,500	90.55	9.8	10.3
16,000	59.50	13.4	23.1
20,100	55.85	13.0	11.0
21,700	118.45	8.8	11.3
44,800	101.20	13.0	8.2
53,200	182.75	14.9	7.2
62,100	104.95	10.1	7.9
95,100	137.95	13.5	4.3

OFFICE BUILDINGS

Project Size Gross S.F.	Project Cost $/S.F.	% Cost Mechanical	% Cost Electrical
2,600	125.30	17.2	9.4
3,400	102.55	10.5	11.3
3,800	100.95	16.4	11.8
4,400	99.40	12.8	8.5
4,500	85.50	13.0	7.0
5,100	62.05	16.6	10.2
5,200	80.95	8.0	5.7
6,700	120.00	16.8	10.4
7,500	130.25	10.1	8.0
7,900	100.95	17.4	9.0
8,100	156.60	10.6	11.0
10,600	90.15	9.9	9.7
10,900	50.45	13.8	10.8

OFFICE BUILDINGS (Cont.)

Project Size Gross S.F.	Project Cost $/S.F.	% Cost Mechanical	% Cost Electrical
11,300	103.20	17.0	6.0
13,000	70.85	15.0	9.0
14,400	93.00	19.8	12.9
14,500	69.80	17.3	12.5
17,000	100.90	14.7	7.9
17,800 A	54.85	10.4	11.0
18,100	106.25	22.5	11.6
19,300	54.10	11.1	8.1
24,600	48.80	18.5	14.1
27,700	62.25	19.6	5.5
27,800	116.00	12.7	5.1
27,800	62.50	17.8	10.3
32,500 R	106.40	13.9	6.4
35,400	61.25	15.0	12.0
36,500	50.00	10.2	10.2
42,300	66.25	10.3	7.7
44,400	91.45	23.5	14.7
44,400	47.10	11.0	5.0
44,500	66.10	11.5	3.1
45,400	56.40	19.1	13.2
47,300	57.20	18.5	8.0
49,700	97.65	22.1	7.2
50,000	92.80	19.4	15.6
50,400	100.65	23.2	7.8
52,200	65.70	18.3	7.8
52,900	76.95	4.4	3.9
53,700	118.20	15.4	9.0

A = Additional   R = Remodel

SQUARE FOOT TABLES

OFFICES (Cont.)

OFFICE BUILDINGS (Cont.)

Project Size Gross S.F.	Project Cost $/S.F.	% Cost Mechanical	% Cost Electrical
54,000	47.10	14.4	2.6
56,000	53.85	10.6	6.2
56,500	73.15	19.1	11.0
72,000 R	24.70	12.9	20.2
74,000	49.60	13.5	6.8
80,800	51.90	12.0	6.0
81,800	71.35	21.1	9.5
81,900	61.05	19.9	9.0
82,000	83.60	14.0	4.2
83,100	98.70	16.0	8.6
85,400	86.55	18.7	8.6
86,200	67.70	14.7	11.0
99,900	78.30	16.6	6.6
100,000	87.55	10.6	12.1
100,000	52.20	16.3	10.4
116,400	83.60	12.6	10.0
134,500	169.40	14.1	12.4
140,000	132.00	20.1	11.2
155,700	164.55	11.1	6.1
171,000	76.95	24.0	7.7
174,300	93.90	12.3	9.1
203,300	128.35	19.0	13.2
265,800	177.70	11.0	10.0
287,300	45.30	10.7	4.1
319,800	77.10	13.4	5.6
350,000	113.90	20.0	13.0
360,900	69.05	14.2	11.1
394,000	54.40	22.5	8.7

OFFICE BUILDINGS (Cont.)

Project Size Gross S.F.	Project Cost $/S.F.	% Cost Mechanical	% Cost Electrical
430,000	82.00	21.8	9.3
490,000	87.35	11.2	7.5
588,400	179.40	15.0	9.0
606,000	71.65	9.4	8.8
620,000	200.60	12.6	12.8
733,500	50.20	10.8	4.2

RECREATIONAL

ARENA

Project Size Gross S.F.	Project Cost $/S.F.	% Cost Mechanical	% Cost Electrical
315,200	205.45	10.4	8.00
385,800	115.25	13.2	8.80
727,000	103.15	14.9	7.40

HEALTH CLUB

15,900	46.25	8.7	9.7
21,800	79.05	10.3	10.3
30,100	122.05	11.4	19.4
66,400	58.90	10.7	8.5

RECREATIONAL CENTER

9,900	110.05	6.6	9.8
14,000	72.20	11.2	5.0
14,000	87.80	11.3	16.7
15,700	62.00	17.8	14.3
20,000	151.55	19.1	8.9
21,200	84.90	16.0	9.6

A = Addition   R = Remodel

SQUARE FOOT TABLES

RECREATIONAL (Cont.)

RECREATIONAL CENTER (Cont.)

Project Size Gross S.F.	Project Cost $/S.F.	% Cost Mechanical	% Cost Electrical
26,000	97.00	9.3	7.0
53,400 A	116.00	11.7	6.4
69,800	188.10	17.7	8.8

RELIGIOUS

CHURCH

Project Size Gross S.F.	Project Cost $/S.F.	% Cost Mechanical	% Cost Electrical
4,100	127.85	8.8	13.5
10,400 R	179.70	12.8	8.1
11,100	77.10	10.9	12.7
13,400	116.80	5.5	6.0
14,500	79.80	12.8	7.4
15,200	103.60	18.3	8.
15,700	89.65	14.4	8.4

For more detailed information see **Building News Square Foot 1995 Costbook.**

CHURCH (Cont.)

Project Size Gross S.F.	Project Cost $/S.F.	% Cost Mechanical	% Cost Electrical
16,000	145.75	14.1	9.6
20,900	90.80	16.0	14.0
21,500	99.80	7.3	8.0
22,900	89.95	12.1	9.5
30,600	68.00	15.5	8.0
42,700	77.90	18.6	7.7

MULTI-PURPOSE

Project Size Gross S.F.	Project Cost $/S.F.	% Cost Mechanical	% Cost Electrical
4,400	81.10	11.5	17.5
5,800 A	103.40	8.9	7.5
6,400	141.15	15.6	15.8
9,000	69.10	7.7	5.9
9,000	98.65	16.0	6.6
10,100	65.90	11.1	12.0
12,000	130.15	13.1	10.7
18,400	71.35	10.8	10.1
19,500	125.60	16.0	17.3

A = Addition   R = Remodel

For more information subscribe to **Design Cost & Data**

Help For Your Toughest Job ... Estimating

With Your
BNi Building News
1995 Costbook Purchase You
Can Receive FREE . . .
CONSTRUCTION ESTIMATOR
With Electronic Data
Available
DECEMBER 1994

Over 500 Pages

BNi Building News

REMODELING
1995
COSTBOOK
FIFTH EDITION

(1557011109) ... **$49.95**

Edited by Plan Room Services, Inc.
Quantity Surveying & Cost Estimating
Consultants to the Construction Industry

Co-Sponsored by

Professional Builder.

Perfect For Tenant Improvements!

Includes The Most Difficult Items To Estimate

- *Demolition — Cutting and Patching*
- *Minimum Quantities and Costs*
- *Working Around Existing Conditions*
- *CSI MASTERFORMAT — All Divisions*
- *Man-Hour Data*
- *Geographic Cost Modifiers*

Remodeling is the toughest type of construction to estimate. You need all the help you can get. **Now** you have a resource you can turn to for fast, accurate cost information.

Hidden costs can be easily identified with the help of the CSI MASTERFORMAT, numbering systems and the clear, concise presentation format . . . and you can have access to this information every day of the year — for about A-Dime-A-Day!

Thousands Of Costs Broken Out Into Easy-To-Work-With Categories For Fast, Accurate Figuring

✔ MATERIAL COSTS:

Represent national averages for prices a contractor would expect to pay, plus allowances for handling. Researched as close to publication time as possible for maximum accuracy and timeliness.

✔ LABOR COSTS:

Prevailing rates — appropriate for each trade and type of work. Eliminates costly "guesstimates."

✔ OVERHEAD AND PROFIT:

Commonly applicable taxes, insurance and other factors are included. No confusion about what a number includes and what it doesn't.

Sample Costbook Page

Sample Man-Hour Table

The Complete Estimating Reference Book

FREE Estimating Software And Data NEW FOR 1995

With Your **BNi Building News** 1995 Costbook Purchase You Can Receive FREE . . . CONSTRUCTION ESTIMATOR With Electronic Data

Available DECEMBER 1994

Over 550 Pages

BNi Building News
GENERAL CONSTRUCTION 1995 COSTBOOK
FIFTH EDITION
(1557011095) ... **$49.95**

Edited by Plan Room Services, Inc.
Quantity Surveying & Cost Estimating
Consultants to the Construction Industry

All New For 1995!

● *Accurate* ● *Authoritative* ● *Up-To-Date*

- *Every Price Looked At And Changed As Needed*
- *Over 10,000 Construction Items*
- *Material And Labor Costs Broken Out*
- *CSI MASTERFORMAT — All 16 Divisions*
- *All Trades Covered In Detail*
- *Man-Hour Tables — For Scheduling And Productivity*
- *Geographic Cost Modifiers*
- *Affordable — About-A-Dime-A-Day!*

This extensive resource provides current, accurate construction costs and man-hour data for all aspects of building construction. Costs are based on "contractors' prices" for materials and prevailing rates for labor. Overhead and profit are included in all costs — there's no confusion!

All data is organized in the 16 Division CSI MASTERFORMAT — the industry standard. This simple, easy-to-use format along with the detailed index means you can instantly find the costs you need for fast, reliable estimates.

Helpful "Extras" To Add To The Costbook's Usefulness

- **Easy-To-Follow Instructions** in no nonsense language.
- **A Detailed Index** for quick reference. You can find cost categories fast.
- **Geographic Cost Modifiers** mean you can customize your estimate by location — by multiplying by an adjustment factor.
- **Square Foot Tables,** page after page of them, allow you to make quick budget estimates based on actual projects.
- **Construction Fact Tables Fast Reference.** Allowance and measurement charts, tips, standard sizes, do's and don'ts . . . and more!

Sample Man-Hour Table

Sample Costbook Page

Electrical Construction Covered In Detail

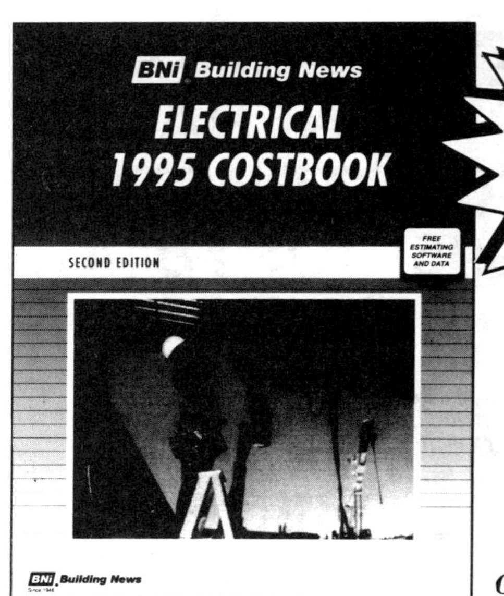

With Your
BNi Building News
1995 Costbook Purchase You Can Receive FREE . . .
CONSTRUCTION ESTIMATOR
With Electronic Data

Over 350 Pages

BNi Building News
ELECTRICAL 1995 COSTBOOK
SECOND EDITION
(1557011141) ... $39.95

Edited by Plan Room Services, Inc.
Quantity Surveying & Cost Estimating
Consultants to the Construction Industry

Available For DECEMBER 1994

Now Includes Low Voltage!

Superbly Indexed And Arranged According To The CSI MASTERFORMAT

- **BASIC MATERIALS**
  - Bus Duct
  - Conduit
  - Underfloor Duct
  - Conductors
  - Boxes & Fittings
  - Receptacles
  - Cable Tray
  - Wireways
  - Wall & Trench Duct
  - Cable
  - Cabinets
  - Man-Holes
- **POWER GENERATION**
  - Generators
  - Transformers
  - Switches
  - Capacitors
  - Circuit Breakers
  - Fuses
- **SERVICE & DISTRIBUTION**
  - Switchboards
  - Panel Boards
  - Transfer Switches
  - Metering
  - Motor Controls
  - Safety Switches
- **LIGHTING**
  - Interior
  - Exterior
- **SYSTEMS**
  - UPS
  - Lighting
  - Clock
  - Signaling
  - Fire Alarm
  - Sound
  - Security
  - Telephone
- **ELECTRIC HEATING** • **CONTROLS**

Accurate, Authoritative Cost Data

From meter to duct, conduit to receptacle, this detailed reference book provides extensive coverage of the most technical aspects of building construction. With thousands of current, reliable electrical costs at your fingertips, you can estimate quickly and accurately. Geographic Cost Modifiers allow you to tailor your estimates to specific areas of the country.

Typical Installation Labor Units For Thousands Of Construction Items

Sample Man-Hour Table

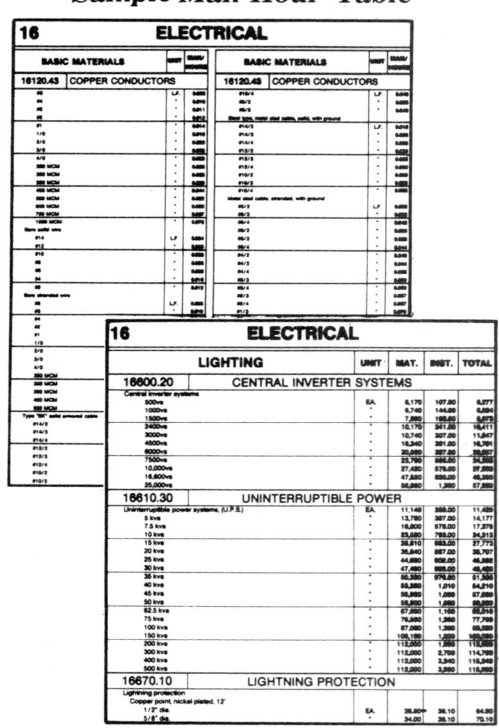

Sample Costbook Page

A Costbook For The FACILITIES MANAGER!

Over 600 Pages

FREE Estimating Software And Data NEW FOR 1995

With Your
BNi Building News
1995 Costbook Purchase You Can Receive FREE . . .
CONSTRUCTION ESTIMATOR With Electronic Data

Available DECEMBER 1994

BNi® *Building News* FACILITIES MANAGER'S 1995 COSTBOOK
FIRST EDITION
(1557011168) ... **$59.95**

Edited by Plan Room Services, Inc.
Quantity Surveying & Cost Estimating
Consultants to the Construction Industry

NEW FOR 1995

Finally, A Comprehensive Costbook For The FACILITIES MANAGER

- **New Buildings**
- **Retrofit Projects**
- **Tenant Improvements**
- **Detailed Costs**
- **Unit Costs**
- **Square Foot Costs**

All New For 1995!

- *Thousands Upon Thousands Of Detailed Unit Costs*
- *Extensive Man-Hour Tables For That Hard-To-Get Information*
- *All In CSI MASTERFORMAT — The Industry Standard*
- *Great For Change Orders And Extra Work*
- *Geographic Cost Modifiers For Many Cities*

This costbook has taken years to develop. It represents thousands of hours of professional thinking and input for the Construction Industry.

The *Facilities Manager's 1995 Costbook* contains the data for owners, architects, engineers and facilities managers to help them develop budgets on commercial and industrial buildings.

This manual should be combined with your own cost data to create a working document for your use. Developed using the CSI MASTERFORMAT, this easy-to-use book can give you — **IN MINUTES** — those hard to get prices that can take hours of your time to find. Our Alphabetical Index in the costbook makes sure of that!

Try This NEW Costbook For 1995 And BUILDING NEWS Will Guarantee Your MONEY BACK!
If Not Fully Satisfied No Questions Asked. We Know You'll Like This Book.

Sample Costbook Page

Sample Man-Hour Table

Finally . . . An Estimating Costbook Designed For The Unique Requirements Of The Home Builder

FREE Estimating Software And Data NEW FOR 1995

With Your **BNi Building News** 1995 Costbook Purchase You Can Receive FREE . . . *CONSTRUCTION ESTIMATOR* With Electronic Data

Available DECEMBER 1994

Over 350 Pages

BNi Building News HOME BUILDER'S 1995 COSTBOOK

THIRD EDITION

(1557011133) ... $29.95

Co-Published with the  HOME BUILDER PRESS

Publishing Arm of the National Association of Home Builders

Edited by Plan Room Services, Inc.
Quantity Surveying & Cost Estimating Consultants to the Construction Industry

- *Man-Hour Data*
  - *New And Remodel*
    - *Foundations to Finish*
      - *Material And Labor Costs*

Published in conjunction with the National Association of Home Builders, the largest construction organization in the U.S., this new costbook is the first in years devoted entirely to the home building industry.

This complete reference is designed for the small to mid-size home builder. It contains material and labor costs and installation times for all components of house construction: concrete, framing, insulation, painting, flooring, plumbing, wiring, etc.

Helpful "Extras" To Add To The Costbook's Usefulness

- **Easy-To-Follow Instructions** in no nonsense language.
- **A Superb, Detailed Index.** You can find cost categories <u>fast.</u>
- **Geographic Cost Modifiers** means you can customize your estimate by location — by multiplying by an adjustment factor.
- **Square Foot Tables,** page after page of them, allow you to make quick budget estimates based on actual projects.
- **Construction Fact Tables For Fast Reference.** Allowance and measurement charts, tips, standard sizes, do's and don'ts . . . and more!

New Pages

Sample Man-Hour Table

Sample Costbook Page

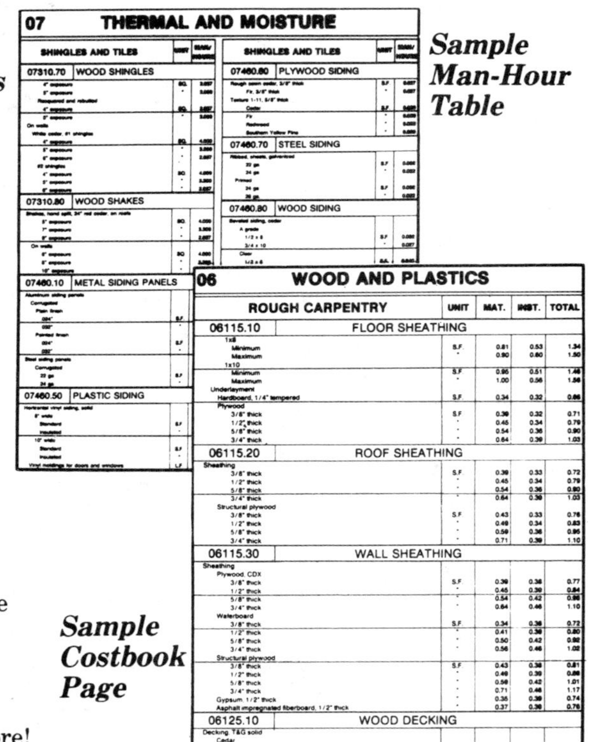

07	THERMAL AND MOISTURE		
SHINGLES AND TILES		UNIT	MAN-HOURS
07310.70	WOOD SHINGLES		
07310.80	WOOD SHAKES		
07460.10	METAL SIDING PANELS		
07460.50	PLASTIC SIDING		
07460.60	PLYWOOD SIDING		
07460.70	STEEL SIDING		
07460.80	WOOD SIDING		

06	WOOD AND PLASTICS			
ROUGH CARPENTRY		UNIT	MAT.	INST.	TOTAL
06115.10	FLOOR SHEATHING			
06115.20	ROOF SHEATHING			
06115.30	WALL SHEATHING			
06125.10	WOOD DECKING			

With ACCURATE ESTIMATOR you can have your choice of any of the Customized Databases listed below:

BNi® *Building News* 1995
Construction Costbook Databases

- *Material And Labor Costs* ■ *Man Hours* ■ *Updated For 1995* ■ *CSI MASTERFORMAT*
- *Data Can Be Adjusted To Local Prices And Labor Rates* ■ *Expanded Geographical Index - 50 States, Over 200 Cities*

Building News
GENERAL CONSTRUCTION
1995 Costbook Database
From Sitework To Painting

Building News
PUBLIC WORKS
1995 Costbook Database
Streets - Traffic - Utilities

Building News
MECHANICAL/ELECTRICAL
1995 Costbook Database
For Facilities Engineers

Building News
HOME REMODELER'S
1995 Costbook Database
¡New For 1995!

Building News
HOME BUILDER'S
1995 Costbook Database
Foundation To Finish

Building News
REMODELING
1995 Costbook Database
Includes Demolition and Asbestos

Building News
ELECTRICAL
1995 Costbook Database
Transformers To Receptacles

Building News
FACILITIES MANAGER'S
1995 Costbook Database
¡New For 1995!

Each Database Comes With The Corresponding
BNi® *Building News* 1995 Costbook ABSOLUTELY FREE

ORDER FORM

PRODUCT	QUANTITY	AVAILABLE	PRICE	TOTAL
ACCURATE ESTIMATOR with Building News 1995 General Construction Costbook Database		1/95	$290.00	
ACCURATE ESTIMATOR with Building News 1995 Home Builder's Costbook Database		1/95	$290.00	
ACCURATE ESTIMATOR with Building News 1995 Public Works Costbook Database		1/95	$290.00	
ACCURATE ESTIMATOR with Building News 1995 Remodeling Costbook Database		1/95	$290.00	
ACCURATE ESTIMATOR with Building News 1995 Mechanical/Electrical Costbook Database		1/95	$290.00	
ACCURATE ESTIMATOR with Building News 1995 Electrical Costbook Database		1/95	$290.00	
ACCURATE ESTIMATOR with Building News 1995 Home Remodeler's Costbook Database		1/95	$290.00	
ACCURATE ESTIMATOR with Building News 1995 Facilities Manager's Costbook Database		1/95	$290.00	

I understand that if I am not satisfied in every way upon inspection, I can return my purchase and receive a full refund, no questions asked.

Make Check Payable To: **BUILDING NEWS** Or
Credit Card Order: ☐ VISA   ☐ MASTERCARD   ☐ AMEX
Card No.______________________________Exp. Date: _________

____
(Authorized Signature)

NAME

COMPANY

ADDRESS

CITY   STATE   ZIP

TELEPHONE

SUBTOTAL....$____________
5% SALES TAX TO MASS. DESTINATIONS....$____________
7.75% SALES TAX TO CALIF. DESTINATIONS....$____________
(ADD 8.25% TO L.A. COUNTY DESTINATION)
ADD SHIPPING/HANDLING CHARGES OF
$5.75 FOR THE FIRST ITEM AND $1.00
FOR EVERY ITEM OVER ONE....$____________
GRAND TOTAL....$____________

Mail your payment to: **Building News**
3055 Overland Ave., Los Angeles, CA 90034

Or Call TOLL FREE **1-800-873-6397**

Square Foot Cost Data Compiled From Actual Projects

BNi Building News
SQUARE FOOT
1995 COSTBOOK
FIFTH EDITION
(1557011117) ... $49.95

**Available
DECEMBER 1994**

With Your

BNi Building News

1995 Costbook Purchase You
Can Receive FREE...
CONSTRUCTION ESTIMATOR
With Electronic Data

Over 475 Pages

Fully Illustrated

Complete Project Costs For 10 Building Types

- *Commercial*
- *Residential*
- *Educational*
- *Hotel/Motel*
- *Industrial/Warehouse*
- *Medical*
- *Public Facilities*
- *Offices*
- *Recreational*
- *Religious*

Hundreds of <u>actual</u> projects. Thousands of <u>actual</u> costs. All data
in this one-of-a-kind book is based on actual building projects
throughout the country.

Each project is broken down into the 16 CSI MASTERFORMAT
divisions. Costs are presented as Cost by Division, Percent of Total
and Cost per Square Foot.

You can use this data with confidence for figuring budgets,
conceptual estimates, cost comparisons, design development . . .
any time you need <u>fast, accurate</u> project costs.

Use With Confidence For Fast, Accurate Figuring

- *Budgets*
- *Conceptual Estimates*
- *Cost Comparisons*
- *Appraisals*
- *Replacement Costs*
- *Design Development*

Compiled By Estimating Experts

Brought to you by *Building Design & Construction* magazine and
Building News, serving your industry with construction/building
information since 1946.

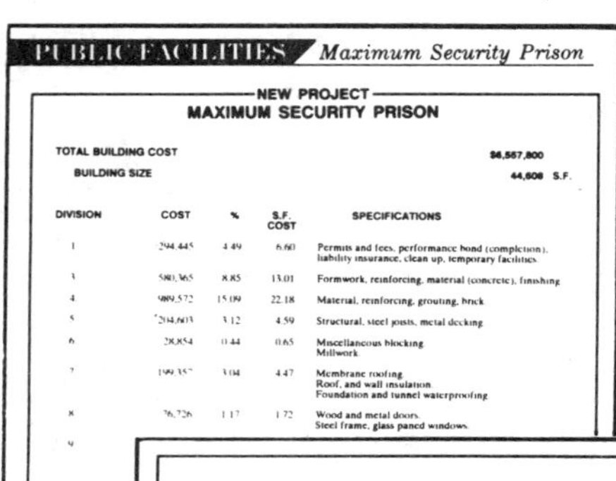

New Project

MAXIMUM SECURITY PRISON

$147.01 Per Square Foot